Eurasian Steppes. Ecological Problems and Livelihoods in a Changing World

PLANT AND VEGETATION

Volume 6

Series Editor: M.J.A. Werger

For further volumes:
http://www.springer.com/series/7549

Marinus J.A. Werger • Marja A. van Staalduinen
Editors

Eurasian Steppes. Ecological Problems and Livelihoods in a Changing World

Springer

Editors

Marinus J.A. Werger
Ecology & Biodiversity
Utrecht University
Padualaan, De Uithof 8
CH Utrecht, The Netherlands

Marja A. van Staalduinen
Ecology & Biodiversity
Utrecht University
Padualaan, De Uithof 8
CH Utrecht, The Netherlands

ISSN 1875-1318 ISSN 1875-1326 (electronic)
ISBN 978-94-007-3885-0 ISBN 978-94-007-3886-7 (eBook)
DOI 10.1007/978-94-007-3886-7
Springer Dordrecht Heidelberg New York London

Library of Congress Control Number: 2012936645

Printed on acid-free paper

Springer is part of Springer Science+Business Media (www.springer.com)

Preface

Eurasian Steppes – Ecological Problems and Livelihoods in a Changing World

Steppes form one of the largest biomes on earth. Eurasia comprises a very large steppe area stretching from China, Mongolia and southern Siberia, across Xinjiang, Kazakhstan, south-western Siberia, European Russia, Ukraine, and Hungary, to Anatolia, Romania, Slovakia, and outlier steppes in Austria and Spain.

Steppe ecosystems are sufficiently old to have provided plants and animals the opportunity to co-evolve as parts of stable ecosystems. Since the last Ice Age the Eurasian steppes first saw gradual changes in plant and animal populations, migrating into the steppe zone from refugia, as the glaciers retreated and temperatures increased. At that time man already appeared in the steppe, living a nomadic life and hunting steppe animals, e.g. antelopes, horses, camels, etc. Soon nomads lightly exploited the steppes as grazing grounds for their sheep and goats, both animals being domesticated at an early date. Later horses, and still later camels were domesticated and bred. They were also used for riding and roaming and waging war, but nomadic pastoralism was the core on which peoples built their livelihoods on the Eurasian steppes. For centuries exploitation pressure was so low that there were hardly, if any, noticeable effects on most of the steppe ecosystems, except very locally around sites of special interest, such as water holes in the driest steppes, or mines of ores and salts. Though obviously, populations of wild large grazers and browsers in the steppe and forest-steppe came under pressure as they experienced the competition from domestic animals.

Changes of increasing use of the steppe came very slowly. In the European steppes, as from the middle of the eighteenth century the rate and intensity of change began to increase, when, beginning in Hungary and Ukraine, man started to plough the steppe on a steadily increasing scale, and grazing started to intensify. In the Asian steppes this occurred considerably later.

Drastic changes began as from 1917, when increasingly larger areas of Eurasian steppes became part of communist states and their governments centralized land ownership, land management and production plans. The Soviet government radically changed the exploitation of the steppe zone, considering those lands a hopelessly under-used wilderness that should be converted into planned productivity. Steppes (including their ecologically marginal areas) were frantically ploughed up, with almost ideological zeal; on a large scale rivers and wetlands were drained; and an extensive infrastructure of roads and canals was constructed. The steppe was settled by large groups of pseudo-industrial farmers and exiles.

The steppe areas coming under the control of the Chinese communist government also changed strongly. A great influx set in, mainly of Han people from the Chinese agricultural provinces, and they ploughed and cultivated sizeable parts of the steppe that allowed modest yields, often moister sites such as valleys, or fields that could be irrigated, while at the same time the steppe area became increasingly fenced in, and nomadism was discouraged. As the population grew in numbers, livestock numbers also drastically increased, leading to severe overgrazing and impressive wind erosion and dune building over very large areas, with intense dust storms reaching much further afield.

Both in the Soviet Union and China these new policies of steppe use implied or explicitly aimed at ending nomadism in the steppe, and thus radically changed the livelihoods of the traditional steppe peoples, while also damaging or destroying very large tracts of steppe land.

Mongolia, however, though allied to the Soviet Union, stayed sufficiently autonomous to follow its own policy on steppe use. The Mongolian communist government also centralized steppe management, but supported services for nomadic pastoralists and took responsibility for accessible public health care, basal schooling, maintenance of watering points, veterinary care, etc. As a result, Mongolia is the only Eurasian country where nowadays substantial numbers of nomadic pastoralists still exist.

After the collapse of the Soviet Union another radical change in the steppe occurred. In ex-Soviet countries the State did largely stop to control steppe land use, and in several post-Soviet countries ownership of large tracts of land was given to the farmers. This soon led to abandonment of large tracts of ploughed steppe, because local farmers did not have the mechanical means nor the labour to work all this land. Huge areas became fallow and steppe regenerated. Several years later, though, frantic production ideas re-merged in some areas and re-ploughing of secondary steppes was undertaken.

In Mongolia, after the fall of the communist government, steppe land stayed public property, but pastoralism was privatized and herders had to take care of their families, their herds and the quality of the grazing grounds and watering points completely by themselves, while governmental provisions for public health, basal education, and veterinary care strongly declined. Moreover, the number of livestock increased enormously in post-communist years, causing overgrazing and degradation of the steppe.

In China, meanwhile, after years of over-exploitation of the steppe, the severe degradation of grazing lands, the creation of immense mobile dune areas, and huge dust storms, new conserving policies and management plans of steppe use were formulated and enforced, in order to drastically improve the situation.

All this again forced important changes on the livelihoods of the people living in the steppes. And on top of all this come the effects of climate change, which prove particularly strong in the northern zone of the Eurasian steppes.

In the late summer of 2010 an international meeting of scientists, policy makers and managers, conservationists and steppe users was organized in Hustai National Park, near Ulaanbataar, Mongolia, in cooperation with IUCN-CEM.

Speakers at that meeting provided valuable and up-to-date overviews of actual changes and associated problems in the steppe zones, covering a wide range of ecological and climatic issues and their consequences for the livelihoods of steppe peoples. They addressed the need for new land use policies and management, and discussed adaptations based on an ecosystem approach.

We decided that, given the importance of the steppe zone and the drastic changes that occurred in the steppe, it was time for a book that would inventory the actual situation in the steppe, review the ecological and climatic changes that are playing important roles in these steppes, discuss the effects of land use policies, and show how livelihoods of people in the steppe zone are affected. We asked a number of the speakers at the Hustai meeting to work out their contributions and invited other authors to contribute additional chapters.

The result is this book. It provides, in English, an extensive and up-to-date overview of steppe status and steppe problems in all of Eurasia, based on the experience of specialists, and it integrates extensive data sources that for a large part so far were only available in a spectre of locally used languages.

We hope that this book will inspire scientists, conservationists, land-users, managers and policy makers concerned with steppes and their people, and who are affecting their future. We also hope that this book, by stimulating interests in steppes and their problems, will ultimately serve to the benefit of the people living in the steppe and improve the ecological status of the steppes of Eurasia.

Utrecht

Marinus J.A. Werger
Marja A. van Staalduinen

Contents

Part II Degradation

Part III Climate Change

Part IV Livelihoods

Contributors

Nadezhda Yu. Arylova Center for Wild Animals of the Republic of Kalmykia, Elista, Russia

Stephan Baas Climate, Energy and Tenure Division FAO, Rome, Italy

Namkhai Bandi Hustai National Park, Ulaanbaatar, Mongolia

S. Bartha Department of Functional Ecology, Centre for Ecology, Hungarian Academy of Sciences, Vácrátót, Hungary

S.N. Bazha A.N. Severtsov Institute for Ecology and Evolution, Russian Academy of Sciences, Moscow, Russia

M. Biró Department of Functional Ecology, Centre for Ecology, Hungarian Academy of Sciences, Vácrátót, Hungary

Inge Bouman Foundation for the Preservation and Protection of the Przewalski Horse, Klaaswaal, The Netherlands

T.M. Bragina Biology and Geography, Kostanay State Pedagogical Institute, Kostanay, Kazakhstan

Natural Sciences, Pedagogical Institute and Institute of Biology of Southern Federal University, Rostov-on-Don, Russia

Zuozhong Chen Graduate University, Chinese Academy of Sciences, Beijing 100049, China

M.M. Cherosov Institute for Biological Problems of Cryolithozone, Russian Academy of Sciences, Siberian Branch, Yakutsk, Russia

A.A. Chibilyov Institute of Steppe, Urals Branch of the Russian Academy of Sciences, Orenburg, Russia

Yu Chu State Key Laboratory of Vegetation and Environmental Change, Institute of Botany, Chinese Academy of Sciences, Beijing, China

Xiaoyong Cui Graduate University, Chinese Academy of Sciences, Beijing 100049, China

E.V. Danzhalova A.N. Severtsov Institute for Ecology and Evolution, Russian Academy of Sciences, Moscow, Russia

Ming Dong State Key Laboratory of Vegetation and Environmental Change, Institute of Botany, Chinese Academy of Sciences, Beijing, China

Yu.I. Drobyshev A.N. Severtsov Institute for Ecology and Evolution, Russian Academy of Sciences, Moscow, Russia

Daniela Dúbravková Institute of Botany, Slovak Academy of Sciences, Bratislava, Slovakia

Homeland Museum in Považská Bystrica, Považská Bystrica, Slovakia

G. Fekete Department of Functional Ecology, Centre for Ecology, Hungarian Academy of Sciences, Vácrátót, Hungary

P.D. Gunin A.N. Severtsov Institute for Ecology and Evolution, Russian Academy of Sciences, Moscow, Russia

Ke Guo State Key Laboratory of Vegetation and Environmental Change, Institute of Botany, Chinese Academy of Sciences, Beijing 100093, China

Mária Hajnalová Department of Archaeology, Faculty of Arts, Constantine the Philosopher University, Nitra, Slovakia

Yanbin Hao Graduate University, Chinese Academy of Sciences, Beijing 100049, China

Tessa Jelley Climate, Energy and Tenure Division FAO, Rome, Italy

Tatiana Yu. Karimova A.N. Severtsov Institute of Ecology and Evolution, Russian Academy of Sciences, Moscow, Russia

G.V. Kazachkov Institute of Steppe, Urals Branch of the Russian Academy of Sciences, Orenburg, Russia

Olga K. Kirilyuk Daursky State Nature Biosphere Reserve, Chita, Russia

Institute of Nature Resources, Ecology and Cryology, Siberian Branch of the Russian Academy of Sciences, Chita, Russia

Vadim E. Kirilyuk Daursky State Nature Biosphere Reserve, Chita, Russia

Institute of Nature Resources, Ecology and Cryology, Siberian Branch of the Russian Academy of Sciences, Chita, Russia

Iryna Korotchenko M.G. Kholodny Institute of Botany, National Academy of Sciences of Ukraine, Kyiv, Ukraine

Harald Kürschner Institut für Biologie, Systematische Botanik und Pflanzengeographie, Freie Universität Berlin, Berlin, Germany

S.V. Levykin Institute of Steppe, Urals Branch of the Russian Academy of Sciences, Orenburg, Russia

Guofang Liu State Key Laboratory of Vegetation and Environmental Change, Institute of Botany, Chinese Academy of Sciences, Beijing, China

Anna A. Lushchekina A.N. Severtsov Institute of Ecology and Evolution, Russian Academy of Sciences, Moscow, Russia

Zs. Molnár Department of Functional Ecology, Centre for Ecology, Hungarian Academy of Sciences, Vácrátót, Hungary

Valery M. Neronov A.N. Severtsov Institute of Ecology and Evolution, Russian Academy of Sciences, Moscow, Russia

Viktor A. Obyazov Institute of Nature Resources, Ecology and Cryology, Siberian Branch of the Russian Academy of Sciences, Chita, Russia

Gerald Parolly ZE Botanischer Garten und Botanisches Museum Berlin-Dahlem, Freie Universität Berlin, Berlin, Germany

Mykyta Peregrym O.V. Fomin Botanical Garden, National Taras Schevchenko University of Kyiv, Kyiv, Ukraine

M.G. Kholodny Institute of Botany, National Academy of Sciences of Ukraine, Kyiv, Ukraine

A.V. Prishcepa A.N. Severtsov Institute for Ecology and Evolution, Russian Academy of Sciences, Moscow, Russia

Jianjiang Qiao State Key Laboratory of Vegetation and Environmental Change, Institute of Botany, Chinese Academy of Sciences, Beijing, China

E.I. Rachkovskaya Institute of Botany and Phytointroduction, Almaty, Kazakhstan

Eszter Ruprecht Hungarian Department of Biology and Ecology, Babeş-Bolyai University, Cluj Napoca, Romania

Helios Sainz Ollero Department of Ecology, Universidad Autónoma de Madrid, Cantoblanco, Madrid, Spain

Ts. Sanjmyatav Argalant Soum Herder and Head of Tost herder group, Tov aimag, Mongolia

Anarmaa Sharkhuu Department of Earth and Environmental Sciences, University of Pennsylvania, Philadelphia, PA, USA

Natsagdorj Sharkhuu Institute of Geography, Mongolian Academy of Sciences, Ulaanbataar, Mongolia

Ilya E. Smelansky UNDP/GEF Project "Improving the Coverage and Management Efficiency of Protected Areas in the Steppe Biome of Russia", Moscow, Russia

NGO Siberian Environmental Center, Novosibirsk, Russia

Savina Tessitore Climate, Energy and Tenure Division FAO, Rome, Italy

Arkadiy A. Tishkov Institute of Geography, Russian Academy of Sciences, Moscow, Russia

Russian Geographical Society, Moscow, Russia

Tatyana E. Tkachuk Daursky State Nature Biosphere Reserve, Chita, Russia

Zabaikalsky State Humanitarian Pedagogical University, Chita 672020, Russia

Jan Treiber Department of Botany, Senckenberg Museum of Natural History, Görlitz, Germany

E.I. Troeva Institute for Biological Problems of Cryolithozone, Russian Academy of Sciences, Siberian Branch, Yakutsk, Russia

Marja A. van Staalduinen Ecology & Biodiversity, Utrecht University, Utrecht, The Netherlands

Department of Ecology, Universidad Autónoma de Madrid, Cantoblanco, Madrid, Spain

Irina B. Vorobyeva V.B. Sochava Institute of Geography, Siberian Branch, Russian Academy of Sciences, Irkutsk, Russia

M.J.A. Werger Ecology & Biodiversity, Utrecht University, Utrecht, The Netherlands

Karsten Wesche Department of Botany, Senckenberg Museum of Natural History, Görlitz, Germany

Piet Wit Foundation for the Preservation and Protection of the Przewalski Horse, Klaaswaal, The Netherlands

Xuehua Ye State Key Laboratory of Vegetation and Environmental Change, Institute of Botany, Chinese Academy of Sciences, Beijing, China

Part I
Steppe Regions

Chapter 1
Abiotic and Biotic Determinants of Steppe Productivity and Performance – A View from Central Asia

Karsten Wesche and Jan Treiber

Abstract With over 13 Mio. km², grasslands of Eurasia form one of the largest continuous terrestrial biomes. They mostly represent environments with low productivity and with a long evolutionary history of natural grazing. Over the last few decades, increasing population sizes and socio-economic changes have subjected these steppes to increasing pressure, and associated degradation. We concentrate on the steppes of Central Asia (Mongolia, northern China and Tibet) and show that land use practice, climate and soil conditions are the most important drivers of change in these grasslands.

Grazing has strongly degrading effects on relatively moist grass and forest-steppes whereas evidence indicates that acute vegetation degradation in semi-arid desert steppes is largely absent. In such environments, precipitation controls community composition and productivity at both the local and regional scales. Recurrent droughts give rise to episodic fodder shortages, which results in animal numbers being maintained at relatively low levels. This may explain the lack of degradation in dry steppes, and supports predictions drawn from the non-equilibrium theory of rangeland science.

On the other hand, soil degradation due to grazing is found across the entire range of hygric conditions without any apparent interaction with precipitation. Soil nutrient contents were recently found to co-limit plant productivity, even at relatively dry sites, indicating that grazing may have indirect effects on steppe performance not predicted by standard theories.

We conclude that moister parts of Central Asia are sensitive to grazing degradation, which directly affects vegetation and soils, while semi-arid parts are mainly and more specifically influenced by soil degradation, which has more indirect effects on plant communities.

K. Wesche (✉) • J. Treiber
Department of Botany, Senckenberg Museum of Natural History, Görlitz, Germany
e-mail: karsten.wesche@senckenberg.de; jan.treiber@senckenberg.de

M.J.A. Werger and M.A. van Staalduinen (eds.), *Eurasian Steppes. Ecological Problems and Livelihoods in a Changing World*, Plant and Vegetation 6,
DOI 10.1007/978-94-007-3886-7_1,

Abbreviations

ANPP	Aboveground (annual) net primary productivity
CV	Coefficient of variation
IMGERS	Inner Mongolian Grassland Ecosystem Research Station
K	Potassium
N	Nitrogen
NEQT	Non equilibrium theory
P	Phosphorus

1.1 Introduction

Together with forests, grasslands and open shrubland constitute the largest terrestrial biomes, with temperate grasslands alone accounting for ca. 8% of land surface (White et al. 2000). They store huge amounts of carbon in their soil, and C-sequestration potential of the world's permanent pastures may be equivalent to 4% of global greenhouse gas emissions (Soussana and Luscher 2007). Asia hosts the largest grassland belt of the world – estimates differ but the most frequently quoted numbers are in the range of 6–9 Mio. km² of (mainly dry) grasslands with broadly similar physiognomy (Photo 1.1), while (mainly eastern) Europe accounts for an estimated 7 Mio. km² (Cressey 1960; White et al. 2000). Together, the Eurasian grasslands may once have formed the largest continuous land biome on the planet, but as in North and South America, they have more recently come under tremendous pressure with vast areas of temperate grassland being converted to arable fields, many of which are now fallow. Moreover, the proportion of former grasslands currently under protection is lower than that afforded to any other major habitat type (Hoekstra et al. 2005; Peart 2008).

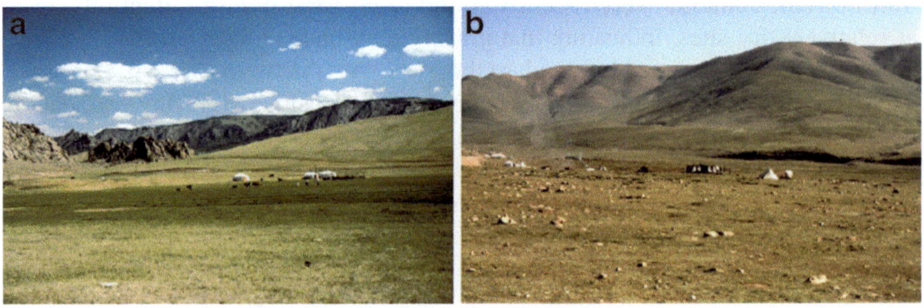

Photo 1.1 Physiognomic resemblance of Central Asian steppes. (**a**) Summer camp in the central Mongolian forest-steppe/grass steppe ecotone. (**b**) Summer camp in the alpine meadows of eastern Tibet (a photos taken by KW unless stated otherwise)

The remaining areas continue to change, and the traditional pastoral grazing systems are successively being replaced. Land use systems are intensifying due to rising human population pressures and new economic constraints that favour sedentary land use systems and larger herds (Sneath 1998; Janzen 2005). This has raised concerns about increasing degradation. However, land use has always been constrained by harsh climates (cold, often dry), and general theory predicts that land use attributes are modified by the overall climate conditions at any given site (Milchunas and Lauenroth 1993; Cingolani et al. 2005; Vetter 2005). This leads to the idea that the relative importance of land use vs. climatic controls change along the aridity gradient (Fernandez-Gimenez and Allen-Diaz 1999; Ellis et al. 2002).

Climate change interacts with land use effects and also introduces new pressures that trigger growing concerns about increasing vulnerability of the Eurasian steppes (Christensen et al. 2004; Angerer et al. 2008). Land use in dry and/or cold environments is especially sensitive to changes in the abiotic conditions (Duraiappah and Naeem 2005), rendering not only the Eurasian grasslands but also their human populations potentially vulnerable. In spite of extensive research on the subject, details of potential effects of changing precipitation and temperatures on ecosystem functioning are far from fully understood (IPCC 2007; Soussana and Luscher 2007). It is even less clear how climate change effects interact with other abiotic controls such as soil conditions. A number of studies have dealt with effects of land use and climate on soil nutrients and carbon contents (see e.g. several chapters in this volume), but their relevance to steppe plant performance has been less commonly studied, particularly for the drier regions of Central Asia.

Research in the grasslands of Asia has traditionally been published in the locally most important languages, i.e. Russian and Chinese, but there are a number of reviews in German and English that provide overviews of the older literature (e.g. Walter 1974; Lavrenko and Karamysheva 1993; Zhu 1993). The last two to three decades have witnessed a rapidly increasing number of publications on various ecosystem functions, with a predominant focus on aboveground net primary production (ANPP, or respective proxies) as the main variable of interest for the prevailing grazing system. Much of this research is now published in easily accessible journals or other outlets, and the older notion that North American prairies are much more comprehensively studied is perhaps no longer valid. Recent reviews covering Central Asia have so far adopted a local or at most part-regional perspective (e.g. Gunin et al. 1995; van Staalduinen and Werger 2006a), while comprehensive evaluations of a given vegetation type, such as the short grass steppes of North America (Lauenroth and Burke 2008), are still lacking. Up-dated overviews and synoptic analyses of the ongoing research in Eurasian grasslands are also scarce.

Compiling the huge literature for all Eurasian grasslands is beyond the scope of a book chapter, and would presumably also require a larger multi-author team. Instead, we present a brief overview of research covering one major part of the Eurasian grasslands, the Central Asian steppes and related communities. While aspects of population ecology and carbon sequestration are merely touched upon in this review and would certainly merit a separate analysis, we focus on the plant community composition and biomass productivity with an emphasis on ANPP.

The main drivers of biomass productivity are discussed with a specific focus on the following three aspects:

- Grazing effects: How does grazing affect plant community composition, and is there evidence for degradation and deteriorating biomass productivity?
- Climate effects: How do precipitation and temperature influence vegetation patterns and ANPP, and are there effects of interannual variability? Do potential climatic constraints influence grazing effects?
- Soil conditions: How does grazing affect soil nutrient conditions, and how are these conditions related to climate? Does soil nutrient availability limit productivity in spite of the strong climatic controls?

We will start with a description of the basic physiogeographic features of the region. The main part of the text will be devoted to a discussion of the specific questions with respect to mostly recent major publications. This is illustrated by already published examples from our own research in the Central Asian drylands and in Tibet.

1.2 Physiogeographic Setting

1.2.1 What and Where Are Central Asian Steppes?

A plethora of literature addressing various aspects of vegetation ecology and grazing ecology reveals both the similarities (Photo 1.1) and the pronounced differences between various types of grassland throughout Eurasia. For the present review, we concentrate on the grasslands of Central Asia, a term which has been somewhat ambiguously used throughout the literature. We adopt the biogeographical approach of Russian scholars rather than taking the simpler topographical view that is now followed by many authors including most publishing in a political context: The region between the Caspian Sea and the western Himalayas, i.e. Kazakhstan and neighbouring countries (Turkmenistan, Uzbekistan, etc.) is in the centre of the Eurasian continent, and is often referred to as Central Asia, terminology which is also found in IPCC reports (IPCC 2007). In contrast, Russian literature traditionally referred to this region as Middle Asia, as opposed to Central Asia, which refers to Mongolia, large parts of China (including most of Tibet) and small adjoining parts of Kazakhstan (Figs. 1.1 and 1.2). Although these regions are no more distant from any ocean than many parts of Middle Asia, they are nonetheless more continental in terms of climate and, consequently, biogeography. The reasons are detailed below and will demonstrate that Middle Asia and Central Asia (and certainly the more oceanic East Asia) should be differentiated, at least if aspects of physical geography are the focus (Cowan 2006).

Grasslands may be referred to as anything from forb-rich grasslands ("meadow steppes") in the forest-steppe ecotone, to typical "grass steppes" with dense swards,

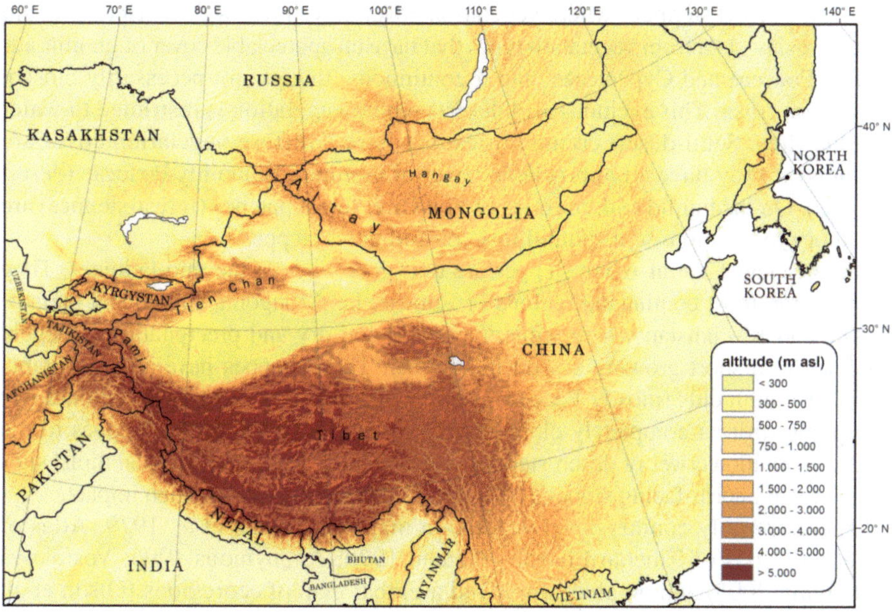

Fig. 1.1 Topographic map of eastern Eurasia (Altitudinal information based on SRTM data)

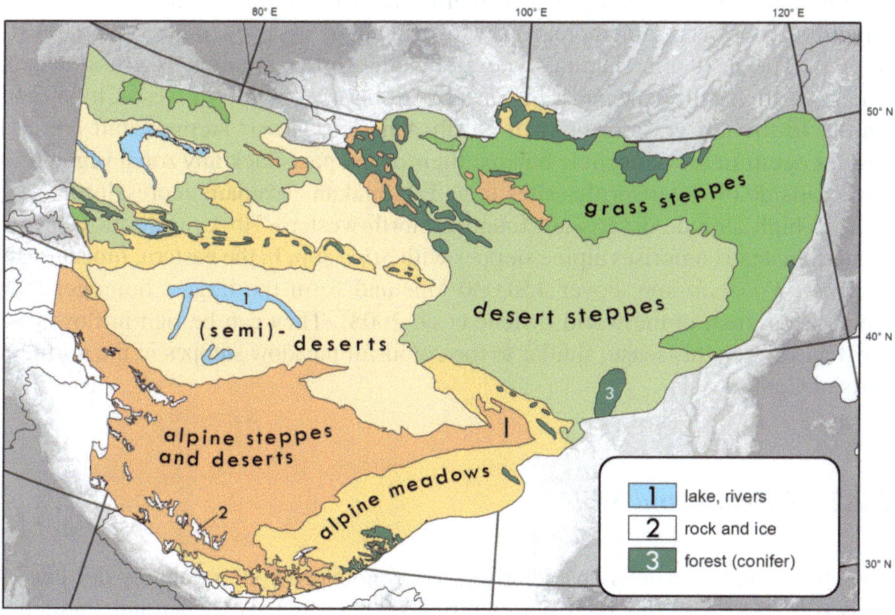

Fig. 1.2 Spatial distribution of the main vegetation types in Central Asia (After Olson et al. 2001, simplified, see Table 1.1 for background environmental data)

or sparse "desert steppes" where vegetation cover is well below 50%. Here, we will mainly review data from vegetation types that have an appreciable cover of graminoids mainly Poaceae and Cyperaceae, and a continuous, though not necessarily closed, vegetation cover. This excludes arid deserts where vegetation is restricted to water surplus sites. Semi-desert communities that have a diffuse vegetation cover are included if they have an appreciable cover of graminoids, mainly Poaceae (desert steppes). On the humid end, mosaics of grass and forest patches (forest-steppes) are transitory to the more semi-humid to semi-arid grass steppes.

In our view, Central Asian steppes are thus grasslands of the mid-latitudes. They occur in the most continental regions of Eurasia, i.e. Mongolia, China, and parts of Russia and Kazakhstan where winters are cold and dry and precipitation is largely confined to a short growing season in summer. These steppes have an appreciable cover of graminoids (mostly Poaceae, some Cyperaceae) and total vegetation cover may range from a completely closed sward in species-rich meadow steppes to the sparse vegetation cover of desert steppes. Trees are restricted to extrazonal moist sites (along rivers, north-facing slopes), be it due to abiotic factors or human impact.

There are only country-level vegetation maps (Lavrenko et al. 1979; Atlas of Tibet en (Atlas of Tibetan Plateau) Plateau 1990; Anonymous 1996; Vostokova and Gunin 2005) and we thus employed a global map of ecoregions (Olson et al. 2001) and simplified this to depict the main biomes (Fig. 1.2). Central Asia is surrounded by drylands and alpine sites in the west and south, and deciduous forests and coniferous forests in the east and north, respectively. Pronounced precipitation gradients control the vegetation belts, leading to a strong north–south differentiation in Mongolia and northern China (Photo 1.2). At their southern limit, boreal forests are increasingly restricted to north-facing slopes (forest-steppe). The next drier habitat type along the aridity gradient is grass steppe, which is often tall, especially in Eastern Mongolia and north-west China, and similar to the tall grass steppes of North America. Desert steppes and (semi-) deserts occur in the endorheic basins; their driest parts lack any zonal vegetation cover outside of water surplus sites (e.g. Taklimakan, Tsaidam depression).

True high altitude deserts are found in north-western Tibet, while most of the western plateau comprises alpine steppes with *Stipa* spp. In the eastern, moister part of Tibet, *Kobresia* spp. cover 450,000 km² and form the largest homogeneous alpine ecosystem in the world (Miehe et al. 2008). They can be rich in flowering forbs and are, in this sense, similar to the colourful meadow steppes in the northern parts of Central Asia.

1.2.2 Climate and Soils

The continental climate is characterised by pronounced cold, relatively dry conditions (Fig. 1.3) and a marked seasonality (Barthel 1983; Weischet and Endlicher 2000). Winters are typically cold with monthly means as low as –40°C, particularly in the northern parts of the region or at the higher altitudes of Tibet. Cold air

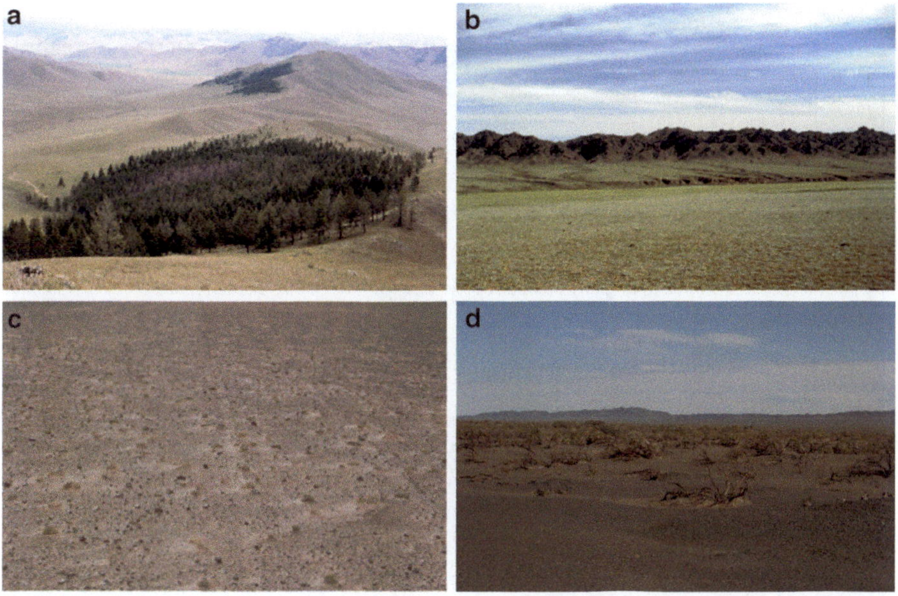

Photo 1.2 Change of plant communities along the zonal precipitation gradient in Mongolia. (**a**) Forest-steppe, i.e. mosaics of forests (north-facing slopes) and meadow steppes near Ulaanbaatar (note defoliation of trees due to outbreak of geometrid moths). (**b**) Grass steppes and mountain steppes on the pediments of the Altay in southern Mongolia. (**c**) Heavily grazed desert steppe in a year of Dzuud. Small nebkhas have formed around bunches of *Stipa glareosa*. (**d**) Semi-desert vegetation dominated by woody Chenopodiaceae in far southern Mongolia, a region that is too dry for livestock grazing

accumulates in the extensive basins while mountain sites are less affected, leading to stable temperature inversions on a local scale. In spring, temperatures rise steeply, but inflow of cold northern air masses can result in occasional frosts up to May. Moreover, strong winds, especially in the Gobi Altay, make living conditions unpleasant. In summer, air partly comes from western circulations but also from the monsoon, resulting in relatively favourable mean temperatures (>20°C) and a frost free period until September.

Precipitation is similarly low as in Middle Asia, but the seasonal distribution is different. Most of the precipitation (typically more than two thirds of the annual mean) falls in the relatively short growing season (Fig. 1.4a). Winter precipitation is very low, and usually springs are also dry (Fig. 1.4b). This contrasts with Middle Asia where moisture availability in spring is much higher, explaining why e.g. geophytes and also evergreen species are more important there. Much of the precipitation is of a convective nature and occurs in spatially and temporally limited events. This results in high spatial and also temporal heterogeneity, where neighbouring sites within a given year, or years at a given site, may differ strongly.

Soils are mostly constituted of fine materials, ranging from typical Loess to almost pure sands. Within substrate groups, soils reflect the availability of moisture

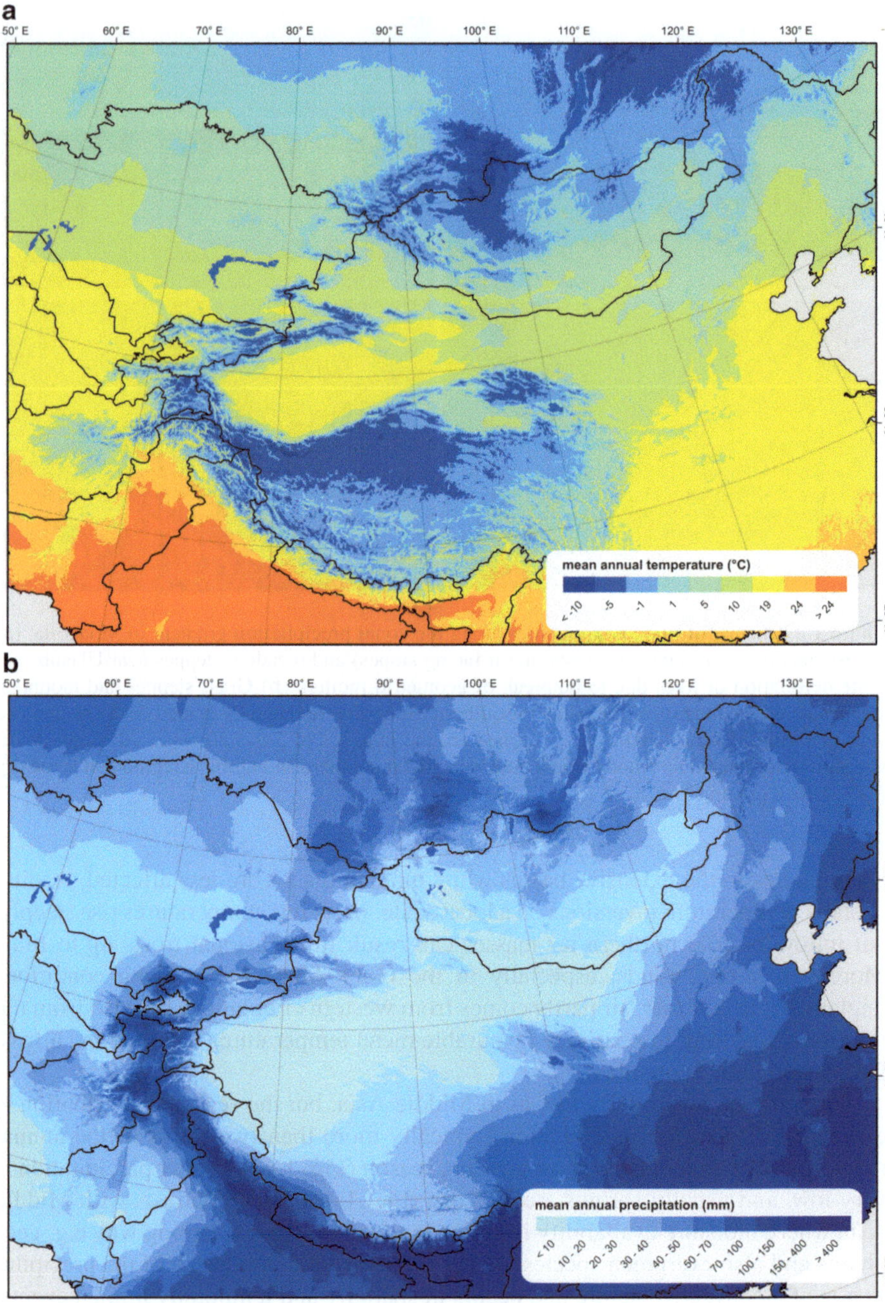

Fig. 1.3 Long-term macroclimate in Central Asia (from the WORLDCLIM model, Hijmans et al. 2005). (**a**) Mean temperatures. (**b**) Mean annual precipitation totals

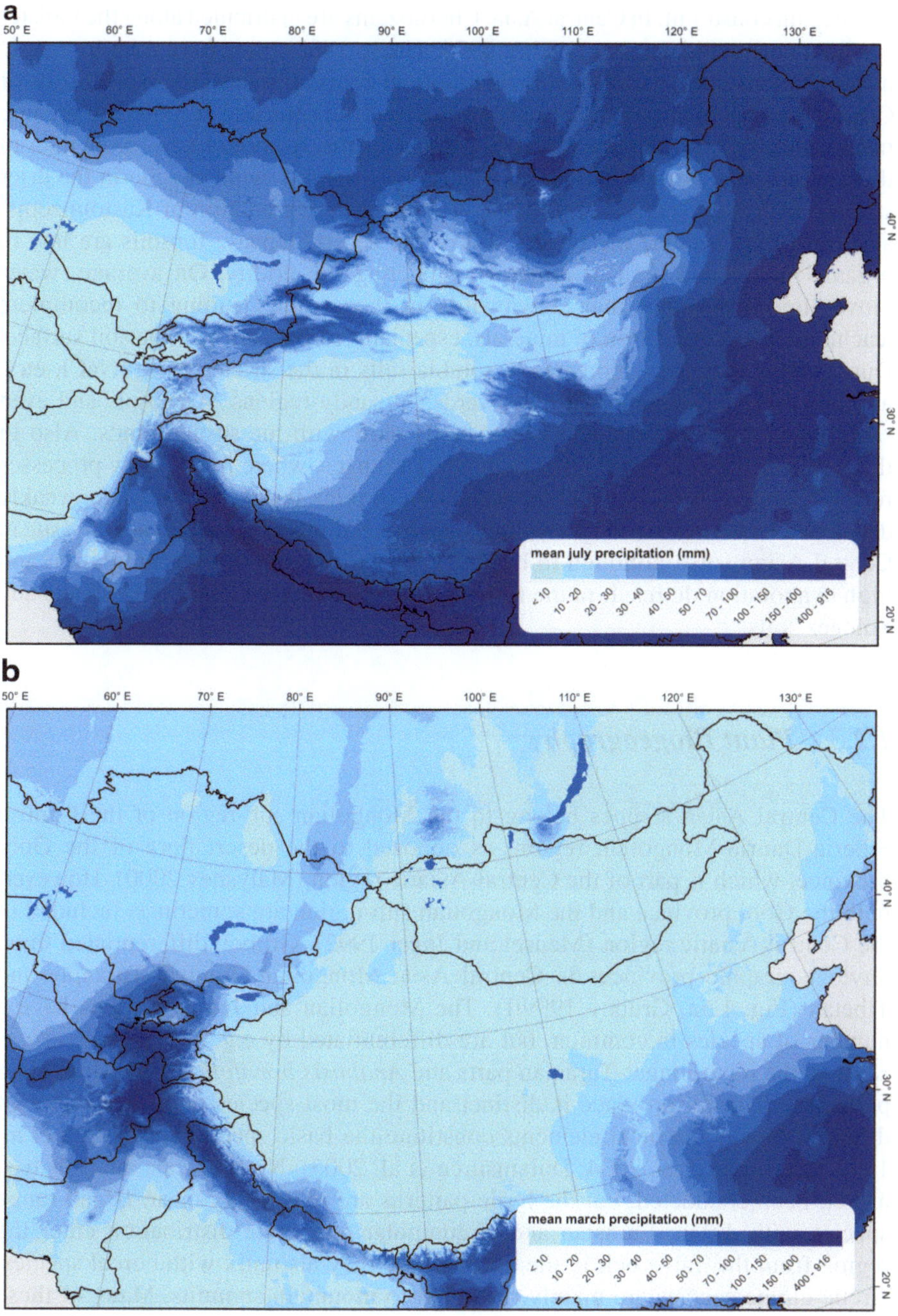

Fig. 1.4 Seasonality in precipitation (Data from Hijmans et al. 2005). (**a**) Mean July precipitation. (**b**) Mean March precipitation

as the main constraint. In Central Asia, Chernozems are distributed along the moister northern and southern parts (Walter 1974; Haase 1983); they have a high organic matter content and typically favourable nutrient concentrations. The typical soils of Central Asian grass steppes are Kastanozems, with an appreciable content of organic matter, though both thickness and C-content of the A-horizon are lower than in the Chernozems. Aridity and permanent, mostly natural, wind erosion in the drier steppes with their sparse vegetation cover result in degradation of Kastonozems, and Burozems become the typical soil of the desert steppes. Topsoils are low in organic matter and high in minerals, reflecting the aridity. On average, water moves upwards rather than downwards in the profile, leading to incomplete leaching and accumulation of minerals, especially carbonates near the soil surface. During rainy summers, more easily soluble salts in the stricter sense (NaCl, etc.) are, however, still transported downwards. In sandy regions, Arenosols and even sand dunes may develop, which constantly shift with the strong winds. Also in these soils salts are washed out. On mountain ranges, where freeze-thaw processes result in heavy weathering and the formation of thick layers of rock scree, weakly developed Leptosols are the prevailing soil type. Salinity is generally not an issue in Central Asian soils, although in basins, pans and oases water accumulation and high evaporation do result in the formation of spatially restricted Solonchaks and Solonez soils.

1.2.3 Plant Biogeography

The Central Asian steppes belong to the Mongolian sub-region of the Central Siberia-Dauria-Mongolian region, as opposed to the desert flora of the Gobi province, which is part of the Central-Asiatic region (Malyshev 2000). However, both the Gobi province and the Mongolian sub-region are sometimes included in the Central-Asiatic region (Meusel and Jäger 1992). Grubov differentiated three phytogeographic provinces in Central Asia: Mongolian, Junggar-Turanian and Tibetan (Fig. 1.5a, Grubov 1999ff). The Mongolian and Jungar regions have a number of species in common, but are differentiated by e.g. *Nanophyton erinaceum* in the recte Jungar-Turanian parts and *Anabasis brevifolia* in the Mongolian parts. The Tibetan province is distinct and the most species-poor. A number of different biogeographical elements constitute the basic species set (Hilbig et al. 1999, 2004; Xie et al. 2004; Dulamsuren et al. 2005). No full-scale analysis has, as yet, been conducted, but the basic patterns are reasonably clear. Broad-range species with Asian, Eurasian and circumpolar temperate distribution enter the region from the Taiga belt in the north where they also mix with boreal species. Arctic-alpine species are mainly restricted to mountain summits. Many of these can tolerate continental temperatures but fade out with increasing dryness. Eastern Asian species extend westwards into northern and eastern Central Asia wherever summer precipitation and temperatures are sufficient. A prominent example is *Ulmus pumila* (Fig. 1.5b), which extends well into the drylands of the central

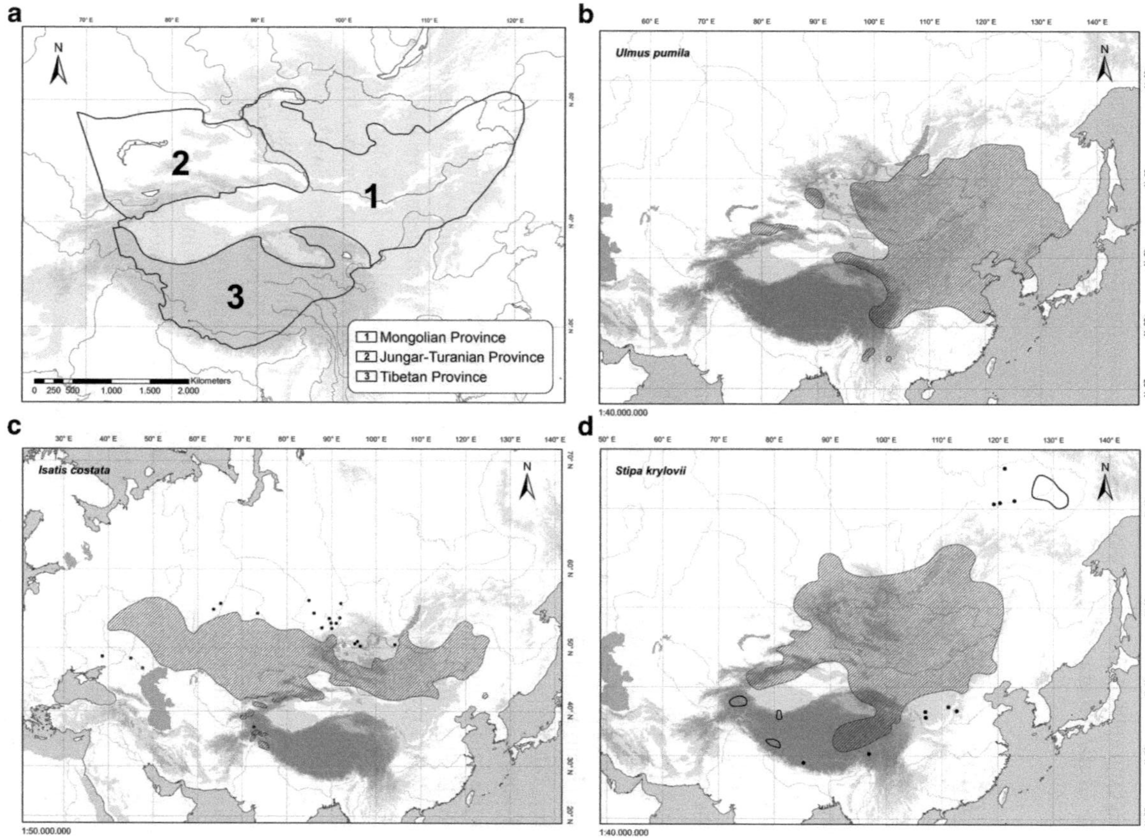

Fig. 1.5 Important biogeographical elements in the flora of Central Asia. (**a**) Outline of the biogeographical regionalization of Central Asia (as delimited by Grubov 1999ff). (**b**) *Ulmus pumila* as an eastern Asian species. (**c**) *Isatis costata* as a Middle and Central Asian species. (**d**) *Stipa krylovii* has a typical Central Asian distribution covering Mongolia, northern China and parts of Tibet

Gobi, but is ultimately replaced by *Populus euphratica*, which is a common tree of more western oases (Wesche et al. 2011). *Isatis costata* is an example of a species with Middle Asian and Central Asian distribution (Fig. 1.5c). One of the region's most common feather grasses *Stipa krylovii* has a typical Central Asian distribution (Fig. 1.5d) that also includes some mountain sites, indicating the close biogeographical links between the Mongolian/northern Chinese grasslands and those of the Tibetan plateau.

1.2.4 Flora – The Species Pool

Poaceae are the most typical constituents of Central Asian steppes. Feather grasses are represented with several sections, with *Stipa krylovii* (section *Leiostipa*) being an example of a species with a wide occurrence in all but the driest steppes (Fig. 1.5d). In Tibet, *Stipa* steppes occur in the drier western parts (Miehe et al. 2011), where *S. purpurea* represents the sect. *Leiostipa*. Other important grass taxa in Central Asia include *Leymus/Elymus*, especially *L. chinensis*, which is very abundant in the tall grass steppes of Mongolia and Inner Mongolia. It is often accompanied by *Agropyron cristatum*, which has a wide range along the hygric gradient, perhaps making it the most important forage grass of Central Asian steppes. It has also been introduced to North America, where it partly outperforms the native vegetation (Hansen and Wilson 2006). Cyperaceae rarely form dominance stands at zonal sites, but sedges of the genus *Kobresia* commonly form dense mats in montane and alpine regions. Forbs are diverse, with many genera, and some species occur in the mid-latitudes all over Eurasia and often even North America. Typical taxa include *Potentilla* spp., *Pedicularis* spp. and a huge diversity of legumes. *Astragalus* and *Oxytropis* are the most diverse higher plant genera in the region, with species numbering into the hundreds in Central Asia alone. They are, however, not exceedingly important in terms of vegetation cover. An example comes from the Gobi Gurvan Saykhan region of southern Mongolia that comprises zonal dry steppe communities, while extrazonally moist conditions in the mountains facilitate the occurrence of grass steppes and even meadow steppes (Wesche et al. 2005). In the Gurvan Saykhan, Fabaceae constitute 6% of the average species richness per plot but account for only 2% of its mean vegetation cover. Poaceae in turn represent 17% of the average species numbers, and have a 30% share in the total cover.

Allium species are also typical for Central Asia and their relative importance increases under drier climatic conditions; they may have the greatest proportional cover in many desert steppes. Chenopodiaceae have a diversity centre in Central Asia and their relative share of cover increases with aridity. Large parts of the semi-deserts of Central Asia are covered with woody Chenopodiaceae, which are accompanied by drought-tolerant Zygophyllaceae and Solanaceae.

Central Asia hosts a range of wild ungulates and small mammal herbivores, many of which are (now) endemic to the region (Bactrian Camel, Przewalski Horse, Schaller 1998; Gao et al. 2011), and steppes have clearly been grazed over

Photo 1.3 Grazing tolerance and avoidance. (**a**) The dwarf sedge *Carex tangulaschanensis* keeps its fruits hidden in the rosette (arrow). (**b**) *Stellera chamaejasme* (Thymelaeaceae) grows as a grazing weed in overgrazed *Kobresia* mats of Tibet and in grass steppes of Central Asia

evolutionary time-scales. The local species pool has been selected under almost omnipresent grazing pressure, and the extant species are, by majority, either tolerant to grazing or they exhibit effective defence strategies. Examples include several Tibetan Cyperaceae, where dwarf species with inflorescences hardly exceed 3 cm in height. Photo 1.3a shows the tiny *Carex tangulaschanensis*, whose fruits are practically hidden in the rosettes and thus beyond the reach of most grazers. Examples of species with an effective defence strategy are *Iris* spp., or *Stellera chamaejasme* (Photo 1.3b), which are poisonous to livestock and may become dominant weeds if steppes are overgrazed. Onions (e.g. *Allium polyrrhizum*) may cause problems if foraged in high quantities, and the Poaceae *Achnatherum splendens* is also mildly toxic. Nevertheless, most of the abundant species, including almost all Poaceae, are of high nutritional value (Jigjidsuren and Johnson 2003).

1.2.5 Land Use

The traditional land use system in Central Asia is nomadic pastoralism, where families migrate with their herds according to availability of livestock forage. Nomadism can be seen as a strategy to buffer temporal variability in climate and forage availability by utilising spatial heterogeneity and moving to less affected regions. Archaeological evidence implies that humans started to roam the steppes at least 6,000 BP (Aldenderfer 2007). Though animal products form the main source of energy, nomads also consume grains traded with sedentary farmers. Nomadic pastoralism in its current form is thus younger than the development of sedentary agriculture in the region, but it may still date back several millennia. Palynological evidence implies that many steppe sites have not changed much in the last millennia (Herzschuh et al. 2004; Miehe et al. 2009), implying that today's rangelands have been grazed since ancient times, with modern land use often merely replacing the natural grazing by wild ungulates and leaving limited impact on the general vegetation physiognomy.

More recently, Central Asia has undergone major social and economic changes. In China, rapid population growth and the rise of sedentary farming practices, even in some of the semi-arid regions, have resulted in tremendous agricultural intensification over the last decades/centuries. Rangelands have partly been degraded resulting in erosion, which has become a problem in some regions where climatic conditions would normally allow for a relatively closed vegetation layer. In the last years, the Chinese government has started to control grazing and has implemented large-scale restoration schemes which have slowed down desertification processes, although not in all territories (Runnström 2000). Mongolia has also experienced rapid population growth in the twentieth century, but overall livestock numbers have remained largely constant at ca. 20 Mio. animals (Janzen 2005). The early 1990s brought a rapid transition to a freer market economy, which caused unemployment to soar, making Mongolia one of the few countries in the world where the number of nomadic pastoralists has risen. The Mongolian economy is still based on agriculture, which is in stark contrast to Middle Asian countries like Kazakhstan where the economy rapidly shifted towards the export of fossil fuels. Fast urbanisation resulted in large tracts of rangeland being abandoned and much less intensive grazing than before the 1990s (Coughenour et al. 2008).

In Mongolia, livestock numbers had risen to 30 Mio. by the end of the 1990s, then collapsed to 20 Mio. in the droughts of 2001/2002 (Janzen 2005; Reading et al. 2006). By 2009, numbers had recovered to >30 Mio., and severe winters in 2009 and 2010 again led to severe losses (up to 50% in some regions; see also Chap. 20 by Baas et al., this volume). We thus seem to have an overall trend of intensification, although heavy fluctuations due to climatic variability partly override this trend. Disentangling the relative importance of socio-economic effects and climatic/abiotic constraints is one of the aims of the main part of this review.

1.3 Factors Controlling Grassland Species Composition and Productivity

1.3.1 Biotic Controls: Grazing Effects on Vegetation

Recent and even ongoing rangeland degradation has been described from a huge number of sites in Mongolia, northern China and Tibet. Authors used a range of indicators for assessing desertification that are not all equally suitable. Lower biomass on grazed sites should not be simply taken as evidence for grazing degradation, as grazing by definition is biomass removal. Lower biomass on grazed rangelands compared to that in exclosures merely confirms the presence of grazing rather than allowing for any conclusions on degradation. In addition, biomass, and also the correlated vegetation cover, is highly variable over the years making inferences on long-term trends difficult. The presence of bare soil alone is also not evidence for degradation, as an assumed closed sward is hardly a suitable reference for a semi-arid rangeland that has naturally low vegetation cover.

Photo 1.4 Effects of grazing exclusion in Central Asia. (**a**) Seven years of grazing exclusion in semi-humid upper montane *Kobresia* mats of northern Tibet (province Qinghai) resulted in replacement of Cyperaceae by tall grasses (mainly *Stipa* spp., right side of the fence, photo by E. Seeber). (**b**) Two years of grazing exclusion in southern Mongolian desert steppes has negligible effects. (**c**) As (**b**), but in the drought year 2002; biomass inside the exclosure is necromass from previous year

From a vegetation perspective, changes in plant community composition, especially in the perennial species, are more suitable indicators. Here, we regard evidence of pronounced changes in plant community composition, often associated with the replacement of palatable species by less preferred weeds, as indicators of degradation. Rangelands can also be seen as degraded where the forage productivity is declining. This introduces a further set of degradation indicators based on the body conditions or productivity of the grazing animals themselves, which reflect changes in forage quality (Allen et al. 2011). The latter criterion is, however, rarely applied in Central Asia.

Long-term records that cover the entire periods of changes in land use are scarcely available from Central Asia. Both experimental data and remote sensing studies typically cover at most three decades and often less, which is relatively short for a harsh, variable environment dominated by slowly growing long-lived perennials. There is, however, a number of fencing studies or short-term monitoring studies that provide clear evidence of grazing-induced degradation. The perhaps most-well studied site is the Inner Mongolia Grassland Ecosystem Research Station (IMGERS), where effects of grazing exclusion on long grass steppes have been studied over almost three decades. A large, continuously growing body of literature shows that intense grazing there has negative effects on rangeland community composition and forage productivity, and that drought can exacerbate the negative effects of overgrazing. However, modest grazing seems to stabilise the functioning of these steppes (Bai et al. 2004; Gao et al. 2009).

Similar results come from central Mongolian steppes where recent increases in animal numbers coupled with reduced mobility result in replacement of grass steppes by weed communities of low forage value (Hilbig 1995). Effects are particularly strong in the sacrifice zone around camps and wells, where heavy trampling adds to the effect of grazing (Sasaki et al. 2008). Evidence from experiments and large-scale surveys in eastern Tibet also points to increasing degradation in the last years. The prevailing *Kobresia* mats are tolerant to traditional Yak grazing, and total grazing exclusion may even lead to a replacement of mats by taller grassland communities (Photo 1.4a). On the other hand, more intensive land use in the last decades has resulted in the gradual breakup of the dense mats and the eventual formation of bare soil that is colonised by often impalatable weeds (Wang et al. 2006; Miehe et al. 2008).

The basic pattern in the examples from the moister region is similar: the current, often zonal communities are composed of perennial hemicryptophytes that tolerate substantial grazing pressure. Overgrazing has occurred in the last decades, leading to a reduction in plant biodiversity, replacement of perennials by short-lived weeds, and deteriorating forage production. ANPP and forage quality decline, with invading weeds (e.g. Chenopodiaceae), having low forage value or being even toxic. The prime example is *Stellera chamaejasme* (Thymelaeaceae), which has become a rapidly spreading grazing weed in grasslands from Mongolia to Tibet. *Stellera* is an exception insofar as it is perennial. Encroachment of larger shrubs, which is a problem in many overgrazed rangeland in other parts of the world, is not important in Central Asian rangelands. Instead, mainly semi-shrubs and dwarf shrubs benefit from increased grazing (e.g. *Artemisia frigida*). Taller shrubs gain dominance at sites with some soil disturbance, especially on small mammal burrows, which is discussed below.

Goats have increased in both relative share and absolute numbers almost everywhere in northern China and Mongolia, and even Tibetan herders increasingly keep goats to produce the relatively high-value Cashmere. In one of the few studies that are specifically devoted to changes in herd species composition, Tsagaan Sankey et al. (2006) showed that goats caused a reduction of tree and shrub growth in the grazed forest-steppes of northern Mongolia. Unfortunately, comparative studies on the effects of goats vs. effects of other animals, such as the traditionally kept camels, are too limited to draw any general conclusions as yet.

Authors working in the drier parts of Central Asia often report contradictory evidence on grazing impact. Li et al. (2011c) describe that grazing exclusion in Saxaul (*Haloxylon ammodendron*) woodlands resulted in increased diversity and carbon sequestration in northern China. In contrast, studies form the Mongolian drylands found minor effects of grazing exclusion. Grazing effects seem to decline with increasing dryness of the sites (Fernandez-Gimenez and Allen-Diaz 2001; Zemmrich et al. 2010; Sternberg et al. 2011), and continued observation of grazed vs. fenced (non-grazed) sites showed that grazing effects were small compared to the effects of annual changes in precipitation (Photo 1.4b, Wesche et al. 2010).

This pattern becomes clearer when a larger number of studies is analysed. We surveyed our own literature database for studies on grazing effects in Central Asian rangelands (numbering several hundred). We took care to report only one sample from any given site (i.e. the mentioned IMGERS site was included once), and we also excluded studies where the "grazing" effect was mainly restricted to the heavily trampled inner sacrifice zone of a piosphere (e.g. Sasaki et al. 2008). We considered only fencing experiments or studies with a sound sampling design along grazing gradients. Studies were checked for evidence on lasting effects on vegetation composition or productivity (for soils see below), which were classified on a simple three-level scale: no/negligible effects; modest effects, severe effects. We simplified further by disregarding more subtle effects that may only be apparent on the population scale. In woody species, effects of grazing may widely differ depending on the given demographical stages, and can have, e.g. detrimental effects on seedlings and positive effects on juveniles (Li 2010). Zemmrich (2007) showed that the

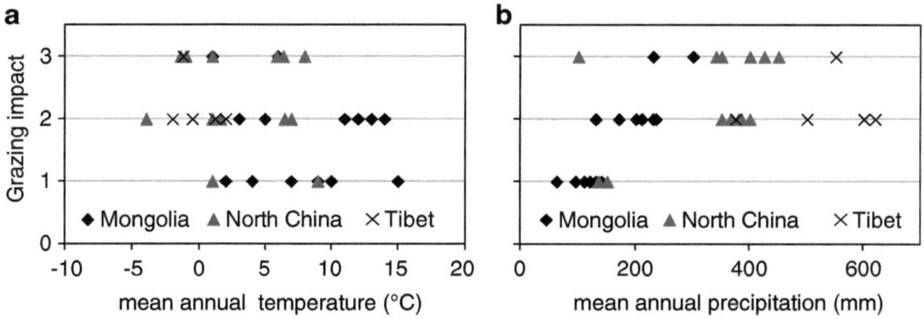

Fig. 1.6 Grazing impact on plant community composition and diversity in Central Asian rangelands. Effects were classified on a simple, three-level scale (*1*: no/negligible, *2*: modest, *3*: severe) and plotted against modelled climate data for a given site (Hijmans et al. 2005). (**a**) Grazing effects plotted against mean annual temperature. (**b**) Grazing effects plotted against mean annual precipitation

demographical structure of the rangeland species *Artemisia xerophytica* may reflect detrimental effects of grazing, while community level species composition and diversity are not affected. Similarly, the production of propagules often suffers heavily from grazing, even though communities have shown no responses as yet (Bläß et al. 2008). Our assessment is therefore probably somewhat conservative.

We plotted data against mean annual temperatures and found no relation to grazing effects (Fig. 1.6a). Annual precipitation of the given site served as a proxy for moisture availability (Fig. 1.6b), which covered a gradient in mean annual precipitation from <100 to 600 mm. Even this simple classification of the grazing impact produced a clear pattern. Studies that report no or negligible effects were restricted to dry regions of Mongolia and northern China, where precipitation is well below 200 mm. Conversely, severe grazing impact was restricted to regions with >300 mm mean annual precipitation. Modest grazing impact is found over the full range of hygric conditions. The sole exception to this pattern is the already mentioned study on *Haloxylon* woodlands in northern China (Zou et al. 2010), where 26 years of grazing exclusion resulted in a pronounced increase in diversity at a mean annual precipitation of ca. 100 mm. The site is somewhat untypical as the majority of species, and all those that were responsive, were short-lived, while the dominant Saxaul trees showed limited responses.

The example nonetheless raises the question as to whether effects of grazing reduction/exclusion are simply missed in many of the other experimental studies from drylands that typical lasted 10 years or often less. There are, however, a number of studies that are not based on fences but on transects that radiate away from permanent water sources or camps (Fernandez-Gimenez and Allen-Diaz 1999; Stumpp et al. 2005; Zemmrich 2007; Sasaki et al. 2009). The high-impact ends of these transects have typically been used for extended periods of time (decades). Thus, even slowly responding species should have had sufficient time to respond. The lack of grazing related patterns in these studies adds some confidence to the results described above.

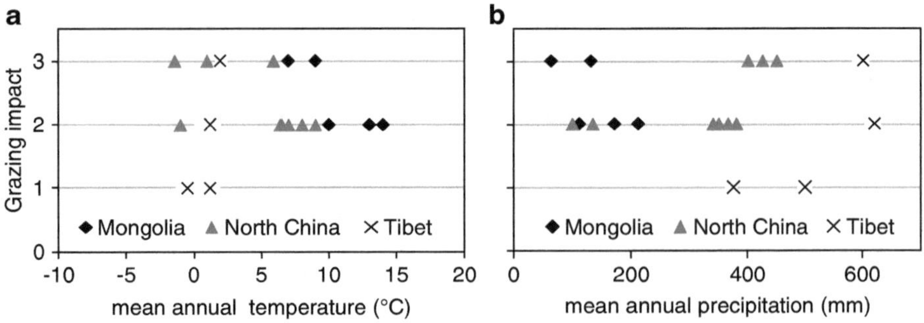

Fig. 1.7 Grazing effects on soil contents of nutrients and carbon in Central Asian rangelands. Effects were classified on a simple, three-level scale (*1*: no/negligible, *2*: modest, *3*: severe) and plotted against climate data for a given site (Hijmans et al. 2005). (**a**) Grazing effects plotted against mean annual temperature. (**b**) Grazing effects plotted against mean annual precipitation

1.3.2 Biotic Controls: Grazing Effects on Soil Conditions

A number of studies screened for Fig. 1.6 also reported grazing effects on soil nutrient contents, but response patterns differed considerably compared to vegetation. Of those that commented on soil conditions, almost all studies described at least modest and often severe losses of soil fertility with increased grazing intensity (Fig. 1.7). Surprisingly few authors comment on positive effects of grazing on soil nutrient availability. Comparison of wild ungulate vs. domestic livestock grazing in the Indian Transhimalaya indicated that wild herbivores facilitated biomass turnover and nutrient availability while the introduction of livestock resulted in losses and reduced availability (Bagchi and Ritchie 2010). In Inner Mongolian steppes, livestock grazing also increased N-turnover and N-availability; the effect was, however, most pronounced at intermediate grazing intensities (Xu et al. 2007). Grazing also increased N in standing crop of central Mongolian steppes, but this was apparently related to increased uptake rather than directly improved availability (van Staalduinen et al. 2007).

There was no apparent relation between the severity of effects and mean annual temperatures (Fig. 1.7a), nor between grazing effects and mean precipitation totals (Fig. 1.7b). There are nonetheless other principal patterns. Nutrient depletion in the course of degradation is sometimes associated with increased heterogeneity and formation of fertile islands under shrubs and other larger perennials (Pei et al. 2006; Cheng et al. 2007). This may at least partly reflect increased erosion as a consequence of reduced plant cover under intense grazing (Li et al. 2005). Relatively few studies report on losses in potassium, but reduced levels of (plant-available) nitrogen and phosphorus are commonly described (Su et al. 2005; Cheng et al. 2007; Zemmrich 2007). Losses in top soil due to erosion inevitably result in losses of carbon, but also nitrogen and phosphorus, which are often concentrated in the upper parts of the profile. Direct grazing impact exacerbates these losses because

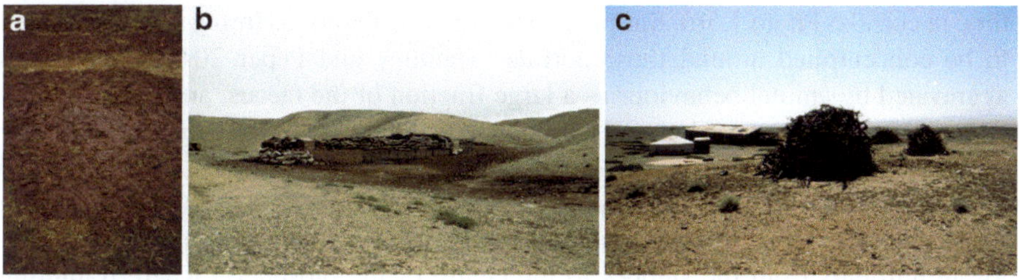

Photo 1.5 Fuel sources in Central Asia. (**a**) In the semi-humid alpine meadows of Tibet, Yak dung is smeared on the sward to dry. (**b**) At Mongolian winter places, dung of small livestock accumulates in corrals and can be later cut as fuel. (**c**) In the drylands of Mongolia, Saxaul trees (*Haloxylon ammondendron*) grow at sites with available groundwater and are heavily exploited as fuel wood (pile in the foreground)

animals take up nutrients with their forage. Under grazing, many, but not all, Central Asian rangeland plants show compensatory growth and shift the belowground-aboveground ratio towards higher aboveground productivity. Nutrients in the tissues are largely taken from the soil – or – as is the case in carbon, can at least not be allocated belowground, resulting in lower soil stocks and also reduced carbon sequestration potential (Gao et al. 2008).

Net effects of grazing on carbon source/sink dynamics, may, however, vary. In Tibetan alpine meadows, grazing favours sedges of the genus *Kobresia*, which tend to allocate the majority of their biomass belowground ultimately forming very dense turfs that store huge amounts of carbon. Grazing exclusion may result in the suppression of these sedges by taller Poaceae, which have a larger fraction of their biomass aboveground, effectively reducing carbon sequestration potential of ungrazed sites (Becker et al. submitted).

The extraction of livestock products, such as milk, meat and wool, causes nutrient loss from the rangelands, which are not closed systems as products are exported or consumed by humans and are thus not re-dispersed over time. Perhaps even more important is the extremely widespread practice of dung collection. Dung has been the main fuel source in large parts of treeless Central Asia, and the importance of dung production for pastoral economies is enormous (Photo 1.5a). For Tibetan pastoralists, the provision of dung by Yaks may be as important as the provision of milk (Rhode et al. 2007). Today, many people on the plateau have stopped animal husbandry, but still rely on dung as the main source of energy (pers. observation).

Livestock usually discards phosphorus in faeces and not in urine, while nitrogen is excreted in both urine and faeces (Clark and Woodmansee 1992). Collection of faeces, especially of larger livestock species, therefore results in losses of P and partly of N. Dung of smaller ruminants (sheep, goat) is hardly collected and nutrient withdrawal due to this pathway is probably limited. Transect studies have, however, been performed in a number of grazing systems that are dominated by small ruminants, and these usually find pronounced spatial patterns of decreasing concentrations of P and (sometimes less clearly) N (Fernandez-Gimenez and Allen-Diaz 2001; Stumpp et al. 2005; Holst et al. 2007). Small ruminants are usually herded at night time in corrals (Photo 1.5b). Simple geometric considerations imply that faeces tend to be concentrated around these corrals (Manthey and Peper 2010), and this is aggravated by animal behaviour as a large fraction of the faeces, and also of urine, is preferentially deposited at night time. They are concentrated in a spatially very restricted spot from where a fraction is lost as N_2O; a larger fraction is finally burned by farmers, and very little is redistributed to the steppes (Holst et al. 2007). Operating on a scale of a few hundred meters to a few kilometres, small ruminants thus translocate nutrients from the steppes to the centre of the piosphere, where trampling prevents plant growth rendering the accumulated nutrients effectively unavailable for forage production.

Figure 1.7 implies that grazing may have strong effects on soils, even where the aboveground vegetation shows no direct response. This renders assessments of grazing degradation that are solely based on aboveground vegetation questionable, because they may fail to detect grazing effects. Our data imply that this problem could be particularly severe in drier regions where vegetation responses to grazing are limited. The very fact that changes in soil nutrient contents are not reflected in the vegetation composition suggests an alternative view. In dry regions, growth should be limited by water availability rather than by nutrient supplies, rendering potential changes in terms of soil nutrient contents irrelevant for ecosystem functioning. This question will be addressed under "abiotic controls" later in this review, but first we have to consider small mammals as the second most important group of herbivores.

1.3.3 Biotic Controls: Effects of Other Herbivore Groups

Small mammals are abundant in steppes worldwide, and those of Central Asia are no exception. Rodents and, to a lesser extent, lagomorphs occur with a considerable diversity and colonise Central Asia from the desert lowlands to the alpine *Kobresia* mats. Species numbers tend to be higher in the more productive environments such as the tall grass steppes (Wang et al. 1999). Some species show high densities and rapid population increases; the microtine vole *Microtus brandtii* is perhaps the most studied example. Brandt's vole occurs in grass steppes of Mongolia and north-eastern China, where populations show boom and bust cycles that may be triggered by large-scale climate phenomena (Zhang et al. 2003b). Outbreaks have increased in frequency, presumably associated with increased grazing, because voles prefer habitats with short vegetation. During outbreaks, vole burrows can rapidly cover the larger part of the steppe surface. Voles consume biomass, and cause massive soil disturbance, eventually leading to the suppression and even disappearance of the otherwise dominant perennial hemicryptophytes (Zielinski 1982; Peterson 1994; Samjaa et al. 2000). Vole burrows carry a small cover of poorly palatable short-lived species. Burrows are abandoned after a few weeks or months and are subsequently colonised by shrubs, which benefit from the lack of competition by hemicryptophytes. Shrubs have comparatively deep roots

Photo 1.6 Pikas (*Ochotona* spp., Lagomorpha) in Central Asia. (**a**) Tibetan pika (*O. curzoniae*) in a disturbed *Kobresia* meadow, characterised by *Leontodon* and *Anaphalis* spp. with low forage value. (**b**) Burrow of the Mongolian pika (*O. pallasi*) characterised by a high abundance of shrubs (*Artemisia* spp.) and annuals (Chenopodiacae). (**c**) Adjacent desert steppe in the southern Mongolian Gobi dominated by herbaceous perennials (*Agropyron cristatum*, *Stipa krylovii*, *Allium* spp.)

(Kutschera et al. 1997; Liu et al. 2003) enabling them to use water that moves rapidly down in the disturbed burrow soils. *Microtus brandtii* may convert relatively productive grassland into stands of weeds or into scrub, which have limited value as pasture. Not surprisingly, the species is seen as a pest and control schemes have been implemented, including large-scale aerial poisoning (Shi et al. 2002), which has severe side-effects on other organisms.

Marmots (*Marmota* spp.) are also abundant in grass steppes, but they have much slower population dynamics and larger burrows that are less easily abandoned and more permanent. Marmots do not demonstrate outbreaks and never devastate entire regions through burrowing activity (van Staalduinen and Werger 2006b). Plant diversity is lower on burrows compared to open steppe, but even active burrows are partly covered by perennial grasses (*Stipa krylovii*, *Leymus chinensis*). Plant tissues have a higher concentration of nitrogen and phosphorus, ANPP is also higher, and burrows thus present valuable forage. Zokors (*Myospalax* spp.) are the main fossorial rodents on the Tibetan plateau, and have similar mixed effects on rangelands. The diversity of plants is lower on burrows, but biomass productivity and nutrient availability is higher (Zhang et al. 2003a; Wang et al. 2008). Burrows also provide special microhabitats for a number of organisms, and zokors are sometimes seen as ecosystem engineers rather than pests.

Pikas (*Ochotona* spp., Lagomorpha) are widespread from the drylands of southern Mongolia to the alpine meadows of Tibet. They have been considered as pests because they compete with livestock for forage and dig burrows like voles. In Tibet, they tend to be associated with disturbed *Kobresia* mats (Photo 1.6a), but whether they cause the fragmentation of sods or just benefit from sod disturbance by livestock or climatic conditions is not clear (Pech et al. 2007; Miehe et al. 2008). Pikas may even have positive effects on rangeland conditions. In Mongolia, pikas do not show population outbreaks and they live in permanent burrows that usually cover only a fraction of the surface (Nadrowski 2006). Diversity on burrows may be higher or lower than on the surrounding steppes depending on the soil substrate (Photo 1.6b, c, Wesche et al. 2007). Pikas are important herbivores and they consume

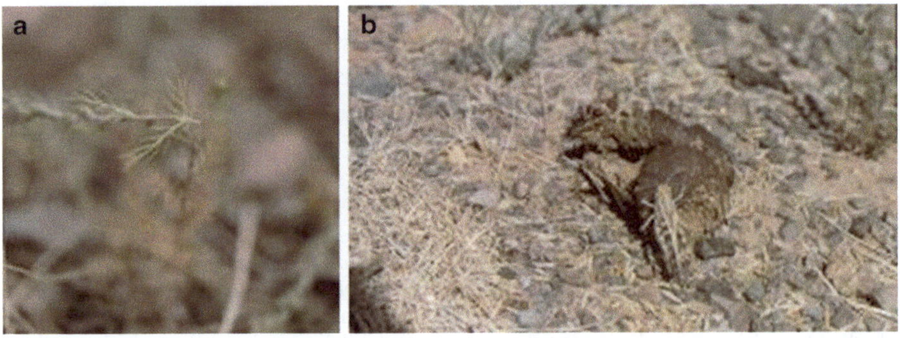

Photo 1.7 In years of outbreaks, Orthoptera can be important herbivores in southern Mongolian desert steppes. (**a**) Tiny stem of the annual *Artemisia* (*Neopallasia*) *pectinata*, that is avoided by livestock but has been cut by grasshoppers. (**b**) During outbreaks, herbaceous forage becomes so scarce that grasshoppers feed on dung

a similar share of the ANPP as livestock (Lai and Smith 2003; Retzer 2007). In Mongolia, in addition to storing forage for winter, pikas concentrate faeces on their burrows, thereby counteracting the large-scale nutrient withdrawal by livestock described above. As a consequence, pika burrows have higher plant available soil nutrient contents, and plants on burrows are more nutritious and productive. Burrows therefore represent preferred forage sources for livestock (Wesche et al. 2007), independent of soil associated moisture. Pikas and livestock can co-exist in the long run (Komonen et al. 2003; Retzer and Reudenbach 2005). Given that burrows also offer microhabitats for other biota, pikas should be seen as ecosystem engineers, that occur under particularly dry (southern Mongolia) and cold (Tibet) conditions (Smith and Foggin 1999). Large-scale poisoning is thus inappropriate and has largely been stopped now.

Much less well studied are other herbivore groups such as insects. Ongoing research is still largely devoted to taxonomy and distribution patterns, while effects on pasture conditions are only known by anecdotal evidence. Even under the generally cold conditions of Central Asia, the warm summer conditions allow for considerable insect diversity. In particular, Orthoptera occur with great diversity and in surprisingly large sizes (Photo 1.7). Abundance levels also appear to be high, but quantitative data are lacking. In southern Mongolian desert steppes, Orthoptera were encountered each year between 2000 and 2008, but feeding effects seemed mostly negligible (pers. obs.). In 2004, we witnessed one of the rare outbreaks, when Orthopterans numbered dozens/m^2 with considerable impact on the steppe vegetation. Grasshoppers clip vegetation even closer to the ground than pikas (Photo 1.7), resulting in the almost complete removal of above-ground forage and inflorescences. Sound measurements are not available and will be hard to obtain due to the irregular nature of such outbreaks, but it seems certain that insect herbivory can have an effect, at least in some years (Zhang et al. 2011).

1.3.4 Abiotic Controls – Climatic Constraints

It is established that herbivores have a strong impact on site conditions, particularly on soil nutrient availability, while direct effects on vegetation composition are mediated by the climate, especially precipitation. Differences in temperatures were of limited importance, partly because entire Central Asia has low mean temperatures (Table 1.1) and cold winters. Rapidly dropping temperatures in September and early October usually terminate the growing season (von Wehrden et al. 2010). The time at which C4 grasses commence growth is dependent on sufficiently high temperatures being reached, which may not be until summer during years with cooler springs (Liang et al. 2002). The dominant C3 species may respond even negatively to warm spring temperatures (Yuan et al. 2007; Yu et al. 2010), presumably due to lack of water in spring. Snow depths have increased in most of the Central Asian grasslands over the last decades (Peng et al. 2010) and analysis of satellite images implies that this results in enhanced growth in summer. This highlights the critical importance of early spring moisture availability. In fact, growth in spring is usually delayed until the onset of summer rains in May/June, although temperatures are well above zero from late April onwards. Together, precipitation and temperature restrict the growing season in much of Central Asia to 3 months or even less.

Precipitation is the crucial factor for all biota in Central Asia, which can be seen at various levels. Large-scale precipitation patterns control the spatial distribution of major vegetation belts in Mongolia and northern China (Table 1.1, Fig. 1.2). Temperature is only a mediating factor, even in Tibet where the *Kobresia* meadows of the moist western half cover an altitudinal range of almost 3,000 m, corresponding to 15–18 K difference between the lowest and the highest occurrence. Still, the vegetation of these eastern Tibetan mats is astonishingly homogenous. The western parts of Tibet have a very different vegetation, dominated by alpine steppes with *Stipa* and *Allium*, but this is related to lower precipitation rather than to different temperatures (Miehe et al. 2008, 2011).

Precipitation patterns are much more continental than in neighbouring Middle Asia, but also compared to North American prairies with their occasional winter precipitation. The prairies are less cold and have a longer growing season (Lauenroth and Milchunas 1992; Lauenroth 2008). A short grass steppe in North America receives around 320 mm/a as compared to 120 mm mean precipitation in the southern Mongolian desert steppes, that also have a continuous but sparse cover of short grass (data from station Dalandzadgad, National Meteorological Service Mongolia). Rain use efficiency (RUE) – in this context the ratio between ANPP and annual precipitation total – is therefore clearly higher in Central Asian steppes (Xiao et al. 1996). This is at least partly an effect of seasonality, as shown by data from the Dalandzadgad station. On average, 90 mm (75%) of the precipitation falls between May and August, resulting in an average of 20–30 mm per month for the growing season. This is not exceedingly low compared to many semi-deserts and explains why appreciable growth and land use practices persist in these desert steppes.

Table 1.1 Overview of spatial extent and altitudinal distribution of major biomes (see Fig. 1.2), and modelled mean annual climatic values (Hijmans et al. 2005)

	Area	Altitude			Temperature (°C)			Precipitation total (mm)		
Biome	(Mio. km²)	Mean	Lowest cell	Highest[a] cell	Mean	Coldest cell	Warmest cell	Mean	Driest cell	Wettest cell
Desert steppe	1.953	1,280	250	3,970	4.2	−9.2	10.9	170	25	695
(Semi-) desert	1.559	1,310	−150	5,200	7.6	−10.5	14.8	105	10	505
Alpine steppes and deserts	1.266	4,540	670	6,550	−3.8	−17.8	11.4	195	230	1,585
Grass steppes	1.153	1,220	500	3,510	−0.4	−10.9	11.1	280	100	715
Alpine meadows	0.865	3,760	190	6,430	−2.2	−17.6	11.4	345	40	1,740
Forest	0.305	2,080	560	5,370	−1.4	−11.7	13.1	275	60	2,375
Lake, rivers	0.101	810	250	1,730	9.8	4.6	12.6	110	20	405
Rock and ice	0.059	5,570	3,830	6,970	−7.2	−15.5	4.4	225	40	1,035

[a]Figures were extrapolated from a relatively crude biome map and extreme values should be treated with care

Precipitation also exerts a strong influence at the local scale, where plot-level species richness (α-diversity) and species composition are correlated to mean annual precipitation (Bai et al. 2007; von Wehrden and Wesche 2007; Ma et al. 2010). This reflects the fundamental diversity – productivity correlation (Wang et al. 2011). According to Ni (2004), aboveground net primary production in northern China ranges between 15 and 1,647 g/m^2. The latter is clearly an extreme value showing that water surplus sites such as saline meadows were included. The precipitation range for these data was approximately 80–620 mm annual mean, and the correlation between ANPP and precipitation was $r=0.32$. The correlation was even stronger in an analysis that covered biomass measurements from all Chinese grasslands including Tibet, but which excluded untypical water surplus sites (ANPP range 10–360 g/m^2, precipitation 100–500 mm, Bai et al. 2007). Mean annual precipitation explained ca. 36% of the differences in ANPP (mean annual temperature only ca. 3%) and 44% of the variation in species richness. Correlations differed in detail between major grassland types, but ANPP/precipitation and species richness showed linear relationships with no evidence of hump-shaped patterns or levelling at the higher precipitation levels. Even moister regions of Central Asia have still limited productivity compared to e.g. monsoonal eastern Asia, and may still represent the rising part of the productivity-diversity relationship (von Wehrden and Wesche 2007).

Mean values do, however, poorly reflect the huge interannual variability in precipitation, which is an inherent feature of most drylands. Coefficients of variation (CV, interannual standard deviation expressed as percentage of the annual mean) are typically high in the semi-arid parts of Mongolia and China (Zhou et al. 2007). Climate stations in southern Mongolia have interannual CVs of 35–55%, and growing season variability may be even higher (von Wehrden et al. 2010). In a given year, the aforementioned Dalandzadgad station may report anything between 30 and 150 mm precipitation (mean 120 mm). With water as the main limiting factor, ANPP shows related changes, as is evidenced by plot-level studies (Hilbig et al. 1993; Xiao et al. 1996; Wang 2004; Munkhtsetseg et al. 2007) but also in larger-scale remote-sensing based approaches (Piao et al. 2004; von Wehrden and Wesche 2007; Zhou et al. 2007). The example in Fig. 1.8 shows the development of standing crop in an exclosure experiment conducted in southern Mongolia (Wesche and Retzer 2005; Retzer 2007, supplemented). At ungrazed sites, standing crop varied between ca. 25 (2001) and 55 g/m^2 (2004). 2001 and 2002 were drought years with very low ANPP, while 2003 and the following years were relatively moist. ANPP thus showed a moderately strong correlation with summer precipitation at the neighbouring Dalandzadgad station ($r=0.46$). Precipitation and ANPP are often not perfectly correlated, even in drylands (Fernández 2007), indicating that plants may need time to recover or are intermittently constrained by other limitations. Moreover, there is also evidence that a few more pronounced rain events may be more important than a higher number of lighter rains (Ronnenberg et al. 2008); an issue which is addressed in the fertilization studies discussed later in this review.

Standing crop at sites grazed by the two main herbivore groups (livestock, pikas) was also highly variable, indicating that both herbivore groups were not able to

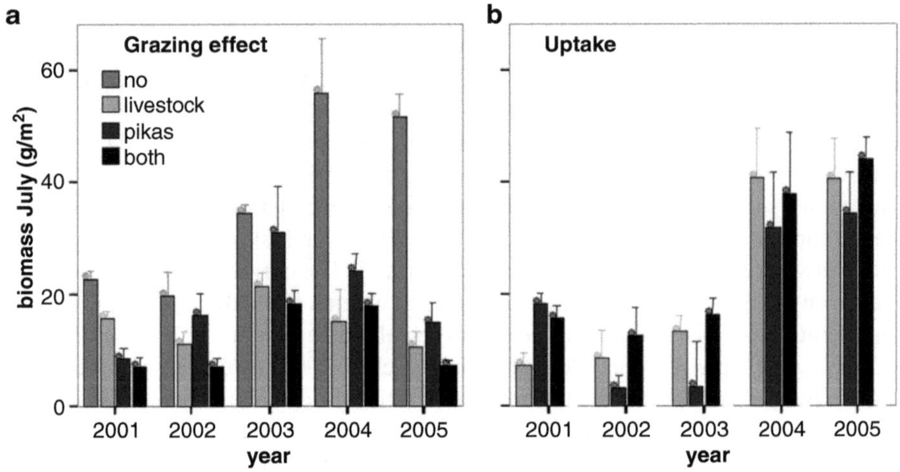

Fig. 1.8 Fencing experiment in desert steppes of southern Mongolia. The experiment comprised four blocks with differential exclusion of the two main herbivore groups (livestock – mainly sheep/goat; pikas – *Ochotona* spp.). (**a**) Amount of aboveground standing crop in July. (**b**) Uptake, estimated as the difference between grazing treatments and the respective ungrazed control (mean, standard deviation, after Wesche and Retzer 2005, supplemented)

track interannual changes in forage availability. Uptake was low during droughts in the first 2 years but remained low in the more favourable year of 2003 (Fig. 1.8b). It took a further year for herbivores to increase uptake and utilise the larger part of the available forage again. The patterns differed between groups: Pikas were able to maintain higher feeding pressure than livestock in the first year of the drought while livestock became the more active group in the following years. This is most easily explained by changes in numbers (Retzer and Reudenbach 2005): Herders moved away from the drought region in 2001, experiencing heavy losses but saving at least part of their herds. Pikas are sedentary but were able to survive quite well during the first year (partly because they had reserve fodder in their burrows). Resources were consumed in the second year and the sedentary pika populations collapsed (Nadrowski 2006).

The correlation between livestock numbers and weather conditions is also evident in an analysis of governmental livestock statistics for the region (Fig. 1.9, Begzuren et al. 2004; Retzer and Reudenbach 2005). Correlation analyses showed that occasional snow cover in winter may cause severe losses (Begzuren et al. 2004). Harsh winters of 2002, and more recently 2009/2010, resulted in massive losses (up to >30%, national livestock statistics), simply because snow cover prevented access to forage on winter pastures (see Chap. 20 by Baas et al., this volume). Winter snow may occasionally hit all sites in Central Asia, including the drier parts (Batjargal et al. 2002; Tachiiri et al. 2008). Such catastrophic conditions are called "Dzuud" or "Zud" by Mongolians. Collapses due to summer drought (called "char Dzuud" – black Dzuud) are mostly restricted to the drier parts of Central Asia, such as southern

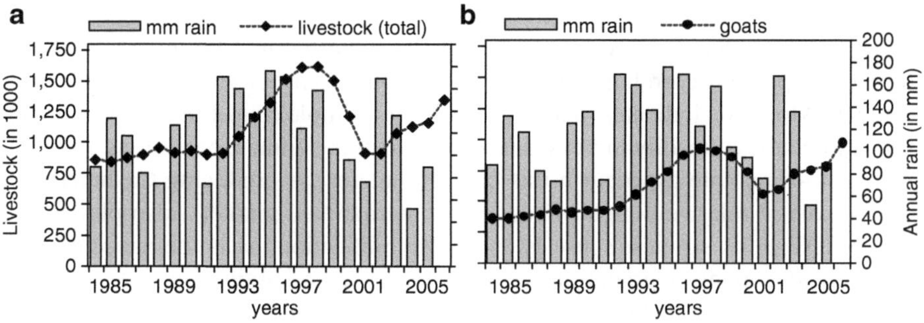

Fig. 1.9 Relationship between animal numbers and precipitation in the South Gobi Aymak, southern Mongolia (data National Statistical Yearbook). (**a**) Total livestock numbers (*dotted line*). (**b**) Goat numbers only (note that precipitation showed no long-term trend over the years; $r^2 < 0.01$)

Mongolia or the alpine steppes of Tibet. The moister tall grass steppes, forest and alpine meadows have more predictable precipitation patterns (interannual CV <25%), resulting in a more equal ANPP. This allows livestock herds to build up without suffering from regular droughts, eventually leading to high numbers and, often, overgrazing.

Such phenomena offer an explanation for the patterns observed in Fig. 1.6. Strong impact of grazing on vegetation is more commonly found under relatively moist conditions, typically >200 mm/a in Central Asia (Ellis et al. 2002). At drier sites livestock numbers are kept relatively low due to recurrent droughts, so that grazing in these steppes often only has a negligible impact on the vegetation. This is described by the non-equilibrium theory (NEQT) of rangeland science: Herds show strong fluctuations as a consequence of highly variable precipitation and thus forage availability. Relatively frequent collapses result in relatively small populations and a lower risk of grazing degradation (Ellis and Swift 1988; Sullivan and Rohde 2002; Vetter 2005). In contrast, moister sites have less variability and allow for more or less permanently high livestock numbers, which may eventually exceed carrying capacity and trigger degradation. This conceptual model describes conditions in large parts of Central Asia reasonably well and has consequently been embraced by a number of authors (Fernandez-Gimenez and Allen-Diaz 1999; Zemmrich 2007; Wesche et al. 2010).

There are, however, unsolved questions, such as the relatively limited impact of grazing in low productive rangelands with a long evolutionary history of grazing (Milchunas and Lauenroth 1993; Cingolani et al. 2005; Milchunas et al. 2008), which are often explained in terms of evolutionary adaptations without any emphasis being placed on stochastic events. The discussion among schools of thought remains intense, particularly because of the far-fetching implications of the NEQT for the evaluation of human impact in dry rangelands such as e.g. the Sahel. No consensus has been reached as yet (Gillson and Hoffman 2007; Hambler et al. 2007). From a Central Asian perspective, there are also contradictory examples including the apparent grazing degradation in dry Saxaul stands described under grazing effects above (cf. Photo 1.5c). An example of a larger-scale analysis is given by Ho (2001), who analysed livestock statistics in northern Chinese grasslands and found no apparent correlation with precipitation. Even our simple example from southern Mongolia indicates that the NEQT is too simple (Fig. 1.9b): Goat numbers have more or less steadily increased, irrespective of the climate conditions, implying that socio-economic aspects (in this case the interest in producing Cashmere that can be sold at relatively high prices to the international market), also play a role. The NEQT in its strictest sense is an oversimplification, as is the assumption of widespread grazing degradation everywhere in Central Asia. The true nature of such trends lies somewhere in between, with NEQ and steady-state conditions representing only two extremes in an overall more complex system.

Perhaps more importantly, much of the discussion on the NEQT ignores belowground conditions. The relation between precipitation patterns and belowground productivity/total productivity is not necessarily close in Central Asia (Ni 2001). Both remote sensing data and actual measurements in Chinese grasslands show that carbon stocks in soils show moderately strong correlations with precipitation patterns (Ni 2001; Yang et al. 2010b). Plants may allocate 50–98% of their growth belowground, with decreasing precipitation resulting in more belowground allocation within a given habitat type such as steppe (Fan et al. 2008; Yang et al. 2009a). The precipitation effect on allocation disappears if all grassland types in China are analysed jointly, which results in specific biological traits becoming more relevant. *Kobresia* spp. in alpine meadows of Tibet have very low aboveground-belowground ratios in phytomass, and build up huge soil carbon sinks. They occur mostly at high altitudes and consequently at low temperatures, explaining why mean annual temperature and altitude emerge as a decisive factor for sizes of soil carbon pools in large-scale comparisons in Chinese grasslands (Ni 2001; Piao et al. 2004; Dai and Huang 2006; Yang et al. 2010b). Clearly, below-ground conditions are not solely controlled by precipitation.

1.3.5 Abiotic Controls – Nutrient Availability

In our context, the main question is whether soil conditions matter at all for steppe plant performance, especially for the crucial forage productivity. Most of Central Asia is cold and dry, implying that plant growth should be limited by climate rather than by soil conditions. The NEQT, but also key reviews on dryland ecology (Noy-Meir 1973; Breman and De Wit 1983), emphasise the overwhelming role of precipitation, regarding e.g. soil nutrient availability as a secondary factor for growth in drylands. More recent reviews have, however, shown that soil nutrients may limit plant growth over a wide range of precipitation (and temperature) conditions (Breman and De Wit 1983; Hooper and Johnson 1999; LeBauer and Treseder 2008). Nitrogen has most commonly been assessed as a limiting factor, but where tested, P is usually found to be co-limiting. In the moister grass steppes (mean annual precipitation >300 mm) experimental nitrogen fertilization under ambient precipitation

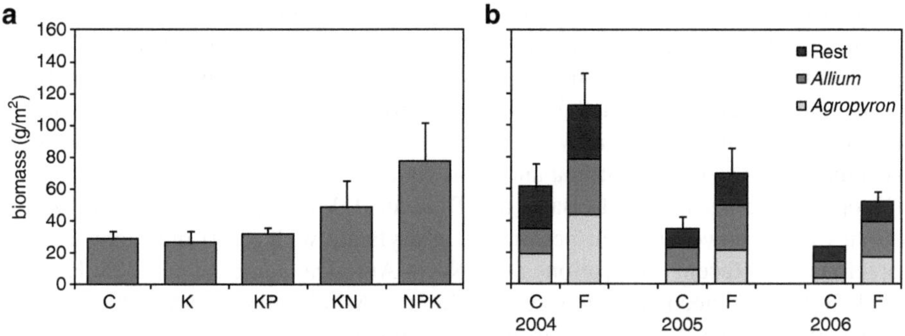

Fig. 1.10 Effects of fertilization in Central Asian grasslands. (**a**) Effects of fertilization in Tibetan *Kobresia pygmaea* mats (n=4; station Kema, 4,450 m asl, Seeber, Miehe and Wesche unpublished). Macronutrients were applied at annual rates of 10 g N, 5 g P and 34 g K per m², and under ambient precipitation. (**b**) Effects of fertilization in southern Mongolian desert steppes (after Wesche and Ronnenberg 2010 changed). NPK fertilizer was applied at equivalents of 10 g N, 3.3 g P and 33 g K per m², plus Mg and various micronutrients at an annual base (*F* fertilized, *C* controls, n=9)

usually results in a significant increase in ANPP (Pan et al. 2005; Bai et al. 2009; Fanselow et al. 2011), improved tissue N contents (Liu et al. 2007; Huang et al. 2008) but also slow changes in dominance patterns, due to differing responses to fertilization among species. Patterns in belowground productivity are less widely studied but may even be opposite (Li et al. 2011a).

Positive effects of N and/or P addition in the moister steppes of Central Asia are not surprising from a theoretical point of view, and they are important in view of the nutrient losses under heavy grazing discussed above. Together, these observations imply that land management should consider conservation of soil nutrient stocks in the Central Asian steppes. Otherwise, the slow degradation of soils will worsen the direct degradation effects apparent in the above-ground vegetation, resulting in severe and lasting losses in steppe productivity.

Less clear is the importance of soil nutrient availability under more severe climatic conditions, such as those found at high altitude sites in Tibet. Studies from other alpine sites (Theodose and Bowman 1997) and data from Central Asia described above suggest that annual mean temperatures are not necessarily decisive for growth. Nutrient shortage may affect plant growth in high altitude Central Asia, and this is supported by fertilization experiments (Fig. 1.10a). Because initial studies pointing to a certain macro-nutrient as the most limiting factor were not available, we tested full NPK fertilization in *Kobresia* meadows of eastern Tibet (ambient precipitation 400–500 mm). We found strong evidence of N-limitation, and biomass productivity was increased even further when NPK were jointly added. Alpine meadows thus suffer from nutrient constraints (N, followed by P), which may have been aggravated by traditional land use and associated nutrient withdrawal. This is also supported by a large-scale comparison of rain use efficiency (unit aboveground biomass productivity per unit of precipitation) in Tibetan grasslands that also demonstrated positive effects of more fertile soils on productivity (Yang et al. 2010a). Studies on the drier alpine steppes of Tibet are, unfortunately, not available.

Not only low temperatures but also low water availability may interact with nutrient constraints. There is contrasting evidence as to whether the relative importance of nutrient availability, especially N, declines with increasing dryness of the sites (Breman and De Wit 1983; Xia and Wan 2008) or is independent of precipitation (Hooper and Johnson 1999; LeBauer and Treseder 2008). Surprisingly few fertilization studies have, however, been performed in drylands with <200 mm annual mean precipitation. Recent publications from North America (James et al. 2005; Harpole et al. 2007) and anecdotal evidence from Central Asia (Slemnev et al. 2004) implied that nutrients may limit plant growth even under dry conditions. This was also supported by our own fertilization experiments in southern Mongolian desert steppes (Wesche and Ronnenberg 2010). Again, we had to opt for NPK fertilization performed under ambient precipitation (120–180 mm). There was clear evidence of nutrient limitation as fertilization more than doubled ANPP (Fig. 1.10b). The valuable forage grass *Agropyron cristatum* showed a particularly strong response. Tissue contents of N, and therefore raw protein contents, increased and flowering was also facilitated. The absolute magnitude of effects was higher in the moist year of 2004 (180 mm precipitation) than in the more average years of 2005 and 2006 (110, 130 mm), indicating co-limitation by water availability.

In a second experiment at the same sites, fertilizers were applied in factorial combination with irrigation equivalent to 100 mm, which is more than the average summer precipitation at the site (Ronnenberg and Wesche 2011). Fertilization again showed strong effects on ANPP, while irrigation had no significant effect. This is in contrast to findings of relatively stronger irrigation effects in moister Inner Mongolia (Fanselow et al. 2011; Li et al. 2011a), but may simply indicate that timing is important. In our experiment, water was only applied in summer (July, August) and in relatively small daily doses (5 mm). The discussion on winter and spring precipitation above implies that precipitation in late summer may have generally limited effects in the dry grasslands. Moreover, there is also evidence that a few large precipitation events may be more relevant than a number of small events (Ronnenberg et al. 2008), a pattern that has recently been described for North American dry steppes as well (Knapp et al. 2008; Heisler-White et al. 2009).

While cations are usually present in sufficiently high concentrations in the grassland soils, it remains unclear whether P or N is the more crucial limiting factor. To our knowledge, differential fertilization experiments have hardly been performed in dry Central Asia, and there is thus only indirect evidence. Plant tissue NP ratios in southern Mongolia were >20–30, which is high compared to measurements on plants all over China (Han et al. 2005; He et al. 2008). Wide NP ratios point to P rather than N as the limiting factor, although this is only an initial indication (Drenovsky and Richards 2004; Güsewell 2004). The pronounced effects of NPK fertilization on flowering activity may also point towards P limitation (Marschner 1995), and low numbers of legumes in the vegetation (see introduction on flora above) suggest that factors other than N limit growth. Indeed, the generally high pH values in the desert steppes will result in quick immobilisation of P, aggravating already low contents of P in bedrock in several parts of Central Asia (e.g. Inner Mongolia, He et al. 2008). These inferences are, however, speculative, and there is evidence in the literature that the addition of N alone may have positive effects on plant growth in dry Central Asia (Slemnev et al. 2004; Li et al. 2011b).

In any case, the data demonstrate clearly that land-use mediated loss of nutrients may constrain steppe productivity, even at relatively dry and/or cold sites. Due to the productivity-diversity relationship discussed above, low levels of nutrients may also affect species richness in the long run (Bai et al. 2007). Large-scale fertilization of steppes is certainly not feasible (due to very low nitrogen use efficiency; Wesche and Ronnenberg 2010), but future studies should address how different forms of grazing such as sedentary vs. nomadic, or goat-dominated herds vs. traditional camels, affect nutrient flows in the landscape. The comprehensive data from the well researched IMGERS station (including the MAGIM project) in Inner Mongolia shows that land use has tremendous effects on soils and vegetation in the landscape, at population and individual levels. Similar data are needed from other climatically different sites, because nutrient constraints have to be taken into account if realistic scenarios on future land use and the effects of climate change in Central Asia are to be developed.

1.4 Conclusion – Central Asia and Global Change

We have restricted our discussion to relatively simple response variables such as productivity and growth, but even these have revealed complicated patterns. These differed strongly between sites, in spite of the general similarity in physiognomy of the Central Asian grasslands. Grazing, precipitation and temperature may all have an influence, be it in isolation or through interaction. As a consequence, predictions in respect of ongoing land use and climate change remain difficult.

A main problem is that trends in key aspects of global change are not clear. In terms of land use change, intensity has increased over the last few decades and is likely to increase further with respect to current population growth in Mongolia and in the ethnic minorities of China. Animal numbers are already high and will probably increase, at least locally; although strong interannual fluctuations as shown for Mongolia render detection of trends difficult. Sedentarisation complicates the picture even further: Movements have largely ceased in Inner Mongolia, sedentarisation has commenced in Tibet, and prospects in Mongolia depend on outcomes of ongoing political discussions (Sneath 1998; Janzen 2005; Long et al. 2008; Ptackova 2011). With respect to the inherently high variability of conditions, sedentarisation is a risky strategy, and may cause higher ecological and, potentially, economic vulnerability. In particular, dry steppes, with their unpredictable conditions, deserve attention in this respect. Management schemes based on steady-state conditions and local carrying capacities are of questionable value for these regions. On the other hand, the assumption of continuous increases in herd numbers and associated degradation certainly is too simple.

Global climatic change is resulting in rising temperatures being reported by the majority of weather stations in the region, and this is also supported by models (Klein Tank et al. 2006; IPCC 2007; Dagvadorj et al. 2009). Rates of increase have widely been >2 K in the last century, triggering concerns about increasing (physiological) aridity. This is surprising in view of the discussion above that provides no evidence for an unequivocal importance of temperature controls. A warming experiment in northern China also showed that precipitation is more important than temperature alone (Niu et al. 2008). An experiment in northern Tibet described severe negative effects of warming (Klein et al. 2004), but this was conducted at the lower altitudinal (and therefore upper thermal) limit of the dominant *Kobresia* spp. More data from the core alpine sites are needed, especially in combination with water addition. Yu et al. (2010) claim that warm temperatures in Tibetan grasslands lead to a lack of winter chilling and an associated delayed onset of the growing season. This may, however, also be related to precipitation patterns: We correlated values reported by Yu et al. (2010) on the start of the growing season with precipitation in March and found a conspicuous trend (Spearman $rho = -0.41$, $p < 0.05$). Simulation model studies also imply that trends in aridity in steppe vegetation largely depend on trends in precipitation (Christensen et al. 2004) and rising CO_2 reduces drought sensitivity in grassland plants (Soussana and Luscher 2007). With no extra rain, increased temperatures may offset positive effects of higher CO_2 availability and thus higher water use efficiency; but again, moisture availability rather than temperature alone is the key variable.

This is also in line with results described above, where precipitation emerged as the perhaps single most important factor in the region. Unfortunately, trends in precipitation patterns are much less well understood than those for temperature, and congruence between model simulations over Central and Middle Asia is limited (Lioubimtseva et al. 2005; IPCC 2007). Most stations in Central Asia have reported no change or increased annual precipitation totals over the last decades (Wei et al. 2005; Klein Tank et al. 2006; Shi et al. 2007). In Mongolia, some stations show decreased precipitation (Dagvadorj et al. 2009; Dulamsuren et al. 2010), but the spatial heterogeneity is huge and, on average, total precipitation seems to be stable (Angerer et al. 2008). Future reductions in the intensity of the Indian monsoon due to warming and increasing air pollution in South Asia have been predicted for parts of the Tibetan plateau (May 2004; Zickfeld et al. 2005), while other studies imply increases of precipitation in (eastern) Tibet (Cui and Graf 2009; Yu et al. 2010). Taken together, general scenarios on moisture availability over Central Asia should therefore be taken with caution.

There is a trend towards increasing winter precipitation and snow cover (Peng et al. 2010). This may have disastrous effects on livestock herds if access to winter pastures is prevented (Dzuud), but may have positive consequences if spring moisture availability is improved. In fact, early summer growth shows positive trends to increasing snow depths in northern Chinese grasslands, while effects in Tibet are not significant, possibly due to cold conditions (Peng et al. 2010). Several authors also report increased variability and higher frequency of extreme events in precipitation (Klein Tank et al. 2006; Dagvadorj et al. 2009). This is usually regarded as critical as severe rains result in more runoff. While effects of more extreme rain events are indeed usually negative in relatively moist regions, they may be positive in dry environments, where small water doses may evaporate quickly and not even become effective for plant growth (see above, Knapp et al. 2008). Therefore, at present, neither knowledge on trends in timing of precipitation patterns, nor understanding of the potential effects seems sufficient in our region.

Finally, effects of climate and land use change on productivity and plant performance are further complicated by soil nutrient effects. We have seen that nutrient availability is likely to constrain growth in dry and moist sites alike. Atmospheric nitrogen deposits will increase over Central Asia, but rates may be too small to exert any major impact (Galloway et al. 2004; Kinugasa et al. 2008). Matter fluxes due to livestock grazing may be much more important, but this is a topic that has been explored at very few sites so far, and trends in the future are unclear.

Carbon stocks in soils are not only relevant for local phenomena but also in terms of sequestration and global carbon cycles. Recent publications imply that carbon stocks have remained largely stable in soils of northern China and Tibet (Yang et al. 2009b, 2010b). There is, however, spatial heterogeneity, and there are reports of contradictory patterns (Cao et al. 2003; Xie et al. 2007), but there is at least no unequivocal evidence for declining carbon pools in Central Asian soils. The same holds true for vegetation, where recent studies describe increasing cover in Central Asia (Piao et al. 2004; Yang et al. 2009a), while there is also evidence for severe declines (Wei et al. 2005; Angerer et al. 2008; Dagvadorj et al. 2009). Whether these are due to climate change or land use effects is still open to debate.

It is clear that knowledge on the ecology of Central Asian grasslands has vastly improved in the last two to three decades. It is, however, equally clear that we still lack a comprehensive understanding of most of the involved processes and the potential drivers for changes in plant community composition or productivity. The successively published and surprisingly often negative scenarios on future development in the course of global land use and climate change should thus be viewed with extreme caution. Central Asia is a huge region and it would be very surprising if it did not show complicated and often unexpected responses.

Acknowledgements We heartily thank M.A. van Staalduinen and M.J.A. Werger for their patience with our manuscript, which came far later than promised. Marinus Werger is also thanked for giving helpful suggestions on the text. We are well aware that reviews like the present one depend on published data by a huge number of people that moved to the field, and we are grateful for all their efforts. Our own fieldwork in Mongolia and China was directly supported by a number of colleagues that cannot be listed here (see acknowledgements in the quoted papers). Special thanks to E. Seeber for her photo from the grazing exlcosures in Qinghai and unpublished fertilization data from Tibet, and to H. von Wehrden for providing unpublished chorological data from Mongolia. Over the years, funding for our different projects came from the German Science Foundation, the German Academic Exchange Service, the German Ministry for Education and Science, the giz (formerly gtz), the Austrian Science Fund and the Schimper Foundation. Daniel McCluskey kindly checked the style.

References

Aldenderfer M (2007) Modeling the Neolithic on the Tibetan Plateau. Dev Quaternary Sci 9:151–165

Allen VG, Batello C, Berretta EJ, Hodgson J et al (2011) An international terminology for grazing lands and grazing animals. Grass and Forage Sci 66:2–28

Angerer J, Han G, Fujisaki I, Havstad KM (2008) Climate change and ecosystems of Asia with emphasis on Inner Mongolia and Mongolia. Rangelands 30:46–51

Atlas of Tibet en (Atlas of Tibetan Plateau) Plateau (1990) Institute of Geography – Chinese Academy of Sciences, Beijing

Anonymous (1996) Digitized vegetation map of China. Institute of Geography, Chinese Academy of Science, Beijing

Bagchi S, Ritchie M (2010) Herbivore effects on above- and belowground plant production and soil nitrogen availability in the Trans-Himalayan shrub-steppes. Oecologia 164:1075–1082

Bai Y, Han X-G, Wu J, Chen Z, Li L (2004) Ecosystem stability and compensatory effects on the Inner Mongolian grassland. Nature 431:181–184

Bai Y, Wu J, Pan Q, Huang J et al (2007) Positive linear relationship between productivity and diversity: evidence from the Eurasian Steppe. J Appl Ecol 44:1023–1034

Bai Y, Wu J, Clark CM, Naeem S et al (2009) Tradeoffs and thresholds in the effects of nitrogen addition on biodiversity and ecosystem functioning: evidence from inner Mongolia Grasslands. Global Change Biol 16:889–889

Barthel H (1983) Die regionale und jahreszeitliche Differenzierung des Klimas in der Mongolischen Volksrepublik. In: Barthel H, Brunner, H, Haase G (eds) Physischgeographische Studien in Asien. (= Studia Geographica 34), Ceskoslovenská Akademie Ved, Geografický ustav. Brno, pp 3–91

Batjargal Z, Oyun R, Togtokh N, Sangidansranjav S (2002) Lessons learnt from the dzud 1999–2000. In: Chuluun T, Ojima D (eds) Fundamental issues affecting sustainability of the Mongolian steppe. IISCN, Ulaanbataar, pp 73–98

Becker L, Seeber E, Wesche K, Spielvogel S et al (submitted) Non-uniform grazing effects on plant biodiversity, fodder quality and soil organic matter in upper montane grasslands of the Tibetan Plateau. Basic and Applied Ecol

Begzuren S, Ellis JE, Ojima D, Coughenour MB, Chuluun T (2004) Livestock responses to drought and severe winter weather in the Gobi Three Beauty National Park, Mongolia. J Arid Environ 59:785–796

Bläß C, Ronnenberg K, Hensen I, Wesche K (2008) Grazing impact on seed production in southern Mongolia. Mongolian J Biol Sci 6:3–9

Breman H, De Wit CT (1983) Rangeland productivity and exploitation in the Sahel. Science 221:1341–1346

Cao M, Prince SD, Li K, Tao B et al (2003) Response of terrestrial carbon uptake to climate interannual variability in China. Global Change Biol 9:536–546

Cheng X, An S, Chen J, Li B et al (2007) Spatial relationships among species, above-ground biomass, N, and P in degraded grasslands in Ordos Plateau, northwestern China. J Arid Environ 68:652–667

Christensen L, Coughenour MB, Ellis JE, Zuo ZC (2004) Vulnerability of the Asian typical steppe to grazing and climatic change. Clim Change 63:351–368

Cingolani AM, Noy-Meir I, Diaz S (2005) Grazing effects on rangeland diversity: a synthesis of contemporary models. Ecol Appl 15:757–773

Clark FE, Woodmansee RG (1992) Nutrient cycling. In: Coupland RT (ed) Natural grasslands, Ecosystems of the world 8A. Elsevier, Amsterdam/New York/Tokyo, pp 137–149

Coughenour MB, Behnke R, Lomas J, Price K (2008) Forage distributions, range condition, and the importance of pastoral movement in Central Asia – a remote sensing study. In: Behnke R (ed) The socio-economic causes and consequences of desertification in Central Asia. Springer, Heidelberg, pp 45–80

Cowan PJ (2006) Geographic usage of the terms Middle Asia and Central Asia. J Arid Environ 69:359–363

Cressey GB (1960) The deserts of Asia. J Asian Stud 19:389–402

Cui X, Graf H-F (2009) Recent land cover changes on the Tibetan Plateau: a review. Clim Change 94:47

Dagvadorj D, Khuldorj B, Aldover RZ (eds) (2009) Mongolia assessment report on climate change 2009. Ministry of Nature, Environment and Tourism, Ulaanbaatar

Dai W, Huang Y (2006) Relation of soil organic matter concentration to climate and altitude in zonal soils of China. Catena 65:87–94

Drenovsky RE, Richards JH (2004) Critical N:P values: predicting deficiencies in desert shrublands. Plant Soil 259:59–69

Dulamsuren C, Welk E, Jäger EJ, Hauck M, Mühlenberg M (2005) Range – habitat relationships of vascular plant species at the taiga forest-steppe borderline in the western Khentey Mountains, northern Mongolia. Flora 200:376–397

Dulamsuren C, Hauck M, Leuschner C (2010) Recent drought stress leads to growth reductions in *Larix sibirica* in the western Khentey, Mongolia. Global Change Biol 16:3024–3035

Duraiappah AK, Naeem S (eds) (2005) Millennium ecosystem assessment. Ecosystems and human well-being. World Resource Institute, Washington, DC

Ellis JE, Swift DM (1988) Stability of African pastoral ecosystems: alternate paradigms and implications for development. J Range Manag 41:450–459

Ellis JE, Price K, Boone R, Yu F (2002) Integrated assessment of climate change effects on vegetation in Mongolia and Inner Mongolia. In: Chuluun T, Ojima D (eds) Fundamental issues affecting sustainability of the Mongolian steppe. IISCN, Ulaanbataar, pp 26–33

Fan J, Zhong H, Harris W, Yu G et al (2008) Carbon storage in the grasslands of China based on field measurements of above- and below-ground biomass. Clim Change 86:375–396

Fanselow N, Schönbach P, Gong X, Lin S et al (2011) Short-term regrowth responses of four steppe grassland species to grazing intensity, water and nitrogen in Inner Mongolia. Plant Soil 340:279–289

Fernández RJ (2007) On the frequent lack of response of plants to rainfall events in arid areas. J Arid Environ 68:688–691

Fernandez-Gimenez ME, Allen-Diaz B (1999) Testing a non-equilibrium model of rangeland vegetation dynamics in Mongolia. J Appl Ecol 36:871–885

Fernandez-Gimenez ME, Allen-Diaz B (2001) Vegetation change along gradients from water sources in three grazed Mongolian ecosystems. Plant Ecol 157:101–118

Galloway JN, Dentener FJ, Capone DG, Boyer EW et al (2004) Nitrogen cycles: past, present, and future. Biogeochemistry 70:153–226

Gao YZ, Giese M, Lin S, Sattelmacher B et al (2008) Belowground net primary productivity and biomass allocation of a grassland in Inner Mongolia is affected by grazing intensity. Plant Soil 307:41–50

Gao Y, Giese M, Han X, Wang D et al (2009) Land use and drought interactively affect interspecific competition and species diversity at the local scale in a semiarid steppe ecosystem. Ecol Res 24:627

Gao XY, Xu WX, Yang WK, Blank DA et al (2011) Status and distribution of ungulates in Xinjiang, China. J Arid Land 3:49–60

Gillson L, Hoffman MT (2007) Rangeland ecology in a changing world. Science 315:54–54

Grubov VI (1999ff) Plants of Central Asia. Science Publishers, Enfield

Gunin PD, Vostokova EA, Dorofeyuk NI, Tarasov PE, Black CC (1995) Ecosystems of Mongolia. Nauka, Moscow

Güsewell S (2004) N:P ratios in terrestrial plants: variation and functional significance. New Phytol 164:243–266

Haase G (1983) Beiträge zur Bodengeographie der Mongolischen Volksrepublik. In: Barthel H et al (eds) Physischgeographische Studien in Asien. (= Studia Geographica 34), Ceskoslovenská Akademie Ved, Geografický ustav. Brno, pp 231–237

Hambler C, Canney S, Coe MJ, Henderson PA, Illius AW (2007) Grazing and "degradation". Science 316:1564–1565
Han W, Fang J, Guo D, Zhang Y (2005) Leaf nitrogen and phosphorus stoichiometry across 753 terrestrial plant species in China. New Phytol 168:377–385
Hansen MJ, Wilson SD (2006) Is management of an invasive grass *Agropyron cristatum* contingent on environmental variation? J Appl Ecol 43:269–280
Harpole WS, Potts DL, Suding KN (2007) Ecosystem responses to water and nitrogen amendment in a California grassland. Global Change Biol 13:2341–2348
He J-S, Wang L, Flynn DFB, Wang X et al (2008) Leaf nitrogen: phosphorus stoichiometry across Chinese grassland biomes. Oecologia 155:301–310
Heisler-White JL, Blair JM, Kelly EF, Harmoney K, Knapp AK (2009) Contingent productivity responses to more extreme rainfall regimes across a grassland biome. Global Change Biol 15:2894–2904
Herzschuh U, Tarasov PE, Wünnemann B, Hartmann K (2004) Holocene vegetation and climate of the Alashan Plateau, NW China, reconstructed from pollen data. Palaeogeogr Palaeoclimatol Palaeoecol 211:1–17
Hijmans RJ, Cameron SE, Parra JL, Jones PG, Jarvis A (2005) Very high resolution interpolated climate surfaces for global land areas. Int J Climatol 25:1965–1978
Hilbig W (1995) The vegetation of Mongolia. SPB Academic Publishing, Amsterdam
Hilbig W, Helmecke K, Schamsran Z (1993) Untersuchungen zur oberirdischen Pflanzenbiomasse von Rasengesellschaften in Gebirgen der Mongolei. Phytocoenologia 23:201–226
Hilbig W, Bastian O, Jäger E, Bujan-Orsich C (1999) Die Vegetation des Uvs Nuur Beckens (Uvs Aimak, Nordwest-Mongolei). Fed Repert 110:569–625
Hilbig W, Jäger E, Knapp HD (2004) Die Vegetation des Bogd-uul bei Ulaanbaatar (Mongolei) – Standortsbindung und pflanzengeographische Stellung. Fed Repert 115:265–342
Ho P (2001) Rangeland degradation in north China revisited? A preliminary statistical analysis to validate non-equilibrium range ecology. J Dev Stud 37:99–133
Hoekstra JM, Boucher TM, Ricketts TH, Roberts C (2005) Confronting a biome crisis: global disparities of habitat loss and protection. Ecol Lett 8:23–29
Holst J, Liu C, Yao Z, Brüggemann N et al (2007) Importance of point sources on regional nitrous oxide fluxes in semi-arid steppe of Inner Mongolia, China. Plant Soil 296:209–226
Hooper DU, Johnson L (1999) Nitrogen limitation in dryland ecosystems: responses to geographical and temporal variation in precipitation. Biogeochemistry 46:247–293
Huang J-Y, Zhu X-G, Yuan Z-Y, Song S-H et al (2008) Changes in nitrogen resorption traits of six temperate grassland species along a multi-level N addition gradient. Plant Soil 306:149–158
IPCC (2007) Climate change 2007: synthesis report. IPCC [Core Writing Team, Pachauri RK, Reisinger A (eds)], Cambridge University Press, Geneva
James JJ, Tiller RL, Richards JH (2005) Multiple resources limit plant growth and function in a saline-alkaline desert community. J Ecol 93:113–126
Janzen J (2005) Mobile livestock-keeping in Mongolia: present problems, spatial organization, interaction between mobile and sedentary population groups and perspectives for pastoral development. Senri Ethnol Stud 69:69–97
Jigjidsuren S, Johnson DA (2003) Forage plants in Mongolia. Admon Publishing, Ulaanbaatar
Kinugasa T, Shinoda M, Tsunekawa A (2008) Increasing nitrogen distribution and grassland productivity in Mongolia. In: Organizing Committee of IGC/IRC Congress (ed) Multifunctional grasslands in a changing world. Guangdongs People's Publishing House, Guangzhou, p 614
Klein J, Harte J, Zhao X (2004) Experimental warming causes large and rapid species loss, dampened by simulated grazing, on the Tibetan Plateau. Ecol Lett 7:1170–1179
Klein Tank AMG, Peterson TC, Quadir DA, Dorji S et al (2006) Changes in daily temperature and precipitation extremes in central and south Asia. J Geophys Res 111:D16105
Knapp AK, Beier C, Briske DD, Classen A et al (2008) Consequences of more extreme precipitation regimes for terrestrial ecosystems. Bioscience 58:811–821
Komonen M, Komonen A, Otgonsuren A (2003) Daurian pikas (*Ochotona daurica*) and grassland condition in eastern Mongolia. J Zool 259:281–288

Kutschera L, Sobotik M, Lichtenegger E (1997) Bewurzelung von Pflanzen in verschiedenen Lebensräumen. 5. Band der Wurzelatlas-Reihe. Stapfia 49, Linz

Lai CH, Smith AT (2003) Keystone status of plateau pikas (*Ochotona curzoniae*): effect of control on biodiversity of native birds. Biodivers Conserv 12:1901–1912

Lauenroth WK (2008) Vegetation of the shortgrass steppe. In: Lauenroth WK, Burke IC (eds) Ecology of the shortgrass steppe. Oxford University Press, Oxford/New York, pp 55–83

Lauenroth WK, Burke IC (eds) (2008) Ecology of the shortgrass steppe. Oxford University Press, Oxford

Lauenroth WK, Milchunas DG (1992) Short-grass Steppe. In: Coupland RT (ed) Natural grasslands. Ecosystems of the world 8A. Elsevier, Amsterdam/London, pp 183–226

Lavrenko EM, Karamysheva ZV (1993) Steppes of the former Soviet Union and Mongolia. In: Coupland RT (ed) Natural grasslands. Ecosystems of the world 8b. Elsevier, Amsterdam/Tokyo, pp 3–59

Lavrenko EM, Yunatov AA et al (1979) Karta rastitelnosti Mongolskoy Narodnoy Respubliki, Moskva

LeBauer DS, Treseder KK (2008) Nitrogen limitation of net primary productivity in terrestrial ecosystems is globally distributed. Ecology 89:371–379

Li S-L (2010) Woody plants in dry sands: life history traits and population dynamics. University of Utrecht, Utrecht, p 114

Li F-R, Kang L-F, Zhang H, Zhao L-Y et al (2005) Changes in intensity of wind erosion at different stages of degradation development in grasslands of Inner Mongolia, China. J Arid Environ 62:567–585

Li J, Lin S, Taube F, Pan Q, Dittert K (2011a) Above and belowground net primary productivity of grassland influenced by supplemental water and nitrogen in Inner Mongolia. Plant Soil 340:253–264

Li LJ, Zeng DH, Yu ZY, Fan ZP et al (2011b) Foliar N/P ratio and nutrient limitation to vegetation growth on Keerqin sandy grassland of North-east China. Grass and Forage Sci 66:237–242

Li C, Li Y, Ma, J (2011c) Spatial heterogeneity of soil chemical properties at fine scales induced by *Haloxylon ammodendron* (Chenopodiaceae) plants in a sandy desert. Ecol Res 26: 385–394

Liang C, Michalk DL, Millar GD (2002) The ecology and growth patterns of *Cleistogenes* species in degraded grasslands of eastern Inner Mongolia, China. J Appl Ecol 39:589–594

Lioubimtseva E, Cole R, Adams JM, Kapustin G (2005) Impacts of climate and land-cover changes in arid lands of Central Asia. J Arid Environ 62:285–308

Liu MZ, Jiang GM, Li YG, Niu SL et al (2003) Leaf osmotic potentials of 104 plant species in relation to habitats and plant functional types in Hunshandak Sandland, Inner Mongolia, China. Trees 17:554–560

Liu P, Sun OJ, Huang J, Li L, Han X (2007) Nonadditive effects of litter mixtures on decomposition and correlation with initial litter N and P concentrations in grassland plant species of northern China. Biol Fert Soils 44:211–216

Long R, Ding L, Shang Z, Guo XH (2008) The yak grazing system on the Qinghai-Tibetan plateau and its status. Rangeland J 30:241–246

Ma W, He J-S, Yang Y, Wang X et al (2010) Environmental factors covary with plant diversity–productivity relationships among Chinese grassland sites. Global Ecol Biogeogr 19:233–243

Malyshev LI (2000) Floristic division on the quantitative basis: Baikalian Siberia, Tuva and Outer Mongolia. Flora 195:330–338

Manthey M, Peper J (2010) Estimation of grazing intensity along grazing gradients – the bias of nonlinearity. J Arid Environ 74:1351–1354

Marschner H (1995) Mineral nutrition of higher plants. Academic Press, San Diego

May W (2004) Simulation of the variability and extremes of daily rainfall during the Indian summer monsoon for present and future times in a global time-slice experiment. Climate Dyn 18:183–204

Meusel H, Jäger E (1992) Vergleichende Chorologie der zentraleuropäischen Flora. Bd. III Text und Kartenband. G. Fischer, Jena, Stuttgart

Miehe G, Miehe S, Kaiser K, Jianquan L, Zhao X (2008) Status and dynamics of the *Kobresia pygmaea* ecosystem on the Tibetan Plateau. Ambio 37:272–279

Miehe G, Miehe S, Kaiser K, Reudenbach C et al (2009) How old is pastoralism in Tibet? An ecological approach to the making of a Tibetan landscape. Palaeogeogr Palaeoclimatol Palaeoecol 276:130–147

Miehe G, Miehe S, Bach K, Nölling J et al (2011) Plant communities of central Tibetan pastures in the Alpine Steppe/*Kobresia* pygmaea ecotone. J Arid Environ 75:711–723

Milchunas DG, Lauenroth WK (1993) Quantitative effects of grazing on vegetation and soils over a global range of environments. Ecol Monogr 63:327–366

Milchunas DG, Lauenroth WK, Burke IC, Detling JK (2008) Effects of grazing on vegetation. In: Lauenroth WK, Burke IC (eds) Ecology of the shortgrass steppe. Oxford University Press, Oxford/New York, pp 389–446

Munkhtsetseg E, Kimura R, Wang J, Shinoda M (2007) Pasture yield response to precipitation and high temperature in Mongolia. J Arid Environ 70:94–110

Nadrowski K (2006) Life history strategy and ecosystem impact of a small mammal herbivore in a mountain steppe. In: Fac. of Geography. Philipps University, Marburg. Available at: http://archiv.ub.uni-marburg.de/diss/z2006/0142/

Ni J (2001) Carbon storage in terrestrial ecosystems of China: estimates at different spatial resolutions and their responses to climate change. Clim Change 49:339–358

Ni J (2004) Estimating net primary productivity of grasslands from field biomass measurements in temperate northern China. Plant Ecol 174:217–234

Niu S, Wu M, Han Y et al (2008) Water-mediated responses of ecosystem carbon fluxes to climatic change in a temperate steppe. New Phytol 177:209–219

Noy-Meir I (1973) Desert ecosystems: environment and producers. Annu Rev Ecol Syst 4:25–41

Olson DM, Dinerstein E, Wikranamayake ED et al (2001) Terrestrial ecoregions of the world: a new map of life on earth. Bioscience 51:933–938

Pan Q-M, Bai Y-F, Han X-G, Yang J-C (2005) Effects of nitrogen additions on a *Leymus chinensis* population in a typical steppe of China. Acta Phytoecol Sin 29:311–317

Peart B (ed) (2008) Life in a working landscape: towards a conservation strategy for the world's temperate grasslands – compendium of regional templates on the status of temperate grasslands. Conservation and Protection. IUCN/WCPA. Temperate Grasslands Conservation Initiative, Vancouver. www.srce.com/files/App_2_Comp_of_Regional_Grassland_Templates.pdf

Pech RP et al (2007) Population dynamics and responses to management of plateau pikas *Ochotona curzoniae*. J Appl Ecol 44:615–624

Pei S, Fu H, Wan C et al (2006) Observations on changes in soil properties in grazed and nongrazed areas of Alxa Desert Steppe, Inner Mongolia. Arid Land Res Manag 20:161–175

Peng S, Piao S, Ciais P, Fang J, Wang X (2010) Change in winter snow depth and its impacts on vegetation in China. Global Change Biol 16:3004–3013

Peterson J (1994) Der Einfluß der Steppenwühlmaus *Microtus brandtii* (Radde 1891) auf Struktur und Dynamik der Steppenvegetation in der Mongolei. Unpublished PhD thesis, Fac. of Biology, Martin Luther Universität, Halle, p 131

Piao S, Fang J, Ji W, Guo Q et al (2004) Variation of a satellite-based vegetation index in relation to climate in China. J Veg Sci 15:219–226

Ptackova J (2011) Sedentarisation of Tibetan nomads in China: implementation of the Nomadic settlement project in the Tibetan Amdo area; Qinghai and Sichuan Provinces. Pastoral Res Policy Pract 1:1–4

Reading RP, Bedunah D, Amgalanbaatar S (2006) Conserving biodiversity on Mongolian rangelands: implications for protected area development and pastoral uses. USDA For Serv Proc RMRS-P 9:1–17

Retzer V (2007) Forage competition between livestock and Mongolian Pika (*Ochotona pallasi*) in southern Mongolian mountain steppes. Basic Appl Ecol 8:147–157

Retzer V, Reudenbach C (2005) Modelling the carrying capacity and coexistence of pika and livestock in the mountain steppe of the South Gobi, Mongolia. Ecol Model 189:89–104

Rhode DE, Madsen DB, Brantingham PJ, Dargye T (2007) Yaks, yak dung and prehistoric human habitation of the Tibetan plateau. Dev Quaternary Sci 9:205–226

Ronnenberg K, Wesche K (2011) Effects of fertilization and irrigation on *Stipa* steppes in dry southern Mongolia. Plant Soil 340:239–251

Ronnenberg K, Wesche K, Hensen I (2008) Germination ecology of Central Asian *Stipa* spp: differences among species, seed provenances, and the importance of field studies. Plant Ecol 196:269–280

Runnström MC (2000) Is northern China winning the battle against desertification. Ambio 29:468–476

Samjaa R, Zöphel U, Peterson J (2000) The impact of the vole *Microtus brandtii* on Mongolian steppe ecosystems. Marb Geogr Schr 135:346–360

Sasaki T, Okayasu T, Jamsran U, Takeuchi K (2008) Threshold changes in vegetation along a grazing gradient in Mongolian rangelands. J Ecol 96:145–154

Sasaki T, Okubo S, Okayasu T et al (2009) Management applicability of the intermediate disturbance hypothesis across Mongolian rangeland ecosystems. Ecol Appl 19:423–432

Schaller GB (1998) Wildlife of the Tibetan steppe. Chicago Press, Chicago

Shi D, Wan X, Davis S et al (2002) Simulation of lethal control and fertility control in a demographic model for Brandt's vole *Microtus brandtii*. J Appl Ecol 39:337–348

Shi Y, Shen Y, Kang E et al (2007) Recent and future climatic change in northwest China. Clim Change 80:379–393

Slemnev NN, Sanjid D, Khongor T, Tsooj S (2004) The features of desertified steppes development in Mongolia at the gradient of ecotopes' moistoning. Arid Ecosyst 10:172–182

Smith AT, Foggin JM (1999) The plateau pika (*Ochotona curziae*) is a keystone species for biodiversity on the Tibetan plateau. Anim Conserv 2:235–240

Sneath D (1998) State policy and pasture degradation in Inner Asia. Science 281:1147–1148

Soussana JF, Luscher A (2007) Temperate grasslands and global atmospheric change: a review. Grass Forage Sci 62:127–134

Sternberg T, Tsolmon R, Middleton N, Thomas D (2011) Tracking desertification on the Mongolian steppe through NDVI and field survey data. Intern J Digital Earth 4:50–64

Stumpp M, Wesche K, Retzer V, Miehe G (2005) Impact of grazing livestock and distance from water points on soil fertility in southern Mongolia. Mount Res Dev 25:244–251

Su Y-Z, Li Y-L, Zhao W-Z, Cui J-Y (2005) Influences of continuous grazing and livestock exclusion on soil properties in a degraded sandy grassland, Inner Mongolia, northern China. Catena 59:267–278

Sullivan S, Rohde R (2002) On non-equilibrium in arid and semi-arid grazing systems. J Biogeogr 29:1595–1618

Tachiiri K, Shinoda M, Klinkenberg B, Morinaga Y (2008) Assessing Mongolian snow disaster risk using livestock and satellite data. J Arid Environ 72:2251–2263

Theodose TA, Bowman WD (1997) Nutrient availability, plant abundance, and species diversity in two Alpine tundra communities. Ecology 78:1861–1872

Tsagaan Sankey T, Montagne C, Graumlich L et al (2006) Lower forest-grassland ecotones and 20th century livestock herbivory effects in northern China. Forest Ecol Manag 233:36–44

van Staalduinen MA, Werger MJA (2006a) Vegetation ecological features of dry Inner and Outer Mongolia. Berichte der Reinhold Tüxen Gesellschaft 18:117–128

van Staalduinen MA, Werger MJA (2006b) Marmot disturbances in a Mongolian steppe vegetation. J Arid Environ 69:344–351

van Staalduinen MA, During HJ, Werger MJA (2007) Impact of grazing regime on a Mongolian forest steppe. Appl Veg Sci 10:299–306

Vetter S (2005) Rangelands at equilibrium and non-equilibrium: recent developments in the debate. J Arid Environ 62:321–341

von Wehrden H, Wesche K (2007) Relationships between climate, productivity and vegetation in southern Mongolian drylands. Basic Appl Dryland Res 1:100–120

von Wehrden H, Hanspach J, Ronnenberg K, Wesche K (2010) The inter-annual climatic variability in Central Asia – a contribution to the discussion on the importance of environmental stochasticity in drylands. J Arid Environ 74:1212–1215

Vostokova EA, Gunin PD (eds) (2005) Ecosystems of Mongolia – Atlas (general scientific edition). Russian Academy of Sciences, Moscow

Walter H (1974) Die Vegetation Osteuropas, Nord- und Zentralasiens. G. Fischer, Stuttgart

Wang RZ (2004) Responses of *Leymus chinensis* (Poaceae) to long-term grazing disturbance in the Songnen grasslands of north-eastern China. Grass Forage Sci 59:191–195

Wang G, Wang Z, Zhou Q, Zhou W (1999) Relationship between species richness of small mammals and primary productivity of arid and semi-arid grasslands in north China. J Arid Environ 43:467–475

Wang W, Wang Q, Wang H (2006) The effect of land management on plant community compositions, species diversity and productivity of alpine *Kobersia* (sic) steppe meadow. Ecol Res 21:181–187

Wang TC, Xiong YC, Ge JP et al (2008) Four-year dynamic of vegetation on mounds created by zokors (*Myospalax baileyi*) in a subalpine meadow of the Qinghai-Tibet Plateau. J Arid Environ 72:84–96

Wang G, Li H, An M et al (2011) A regional-scale consideration of the effects of species richness on above-ground biomass in temperate natural grasslands of China. J Veg Sci 22:414–424

Wei W, Zhou H, Shi Y et al (2005) Climatic and environmental change in the source areas of dust storms in Xinjiang, China, during the last 50 years. Water Air Soil Pollut 5:207–216

Weischet W, Endlicher W (2000) Regionale Klimatologie. Teil 2. Die Alte Welt. Teubner, Stuttgart, Leipzig

Wesche K, Retzer V (2005) Is degradation a major problem in semi-desert environments of the Gobi region in southern Mongolia? In: Retzer V et al (eds) Erforschung Biologischer Ressourcen der Mongolei. Martin-Luther-University Halle-Wittenberg, Halle, pp 133–146

Wesche K, Ronnenberg K (2010) Effects of NPK-fertilisation in arid southern Mongolian desert steppes. Plant Ecol 207:93–105

Wesche K, Miehe S, Miehe G (2005) Plant communities of the Gobi Gurvan Sayhan National Park (South Gobi Aimag, Mongolia). Candollea 60:149–205

Wesche K, Nadrowski K, Retzer V (2007) Habitat engineering under dry conditions: the impact of pikas (*Ochotona pallasi*) on southern Mongolian mountain steppes. J Veg Sci 18:665–674

Wesche K, Ronnenberg K, Retzer V, Miehe G (2010) Effects of herbivore exclusion in southern Mongolian desert steppes. Acta Oecol 36:234–241

Wesche K, Walther D, von Wehrden H, Hensen I (2011) Trees in the desert: reproduction and genetic structure of fragmented Ulmus pumila forests in Mongolian drylands. Flora 206:91–99

White RP, Murray S, Rohweder M (2000) Pilot analysis of global ecosystems. Grassland ecosystems. World Resource Institute, Washington, DC

Xia J, Wan S (2008) Global response patterns of terrestrial plant species to nitrogen addition. New Phytol 179:428–439

Xiao X, Shu J, Yifeng W et al (1996) Temporal variation in aboveground biomass of *Leymus chinense* steppe from the species to community levels in the Xilin River Basin, Inner Mongolia, China. Vegetatio 123:1–12

Xie Y, MacKinnon J, Li D (2004) Study on biogeographic divisions of China. Biodivers Conserv 13:1391–1417

Xie Z, Zhu J, Liu G et al (2007) Soil organic carbon stocks in China and changes from 1980s to 2000s. Global Change Biol 13:1989–2007

Xu Y, Li L, Wang Q et al (2007) The pattern between nitrogen mineralization and grazing intensities in an Inner Mongolian typical steppe. Plant Soil 300:289–300

Yang Y, Fang J, Ji C, Han W (2009a) Above- and belowground biomass allocation in Tibetan grasslands. J Veg Sci 20:177–184

Yang Y, Fang J, Smith P et al (2009b) Changes in topsoil carbon stock in the Tibetan grasslands between the 1980s and 2004. Global Change Biol 15:2723–2729

Yang Y, Fang J, Fay PA et al (2010a) Rain use efficiency across a precipitation gradient on the Tibetan Plateau. Geophys Res Lett 37:L15072. doi:15010.15129/12010GLO043920

Yang Y, Fang J, Ma W et al (2010b) Soil carbon stock and its changes in northern China's grasslands from 1980s to 2000s. Global Change Biol 16:3036–3047

Yu H, Luedeling E, Xu J (2010) Proceedings of the National Academy of Sciences of the United States of America 107:22151–22156

Yuan W, Zhou G, Wang Y et al (2007) Simulating phenological characteristics of two dominant grass species in a semi-arid steppe ecosystem. Ecol Res 22:784–791

Zemmrich A (2007) Equilibrium or non-equilibrium ecosystems? Scaling effects in rangeland ecosystems of western Mongolia. Erdkunde 61:335–343

Zemmrich A, Manthey M, Zerbe S, Oyunchimeg D (2010) Driving environmental factors and the role of grazing in grassland communities: a comparative study along an altitudinal gradient in Western Mongolia. J Arid Environ 74:1271–1280

Zhang Y, Zhang Z, Liu J (2003a) Burrowing rodents as ecosystem engineers: the ecology and management of plateau zokors *Myospalax fontanierii* in alpine meadow ecosystems on the Tibetan plateau. Mammal Rev 33:284–294

Zhang Z, Pech R, Davis S et al (2003b) Extrinsic and intrinsic factors determine the eruptive dynamics of Brandt's voles *Microtus brandtii* in Inner Mongolia, China. Oikos 100:299–310

Zhang G, Han X, Elser J (2011) Rapid top–down regulation of plant C:N:P stoichiometry by grasshoppers in an Inner Mongolia grassland ecosystem. Oecologia 166:253–264

Zhou R, Yang Y, Fang J (2007) Responses of vegetation activity to precipitation variation on the Tibetan Plateau. Acta Sci Nat Univ Pekin 43:771–775

Zhu T-C (1993) Grasslands of China. In: Coupland RT (ed) Natural grasslands, Ecosystems of the world 8b. Elsevier, Amsterdam/London/New York/Tokyo, pp 61–82

Zickfeld K, Knopf B, Petoukhov V, Schellnhuber HJ (2005) Is the Indian summer monsoon stable against global change? Geophys Res Lett 32. doi:10.1029/2005GL022771: L15707. doi: 15710.11029/12005GL022771

Zielinski J (1982) The effect of the Brandt vole (*Microtus brandtii* Radde, 1861) colonies upon vegetation in the *Caragana* steppe of central Asia. Pol Ecol Stud 8:41–56

Zou T, Li Y, Xu H, Xu G-Q (2010) Responses to precipitation treatment for *Haloxylon ammodendron* growing on contrasting textured soils. Ecol Res 25:185–194

Chapter 2
The Steppe Biome in Russia: Ecosystem Services, Conservation Status, and Actual Challenges

Ilya E. Smelansky and Arkadiy A. Tishkov

Abstract Russian steppe should be recognized as an endangered biome which dramatically declined and degraded during the last two centuries. We summarize the major economical and conservational values of the Russian part of the steppe biome relating it with the main elements of the biodiversity pattern of the biome. The principal factor actually driving the biome dynamics is land use, foremost agriculture. Recently almost all surviving steppes in Russia were used as pastures for cattle, sheep, and horse breeding while specific land-use patterns are characteristic for each main subdivision of the biome.

The conservation status of the steppe ecosystems is strongly insufficient. The steppe is the least protected biome according to all indicators: the least coverage of protected areas, the least biome fraction in national PAs, the least average area of PA, etc. For main subdivisions of the steppe biome we estimate the proportion of the area under protection nationally. It comprises 3–10% and for the total biome and all levels of conservation is not exceeding 5%.

Thus the major part of the biome is still unprotected and faces many actual threats and new challenges caused by land use changes, the main ones being conversion (and re-conversion) into cropland, overgrazing and pasture abandonment, wild fires, mining and fossil hydrocarbons extraction, over-exploitation and poaching, and

I.E. Smelansky (✉)
UNDP/GEF Project "Improving the Coverage and Management Efficiency of Protected Areas in the Steppe Biome of Russia", Moscow, Russia

NGO Siberian Environmental Center, Novosibirsk, Russia
e-mail: oppia@yandex.ru

A.A. Tishkov
Institute of Geography, Russian Academy of Sciences, Moscow, Russia

Russian Geographical Society, Moscow, Russia
e-mail: tishkov@biodat.ru

M.J.A. Werger and M.A. van Staalduinen (eds.), *Eurasian Steppes. Ecological Problems and Livelihoods in a Changing World*, Plant and Vegetation 6,
DOI 10.1007/978-94-007-3886-7_2,

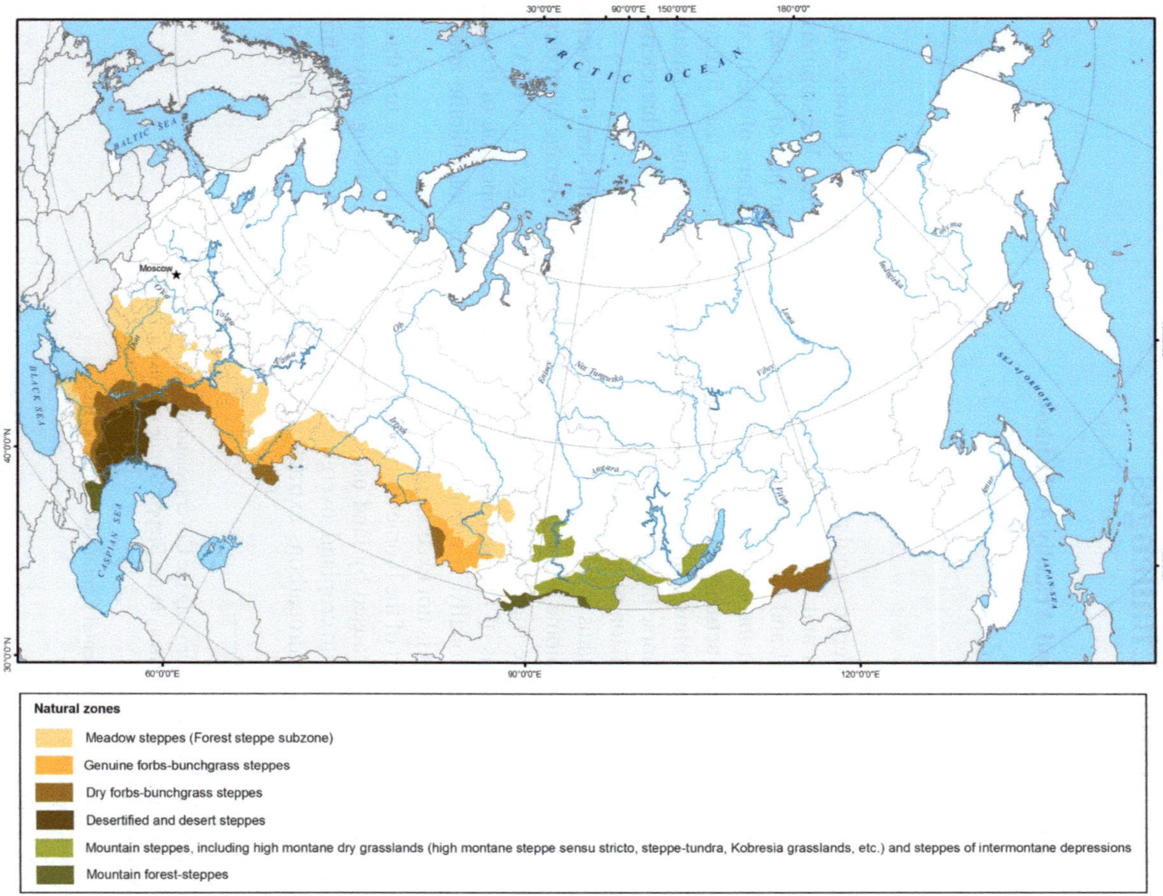

Fig. 2.1 Steppe biome division within the borders of Russia (Based on the map of Ogureeva (1999) with some changes)

afforestation. Almost all threat intensities and specifics vary following market signals. For example, turning secondary steppes into cropland was significantly intensified due to the global biofuel boom and grain crisis.

2.1 Main Divisions of the Steppe Biome in Russia

While Russia commonly seems to be a 'forest country' it actually possesses a huge extent of grasslands. By its acreage and diversity, steppe is the largest temperate grassland type in Russia.

The huge Eurasian steppe biome is divided along latitudinal, altitudinal and longitudinal gradients, leading to many subdivision schemes recognized by different authors (Kucheruk 1959; Lavrenko et al. 1991; Mordkovich 1994; Namzalov 1994; Chibilev 1998; Nikolaev 1999; Korolyuk 2002). Another way to divide the biome is distinguishing edaphic and other special variants. Whatever dividing scheme we accept, Russia encompasses a very significant part of the diversity of the Eurasian steppe.

It is necessary to first explain the basic gradients and characteristics we used (Fig. 2.1).

2.1.1 Latitudinal and Altitudinal Zoning

According to the scheme by Lavrenko (1970a, b), Lavrenko et al. (1991), the most accepted in post-Soviet countries, the steppe zone in the Eurasian plains divides into four latitudinal bands. All of them have analogous zones describing altitudinal changes in the mountains along the southern border of Russia (and outside Russia as well). In general, from north to south all community characters decrease: species richness, total biomass, aboveground/belowground biomass ratio (from 1/8 to 1/30 over the entire gradient), productivity (by a 100 times), vegetation cover, etc. The number and diversity of small burrowing mammals as well as other burrowers increase from north to south as do the share of xerophytic semi-shrubs in vegetation cover and number of species in plant communities.

In this analysis we accept the following categorization of steppe types:

1. **Meadow steppes** are the least dry and the most mesophytic steppe grasslands that typically form the grassland component of the forest-steppe landscape, i.e. intermingle with tree and bush groves even on plain watersheds. Grass cover is dense and colorful, dominated by many species of loose bunch grasses and forbs. The vegetation's species richness (64–67 species per 100 m^2 on average, but up to 70–95 species per 100 m^2 – Korolyuk et al. (2008); Dzybov 2010), evenness, coverage (about 100%), total biomass (15–30 $t.ha^{-1}$), and productivity (18–25 $t.ha^{-1}.year^{-1}$) are high. A relatively large proportion of the biomass stands aboveground. Many blossom waves continue non-stop, one after another,

from April to October. Characteristic grasses are: *Phleum phleoides, Poa stepposa, Helictotrichon schellianum, Calamagrostis epigeios, Stipa pennata, Carex humilis*, etc. The thickest Chernozem soils (Rich Chernozem subtype in the Russian National Soil Classification) are formed under meadow steppes.

2. **Genuine forbs-bunchgrass steppes.** These are dominated by xero-mesophytic graminoids (*Stipa zalesskii, S. tirsa, S. pulcherrima, Helictotrichon desertorum* are characteristic), and have a significant proportion of xero-mesophytic and even mesophytic forbs. Blossom waves are not so numerous. The vegetation's development is interrupted by a short summer pause, though not every year. Typical Chernozem soils are formed under these grasslands.
3. **Genuine (dry) bunchgrass steppes** are even more droughty grassland types. They are dominated by xerophytic and meso-xerophytic graminoids (e.g. *Stipa lessingiana, S. krylovii, Festuca valesiaca, Koeleria cristata, Agropyron pectinatum*), with xerophytic semi-shrubs (*Artemisia* spp., some other Asteraceae, Chenopodiaceae, etc.). The vegetation is not dense; most of the biomass is belowground. The characteristic soils (in the Russian classification which is rather similar to the FAO one) are Southern Chernozem, Dark Kastanozem, and Kastanozem.

   Many authors combine types (2) and (3) into one large **genuine steppes** unit.

4. **Desertified and desert steppes** are the driest types. The vegetation is rather sparse; no more than 25–40% of the ground is covered with plants while belowground root systems are densely packed. Typical dominants are *Stipa sareptana, S. glareosa, S. caucasica, S. gobica, Cleistogenes squarrosa.* The characteristic soils are Light Kastanozem and Kastanozem. Desertified steppes form the grassland component of semi-desert landscapes, where they are intermingled with desert semi-shrub communities, saline habitats, and xerophytic shrub communities. Desert steppe is extremely arid grassland, where tiny xerophytic grasses and semi-shrubs form the usual communities.
5. **Special mountain steppes.** The stony variants of steppes (see below) are characteristic for mountain landscapes but there are also some additional steppe formations that are not at all found in the plains. Especially noteworthy are the cold cryophytic steppes in the Altay and Sayan Mts., Dauria, etc. In these extremely continental arid mountains steppes and alpine tundra grasslands are in direct contact and transgress into each other. Another type is the "subtropical steppe" with *Andropogon ischaemum,* some species of *Elytrygia,* etc., present in the Caucasus Mountains (and much more developed in the mountains of Central Asia).

2.1.2 Longitudinal Zoning

There are two major subregions in the Steppe Region of Eurasia: the Pontic-Kazakh Steppe Subregion and the East Siberian – Inner Asian (Daurian-Mongolian) Steppe Subregion (Lavrenko et al. 1991; Korolyuk 2002). They have their dividing line roughly across the Altay Mountains and Yenisei River. Actually, the Steppe Region of Russia could be divided into five longitudinal sectors (out of seven recognized for the entire continent), but subdivision into two Subregions suffices for this chapter. The two major subregions differ in many characteristics. Their floras differ notably from one another. The distribution area of many plant species is restricted to only one steppe subregion: *Cymbaria, Saposhnikovia, Filifolium, Panzeria, Schizonepeta, Stellera, Lespedeza*, etc. are recorded only in Eastern steppes, while *Trinia, Seseli, Crambe, Salvia, Verbascum, Tulipa, Ornithogalum*, etc. are characteristic for Western ones. Feather-grasses (*Stipa spp.* from the *Stipa* and *Barbata* sections) are typical for the Western steppes but are almost not present in Inner Asia, while all species of *Stipa* dominating the Eastern steppes belong to the needle-grasses (the *Capillatae (=Leiostipa)* and *Smirnovia* sections). Many plant and invertebrate species are endemic for only one subregion. The Pontic-Kazakh Steppe Subregion is rich in spring ephemeroids while late-summer annuals are characteristic for the Daurian-Mongolian steppes due to a very different precipitation regime. Communities dominated by *Filifolium sibiricum, Leymus chinensis, Stipa baicalensis* are not found in the western half of the Steppe Region, and communities in which *Stipa lessingiana, S. zalesskii, S. ucrainica, Artemisia austriaca* and *Artemisia spp.* from the *Seriphidium* subgenus do not occur in the eastern one. Desert steppes occur only in the Eastern Steppe Subregion.

2.1.3 Edaphic Variants

Depending on specific substrates steppes form many types of edaphic variants. Sand (psammophytic) steppes, stony (petrophytic) steppes, different variants of salt steppes, and calcareous steppes are recognized generally. Nearly all variants occur in every latitudinal band and each subregion (only sand steppes are restricted to the Western Subregion). Specific patterns of typical and edaphic variants are characteristic for different steppe landscapes depending on their geology, evolutionary history, and land use.

2.1.4 Shrub Steppes

Shrubs are a common component of steppe vegetation in all latitudinal bands and longitudinal sectors. Many shrub species are characteristic for and not found outside the steppe biome. There are species of *Caragana* (in Russia the main ones are: *C. frutex, C. altaica, C. bungei, C. spinosa, C. scythica*), *Chamaecytisus, Calophaca, Spiraea* (the most common are: *S. crenata, S. hypericifolia, S. trilobata*), *Amygdalus, Prunus, Cerasus, Armeniaca, Ulmus*, etc. Sometimes shrubs have a special, important role and define certain steppe community types – they are recognized as special shrub steppes (usually with *Caragana* and *Spiraea*).

2.1.5 Steppe Landscape Complexes with Other Ecosystems

In most cases steppe grasslands in mosaics with other ecosystem types form the typical landscapes of the Steppe zone. The two most obvious examples are forest-steppe (wooded steppe) and semi-desert, in both of which steppe communities occur in mosaic patterns with other types of vegetation (forest, desert, shrubland, etc.).

Throughout the vast plains of Western Siberia, Dauria, and to a smaller extent the Russian Plain and Middle Siberia, the steppes fringe on salty and freshwater wetlands, saline grasslands, halophytic succulents communities, or are dotted with *Betula*, *Populus*, and *Salix* groves.

Many communities of these landscape complexes are found only or primarily in steppe landscapes – examples are the characteristic ravine "*bairak*" forests and shrubberies, steppe wetlands, rocky (petrophytic) steppic dwarf shrub lands, etc. For conservation purposes these additional "steppe related ecosystems" (i.e. they are 'intrazonal' in the Steppe zone) should be treated together with the steppe ecosystems as a holistic landscape complex.

2.1.6 Succession Series

All kinds of steppe ecosystems are highly dynamic. They tend to experience extensive and strong disturbances. Therefore all steppe types listed above have both degraded and serial variants. Some kinds of disturbances are common and characteristic for the steppe biome, particularly grazing and fire which are intimately interacting. Another characteristic disturbance is the digging activity of burrow-inhabiting colonial mammals.

In the last decades steppe succession on old-fields and formerly overgrazed lands became very common in Russia. Therefore serial communities occur widely and are highly diverse. Disturbed and serial ecosystems make an important contribution to the total steppe area and the actual biodiversity of the Steppe Region. Thus, serial steppes dominated by *Stipa capillata* are probably the commonest type recorded in all latitudinal bands and longitudinal sectors. Several steppe plants and animals occur, as regards the Russian part of their distribution area, only in serial steppes dominated by *Andropogon (=Botriochloa) ischaemum* in the Caucasus foothills.

2.1.7 Steppe Biome Area and Distribution

Steppe grasslands occur in 35 of the 83 administrative provinces of Russia (Federal Subjects) that extend at least partly south of 55° NL (Fig. 2.1). Extremely small pieces of steppe may be also found in several other provinces, not considered here. These 35 provinces cover a massive area of 4,806,295 km^2 but only 2,300,000 km^2 are part of the steppe geographical zone (including forest-steppe and semi-desert).

Accurate inventory of the actually existing steppe area in Russia has not been done yet; neither for the total area of temperate grasslands. An estimate may be obtained on the basis of land-use and land inventory data. For several provinces special investigations on steppe distribution were undertaken and more accurate data is available. Nevertheless here we use only coarse land inventory data for comparability and uniformity among different provinces.

Based on this method the total area of grazing lands (as taken in land inventory data) is 681,249 km^2 (mountainous grasslands are underestimated and arctic tundra is not included) (official data of 2006 – Compendium 2007). In every province where steppe ecosystems occur most of the natural pastures are steppe grasslands. Within 35 cited "steppe" provinces the total area of grazing lands (without old-fields and reserve lands) in the land inventory is 565,823 km^2 (data of 2006 – Compendium 2007). Based on this figure the actual steppe area in these provinces can be estimated at approximately 500,000 km^2 or somewhat less (Table 2.1) constituting about 20% of the whole steppe zone area of Russia.

Thus, four provinces top the list with grazing lands areas up of 30,000 km^2: Kalmyk Republic, Zabaikalskii Krai (Territory), Orenburg Province, and Tyva Republic. These four are scattered along the steppe zone of Russia from its western (Kalmyk R.) to its eastern end (Zabaikalskii Krai). Of these, Kalmyk R. falls entirely in the semi-desert geographical subzone and its grazing lands consist of steppes, semi-deserts, and sand and salt deserts. The steppes take at most half of the total grazing lands area. The areas of grazing lands of Zabaikalskii Krai and Tuva R. are more heterogeneous. They comprise steppes and desertified areas, meadows of the taiga zone, and alpine grasslands (in Tyva especially). The grazing lands of Orenburg Province are almost completely steppe; the province lies in the core of the steppe zone. Thus Orenburg should be considered as a region of highest priority for steppe occurrence and steppe conservation, while the other three top provinces come next. Other important provinces are: Altaiskii Krai, Rostov, Saratov, Volgograd provinces, Bashkortostan, and Buryat Republic.

2.2 Global Importance

2.2.1 UNESCO World Natural Heritage (WHN)

As from 1996 Russia holds eight WNH sites. Among them three properties include some steppe landscapes, all situated in South Siberia: the Golden Mountains of Altai, the Uvs Nuur Basin, and Lake Baikal Basin (Fig. 2.2).

The Golden Mountains of Altai (1998) comprise 1,611,457 ha in total, of which 1,002,000 ha are strictly protected areas while the remaining 609,457 ha have a lower protection status. The following areas are protected (World Heritage Site categories (I), (II), (III) and (IV)):

Altaiskiy Zapovednik, category Ia (Strict Nature Reserve) 872,000 ha
Katunskiy Zapovednik, Ia (Strict Nature Reserve) 130,000 ha

Table 2.1 Occurrence of grazing lands in the Russian provinces (only provinces with steppe ecosystems are considered)

Province (Federal subject)	Grazing lands (km^2)	Province total area (km^2)	Grazing lands as a % of the total province area	Ecosystem cover of the grazing lands	Zonal position	Location
Kalmyk Republic	52,418	74,731	70.1	Steppes + deserts	Semi-desert	South-West Russia
Zabaikalskii Krai (Territory)	44,541	431,892	10.3	Steppes and meadows	Steppe/Forest-steppe/Taiga/Mountains	East Siberia – Baikal
Orenburg P.	39,908	123,702	32.3	Steppes mainly	Steppe/Forest-steppe	South Urals
Tuva R.	34,606	168,604	20.5	Steppes + alpine grasslands	Steppe/Taiga/Mountains	Central Asia
Altaiskii Krai (Territory)	27,985	167,996	16.7	Steppes mainly	Steppe/Forest-steppe/Taiga	West Siberia – Altai
Volgograd P.	26,580	112,877	23.6	Steppes mainly	Steppe/Semi-desert	Volga, Don
R. of Daghestan	25,905	50,270	51.5	Steppes, deserts, and alpine grasslands	Semi-desert/Mountains	South-West Russia/Caucasus
Rostov on Don P.	25,564	100,967	25.3	Steppes mainly	Steppe	South-West Russia/Don
Saratov P.	24,611	101,240	24.3	Steppes mainly	Steppe/Forest-steppe	Volga
Astrakhan Province	24,149	49,024	49.3	Steppes + deserts	Semi-desert	Volga
Bashkortostan R.	23,664	142,947	16.6	Steppes mainly	Steppe/Forest-steppe/Forest	South Urals
Novosibirsk P.	23,193	177,756	13.1	Meadows and saline vegetation mainly, steppes	Steppe/Forest-steppe/Forest	West Siberia
Buryat R.	18,579	351,334	5.3	Steppes mainly, woodlands	Steppe/Forest-steppe/Taiga/Mountains	East Siberia –Baikal
Stavropol Territory	16,286	66,160	24.6	Steppes + alpine grasslands	Steppe/Mountains	South-West Russia/Caucasus
Altai R.	15,257	92,903	16.4	Steppes + alpine grasslands	Steppe/Forest-steppe/Taiga/Mountains	Altai
Chelyabinsk P.	13,628	88,529	15.4	Steppes mainly	Steppe/Forest-steppe/Forest/Mountains	South Urals – West Siberia
Krasnoyarsk Territory	13,275	723,671	1.8	Steppes, meadows, woodlands	Steppe/Forest-steppe/Forest/Taiga/Tundra/Mountains	Central Siberia

Omsk P.	12,669	141,140	9.0	Steppes and meadows	Steppe/Forest-steppe/Forest	West Siberia
Kurgan P.	10,264	71,488	14.4	Steppes and meadows	Forest-steppe	West Siberia
Khakass R.	10,244	61,569	16.6	Steppes mainly	Steppe/Taiga/Mountains	Central Siberia
Tatarstan R.	9,185	67,847	13.5	Steppes mainly	Forest-steppe/Forest	Volga
Samara P.	8,490	53,565	15.9	Steppes mainly	Steppe/Forest-steppe	Volga
Voronezh P.	7,673	52,216	14.7	Steppes mainly	Forest-steppe	Central Chernozem Region
Tyumen P.	7,581	160,122	4.7	Steppes, meadows, woodlands	Forest-steppe/Forest/Tundra	West Siberia
Irkutsk P.	6,409	774,846	0.8	Steppes, meadows, woodlands	Steppe/Forest-steppe/Taiga/Mountains	Baikal
Kemerovo P.	5,924	95,725	6.2	Steppes and meadows	Forest-steppe/Taiga/Mountains	West Siberia – Central Siberia
Chechen R.	5,763	15,647	36.8	Steppes, deserts, and alpine grasslands	Semi-desert/Mountains	South-West Russia/Caucasus
Krasnodar Territory	5,343	75,485	7.1	Steppes and alpine grasslands	Forest-steppe/Mountains	South-West Russia/Caucasus
Penza P.	5,311	43,352	12.3	Steppes mainly	Forest-steppe	Central Chernozem Region – Volga
Belgorod P.	3,996	27,134	14.7	Steppes mainly	Steppe/Forest-steppe	Central Chernozem Region
Ulyanovsk P.	3,929	37,181	10.6	Steppes and meadows	Forest-steppe	Volga
Kursk P.	3,652	29,997	12.2	Steppes mainly	Forest-steppe	Central Chernozem Region
Orel P.	3,421	24,652	13.9	Steppes and meadows	Forest-steppe	Central Chernozem Region
Tula P.	3,004	25,679	11.7	Steppes and meadows	Forest-steppe/Forest	Central Chernozem Region
Lipetsk P.	2,816	24,047	11.7	Steppes mainly	Forest-steppe	Central Chernozem Region
Total	**565,823**	**4,806,295**	**11.8**			

Area data source: Compendium (2007)

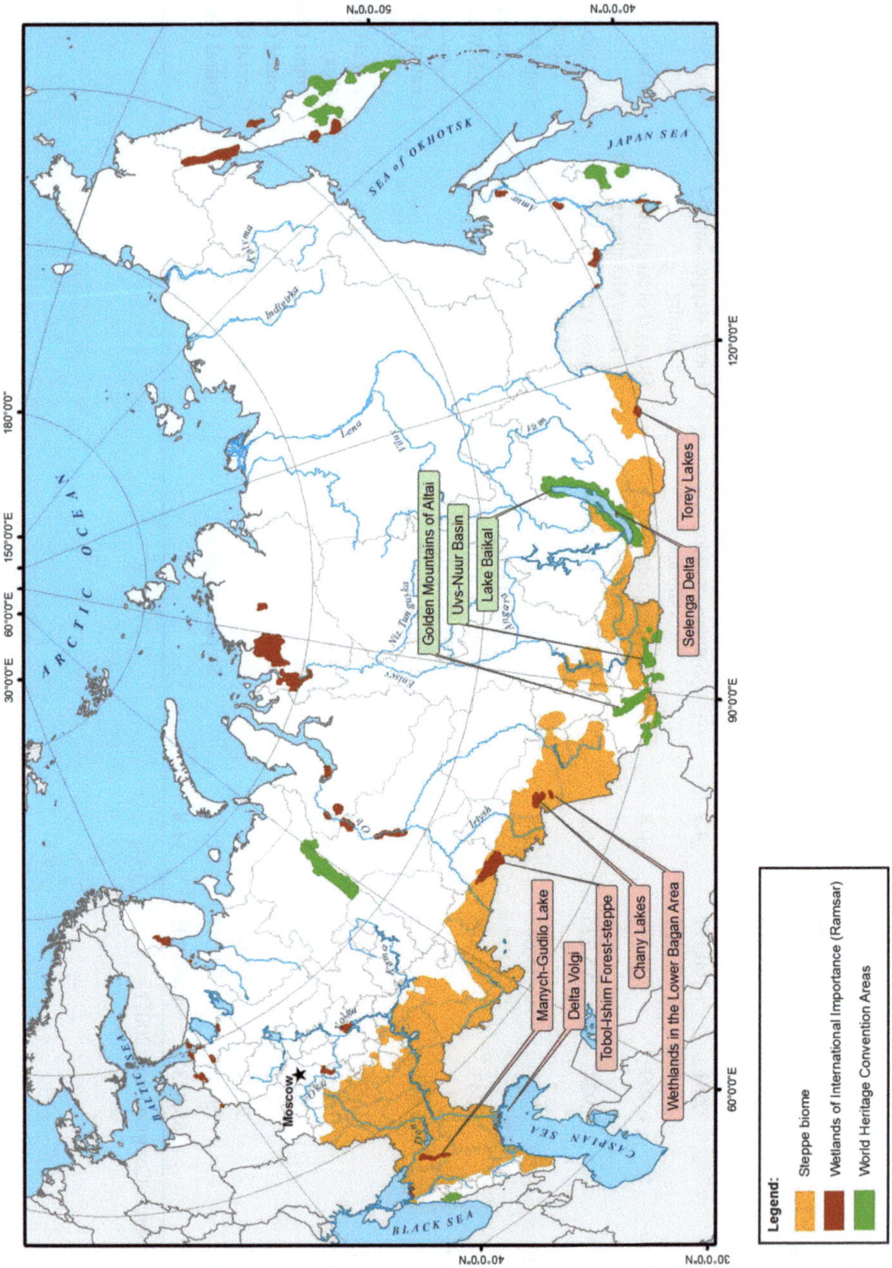

Fig. 2.2 UNESCO World Natural Heritage and Ramsar sites in the Steppe Region of Russia. On the map: WNH site labels are *light blue* while labels of Ramsar sites are filled *pink*

Lake Teletskoe, III (Nature Monument) 93,753 ha
Mt. Belukha, VI (Nature Park) 262,800 ha
Ukok Quiet Zone, VI (Nature Park) 252,904 ha

The site Ukok Plateau comprises mainly mountain (highland) steppes, steppe-like *Kobresia* grasslands, and tundra-steppe ecosystems; small areas of steppe are protected in the Altaiskiy and Katunskiy Strict Nature Reserves (also see below).

The Uvs Nuur Basin (2003, transboundary with Mongolia) comprises 1,068,853 ha. On the Russian side the property includes seven scattered plots of the Ubsunurskaya Kotlovina Strict Nature Reserve:

1. Mongun Taiga, 15,890 ha core area + 84,510 ha buffer area
2. Ubsu-Nur, 4,490 ha
3. Oroku-Shinaa, 28,750 ha
4. Aryskannyg, 15,000 ha + 11,800 ha
5. Yamaalyg, 800 ha + 4,000 ha
6. Tsugeer Els, 4,900 ha + 50,000 ha
7. Ular, 18,000 ha + 20,480 ha

Total: 258,620 ha (87,830 ha core area + 170,790 ha buffer area).

Steppe ecosystems are the main land cover in plots (2), (5), (6), and partially in plots (1), (3), (4) and (7). A high diversity of different types of Central Asian steppe ecosystems and landscapes is characteristic in this WNH site.

Lake Baikal Basin (1996) 88,000 km^2: Lake Baikal, 31,471 km^2, Coastal Protection Zone 56,500 km^2 (19,490 km^2 nationally protected):

Baikalo-Lenskiy Zapovednik, Ia (Strict Nature Reserve) 659,919 ha
Baikalskiy Zapovednik, Ia (Strict Nature Reserve, Biosphere Reserve), 165,724 ha
Barguzinskiy Zapovednik, Ia (Strict Nature Reserve, Biosphere Reserve), 374,423 ha
Zabaikalskiy National Park, II (National Park), 245,000 ha
Pribaikalskiy National Park, II (National Park), 418,000 ha
Frolikhinskiy Zakaznik, IV (Habitat/Species Management Area), 68,000 ha
Kabanskiy Zakaznik, IV (Habitat/Species Management Area, Ramsar Site), 18,000 ha.

From these only the Pribaikalskiy NP includes a large steppe area, Olkhon Island (dry genuine steppe of the Daurian-Mongolian Subregion, mostly sandy and stony variants). Tiny pieces of meadow steppe are protected in some other parks.

2.2.1.1 Ramsar Sites

Even though wetlands seem to be diametrically different from steppe ecosystems, due to their environmental and ecological characters, several Ramsar sites encompass some areas of steppe and/or are important for some steppe species. To date, Russia designated 35 Ramsar sites (as of 13 September 1994) while the Shadow list is much longer (166 sites) (Krivenko 1998, 2000). Only six Ramsar sites contain some steppe areas (Fig. 2.2, Table 2.2). With the exception of the Selenga Delta, all these sites include steppe as part of steppe – wetland patterns and in combination with diverse saline habitats. Generally, halophytic variants of steppes are best developed in these situations.

2.2.2 WWF Terrestrial Global 200

Within the bounds of Russia there are 11 eco-regions listed in the WWF Terrestrial Global 200 (Fig. 2.3). Only one of them is recognized especially for its steppe ecosystems, the Daurian/Mongolian Steppe. Its main part lies in Mongolia and China, with only approximately 10% in Russia. There are several eco-regions marked as "forest" but which actually contain some important steppe areas: Altai-Sayan Montane Forests and Ural Mountains Taiga and Tundra. Unfortunately the WWF Terrestrial eco-region Altai Steppe and Semi-desert is mistakenly placed outside of Russian territory and not recognized as globally important (not listed in Global 200).

Ural Mountains Taiga and Tundra: The southern part of the eco-region corresponds with the mountainous part of Bashkortostan. It includes some steppe tracts while the largest and most important steppe area of Bashkortostan is left outside of the eco-region.

Altai-Sayan Montane Forests: The Russian part of the eco-region includes both the Altai and Tuva Republics, and partially Altaiskii Krai, Krasnoyarsk, Khakassia, Kemerovo, as well as part of Buryatia and Irkutsk. Both current and completed WWF eco-regional projects do not cover the latter ones. The many steppe areas of the Altai and Khakass Republic and all the territory of Tuva are on the border of the eco-region while the steppe areas of Altaiskii Krai fall outside the limits.

Daurian/Mongolian Steppe: The Russian part of the eco-region includes the southern parts of Buryatia and Zabaikalskii Krai (the steppe region known as Dauria).

2.2.3 Recognized IBAs

Recently the Russian Bird Conservation Union (RBCU) recognized in Russia 746 Important Bird Areas (IBAs). 462 IBAs are situated in the steppe zone and about 170 IBAs include steppe ecosystems. Only 88 IBAs contain significant steppe areas and/or are important for characteristic steppe birds, and 25 IBAs could be marked as very important to steppe conservation (Fig. 2.4, Tables 2.3 and 2.4) (BirdLife International 2008; Bukreev 2006, 2009; RBCU IBA Data Base – version June 2008; Antonchikov, personal communication 2010).

Table 2.2 Ramsar sites that are important for steppe conservation

WI site reference	Site name	Ramsar site no.	Coordinates of site center	Total site area (km^2)	Terrestrial area (km^2) (if known)	Province	Sector of the steppe region
3RU009	Lake Manych-Gudilo	673	44°36′N 042°50′E	1,126	335.55	Kalmykia + Rostov	Pontic Steppe
2RU015	Tobol-Ishim Forest-steppe	679	55°27′N 069°00′E	12,170		Tyumen	West Siberian Forest -steppe
2RU016	Chany Lakes	680	55°02′N 077°40′E	3,648.5		Novosibirsk	West Siberian Forest-steppe
2RU017	Wetlands in the Lower Bagan area	681	54°09′N 078°23′E	268.8		Novosibirsk	West Siberian Forest-steppe
2RU018	Selenga Delta	682	52°17′N 106°22′E	121		Buryatia	Khangai-Daurian Forest -steppe
2RU019	Torey Lakes	683	50°05′N 115°32′E	1,725	485	Zabaikalskii Krai	Mongolian Steppe

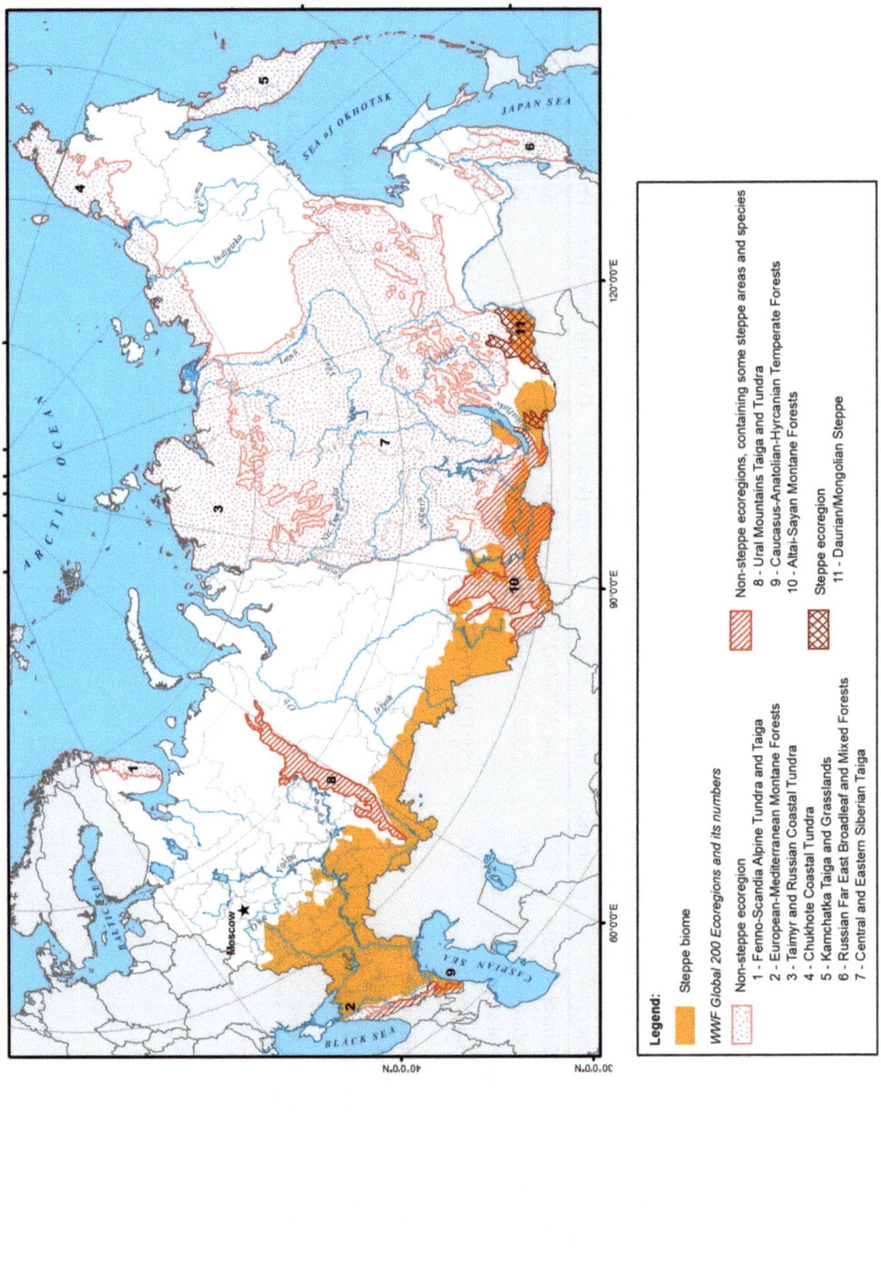

Fig. 2.3 WWF Terrestrial Global 200 eco-regions in Russia and post-Soviet countries. Indication on the map: *Brown* – steppe eco-region; *Dark red* – non-steppe eco-region containing some steppe ecosystems and species; *Light red (pink)* – non-steppe eco-region

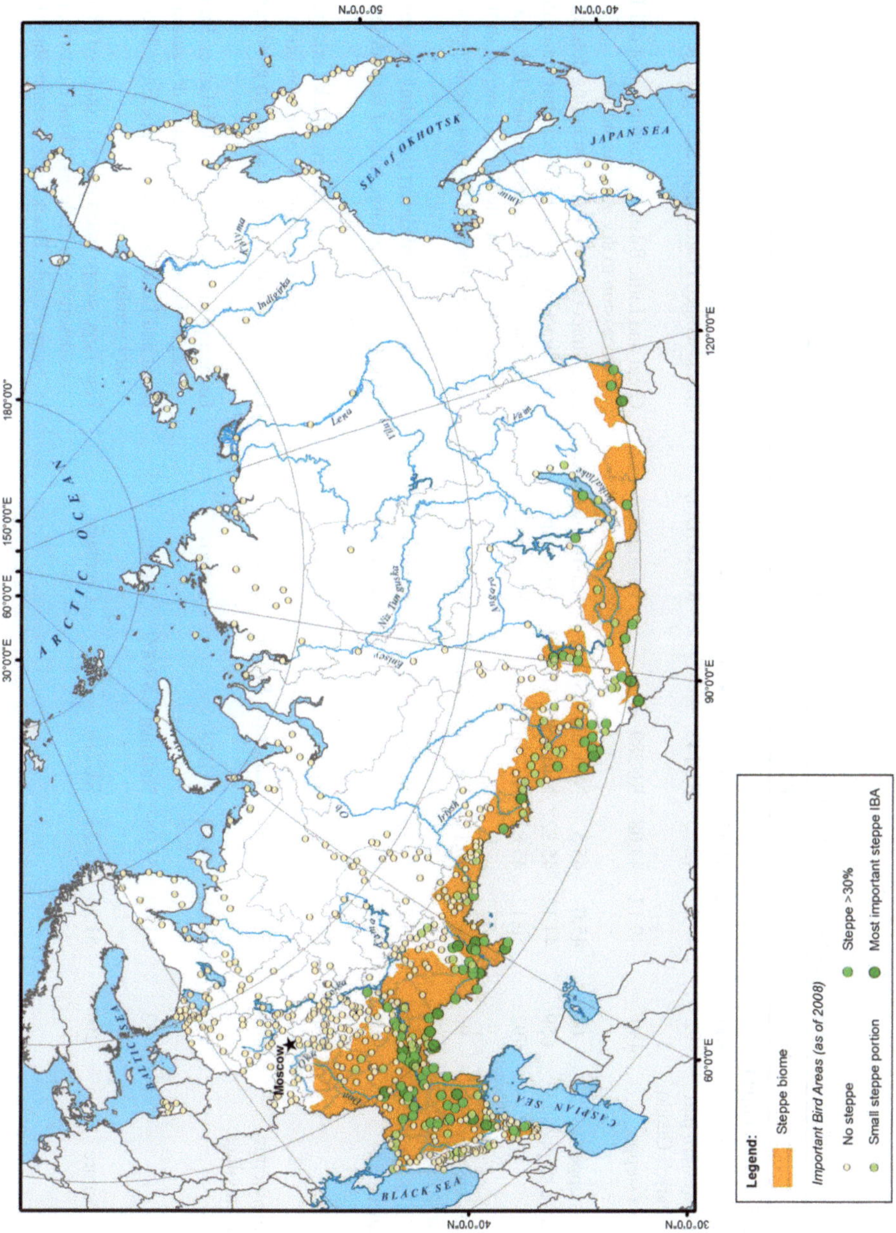

Fig. 2.4 IBAs in Russia as recorded in RBCU data base. The Steppe Region is *coloured*. Steppe IBAs are presented as *green filled dots*, non-steppe IBAs as *white dots*

Table 2.3 Russia's IBAs including a solid steppe area and/or important for characteristic steppe birds or most important for steppe conservation on national level

IBA name	Area, km²	NL	EL	BLDBC or RBCU	Province	Importance for steppe conservation
Western half of the steppe biome						
Dudarevskaya steppe	300	49.92	41.75	National – RO004	Rostov/Don	
Islands in the western part of Lake Manych-Gudilo	192	46.5	42.55	RU143	Rostov/Don	Black-winged Pratincole breeding; large steppe tract
Manoilinskaya steppe	480	49.12	43.08	RU289	Volgograd	Great and Little Bustards breeding; vast old-field area with succession to dry genuine steppe
Prikumskiye steppes	60	45.0	45.5	National – ST012	Stavropolskii Krai	Large tract of semi-desert (including desertified steppe)
Karanogaiskiye steppes	659	44.12	45.82	National – DS030	Dagestan	Large tract of semi-desert (including desertified steppe)
Drofiny area	792	50.12	45.83	RU278	Volgograd	Great and Little Bustards breeding; large tract of semi-desert (including desertified steppe, steppe shrubs)
Steppes in the vicinity of Zeleni Dol village	96	51.33	46.17	National – SR007	Saratov	One of largest tracts of dry genuine steppe in good condition
Oling area	332	46.27	45.20	RU256	Kalmyk	Large tract of semi-desert (including desertified steppe)
Erdniyevskaya area	2,490	46.92	46.4	RU280	Kalmyk	Little Bustard stop; large tract of semi-desert (including desertified steppe)
Algaiski	130	50.13	48.57	RU135	Saratov	Little Bustard, Pallid Harrier, Black Lark breeding; large tract of dry genuine steppe in good condition
Steppes in the vicinity of Kanavka village	64	50.3	48.67	National – SR006	Saratov	Large tract of dry genuine steppe in good condition
Siniye hills	150	51.08	49.48	RU128	Saratov	Great and Little Bustards, Pallid Harrier, Imperial and Steppe Eagles breeding; large tract of dry genuine steppe
Ozinskii	32	51.08	51.08	National – SR009	Saratov	Great and Little Bustards, Steppe Eagle breeding; large tract of dry genuine steppe
Orenburgski Nature Reserve	216	51.25	57.33	RU218	Orenburg	Four large steppe tracts in good condition (from mesic genuine to desertified); Little and Great Bustards, Imperial Eagle, Steppe Eagle, Black Lark, Greater Short-toed Lark breeding

Zilairskoye Prisakmar'ye	150	52.17	57.75	RU322	Bashkortostan	Large tract of meadow and mesic genuine steppe in forest-steppe pattern; Imperial Eagle breeding
Irendyk ridge	1,500	53.33	58.5	RU212	Bashkortostan	Imperial Eagle breeding; part of large steppe tract in good condition (mesic genuine steppe, steppe shrubs)
Cheka hill	227	52.53	59.07	National – ChL008	Chelyabinsk	Large steppe tract in good condition (mesic genuine steppe, stony steppe, steppe shrubs)
Sources of the Bolshaya Karaganka and Syntasty rivers	2,039	52.62	59.82	National – ChL009	Chelyabinsk	Large steppe tract included (mesic genuine steppe, steppe shrubs);
Shalkaro-Zhetykol'ski lake system	813	50.92	60.83	RU217	Orenburg	Large steppe tract in good condition (dry steppe); Pallid Harrier, Black-winged Pratincole breeding; Black Lark wintering
Kurumbel'skaya steppe	1,123	54.42	75.42	National – OM016	Omsk	Demoiselle Crane breeding; large steppe tract in good condition (genuine dry steppe in pattern with salt habitats)
Loktevskaya	528	51.15	81.65	National – AL029	Altaiskii Krai	Large steppe tract in good condition (genuine mesic and dry steppes, stony steppe)
Eastern half of the steppe biome						
Plateau Ukok	2,529	49.3	87.57	National – AT002	Altai	Saker, Steppe Eagle, Black Vulture, Demoiselle Crane breeding; vast tract of mountain steppe in good condition
Talduair mountain	2,061	49.97	89.27	National – AT008	Altai	Saker, Steppe Eagle, Demoiselle Crane breeding; vast tract of mountain steppe in good condition
Ol'khon area	2,200	53.17	107.0	RU046	Irkutsk	Imperial Eagle breeding; large steppe tract in good condition (genuine dry steppe)
Torey lakes	1,725	49.92	115.5	RU055	Zabaikalskii Krai	Eastern Great Bustard breeding; large steppe tract in good condition (meadow, genuine dry, and halophytic steppes)

NL Northern Latitude (degree), *EL* Eastern Longitude (degree), *BLDBC* Bird Life Data Base Code, *RBCU* Russian National Data Base Code

Table 2.4 Russia's IBAs that include a solid steppe area and/or are important for characteristic steppe birds – other areas

IBA name	Area, km^2	NL	EL	Province
Agar-Dag	217	50.28	94.55	Tyva
Aleyskaya	755	50.90	82.17	Altaiskii Krai
Anuyskaya	3,591	51.72	84.50	Altaiskii Krai
Archedinskiye sands	1,500	49.58	43.25	Volgograd
Argun' river	1,000	49.92	118.50	Zabaikalskii Krai
Balaganskaya steppe	1,600	53.83	102.83	Irkutsk
Birsuat	385	52.20	60.35	Chelyabinsk
Blagoveschenskaya (Kulunda lake and vicinity)	1,344	53.00	79.67	Altaiskii Krai
Bogdinsko-Baskunchakski	926	48.17	47.00	Astrakhan
Bograd forest-steppe	1,148	54.17	90.78	Khakass
Bulukhta area	625	49.33	46.17	Volgograd
Charyshskaya	1,598	51.48	83.28	Altaiskii Krai
Cherebayevskaya floodplain	108	50.75	45.67	Saratov
Chonta	680	46.73	44.95	Kalmyk
Dzhirim mountain	86	54.75	90.45	Khakass
Ebeity lake	214	54.65	71.73	Omsk
Forest-steppe Gyulchachak	1,005	52.70	47.62	Ulyanovsk
Golubinskiye Sands	200	48.83	43.67	Volgograd
Gornaya Kolyvan'	528	51.33	82.23	Altaiskii Krai
Gunibskoye plateau	85	42.42	46.90	Dagestan
Kanskaya steppe	2,099	50.90	84.87	Altai
Kayakentski reserve	143	42.33	47.80	Dagestan
Kholmanskie feathergrass steppes	656	51.667	50.5	Saratov
Kissyk area	3	43.73	46.07	Chechen
Kraka Mountain	1,300	53.50	58.00	Bashkortostan
Krasnoschekovskaya	946	51.87	82.73	Altaiskii Krai
Kulaksay lowland	50	50.73	55.83	Orenburg
Kundryuchenskiye sands	175	47.75	41.00	Rostov/Don
Kupy area	20	51.33	53.77	Orenburg
Lake Ayke	100	50.98	61.58	Orenburg
Lake El'ton	300	49.17	46.83	Volgograd
Lake Manych-Gudilo	500	46.18	43.00	Kalmyk
Large liman	400	48.75	45.00	Volgograd
Levo-Dobrinskaya valley	94	48.75	43.00	Volgograd
Lower Eruslan	496	50.30	46.42	Volgograd
Marinskaya cuesta of Skalisti ridge	25	43.80	42.10	Karachaevo-Cherkess
Migulinskiye sands	320	49.67	41.33	Rostov/Don
Mikhailovskii Wildlife Refuge	180	52.05	47.20	Saratov
Novokvasnikovski liman	3	50.53	46.50	Volgograd
Orota depression	47	42.58	46.95	Dagestan
Oruku-Shinaa	130	50.63	93.13	Tyva
Ostrovnoi Reserve (recently part of Donskoi Nature Park)	75	47.58	41.92	Rostov/Don
Outskirts of Arbali village	16	45.12	45.22	Stavropolskii Krai
Privolzhskaya forest-steppe	300	53.00	47.83	Ulyanovsk

(continued)

Table 2.4 (continued)

IBA name	Area, km^2	NL	EL	Province
Priyeruslanskiye Sands	200	50.70	46.72	Saratov
Sarpinskaya lake system	4,500	47.50	45.25	Volgograd
Sengileyevskiye mountain	224	53.95	48.62	Ulyanovsk
Shaitan-Tau ridge	600	51.75	57.42	Bashkortostan
Shcherbakovskaya bend of Volga river	346	50.45	45.75	Volgograd
Stepan Rasin Rock	351	51.00	45.58	Saratov
Steppe valley of Sakmara river	750	51.53	56.92	Orenburg
Sviyago-Kubninskaya forest-steppe	320	55.50	48.37	Tatarstan
Tazhinski liman	96	49.22	45.45	Volgograd
Tazlarovskiye hills	6	52.17	56.72	Bashkortostan
Trekhozerki lakes	5	53.30	90.52	Khakass
Tri Gusikhi	277	52.28	59.08	Chelyabinsk
Tsimlyanskiye sands	1,500	48.00	42.67	Rostov/Don
Urulunguevskaya hollow	500	50.30	117.08	Zabaikalskii Krai
Uttinskaya area	980	46.23	46.10	Kalmyk
Uzkaya steppe	12,020	51.52	80.27	Altaiskii Krai
Varfolomeyevskiye saltmarshes	28	50.17	48.20	Saratov
Zhuravlinaya area	710	45.95	44.07	Kalmyk

2.2.4 *Recognized IPAs*

Recently within the framework of the IUCN/SibEcoCenter/PlantLife project "Plant Conservation Strategy of the Altai-Sayan Eco-region" for the first time in Russia over 80 Important Plant Areas (IPAs) had been identified within the Altai-Sayan Montane Forests eco-region, in Altai, Tuva, and Khakass Republics, Altaiskii Krai, and Kemerovo province. More than a half of all IPAs (47) include some steppe area, and half of them (24) hold relatively large steppe tracts and/or are recognized specifically for steppe species or habitats. Seven IPAs are recognized as very important for steppe conservation areas for their remarkable area or uniqueness of steppe habitat or endemic plant species (Fig. 2.5, Table 2.5).

2.2.5 *Other Types of Recognized Biodiversity Hot Spots*

High Nature Value Farmlands (HNVF). The original steppe area comprises large tracts of lands that were suitable, and used, for crop farming, and steppes also were always extensively used for livestock raising. Thus, presently, the steppe ecosystems strongly depend on agricultural use and HNVF could be a very appropriate format for characterization of the biodiversity of the steppe and for conservation. Until now this type of ranking of natural areas is not implemented in Russia. Recently, Russia became one of several post-Soviet countries where the methodical and scientific base for HNVF promotion was developed in the course of the

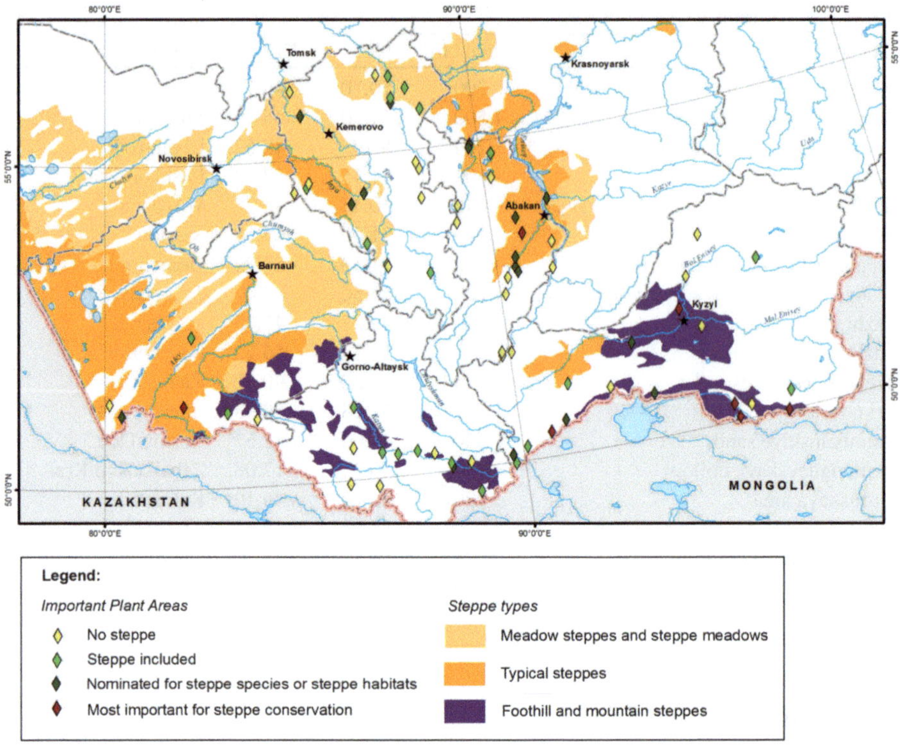

Fig. 2.5 IPAs in Russia identified as a result of a pilot project in the Altai-Sayan eco-region (SibEcoCenter, IUCN Office for Russia and CIS, PlantLife International)

Table 2.5 Russia's IPAs (Altai-Sayan eco-region only) including any steppe area and/or important for characteristic steppe plants

IPA name	Area, km^2	NL	EL	Province	ISC
Ustianka steppe low hills	100	51.28	81.95	Altaiskii Krai	3
Saksary	600	53.50	90.78	Khakass	3
Tsugheer Els sands	46	50.07	95.22	Tuva	3
Lower Katchyk river	74	50.00	96.38	Tuva	3
Erzin river valley	42	50.28	95.15	Tuva	3
Upper Barlyk river	135	50.38	90.73	Tuva	3
Seserlig river and Kamennyi stream watershed	80	51.90	94.33	Tuva	3
Kyzylchin area	8	50.07	88.30	Altai	2
Tassor salt lakes	20	51.15	80.42	Altaiskii Krai	2

(continued)

Table 2.5 (continued)

IPA name	Area, km^2	NL	EL	Province	ISC
Kapchaly – 'Khutor #7'	0.2	53.75	90.68	Khakass	2
Khujur lake	17	53.00	90.48	Khakass	2
IPA Fyrkal	5	54.97	89.78	Khakass	2
Malyy Kobejekov	10	54.92	89.75	Khakass	2
Tepsey Mountain	25	53.97	91.57	Khakass	2
Bondarevo	5	52.92	90.53	Khakass	2
Askiz Cuesta ridge	16	53.15	90.52	Khakass	2
Lower Ustu-Ghimateh river valley	6	50.12	89.75	Tuva	2
Khayirakan mountain	15	51.53	93.02	Tuva	2
Amdaighyn-Khol lake	40	50.70	93.33	Tuva	2
Terektig river bassin	22	50.53	91.12	Tuva	2
Chumaiskie Bukhtui hills	15	55.76	87.82	Kemerovo	2
Artyshta	20	54.27	86.52	Kemerovo	2
Karakan ridge	45	54.42	86.87	Kemerovo	2
Novoromanovo rocks	1	55.68	85.37	Kemerovo	2
Rocks and screes along right bank of Tchuya river near Belgebash mouth	12	50.37	87.50	Altai	1
Limestone rocks at Chuya mouth	35	50.40	86.67	Altai	1
Akkaya	7	51.13	86.10	Altai	1
Upper Ulandryk river	45	49.63	88.90	Altai	1
Sukor mountain	7	50.12	88.27	Altai	1
Limestone rocks on the left bank of Katun near river mouth	12	51.13	86.10	Altai	1
Limestone rocks of Belyi Bom	15	50.35	87.03	Altai	1
Tigirek	175	51.17	83.03	Altaiskii Krai	1
Kolyvanskoe lake	4	52.35	82.20	Altaiskii Krai	1
Krasnen'kie lake	1	54.78	90.30	Khakass	1
Tolaity river bassin	120	50.22	90.12	Tuva	1
Watershed Alty-Ghimateh and Aspaity rivers	100	49.98	89.80	Tuva	1
Upper Naryn and Balyktyg-Khem rivers	280	50.30	96.53	Tuva	1
Shemi river valley	10	51.08	91.30	Tuva	1
Azas lake	3	52.42	96.47	Tuva	1
Sertinskaya Forest-steppe	10	55.97	88.25	Kemerovo	1
Tambar mire	30	55.62	88.58	Kemerovo	1
Shestakovskie mires	30	55.83	87.83	Kemerovo	1
Kokuiskoe mire	1	54.57	85.38	Kemerovo	1
Archekass ridge	16	56.17	87.83	Kemerovo	1
Rocks near Kostenkovo village	1	53.63	86.82	Kemerovo	1
Rocks along Mrassu river	55	53.08	88.33	Kemerovo	1

ISC Importance for steppe conservation, 1 lowest, 3 highest international project ‘Identification of High Nature Value Farmlands for EECCA: results of assessment and recommendations’ in 2005–2006. Nevertheless, the HNVF identification process in Russia has not yet started.

Units of Eeconet. The Russian Federal Government has no policies in the field of Eeconet identification and legal designation. But many provincial and regional Eeconet projects were elaborated by NGOs and (sometimes) provincial authorities, including provinces in the Steppe Region, like Orenburg, Samara, Bashkortostan, Orel, the Altai-Sayan eco-region (several provinces), etc.

2.2.6 Animal Species of Special Concern Inhabiting Steppes

Fragmentation and disturbances of the steppe biome have resulted in threats to many animal species inhabiting steppes and the Russian steppes harbour many endangered and vulnerable species. As usual, it is hard to assess the situation for steppe invertebrates, despite a number publications on the subject. Thus we concentrate on vertebrates only.

2.2.6.1 Mammals

The steppe habitats of Russia harbour 11 species that are of global conservation concern (and two ungulates became extinct in the wild). Two species were formerly listed as Vulnerable, but are now reassessed as Least Concern. On a national level the Federal Red Data Book listed 14 steppe species (two of them are extinct in the wild in Russia) (Table 2.6). Eight species are listed at both national and international levels (including the two extinct in the wild) and ten species at only one of these levels (Low Risk or Least Concern in the global IUCN Red List). On a subnational (provincial) level there are a number of mammal species related to steppe habitats listed (more details are provided for selected provinces only).

2.2.6.2 Birds

Russia is recognized as the most important European country for conservation of steppe birds. The list of bird species whose European population are for 75% or more concentrated in steppe habitats is topped by Russia and Turkey with 21 species (from 27 in total). Summing the numbers of national populations of all steppe bird species in Europe reveals that Russia supports 39% of the total European breeding populations of these 27 species (Turkey 36% and Spain 22%). Ten of these twenty seven breeding steppe species are of global conservation concern; Russia harbours nine of them – more than any other European country (Burfield 2005; Tacker and Evans 1997).

Table 2.6 Mammal species of global and/or national concern inhabiting steppe habitats

Scientific name, English name	Global IUCN Red List	RF Red Data Book[a]	Russian Provinces (total range or maximum concentration only)[b]	LSS[c]; associated steppe ecosystem/habitat	Major identified threats
Saiga tatarica, Saiga Antelope	CR A2acd ver 3.1 (2008)	No	Kalmyk	Western; Desertified steppe	Habitat loss/degradation, poaching, international trade
Equus ferus (= *Equus przewalskii*), Wild Horse	CR D ver 3.1 (2008)	0	In Russia extinct in the wild	Eastern;	
Cuon alpinus, Asiatic Wild Dog or Dhole	EN C2a(i) ver. 3.1 (2008)	1	Altai, Tyva, Buryat, Zabaikalskii Krai	Eastern; Mountain steppe, tundra-steppe, *Kobresia* grasslands in Russian part of the range	Poaching?
Equus hemionus, Asian Wild Ass (Kulan)	EN A2abc+3bd ver 3.1 (2008)	0	In Russia extinct in wild (last record –Zabaikalskii Krai, 1926)	Eastern; Desert and desertified steppes	Habitat loss/degradation, poaching, competition with livestock, interruption of migration routes
Marmota sibirica, Tarbagan Marmot	EN ver 3.1 (2008)	1	Altai, Tyva, Buryat, Zabaikalskii Krai	Eastern; All variants of mountain steppes	Hunting, habitat loss/ degradation
Spalax giganteus, Giant Mole Rat	VU B2ab(iii) ver 3.1 (2008)	3	Dagestan, Chechnya, Stavropol	Western; Sand steppe, desertified steppes	Habitat loss/degradation
Mustela eversmanni ssp. *amurensis* (the sub-species is not distinguished in latest version of the Red List)	VU A2cd ver 2.3 (1994)	2	Amur Province	Eastern; Steppefied meadows ("Amur prairie")	Habitat loss/degradation
Vormela peregusna ssp. *peregusna*, European Marbled Polecat	VU A2c ver 3.1 (2008)	1	Rostov, Saratov, Samara, Orenburg, Krasnodar, Dagestan	Western; Sandy and shrub steppes	Habitat loss/degradation
Otocolobus manul, Pallas Cat	NT ver 3.1 (2008)	3	Altai, Tyva, Buryat, Zabaikalskii Krai	Eastern; Mountain and piedmont dry and stony desert steppes	hunting – Regional/international trade; Changes in native species dynamics – Prey/food base

(continued)

Table 2.6 (continued)

Scientific name, English name	Global IUCN Red List	RF Red Data Book[a]	Russian Provinces (total range or maximum concentration only)[b]	LSS[c]; associated steppe ecosystem/habitat	Major identified threats
Ovis ammon, Argali	NT ver 3.1 (2008)	1	Altai and Tyva	Eastern; Mountain steppe, tundra-steppe, *Kobresia* grasslands	Habitat loss/degradation, poaching, competition with livestock
Spermophilus suslicus, Speckled Ground Squirrel	NT ver 3.1 (2008)	No	Kursk, Voronezh, Belgorod, Orel, Penza, Tula, Tambov, Samara, Ulyanovsk, Stavropol, Krasnodar, Rostov, Volgograd	Western; Genuine and meadow steppes	Habitat loss/degradation, poisoning, grazing decline
Mustela altaica, Altai or Mountain Weasel	NT ver 3.1 (2008)	No	Altai, Altaiskii Krai, Irkutsk, Buryat, Zabaikalskii Krai, Amur Province	Eastern; Mountain steppe and alpine tundra-steppe	Habitat degradation (overgrazing)/pika-control campaigns
Ochotona pusilla, Steppe Pika	LC ver.3.1 (2008) (VU A1cd, C2a ver 2.3 (1994))	No	Saratov, Samara, Orenburg, Bashkortostan, Chelyabinsk, Altaiskii Krai	Western; Shrub steppes (meadow, genuine, and dry types)	Habitat loss/degradation
Spalax microphthalmus, Greater Mole Rat	LC ver 3.1 (2008) (VU D2 ver 2.3 (1994))	No	Kursk, Voronezh, Belgorod, Orel, Penza, Tula, Tambov, Saratov, Samara, Ulyanovsk, Rostov	Western; Genuine and sand steppe	Habitat loss/degradation
Mesechinus dauuricus, Daurian Hedgehog	LC ver 3.1 (2008)	4	Buryat, Zabaikalskii Krai	Eastern; Any types of steppes in Russian part of the range	Fires, pesticides, agricultural processes, predation by dogs
Myospalax psilurus aspilanus, Transbaikal Zokor	LC ver 3.1 (2008)	2	Zabaikalskii Krai	Eastern; Meadow steppes	Habitat loss/degradation
Eolagurus luteus, Yellow Steppe Lemming	LC ver 3.1 (2008)	1	Altai (one site only)	Eastern; Mountain steppe	?

Vormela peregusna ssp. *pallidior*, Semirechye Marbled Polecat (the sub-species is not distinguished in latest version of the Red List)	LR/lc ver 2.3 (1994)	1	Altaiskii Krai (extinct in the wild?) Altai, Tyva	Eastern; Sandy and shrub steppes	Habitat loss/degradation
Procapra gutturosa, Mongolian Gazelle	LC ver 3.1 (2008)	1	Zabaikalskii Krai	Eastern; Any types in Russian part of the range	Fires, hunting, interruptions of migration routes, habitat loss

IUCN (2008) and Russian Red Data Book (2001)

RF Russian Federation (Red Data Book), *LSS* Longitudinal steppe sector (Eastern or Western[c])

[a]Categories of RF Red Data Book used in Table 2.6 (and Table 2.7 as well):

0. *Probably disappeared species*:
   In this Category there are species and populations that were known earlier from the territory (or water area) of the Russian Federation, but their presence in the wild is not confirmed (during the last 50 years)
1. *Under threat of disappearance*:
   Species and populations,
   - Whose number has decreased up to a critical level and they might disappear in the near future;
   - On the verge of disappearance; there are only a few registrations during the last 25 years;
   - Not under threat of disappearance, but being in great risk of loss (because of extremely low numbers and/or of narrowness of distribution area or in an extremely small number of habitat sites).
2. *Declining*:
   Species and populations which are steadily declining in numbers. They soon might enter the category of 'threat of disappearance' if negative factors continue
3. *Rare*:
   Species and populations with normally low numbers, living in limited (land or water) areas or sporadic but widespread in specific (land or water) areas; needing special protective measures to survive
4. *Indefinite status*:
   Species and populations which (probably) fit in one of the previous categories, but there is no detailed information about their present status, or they do not meet completely with the criteria of those categories
5. *Restored and restoring species*:
   Species and populations, which in numbers and distribution have been restored and would not require urgent conservation and restoration measures (their restoration resulted from natural phenomena or from applied conservation measures)

[b]Only those where the species is living in steppe habitats

[c]Within the borders of Russia only

Table 2.7 Bird species of global, European, and/or national concern inhabiting steppe habitats

English name, Scientific name	Global IUCN Red List (LC and LR not shown)	2004 SPEC category[a]	RF	Russian Provinces (nesting range and/or maximum concentration on flyways)[b]	LSS[c]; Keystone steppe habitats	Major identified threat (if known)
Ruddy Shelduck, *Tadorna ferruginea*		3	–	Rostov, Stavropol, Volgograd, Saratov, Samara, Bashkortostan, Altaiskii Krai, Altai, Khakass, Kemerovo, Krasnoyarsk, Kurgan, Novosibirsk, Orenburg, Tuva, Buryat, Chelyabinsk, Omsk, Tyumen	Both; genuine, dry, and desertified steppe dotted with lakes or ponds	
Shelduck, *Tadorna tadorna*		Non-SPEC	–	Rostov, Stavropol, Volgograd, Orenburg, Kurgan, Chelyabinsk, Novosibirsk, Altaiskii Krai	Western; genuine, dry, and desertified steppe dotted with lakes or ponds	
White-headed Duck, *Oxyura leucocephala*	EN A1acde ver 2.3 (1994)	1	1	Kalmyk, Rostov, Chelyabinsk Kurgan, Omsk, Tyumen Novosibirsk, Orenburg, Altaiskii Krai, Khakass, Tuva	Both; steppe lakes or ponds	
Pallid Harrier, *Circus macrourus*	NT ver 3.1 (2001)	1	2	Orenburg, Novosibirsk, Altaiskii Krai, Altai, Khakass, Kemerovo, Krasnoyarsk, Kurgan, Tuva, Buryat, Chelyabinsk, Omsk, Tyumen	Both; different types of steppes and meadows close to water	
Upland Buzzard, *Buteo hemilasius*		–	Monit.	Altai, Buryat, Irkutsk, Zabaikalskii Krai, Khakass, Tuva	Eastern; mountain and hilly steppe with outcrops	
Long-legged Buzzard, *Buteo rufinus*		3	3	Altaiskii Krai, Orenburg, Samara, Saratov, Volgograd, Astrakhan, Chelyabinsk	Western; mountain and hilly steppe with outcrops	
Eurasian Griffon, *Gyps fulvus*		Non-SPEC	3	Stavropol, Dagestan	Both; high mountain steppe with large rocks and cliffs	

Black Vulture, *Aegypius monachus*		–	3	Dagestan, Stavropol, Altai, Tuva, Buryat, Zabaikalskii Krai	Both; high mountain steppe with large rocks and cliffs	
Egyptian Vulture, *Neophron percnopterus*		3	3	Stavropol, Krasnodar, Dagestan	Western; mountain steppe with large rocks and cliffs	Habitat loss/ degradation, grazing decline
Lammergeier, *Gypaetus barbatus*		–	3	Stavropol, Krasnodar, Dagestan, Altai, Tuva	Both; high mountain steppe with large rocks and cliffs	Habitat loss/ degradation, grazing decline
Short-toed Eagle, *Circaetus gallicus*		3	2	Rostov, Stavropol, Tambov, Voronezh, Saratov, Samara, Altaiskii Krai, Altai, Kemerovo, Krasnoyarsk, Kurgan, Novosibirsk, Orenburg, Chelyabinsk, Tumen	Western; steppe dotted with small groves and/or outcrops	
Steppe Eagle, *Aquila nipalensis*		3	3	Rostov, Stavropol, Saratov, Samara, Bashkortostan, Altaiskii Krai, Altai, Buryat, Zabaikalskii Krai, Khakass, Irkutsk, Kemerovo, Krasnoyarsk, Kurgan, Novosibirsk, Orenburg, Tuva, Chelyabinsk, Tumen	Both; hilly short-grass steppe preferred	
Imperial Eagle, *Aquila heliaca*	VU C1 ver 3.1 (2001)	1	2	Rostov, Volgograd, Saratov, Samara, Altaiskii Krai, Altai, Buryat, Zabaikalskii Krai, Khakass, Irkutsk, Kemerovo, Krasnoyarsk, Kurgan, Novosibirsk, Orenburg, Tuva, Chelyabinsk	Both; forest edges opened to steppe, forest-steppe, steppe with solitary trees, etc.	

(continued)

Table 2.7 (continued)

English name, Scientific name	Global IUCN Red List (LC and LR not shown)	2004 SPEC category[a]	RF	Russian Provinces (nesting range and/or maximum concentration on flyways)[b]	LSS[c]; Keystone steppe habitats	Major identified threat (if known)
Lesser Kestrel, *Falco naumanni*	VU A2bce+3bce ver 3.1 (2001)	1	1	Altaiskii Krai, Altai, Buryat, Zabaikalskii Krai, Khakass, Krasnoyarsk Krai, Orenburg, Saratov, Bashkortostan, Tuva, Chelyabinsk	Both; steppe with weathered outcrops or artificial constructions (buildings, tombs, walls)	Habitat loss/ degradation, grazing decline, pesticide use
Red-footed Falcon, *Falco vespertinus*	NT ver 3.1 (2001)	3	Monit.	Buryat, Khakass, Irkutsk, Krasnoyarsk, Novosibirsk, Kemerovo, Kurgan, Omsk Orenburg, Tyumen, Tuva	Both; small forests contacting steppe	Habitat loss/ degradation, pesticide use
Saker, *Falco cherrug*	VU A2bcd+3cd+4bcd ver 3.1 (2001) (re-assessed in 2010)	1	2	Orenburg, Bashkortostan, Altaiskii Krai, Altai, Tuva, Buryat, Khakass, Irkutsk, Zabaikalskii Krai	Both; cliffs, outcrops, single trees within steppe	Poaching – international trade, pesticide use, habitat loss/ degradation
Eagle Owl, *Bubo bubo*		3	2	All provinces in steppe zone	Both; cliffs, outcrops, ravines, shrubs within every type of steppe	
Demoiselle Crane, *Grus virgo*		Non-SPEC	5	Altaiskii Krai, Altai, Buryat, Zabaikalskii Krai, Khakass, Irkutsk, Kemerovo, Krasnoyarsk Krai, Kurgan, Novosibirsk, Orenburg, Tuva	Both	
Great Bustard, western subspecies, *Otis tarda tarda*	VU A3c ver 3.1 (2001)	1	3	Rostov, Volgograd, Dagestan, Astrakhan, Saratov, Samara, Orenburg, Chelyabinsk, Omsk, Novosibirsk, Altaiskii Krai, Altai, Tuva, Khakass	Both	Habitat loss/ degradation, poaching, agrochemicals use

Great Bustard, eastern subspecies, *Otis tarda dubowskii*	VU A3c ver 3.1 (2001)	–	2	Tuva, Irkutsk, Buryat, Zabaikalskii Krai, Krasnoyarsk, Amur	Eastern	Habitat loss/ degradation, poaching
Little Bustard, *Tetrax tetrax*	NT ver 3.1 (2001)	1	3	Stavropol, Kalmykia R., Rostov, Volgograd, Saratov, Samara, Orenburg	Western	Habitat loss/ degradation, poaching
Houbara Bustard, *Chlamydotis undulata*	VU A2bcd + 3bcd ver 3.1 (2001)	1	1	Tuva	Eastern	Habitat loss/ degradation, poaching
Greater Sand Plover, *Charadrius leschenaultia*		3	Monit.	Altai, Tuva	Eastern	
Oriental Plover, *Charadrius veredus*		–	Monit.	Tuva, Zabaikalskii Krai	Eastern	
Sociable Lapwing, *Vanellus gregarius*	CR A3abc ver 3.1 (2001)	1	1	Nesting: Altaiskii Krai, Orenburg, Chelyabinsk; Flyway: Stavropol, Kalmykia	Western;	Land abandonment, habitat loss/ degradation
Slender-Billed Curlew, *Numenius tenuirostris*	CR C2a(ii) ver 3.1 (2001)	1	1	Omsk, Altaiskii Krai; probably: Chelyabinsk, Kurgan, Novosibirsk	Western	
Whimbrel, *Numenius phaeopus alboaxillaris*		Non- SPEC	Monit.	Bashkortostan, Orenburg	Western	
Stone Curlew, *Burhinus oedicnemus*		3	4	Dagestan, Stavropol, Krasnodar, Kalmykia R., Volgograd, Saratov, Orenburg	Western	
Oriental Pratincole, *Glareola maldivarum*		–	Monit.	Zabaikalskii Krai	Eastern	

(continued)

Table 2.7 (continued)

English name, Scientific name	Global IUCN Red List (LC and LR not shown)	2004 SPEC category[a]	RF	Russian Provinces (nesting range and/or maximum concentration on flyways)[b]	LSS[c]; Keystone steppe habitats	Major identified threat (if known)
Black-winged Pratincole, *Glareola nordmanni*	NT ver 3.1 (2001)	1	2	Orenburg, Kurgan, Altaiskii Krai	Western	Land abandonment, Habitat loss/ degradation
Black-bellied Sandgrouse, *Pterocles orientalis*		3	–	Volgograd, Astrakhan, Kalmykia R.	Western;	
Mongolian Lark, *Melanocorypha mongolica*		–	2	Altai, Tuva, Buryat, Zabaikalskii Krai	Eastern	
Black Lark, *Melanocorypha yeltoniensis*		3	–	Astrakhan, Volgograd, Saratov, Orenburg, Omsk	Western	
Calandra Lark, *Melanocorypha calandra*		3	–	Dagestan, Stavropol, Rostov, Volgograd, Saratov, Orenburg	Western	
Short-toed Lark, *Calandrella brachydactyla*		3	–	Altaiskii Krai, Krasnoyarsk	Both	
Lesser Short-toed Lark, *Calandrella rufescens*		3	–	Dagestan, Stavropol, Rostov, Volgograd, Saratov, Orenburg	Western	
Asian Short-toed Lark, *Calandrella cheleensis tuvinica*		–	Monit.	Altai, Tuva, Buryat, Zabaikalskii Krai	Eastern	

Crested Lark, *Galerida cristata*		3	–		Both	
Tawny Pipit, *Anthus campestris*		3	–	Dagestan, Stavropol, Krasnodar, Rostov, Volgograd, Saratov, Orenburg, Novosibirsk, Kemerovo, Altaiskii Krai, Altai, Tuva, Kakass, Krasnoyarsk, Buryat, Irkutsk Zabaikalskii Krai	Both	
Pere David's Snowfinch, *Pyrgilauda davidiana*		–	Monit.	Altai, Tuva, Zabaikalskii Krai	Eastern	
White-Throated Bushchat, *Saxicola insignis*	VU C2a(ii) ver 3.1 (2001)	–	1	Altai	Eastern	Habitat loss/ degradation

IUCN (2008), Antonchikov (2005), Burfield (2005), and Prisyazhnyuk (2004), BirdLife International (2004)

RF Russian Federation Red Data Book, *LSS* Longitudinal steppe sector (Eastern or Western[c])

[a]SPEC (Species of European conservation Concern) category is based on both its global conservation status and its proportion of the global population (or range) in Europe. Each bird species was assigned to one of five categories including 1–3 SPEC categories and two non-SPEC ones (Burfield 2005). The SPEC categories are:

SPEC 1 – European species of global conservation concern (i.e. classified as CR, EN, VU, NT, or DD in the global IUCN Red List),

SPEC 2 – species whose global populations are concentrated in Europe, and which have an unfavorable conservation status in Europe,

SPEC 3 – species whose global populations are not concentrated in Europe, but which have an unfavorable conservation status in Europe.

'Monit.' is a special status in the RF Red Data Book: The species should be monitored while actually it has no listing under any category of the RD Book.

[b]Only those where the species is living in steppe habitats

[c]Within the borders of Russia only

On the national level the Russian Federation Red Data Book (2001) lists 126 bird species and subspecies, 14 of which are typical steppe birds (Antonchikov 2005).

It is important to point out that, as Russia holds several relatively intact steppe landscapes, there are some so-called 'intrazonal' non-steppe habitats intimately related with these landscapes (special steppe wetlands, rocky habitats in mountain steppes, special small forests associated with steppe, etc.). Some characteristic bird species inhabit these habitats. All these species as well as the proper steppe species should be taken into consideration. Thus, we find in the RF Red Data Book 30 species and subspecies inhabiting steppe habitats or habitats closely related with steppe landscapes. Accordingly, our complete list includes 42 species (Table 2.7).

The total number of 61 vertebrate species of special concern is almost equally divided over the Eastern and Western subregions of the steppe biome (40 species each and 37 species inhabiting both subregions and thus are scored twice). Mammals concentrate more in the Eastern half of the steppe biome: 13 species against 6 in the Western half and none over the entire biome (present in both subregions). As regards birds, 10 species occur in the Eastern steppes only, 14 in the Western steppes only, and 17 in both subregions.

Suitable habitats for many mammals (ungulates and carnivores) and some birds (carrion feaders) are predominantly large steppe tracts in semi-desert or mountain regions. Saving these species is rather impossible without designing large PAs. Actually, large PAs are the only adequate and appropriate legal instrument for the conservation of these steppe species. On the other hand, some species listed above inhabit mainly farmland habitats and strongly depend on agricultural practice. The best known examples are western Great Bustard, Demoisell Crane, Sociable Lapwing, Lesser Kestrel, Steppe and Imperial Eagles (in the Russian part of their ranges). The survival of these species is not secured by conservation in strict reserves only. They need special protection on agricultural land outside of PAs.

2.3 Land Use Modes and Actual Threats

Historically, all steppes of Russia have been extensively used by mobile (nomadic) herders for sheep, horse, cattle and camel breeding. Modern crop-based agricultural development in European style started in the north-west corner with the most productive East-European meadow steppes in the seventeenth century (in the area which is known as the Central Chernozem Region of Russia and the adjacent part of Ukraine). Croplands were gradually extended to the East and South to reach the desertified steppes near the Caspian Sea by the end of the nineteenth century and the extremely arid desert steppes of Tuva only by the 1940s–1950s. The recent situation is presented in Table 2.8 (data are presented for those provinces of Russia which have some areas of steppe within their boundaries).

Table 2.8 Agricultural lands important for steppe conservation in Russia

Province	Total Province area, km²	Agricultural land, km²			Arable land, km²	Arable land as % of total area	Arable land as % of agricultural area	Old-field, km²	Old-field as % of former arable area	Arable lands decrease, km²
Data of		1990	1990	1998	2007	1990	1990	2007	2007/1990	2007/1990
Altai	92,903	13,876	1,488	1,405	1,432	1.6	10.7	23	1.6	56
Altaisky Krai (= Altai Territory)	167,996	107,759	70,903	65,906	64,013	42.2	65.8	5,635	7.9	6,890
Astrakhan	49,024	30,152	3,596	2,859	3,461	7.3	11.9	86	2.4	135
Bashkortostan Republic	142,947	72,085	48,491	44,631	36,788	33.9	67.3	0	0.0	11,703
Belgorod	27,134	21,495	16,657	16,175	16,528	61.4	77.5	1	0.0	129
Buryat Republic	351,334	26,556	9,654	7,885	8,326	2.8	36.4	616	6.4	1,328
Chechen Republic	15,647				3,375			1		
Chelyabinsk	88,529	48,611	31,636	30,423	30,621	35.7	65.1	643	2.0	1,015
Dagestan Republic	50,270	32,966	5,023	4,921	5,248	10.0	15.2	48	1.0	–225
Irkutsk	774,846	34,739	24,364	22,278	17,341	3.1	70.1	33	0.1	7,023
Kalmyk Republic	74,731	59,462	9,086	8,338	9,134	12.2	15.3	154	1.7	–48
Kemerovo	95,725	24,976	15,821	15,066	15,563	16.5	63.3	1	0.0	258
Khakass Republic	61,569	17,636	7,303	6,803	6,877	11.9	41.4	404	5.5	426
Krasnodar Territory	75,485	45,475	39,303	38,804	39,896	52.1	86.4	2	0.0	–593
Krasnoyarsk Territory	723,671	52,802	32,826	31,096	31,269	4.5	62.2	1,410	4.3	1,557
Kurgan	71,488	44,462	30,051	27,595	25,259	42.0	67.6	3,387	11.3	4,792
Kursk	29,997	24,343	19,652	19,227	19,442	65.5	80.7	7	0.1	210
Lipetsk	24,047	19,440	16,436	15,728	15,544	68.4	84.6	1	0.0	892
Novosibirsk	177,756	82,341	39,166	36,622	37,624	22.0	47.6	857	2.2	1,542
Omsk	141,140	66,955	43,541	41,668	41,616	30.9	65.0	1,733	4.0	1,925
Orel	24,652	20,830	16,661	15,807	15,691	67.6	80.0	598	3.6	970
Orenburg	123,702	106,740	62,233	61,011	61,373	50.3	58.3	0	0.0	860
Penza	43,352	30,518	25,040	23,156	22,420	57.8	82.1	1,787	7.1	2,620

(continued)

Table 2.8 (continued)

Province	Total Province area, km²	Agricultural land, km²			Arable land, km²	Arable land as % of total area	Arable land as % of agricultural area	Old-field, km²	Old-field as % of former arable area	Arable lands decrease, km²
Data of		1990	1990	1998	2007	1990	1990	2007	2007/1990	2007/1990
Rostov	100,967	85,154	60,805	59,626	58,166	60.2	71.4	0	0.0	2,639
Samara	53,565	39,934	31,079	30,627	30,165	58.0	77.8	354	1.1	914
Saratov	101,240	85,039	63,731	57,925	59,407	63.0	74.9	0	0.0	4,324
Stavropol Territory	66,160	56,854	40,774	38,883	39,943	61.6	71.7	147	0.4	831
Tatarstan Republic	67,847	45,511	37,364	34,802	34,625	55.1	82.1	6	0.0	2,739
Tula	25,679	19,654	15,668	14,675	15,571	61.0	79.7	77	0.5	97
Tuva (= Tyva) Republic	168,604	35,779	4,326	1,894	2,350	2.6	12.1	1,259	29.1	1,976
Tyumen	160,122	37,720	17,432	15,042	14,801	10.9	46.2	2,439	14.0	2,631
Ulyanovsk	37,181	22,079	18,090	1,720	16,767	48.7	81.9	860	4.8	1,323
Volgograd	112,877	87,361	58,705	57,045	58,486	52.0	67.2	48	0.1	219
Voronezh	52,216	40,986	32,350	30,668	30,593	61.9	78.9	441	1.4	1,757
Zabaikalskii Krai (=Territory)	431,892	80,588	25,274	9,536	5,634	5.9	31.4	9,031	35.7	19,640

Data source: Compendium (2007) and Regions (1999)

For the whole steppe biome in Russia the greatest threat is conversion of steppe into cropland and other forms of cultivation (for instance pasture and hay grasslands "improvement"). Other important threats (but to a lesser extent) are mining (including conversion of mining lands into dumps, concentration plants, infrastructure, etc.), urban encroachment, development of infrastructure, and afforestation. The most common threats, though not so destructive, are overgrazing, hay cutting and burning. The intensity of fires in most places increases with the decrease in grazing pressure. Therefore, the effects of recently observed decreases in livestock numbers are very controversial for such landscape types.

Several characteristic steppe species are endangered because of overharvesting for international and/or domestic trade which is commonly illegal (Saiga Antelope, Saker Falcon, several species of Marmots, several species of early spring wildflowers like *Crocus* spp., *Bulbocodium versicolor, Pulsatilla* spp., *Tulipa gesneriana*, etc.).

New challenges for Russian steppes emerged recently as a result of the global food crisis and biofuel boom of 2007/2008. A significant threatening factor at the national level is the increase of investment into agriculture in Russia. The increased land use profitability leads to increased pressure on arable land and more and more old-fields, forage crop fields and even virgin steppe grasslands are being converted into arable lands.

In Russian markets the price for wheat has risen twofold from 2007 to 2008 (€111 per ton in May 2007 against €246+ in May 2008 for 3rd class wheat) (data of the Russian Grain Union, http://grun.ru/). In the same period, the price for diesel oil has risen about 20% only. As a result the 2.3 million hectares of old-fields and perennial forage crops were ploughed up anew in Russia in 2008. It made the crop area increase by 5.2%. A significant part of these re-ploughed areas was on account of the spontaneously regenerated steppe and steppe-like vegetation. In 2009 the process of re-ploughing was continued. It was interrupted only in 2010–2011 as a result of the 2010 crop failure due to heavy drought in the Western Sector of the Steppe Biome in Russia (also see below and Fig. 2.9).

The modes of land use differ according to the latitudinal and longitudinal positions of the type of steppe ecosystems:

Meadow steppes and the more mesic parts of genuine steppes (types (1) and (2) listed above) were massively turned into arable land for production of crops and fodder as early as the end of nineteenth century. The forest-steppe zone also has the highest population density in Russia and is most affected by infrastructural, industrial and urban encroachment.

Nowadays almost the entire area west of the Volga River has been turned into arable lands, with a noticeable share of forest plantations, settlements, and infrastructure. The steppe remnants are small, highly fragmented, and dispersed among arable lands. They survive only on sites which are inconvenient for ploughing because of soil or relief characteristics. Steppe remnants are used for both cattle grazing and hay making. Recreation and collection of wild plants and fruits are additional modes of use. In general, meadow steppes are more often ploughed to cultivate fodder crops than used as pastures. These steppe types are strongly affected by afforestation, but also by settlements and suburb expansion.

East of the Volga River the situation is a bit different. The steppes survive here in much larger areas. Probably it is not so much because of belated European colonization than due to much more complicated relief and soil patterns. Relatively large meadow steppe and mesic genuine steppe tracts remain in the Ural and the foothills of the Altai Mts. (surviving due to relief features) as well as in the vast West Siberian and Daurian Plains where grasslands are interconnected with wetlands, salt habitats and extensive birch (*Betula*) groves (surviving due to soil complexity). Nevertheless these steppe types are the most destroyed parts of the eastern longitudinal steppe sector. Land use is almost the same as in the European part but it has some specific features. Burning of grasslands is much more common here. On the other hand, settlement encroachment has less impact. In recent years mining has become a significant threat, in particular in the mountains and their foothills.

Genuine dry bunchgrass steppes (type 3) were used as rangelands for centuries and massively reclaimed into croplands only in the 1950s (while some local areas were ploughed earlier). The history of this massive land development is highly dramatic and closely related with the political history of Russia. The total area of steppes converted to arable lands in the USSR (not Russia only) during 1954–1963 was 43 million hectares. The main part of these steppe grasslands was turned into arable lands for merely ideological reasons not economic ones. Not surprising, the use of these new croplands was not sustainable due to climate fluctuations, relatively low soil fertility, aridity, widespread salinity, etc. Therefore these croplands started to be abandoned as early as at the end of the 1960s. Ideological constraints were crushed in 1991, and widespread abandonment occurred after that. Since the early 1990s the former area of dry steppes became a scene of old-field succession. Recently part of these lands returned to arable farming while the share of old-fields is still significant.

This type of steppes is used widely again as rangelands for cattle and sheep, locally for horses. Dry steppes usually are not mown; floodplain meadows and fodder crops are commonly used here for hay production. Fires have an important role only in ungrazed lands. Afforestation and urbanization are less important factors. Some valuable steppe tracts are imperilled because of oil and gas production (in Trans-Volga, South Urals, north of the Caspian Sea, east of the Azov Sea) while others are threatened because of mining (in South Urals, Altai, Buryatia, Dauria).

Livestock numbers in these steppes decreased after 1991: cattle twofold and sheep tenfold. Therefore, during the last decades everywhere in these dry steppes grazing pressure has been relatively low.

Desertified and desert steppes could become croplands only through costly irrigation. In Russia they are mostly used as rangelands for seasonal grazing. Semi-nomadic mobile pastoralism still continues in some South Siberian mountains (South-Eastern Altai, Tuva). Desert steppes are used here along with the high-mountain cryophytic steppes within the common transhumance pattern.

High-mountain cryophytic steppes are used as seasonal pastures only. Stock rearing is the main land use here and herders follow the traditional transhumance pattern. These steppes have never been plowed at all. Recently mining became one of the main threats for these steppes. Mining affects relatively small pieces of area but every mining place entails a massive infrastructural development and a general increase in the accessibility of the area.

Edaphic and other variants. Stony slopes, sand steppes and salt habitats cannot be converted into croplands and often remain the only untouched pieces of steppe grassland in the regions of meadow steppe and genuine steppe. At the same time all these edaphic variants (compared with the typical variants) are very vulnerable to overgrazing causing significant erosion.

Water infrastructure affects steppe wetlands and surrounding areas as water is diverted for irrigation and reclamation. It is the biggest threat, especially in dryer areas where water is scarcer and irrigation/water diversion has a more dramatic effect on ecosystems. The best-known Russian example is the massive irrigation of the Black Lands region, Kalmyk Republic, undertaken in the 1980s; it resulted in a local environmental catastrophe and dramatic desertification. A more recent case is the actual threat of a Chinese plan of a massive water transfer from the Argun River into Dalai Nor (Dalai Lake) in the transboundary region of Dauria which will strongly affect the Russian part of the Argun River Basin (Tkachuk et al. 2008).

Shrub steppe communities are very vulnerable to fire in all zones (while steppe shrubs have specific adaptations to resist fire damage). The same threat is even more severe for forest-steppe and steppe landscapes with tree and shrub groves. In general, the intensity of wild fires increases when grazing decreases. The effects of the recently observed decrease in livestock numbers are controversial for such types of steppe landscape (several known examples include Altaisky Krai, Khakassia and Dauria where steppe fires became extremely frequent and heavy in the last decade).

Serial steppe communities on abandoned lands are the result of old-field succession. Fires and grazing affect them as well, but these disturbances actually may suppress ruderal vegetation and lead to succession towards steppe (however in most cases the climax stage is unreachable).

In general, agricultural activities selectively affect certain steppe types adversely, while they only moderately alter other ones.

It is important to point out that the quantitative data presented in Tables 2.8 and 2.9 on the official land inventory and livestock survey data should be used carefully. The official land inventory data on the old-fields area is inaccurate because of (1) many lands are actually set aside but not marked as old-field in the land cadastre (still registered as arable land), (2) the figure on the old-field area is strongly depending on provincial government policy (thus some provinces report very low values or no old-field at all, not because it actually is absent but for a politically-induced refusal to register abandoned lands as old-field). And vice versa, short-term farming on old-fields is not uncommon; after 1–2 years of farming the field becomes abandoned again similar to the historical "steppe perelog" farming regime, a long-fallow farming practice in steppe areas which was very common in the era of the Russian colonial expansion into steppe regions. It was a specific land use mode, practiced when new settlers had extremely rich land in abundance. They were able to plow new fields for 1–3 years and then shift to another plot abandoning the former one for spontaneous restoration of 10–15 years or more. Thus very good yields were produced with only minimal cultivation and no fertilizers, based on natural soil fertility built by the steppe ecosystem at no cost. (3) Furthermore, parts of abandoned arable lands were turned into grazing lands in the land inventory. In fact, conserving these areas as grasslands and allowing succession towards the secondary steppe stage would be the best approach. In any case these official data should be regarded as approximations (Table 2.8). A more correct way to assess the extent of abandoned arable lands is to compare the values for actual arable area against the values in 1990 (1990 is the starting year because there has not been any formal change in the acreage of agricultural lands from the early 1970s to the collapse of the Soviet Union in 1991).

2.3.1 Rural Economics in the Steppe Biome in Russia

A major main part of the steppe areas in Russia falls in the category of agricultural lands and is used mainly for agriculture. However, the principal branch of agriculture in Russia is crop farming. Crop farming extensively makes use of some steppe ecosystem services like soil fertility (steppe-produced soils, Chernozems and Kastanozems are among the most fertile on Earth), prevention of erosion, wild relatives of crop plants, providing restoration facilities for abandoned croplands, natural enemies of insect and rodent pests, etc. But as steppe is transformed into croplands most of these functions are eliminated. Ploughing and other farming techniques are the most damaging threats to steppe.

Only one important agricultural sector is favourable to steppe ecosystems and that is livestock breeding. As we discussed above the steppe areas in Russia are predominantly grazing lands and not so much hay grasslands. Not all forms of livestock-breeding need the steppe but only those that are based on grazing lands. Conventional cattle-breeding is based on fodder production on arable lands and therefore cannot be considered as steppe-friendly.

The main domestic animals using temperate grasslands in Russia are beef cattle, sheep, goats, horses, to a lesser extent camels, yaks, reindeer (in the mountains of South Siberia), and Siberian deer. Pigs are raised in grasslands only occasionally. The first four (especially cattle and sheep) compose the most important livestock raised in steppe (Table 2.9, Fig. 2.6).

The "friendliness" of livestock grazing to steppe is based on ecological specifics of the steppe ecosystem, including the prevalence of herbivore-based food-webs, wild gregarious nomadic ungulates as ecological edificators (or keystone species), high rates of primary production, high ratios of belowground/aboveground phytomass. Domestic livestock in the steppe takes the place of wild ungulates that became extinct in the major part of the steppe biome. Within the steppes of Russia wild gregarious nomadic ungulates (Saiga Antelope, Argali, and Mongolian Gazelle) now play a noticeable role only in very limited areas of Kalmykia, South-Eastern Altai, and Dauria. And these steppe areas maintain only incomplete sets of ungulate species surviving in low density and affected strongly by competition with livestock.

Table 2.9 Livestock numbers in Russia (live animals), FAO data

	Cattle	Sheep	Goats	Horses	Camels	Buffaloes	Asses
1992	54,676,704	52,194,600	3,060,000	2,590,000	10,000	23,000	21,500
1993	52,226,000	48,182,500	3,186,000	2,556,000	10,300	22,000	21,500
1994	48,914,000	40,616,000	3,097,000	2,500,000	10,400	17,900	25,549
1995	43,296,000	31,818,000	2,722,400	2,431,000	10,500[a]	23,500	26,000[a]
1996	39,696,000	25,344,600	2,682,000	2,363,000	10,800[a]	23,548	26,500[a]
1997	35,102,800	20,327,000	2,445,400	2,197,000	11,000[a]	20,840	26,000[a]
1998	31,519,900	16,482,700	2,291,300	2,013,000	11,000[a]	16,586	25,000[a]
1999	28,480,800	13,412,500	2,143,900	1,801,000	11,000[a]	16,141	25,000[a]
2000	28,032,300	12,603,000	2,147,500	1,683,000	10,000[a]	16,500[a]	22,000[a]
2001	27,293,500	12,560,800	2,211,700	1,619,000	9,000[a]	17,370	20,000[a]
2002	27,106,902	13,035,009	2,291,696	1,578,000	8,000[a]	14,004	19,000[a]
2003	26,524,360	13,728,497	2,322,143	1,499,000	7,000[a]	15,863	18,000[a]
2004	24,935,140	14,669,420	2,360,992	1,498,500	6,792	17,943	17,000[a]
2005	22,987,700	15,494,014	2,277,366	1,409,261	6,654	17,288	22,000[a]
2006	21,473,926	16,074,449	2,138,202	1,319,358	6,289	14,362	21,929
2007	21,466,217	17,508,132	2,166,536	1,303,837	6,385	15,654	20,847
2008	21,546,000	19,290,400	2,212,870	1,321,340	6,356	14,363	19,934
2009	21,038,000	19,602,300	2,167,940	1,353,320	6,222	14,500[a]	19,274

No data available for yaks, reindeer and Siberian deer
Source: FAO Statistics Division (2008)
[a]FAO estimate

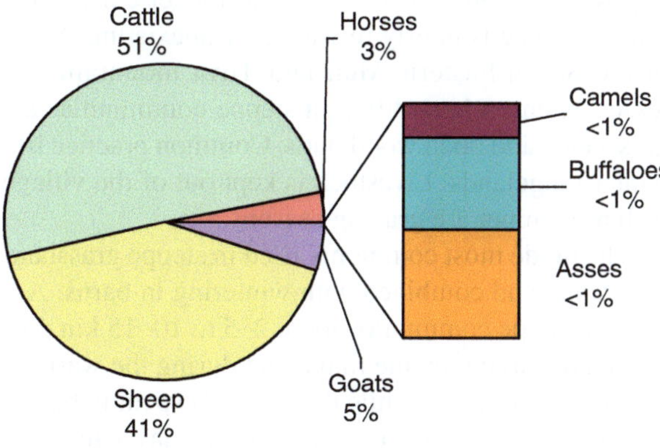

Fig. 2.6 Proportional division of livestock in Russia in 2007, based on FAO data (see Table 2.9)

Livestock does not use the steppe grasslands in the same manner as the wild steppe ungulates but the traditional grazing style is the best available substitution for the disappeared wild ungulates.

Different species of domestic animals, different breeds, and different modes of grazing have unequal impacts on the steppe ecosystem. In general, most steppe-friendly are those species, breeds and grazing modes which use steppe grassland in a similar way as the (extinct) wild ungulates.

Grazing of sheep and goats is rather destructive for steppe plants and soil cover, and easily results in vegetation degradation and soil erosion. Both these species share some common characters: feeding animals pinch off plants very close to the ground damaging vitally important parts of the plants (buds hidden under leaf bases in bunchgrass and rhizomes in forbs); their hoofs are pointed and cover a relatively small area resulting in relatively high pressure on the ground; their characteristic behaviour includes digging with their frontal legs, thus fragmenting the turf and reaching the roots; on pasture they move in thick flocks that results in a very concentrated grazing impact; sheep and goat excrements decompose slowly. For these reasons sheep and goat grazing can be considered as steppe-unfriendly, except when it is practised in relatively low stocking densities resulting in low grazing pressure. Cattle grazing is more acceptable. But the most steppe-friendly grazing animal is the horse. In the fairly recent past, the wild horse was one of the most important wild ungulates in the steppes of Russia (along with the Saiga and Steppe Aurochs). The domestic horse shares the same morphological and physiological characters as the wild one. Especially local indigenous breeds (Bashkirian, Altai, Kalmykian horses, etc.) are close relatives of the extinct Wild Horse. Recently several successful attempts to reintroduce the Przewalski Horse into the wild were undertaken in Mongolia and China. Russia also has a plan to establish a semi-wild population of Przewalski Horse in the next years. This is important for the conservation of a characteristic steppe species, but these populations are very small and local, and do not affect the biome as a whole. Thus the domestic horse still remains the most important ungulate for the steppe biome.

In Russia different grazing modes (grazing methods and grazing systems) are in use. Real nomadic grazing does no longer exist in the Russian steppes. Horizontal transhumance is utterly restricted or unexisting. Vertical transhumance is still vital in the South-Eastern Altai and Tuva mountains (and in the Caucasus to a far lesser extent) where different steppe communities are used in rotation with alpine grasslands and open woodlands. Common practice here is division into summer and winter rangelands. Livestock is kept out of the village all year round and winter as well as summer are grazing seasons.

The mode most commonly used in steppe grasslands is extensive grazing in summer rangeland combined with wintering in barns. As a rule summer rangeland is a natural steppe community about 2–5 to 10–15 km distant from the village. Cattle or sheep are staying on the rangeland during the warm season (from early May to late October on average) while the winter is spent in barns within the village. During the warm season animals are grazed every day while in winter they are supplied with hay, silage, and crop forage. Unlike cattle and sheep, horses can use the steppe rangeland in a semi-wild regime without everyday herders attending to them. Horses also more often are grazed in winter.

Thus, vertical transhumance grazing and extensive grazing in distant pastures are the grazing modes best compatible with the steppe ecosystem. Horses are the most steppe-friendly stock, cattle comes the next, and grazing sheep and goats requires special care. Multi-species use of the steppe rangeland is preferable.

These grazing modes also are the most compatible with meat and wool production. The production of milk and dairy products fits these modes to a lesser extent (except horse milk processing).

Some other agricultural branches could be considered more or less steppe-friendly but they are rather side activities adding to crop farming or animal breeding. Those are bee keeping, collecting wild medicinal and aromatic plants, and rural tourism.

2.3.2 *Steppe Ecosystem Services and Values*

Nowadays the steppes on chernozem soil bring the highest economic value, serving as the base of the Russian crop production, especially of cereals and oil-bearing seeds. The average annual yield of grain in the period 2005–2008 was 112,300 kilo-ton; of wheat it was 49,368–61,694 kilo-ton in the period 2007–2009, ranking 4th in the world; sunflower production was 7,350–6,425 kilo-ton in the period 2008–2009, ranking 1st in the world; it engaged approximately 6,000,000 persons. All top 20 provinces on the list of national grain production are “steppe provinces”, and together the 35 Russian provinces which comprise steppes contribute 86% to the national total (for 2008) (Fig. 2.7).

Steppe rangelands also are the main base of beef, sheep, and horse meat production as well as other products of cattle and sheep: beef production amounted to 1,828 kilo-ton in 2007 and 2,733 kilo-ton in 1995; sheep 160 kilo-ton in 2007. Of the 10 top provinces of cattle meat production 9 are situated in the Steppe Region; and of the 20 provinces with the highest numbers of cattle only 2 are lying outside the Steppe Region. The 35 “steppe provinces” total 70+% of the total number of cattle in Russia (Fig. 2.8; Table 2.9), and 90+% of its wool yield (in 2008).

An estimation of the values of various ecosystem services of the Russian steppe is given in Table 2.10. Based on data from the national land inventory, the national area of steppe grasslands is about 50,000,000 ha, of which arable lands in the steppe region comprise 89,534,900 ha, and PAs containing steppe measure 1,774,860 ha. Based on these figures and the per hectare value, for the whole steppe region of

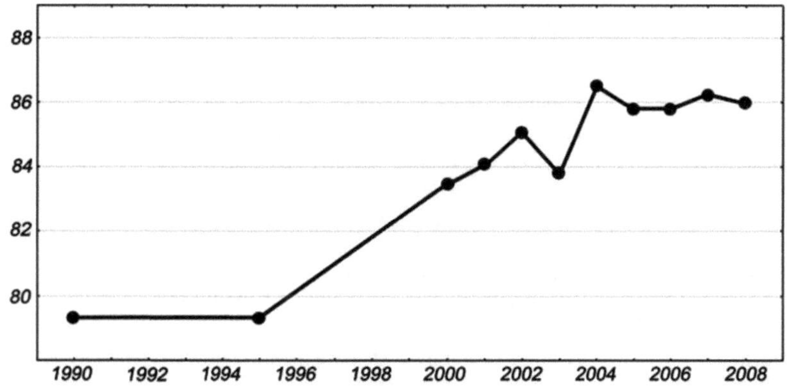

Fig. 2.7 Share of the Steppe Region in the total national grain production of Russia (35 provinces), (as% of the total) during the period 1990–2008

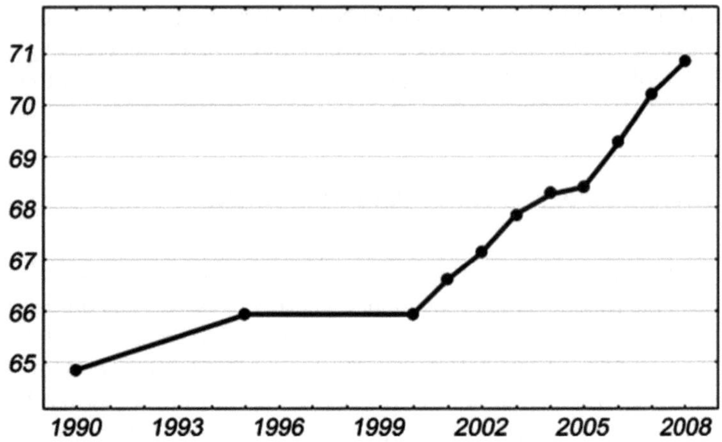

Fig. 2.8 Share of the Steppe Region (35 provinces) in the total number of cattle in Russia, in%, during the period 1990–2008

Table 2.10 Estimation of main values of ecosystem services in the steppe biome of Russia

Ecosystem service	Evaluation method	Specific value, $ per hectare per year (price level 2009)
Climate regulation	Expected loss of yield through climate factors and average values of main crops	2–3
Water regulation	Compensation costs for decreasing rivers run-off resulting from conversion of steppe grasslands to arable lands	3–5
Carbon sequestration	Prices of Carbon Market and (both field and model) data on carbon accumulation rates, 1,500 kg/ha·year (Kurganova et al. 2008; Belelli Marchesini et al. 2007)	7.5–75
Erosion control	Costs of engineer work to protect slopes and to control erosion on degraded lands based on insurance money (0.1–2.5%)	5–67
Pollution control	Direct costs of maintenance of environmental standards (waste treatment, aftermath elimination)	7–13
Biological production (soil fertility)	Direct farming costs to produce, on damaged or other poor soils, a crop of equal value as the average crop value produced on fertile soil (assessed as costs to produce crop on soil-less dumps in steppe landscapes)	200–270
Biological resources	Value of sustainably yielded steppe natural products (hay, fruits, medicinal & aromatic plants, wildlife, etc.)	1.5–3
Maintenance of biodiversity	Average operating costs for existing steppe reserves ('zapovedniks')	3.3

(continued)

Table 2.10 (continued)

Ecosystem service	Evaluation method	Specific value, $ per hectare per year (price level 2009)
Maintenance of human well-being	Decreasing the sickness rate after recreation in steppe landscapes assessed by average compensation rate	0.3–0.5
Recreation (paid recreational services)	Average revenue from recreation on steppe protected areas	1.7–2
Hedonistic value (not for profit use, non-use values)	Decreasing realty value after degradation of adjoining steppe area, visitors "ready to pay" and transportation costs to reaching steppe PAs	2.7
Total value		**234–444.5**

Russia we estimate a national steppe flow value of 29.3–46.0 billions (thousand milliard) $$ per year.

We can compare these figures with known global estimations of Costanza et al. (1997). In their summary of the average global value of annual ecosystem services the authors estimated the specific value of all grasslands at $232 per ha per year which amounts to $\$906 \cdot 10^9$ per year for the global grasslands area ($3{,}898 \cdot 10^6$ ha). We estimate that the Russian steppe grasslands area is ca. 1% of Constanza et al.'s global area. Thus we see that according to our estimates the value of the Russian steppe grasslands amounts to 3–5% of the global figure.

2.4 Conservation Status of Steppe Ecosystems

2.4.1 Protected Areas

2.4.1.1 National

There are three levels of PAs in Russia: national (federal), provincial, and local (municipal) ones. Quite accurate data on acreage and localization of PAs is available for federally protected areas only. Steppe ecosystems occur in 26 of the 101 Federal State Strict Nature Reserves (Zapovedniks, IUCN category Ia), 7 of the 39 National Parks (IUCN category II), 15 of the 69 Federal Wildlife Refuges (Federal Zakazniks, IUCN categories IV and VI), and 1 of the 39 Federal Natural Monuments (IUCN category III). Among the 'Zapovedniks' containing steppe ecosystems 11 have the status of Biosphere Reserve (Russia has a total of 37 Biosphere Reserves).

Information on the national level PAs that comprise steppe ecosystems is given in Table 2.11 (also see Fig. 2.9). The table is based both on official data on protected

Table 2.11 Steppe in nationally (federally) protected areas in Russia

Nationally (Federally) protected areas	Land area, ha	Province	Steppe ecosystem type	Longitudinal sector (Eastern or Western)
Federal state strict nature reserves (Zapovedniks), IUCN category Ia				
A. Protecting steppe grasslands in more than 25% of the reserve area				
Belogorye	2,131	Belgorod	Stony and typical meadow steppes	Western
Daurskiy (Biosphere Reserve since 1997, Dauria Transboundary Reserve – Dauria International Protected Area)	45,790	Zabaikalskii Krai	Meadow and genuine steppes – typical, sandy, salty, and stony variants	Eastern
Orenburgskiy	21,653	Orenburg	Genuine and dry steppes – typical, salty, and stony variants	Western
Privolzhskaya Lesostep'	8,373	Penza	Meadow steppe in forest-steppe landscapes	Western
Rostovskiy (Biosphere Reserve since 2008)	9,532	Rostov/Don	Genuine and dry steppe – typical and salty variants	Western
Ubsunurskaya Kotlovina (Biosphere Reserve since 1997, UvsNuur Transboundary Reserve)	323,198	Tyva	Dry and desert steppes – typical, salty, sandy, and stony variants	Eastern
Khakasskiy	267,565	Khakass	Meadow, genuine, dry steppes – typical, salty, and stony variants	Eastern
Centralno-Chernozemny (Tsentral'nochernozem Biosphere Reserve since 1978)	5,287	Kursk	Meadow steppe in forest-steppe landscapes	Western
Chernye Zemli (Biosphere Reserve since 1993)	121,482	Kalmyk	Dry and desertified steppes – different variants	Western
Galichya Gora	231	Lipetsk	Stony and typical meadow steppes	Western
B. Protecting only small pieces of steppe grasslands (<10% of the reserve area)				
Azas	333,884	Tyva		Eastern
Altaiskiy	871,212	Altai		Eastern
Astrakhanskiy (Astrakhanskiy Biosphere Reserve since 1984)	56,619	Astrakhan		Western
Baikalo-Lenskiy	660,000	Irkutsk		Eastern
Bashkirskiy	49,609	Bashkortostan		Western

Bogdinsko-Baskunchakskiy	18,525	Astrakhan		Western
Voroninskiy	10,320	Tambov		Western
Zhigulevskiy ('Middle-Volga Integrated' Biosphere Reserve since 2006)	23,157	Samara		Western
Katunskiy (Biosphere Reserve since 2000)	151,678	Altai		Eastern
Prioksko-Terrasnyi (Biosphere Reserve since 1978)	4,945	Moscow Province		Western
Prisurskiy	9,148	Chuvash		Western
Sayano-Shushenskiy (Biosphere Reserve since 1984)	390,368	Krasnoyarsk		Eastern
Sokhondinskiy (Biosphere Reserve since 1984)	210,985	Zabaikalskii Krai		Eastern
Shulgan-Tash	22,531	Bashkortostan		Western
Jerghinskiy	238,088	Buryat		Eastern
Tighirekskiy	41,415	Altaiskiy Krai		Western
Total steppe Zapovedniks (A group)	***764,031***			
Total area of Zapovedniks protecting small steppe plots (B group)	***3,092,484***			
National parks, IUCN category II				
A. Protecting steppe grasslands in more than 25% of the reserve area				
Pribaikalskiy	418,000	Irkutsk	Dry steppes, mainly stony	Eastern
B. Steppe grasslands <10% of the park area				
Samarskaya Luka	134,000 (127,200 in other source)	Samara		Western
Khvalynskiy	25,514	Saratov		Western
Buzulukskiy Bor	106,788	Orenburg		Western
Bashkiria	82,300	Bashkortostan		Western
Zabaikalskiy	267,177	Buryat		Eastern
Tunkinskiy	1,183,662	Buryat		Eastern
Alkhanai	138,234	Zabaikalskii Krai		Eastern
Total area of NPs protecting steppe plots	***109,037,167***			

(continued)

Table 2.11 (continued)

Nationally (Federally) protected areas	Land area, ha	Province	Steppe ecosystem type	Longitudinal sector (Eastern or Western)
Federal wildlife refuges (Federal Zakazniks), IUCN categories IV and VI				
A. Protecting steppe grasslands in more than 25% of the refuge area				
Tsimlyanskiy	44,998	Volgograd	Sandy genuine steppes	Western
Kharbinskiy	163,900	Kalmyk	Dry and desertified steppes	Western
Sarpinskiy	195,925	Kalmyk	Dry and desertified steppes	Western
Mekletinskiy	102,500	Kalmyk	Dry and desertified steppes	Western
Saratovskiy	44,302	Saratov	Genuine and dry steppes	Western
B. Steppe grasslands <10% of the refuge area				
Tsasucheiskiy Bor	57,867	Zabaikalskii Krai		Eastern
Stepnoy	75,000	Omsk		Western
Starokulatkinskiy	20,166	Ulyanovsk		Western
Kurganskiy	31,846	Kurgan		Western
Kirzinskiy	119,808	Novosibirsk		Western
Kamennaya Step'	5,232	Voronezh		Western
Beloozerovskiy	17,850	Tyumen		Western
Bairovskiy	64,831	Omsk		Western
Altacheyskiy	60,000	Buryat		Eastern
Kabanskiy	12,100	Buryat		Eastern
Total steppe Wildlife Refuges (A group)	***551,625***			
Total area of Wildlife Refuges protecting small steppe plots (B group)	***464,700***			
Federal natural monuments, IUCN category III				
Protecting steppe grasslands in more than 75% of the area				
Janybek' biological station	228	Volgograd		Western

areas (from the State Report 'On the state and protection of the environment in the Russian Federation in 2006', data is presently still relevant) and on expert assessment of the proportion of steppe in the specific PAs.

2.4.1.2 Provincial

At present we have no possibility to do the same assessment for PAs of provincial level for the whole country. It still is a task of high priority to make a complete list of steppes protected as provincial PAs. We can provide such a list for only one category of provincial PAs, Nature Parks (IUCN category V) (Table 2.12). According to Russian legislation a Nature Park is a PA category analogous to a National Park but it is established and controlled at the provincial level.

2.4.1.3 Other Forms, Different Levels

An additional form of protected areas under Russian legislation is 'Landshaftny Muzey-Zapovednik' (Landscape Museum-Reserve, IUCN category V) designed to conserve cultural and historical heritage in its natural landscape. Museum-Reserves are managed by federal or provincial Ministries of Culture and are formally not a category of "nature protection area" in terms of conservation legislation. Only partial information is available on this kind of PAs. Only 9 from 96 existing Museum-Reserves include some areas of steppe (Table 2.13).

2.4.1.4 Proportion of the Steppe Biome Protected in Russia

Based on the steppe biome division used above we can try to estimate the proportion of each unit of the biome that is protected (listed as 'protected area coverage', i.e. the percentage of area within the PAs covered by that particular biome). Of course this is just a rough estimation only (Table 2.14). The area formally protected at federal (national) and local (provincial) level is calculated here as area that is left, not as the area historically covered by the ecosystem before settlement.

In general only 0.11% of the steppe biome area is formally protected in protected areas, and this is the lowest value of PAs coverage among all biomes of Russia (Nikolsky and Rumyantsev 2002). But inside even those few protected areas, different non-steppe ecosystems occupy most of the acreage. Recently, WWF recognized steppe as the worst protected terrestrial biome in Russia (the least covered by federally protected areas).

Only 10% of the Zapovedniks (federal strict nature reserves, IUCN category Ia; Russia has 101 of these) include some significant area of steppe (see Table 2.11). At least some small steppe areas are found in 25% of the Zapovedniks and 16% of the National Parks. Note furthermore that the average area of steppe-containing PAs is significantly smaller than the average area of all PAs in Russia. Thus only 1% of the total acreage of the system of national PAs is covered with grasslands of all types.

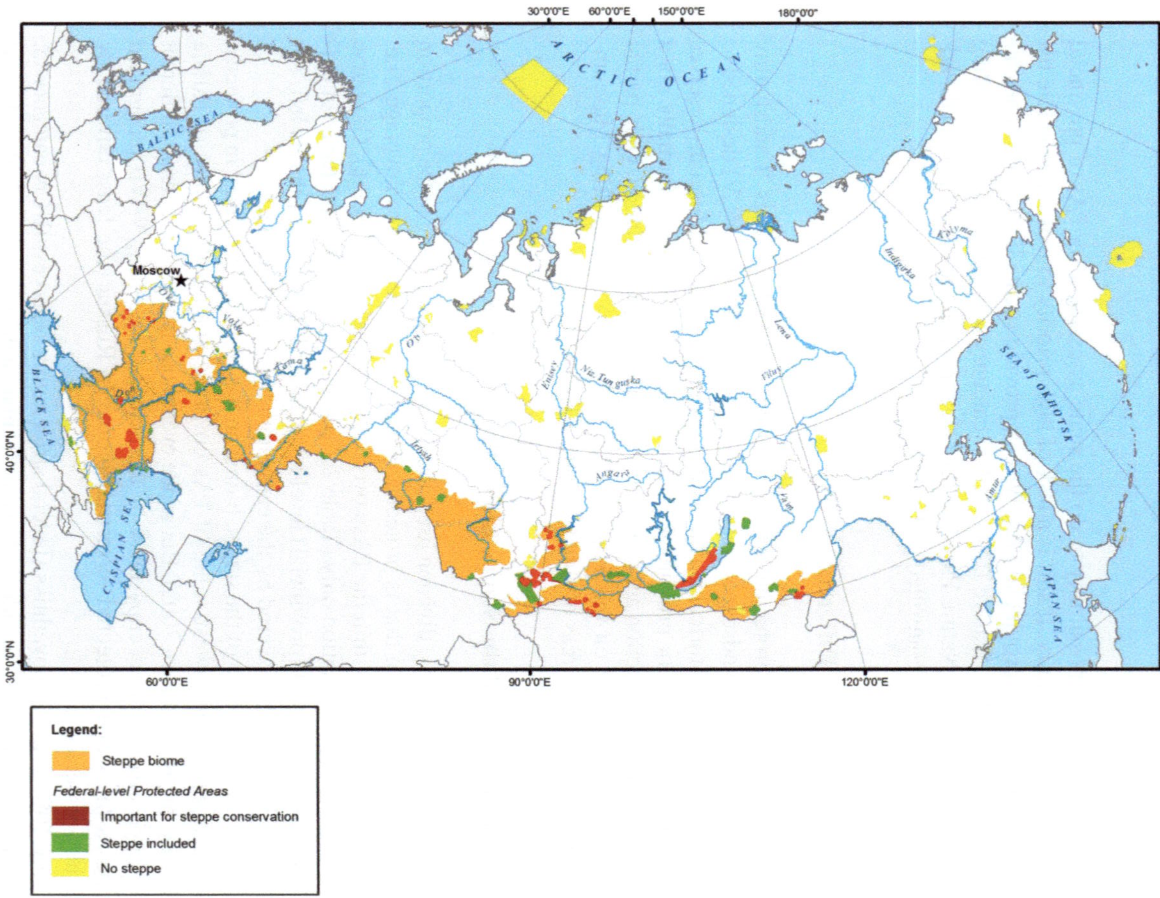

Fig. 2.9 Areas within Russia protected at federal level. The Steppe Region is *brown*. PAs which are especially important for steppe conservation are *red*, less important are *green*, and unimportant ones are *yellow*

Table 2.12 Steppe in nature parks (protected areas at provincial level) in Russia, 2010

Nature park	Land area, ha	Since	Province/Territory	Longitudinal sector (Eastern or Western)
Steppe grasslands >25% of the park area				
Uch Enmek[a]	65,000	2001	Altai	Eastern
Ukok	254,000	1994/2005	Altai	Eastern
Chuy Oozy[b]	9,538	2001	Altai	Eastern
Kandry-Kul	8,500	1995	Bashkortostan	Western
Bamb Tsetsg[a]	529	1991	Kalmykia	Western
Rovenskiy	1,300	1998	Belgorod	Western
Donskoy	40,955	2006	Rostov	Western
Donskoy	17,600	2001	Volgograd	Western
Eltonskiy	132,000	2000	Volgograd	Western
Nizhnekhoperskiy	231,206	2003	Volgograd	Western
Scherbakovskiy	20,000	2003	Volgograd	Western
Tsimlyanskie Peski (Tsimla' Sands)	66,951	2003	Volgograd	Western
Steppe grasslands <10% of the park area				
Argut[a]	20,500	2001	Altai	Eastern
Muradymovskoe Uschelye (Muradymovo Canyon)	23,586	1998	Bashkortostan	Western
Aya[c] (boundaries not legally approved)	1,109	2003	Altaiskii Krai	Western
Ust-Medveditskiy	51,200	2005	Volgograd	Western
Malaya Izluchina Dona (The Little Bend of the Don R.)	4,000	1993	Volgograd	Western
Nature park of Kalmyk Republic (also referred as Volgo-Akhtubinskoe Mezhdurechye (Volga-Akhtuba Watershed))	4,323	1995	Kalmykia	Western
Total area of NPs conserving significant steppe plots	***81,387***			
Total area of NPs conserving small steppe plots	***104,718***			

[a]Formal status only, no legal entity, staff, and budget
[b]Revoked in 2011; had formal status only, no legal entity, staff, and budget
[c]Formal status only, no fixed borders

Table 2.13 Steppe in protected cultural areas in Russia

Museum – landscape reserve	Land area, ha	Since	Province/Territory	Managed at level	Longitudinal steppe sector (Eastern or Western)
Protecting steppe grasslands at more than 75% of the museum area					
Irendyk Historical-Archeological and Landscape Museum-Reserve	29,550	2002	Bashkortostan	Provincial	Western
Historical-Cultural Provincial Reserve "Arkaim"	4,415	1991	Chelyabinsk	Provincial	Western
Steppe grasslands <10% of the museum area					
Razdorskiy Ethnographic Museum-Reserve	3,650	1988	Rostov/Don	Provincial	Western
Mikhail Sholokhov Museum-Reserve	38,236; including 3,820 ha protected natural landscape	1984	Rostov/Don	National	Western
Tanais State Museum-Reserve	8,500; including only 126 ha steppe area	1955	Rostov/Don	National	Western
Archeological and Natural Museum-Reserve "Tatarskoe gorodische" [Tatar Ancient Settlement Site] as part of	200	1992	Stavropol	Provincial	Western
G.N. Prozritelev and G.K. Prave Stavropol State Historical, Cultural and Natural Landscape Museum-Reserve					
Divnogorie Natural Architectural and Archeological Museum-Reserve	1,100	1991	Voronezh	National	Western
"Kulikovo Pole" State Military Historical Museum	nearly 16,000	1996	Tula	National	Western
Kazanovka Archeological Museum-Reserve	18,400	1996	Khakass	Provincial	Eastern

Table 2.14 PAs (both national and provincial) coverage in different types of steppe in Russia

Main units of the steppe biome	Percent of steppe area formally protected by law (protected area coverage)
At the national level in total	Federally protected <1%, totally protected ≤ 5% (not sufficient data)
Meadow steppes	
Typical	<3%
Rocky (petrophytic) (+ "rocky steppe" dwarf shrubs)	<10%
Sand (psammophytic)	<10%
Genuine forbs-bunchgrass steppes	
Typical	<3%
Rocky (petrophytic) (+ "rocky steppe" dwarf shrubs)	<10%
Sand (psammophytic)	<10%
Genuine (dry) bunchgrass steppes	<3%
Desertified and desert steppes	<5%
Cryophytic steppes (mountainous and sub-arctic)	3%
Steppes of subtropical mountains	<5%? (not sufficient data)

Sources: Compendium (2007) and MNR (2007)

The area of protected steppe is not proportional to the extant steppe area. Most of the steppe biome remains unprotected. In fact, the Russian steppe survives primarily on agricultural lands where it is used as pasture in stock breeding and, to a small extent, as hay grassland.

At the same time some protected areas could be important centers for managing steppe conservation in the much larger areas beyond the borders of the protected areas. The main ones are the Rostovskiy Biosphere Reserve (BR) for Rostov Province, the Centralno-Chernozemny BR for Kursk Province, the Orenburgskiy Zapovednik (SNR) for Orenburg Province, the Ubsunurskaya Kotlovina BR for Tuva Republic, the Khakasskiy SNR for Khakass Republic, and the Daurskiy BR for Zabaikalskii Krai.

2.5 New Challenges and Actual Threats

2.5.1 Turning into Croplands

Historically the greatest threat for steppes was ploughing (Chibilev 1998; Moon 2008). The total steppe area turned into arable land is assessed at almost one million km^2 nationally (for 1990, before the Soviet Union had crashed). But in fact it does not seem to be so severely threatened on a national level. After the collapse of the Soviet Union the area of arable land decreased significantly; over 26 millions hectares were abandoned in the steppe region up to 2007 (Fig. 2.10). These were mainly (not all) low-productive lands which were formerly developed under the specific conditions of the Soviet style economy, frequently for ideological reasons only.

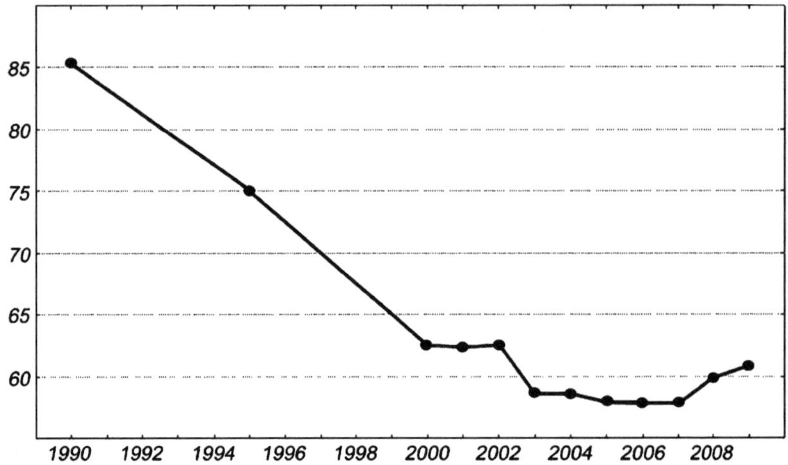

Fig. 2.10 Crop area in the Steppe Region of Russia, in millions of ha, during the period 1990–2008

The tendency was broken in 2007, however. Since 2007 almost one million hectares have been re-converted to cropland. This concerned mainly secondary steppe and probably also some natural steppe grasslands. There is no data on this. More information comes from the sub-national level. According to national land statistics 8 (of the 35) "steppe provinces" have, since 2007, increased their crop area by 140,000–340,000 ha. On top of the list are Orenburg and Voronezh Provinces and the Altai Territory. In the steppe provinces many officials proclaimed that no abandoned lands existed in their provinces in 2009. The real situation is apparently not so bad but remains rather unclear. After the bad harvests of 2010 officials and markets appear to have high expectations to again raise the crop production in Russia. Some market analysts consider that a new boom can be created by re-turning grassland areas into croplands.

2.5.2 *Over-Grazing and Under-Grazing*

In the 1960s–1980s over-grazing was widespread and damaging. After the agricultural collapse, the livestock number decreased dramatically in the early 1990s. By 1999 cattle numbers had dropped to half of the 1990 figure, while sheep and goats stabilized at a third of that level (ca. 1,900,000 in 2006–2009 as against 5,150,000 in 1990) (Fig. 2.11). Due to this decline over-grazing diminished and it became largely restricted to merely the local level, mainly around settlements. But over-grazing still is an important threat in the dry and desert steppes of the Caucasus region, because Daghestan, Kalmykia, and Stavropol are at the top of the national list of sheep and goat numbers, accounting for ca. 45% of the national figure.

Quite to the contrary, as in many areas the livestock number dropped under-grazing became very common. Because of this, during the last decade, it has let to

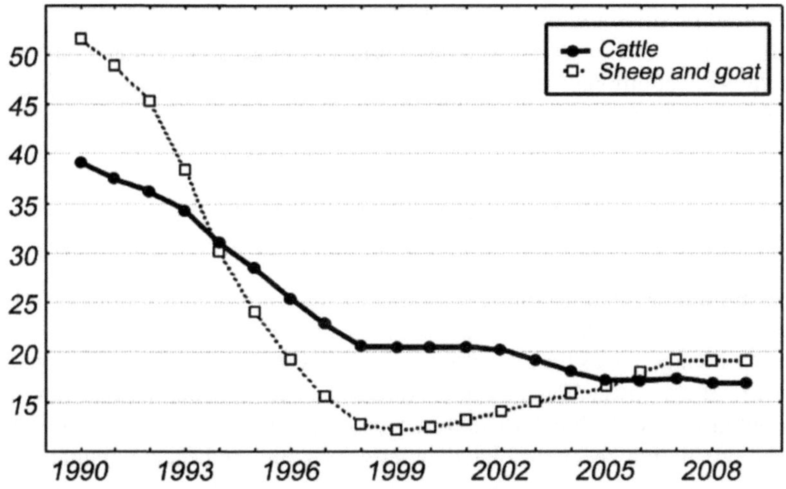

Fig. 2.11 Number of cattle (*filled circles*), and sheep and goats (*transparent squares*), in thousands of heads, during the period 1990–2008

the degradation of many steppe tracts due to the spontaneous spreading of trees and shrubs as well as weeds and alien species.

2.5.3 *Wild Fires*

Unfortunately there are no national statistics for grassland wild fires (only for forest). Nevertheless, the Russian Federation has been listed as the country with the world's largest forested and non-forested areas in which large natural and human-caused wildfires occur (FAO 2006). It has been assessed that Russia contributed 78–84% of the world's springtime black carbon from agricultural fires between 2004 and 2007 (Pettus 2009; Gordon 2010). But reliable information about agricultural fires in Russia does not exist. A FAO report of 2006 estimates that 300,000 km^2 is burning annually (arable land and grassland together, total national territory). At least 1/3 of this should be steppe tracts, which implies that a minimum of 20% of the national steppe area burns annually. In Orenburg Province and the Altai Territory every steppe tract is burnt every 2–5 years.

2.5.4 *Mining, Oil and Gas Development*

Mining projects (for metal ores, especially copper, cobalt, precious metals, etc.) affect both the western and eastern sectors of the Steppe Region: the Russian Plain, South Urals, Altai, and Trans-Baikalia. Oil and gas development threatens steppe ecosystems in Kalmykia, and the Middle and Lower Volga. A specific threat there is that these projects are localized in the best and last remnants of steppe persisting in hilly country and foothills. And any project in mining and oil and gas development involves additional infrastructural works and settlements encroaching on the steppe and causing damage.

2.5.5 Over-Exploitation and Poaching

Over-exploitation and poaching play a role for only a few steppe species in Russia. Probably, the Saiga antelope (*Saiga tatarica*) and the Saker falcon (*Falco cherrug*) are most endangered, as they are affected by poaching for illegal international trade. Kalmykia in the Lower Volga harbours one of the fife main populations of the saiga. Its population size declined 15-fold in 5 years (1997–2001) and still is extremely low; the population became isolated, and its sex structure became dramatically disturbed. The main aim for poachers are horned males and that resulted in an extremely low percentage of males in the adult population, namely 6–10% in 2005–2008 (Melnikov and Sidorov 2009; Neronov et al. 2010; Chap. 12, this volume).

During the past 20 years the Saker population in Russia decreased from more than 9,000 individuals to 2,400, a decline of more than 70%. Some nesting populations have completely disappeared. The species became extinct in European Russia and dramatically decreases in the Siberian steppes (Karyakin 2008; Moshkin 2010).

Also some Daurian and Caucasian medicinal steppe plants are affected by over-exploitation for BAS production.

2.5.6 Afforestation

Grasslands conversion by woody plant encroachment (afforestation) is a long tradition in Russia as a measure for "improvement" of the environment in agricultural landscapes. The last decade it is regarded also as a way of carbon sequestration under the Kyoto framework. Nationally 36,000 ha have to be afforested to 2012 under just the Federal Plan for Kyoto forests and provincial governments are planning even much more acreage: Stavropol ca. 56,000 ha until 2020 (this is more than 1% of the province area); Samara planned 140,000 ha for 2006–2015 (a 20% rise of the province's forest area) but the program ended in complete failure and was cancelled in 2009. Unfortunately steppe sites are considered the most suitable for afforestation because of their low price and easy availability.

2.5.7 Biofuel Production

At the moment biofuel production is not a specific threat but it is an additional driver of turning steppe grasslands into arable land. Up to now the main biofuel crop in Russia is rapeseed (*Brassica napus*); since 2005 its cultivation area rocketed

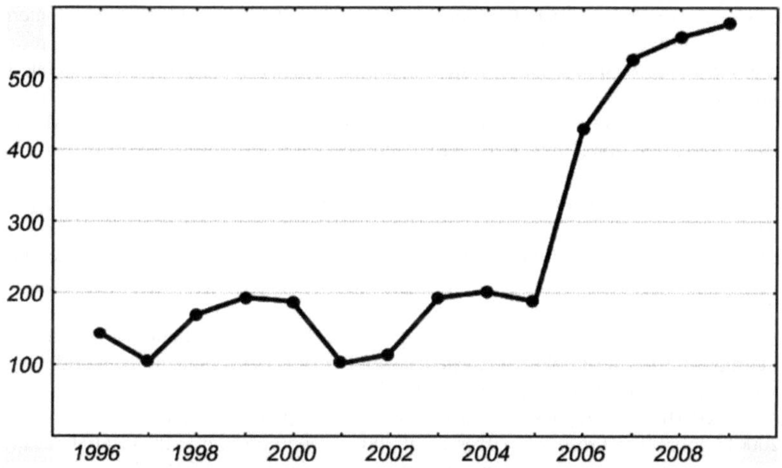

Fig. 2.12 Area of rapeseed crops in the 35 steppe provinces, in thousands of ha, during the period 1996–2008

threefold (Fig. 2.12). Its cultivation area increased by re-developing abandoned lands rather than by changing to rapeseed instead of other crops, but this means that croplands are enlarged at the expense secondary steppe ecosystems.

Acknowledgement We are very grateful to Dr. Evgeny Simonov (Daurian International Protected Area), Mr. Philip and Mrs Julia Johnson, Dr. Jozef Bednar and Mrs Svetlana Belova-Bednar who so much helped us to make an acceptable English text and made many important comments to the manuscript. Our warm gratitude to Mr. Konstantin Kobyakov and Mrs Svetlana Titova-Kobyakova, Mrs Alexandra Egorova, and Mrs Anna Barashkova for their invaluable help in GIS work and map preparing. Special thanks to Dr. Marinus J.A. Werger for his careful and fruitful editing of the manuscript.

The manuscript is based on our analyses prepared for the RSPB/EBRD Project "Developing Pro-Biodiversity Business in the Steppe Zones of the Eurasia region" (2008) and UNDP/GEF Project "Improving the Coverage and Management Efficiency of Protected Areas in the Steppe Biome of Russia", preparatory phase (2009).

References

Antonchikov AN (2005) A review of the conservation status of steppe birds of the Northern part of the Eastern Paleartic. In: Bota G, Morales MB, Mañosa S, Camprodon J (eds) Ecology and conservation of steppe-land birds. Lynx Edicions & Centre Tecnològic Forestal de Catalunya, Barcelona, pp 103–118

Belelli Marchesini L, Papale D, Reichstein M, Vuichard N, Tchebakova N, Valentini R (2007) Carbon balance assessment of a natural steppe of southern Siberia by multiple constraint approach. Biogeosciences 4:581–595

BirdLife International (2004) Birds in Europe: population estimates, trends and conservation status, Conservation series No.12. BirdLife International, Cambridge

BirdLife International (2008) BirdLife's online world bird database: the site for bird conservation. Version 2.1. BirdLife International, Cambridge, UK. Available http://www.birdlife.org

Bukreev SA (Ed.) (2006) Important Bird Areas in Russia. Vol. 2. Important Bird Areas of international significance in Western Siberia. Russian Bird Conservation Union, Moscow, 336 pp
Bukreev SA (Ed.) (2009) Important Bird Areas in Russia. Vol. 3. Important Bird Areas of international significance in Caucasus Ecoregion. Russian Bird Conservation Union, Moscow, 302 pp
Burfield IJ (2005) The conservation status of steppic birds in Europe. In: Bota G, Morales MB, Mañosa S, Camprodon J (eds) Ecology and conservation of steppe-land birds. Lynx Edicions & Centre Tecnològic Forestal de Catalunya, Barcelona, pp 119–139
Chibilev AA (1998) Basics of steppe science. Publisher House DIMUR, Orenburg, 120 pp (in Russian)
Compendium (2007) Compendium of information on the state and use of lands in Federal Districts of the Russian Federation in 2006. Federal Agency Real Estate Cadastre Agency, Moscow (in Russian)
Costanza R, d'Arge R, de Groot R, Farber S, Grasso M, Hannon B, Limburg K, Naeem S, O'Neill RV, Paruelo J, Raskin RG, Sutton P, van den Belt M (1997) The value of the world's ecosystem services and natural capital. Nature 387:253–260
Dzybov DS (2010) Agro-steppes. AGRUS, Stavropol, 256 pp (in Russian)
FAO (2006) Global forest resources assessment 2005 – report on fires in the Central Asian Region and adjacent countries. Fire management working paper 16. FAO, Rome. www.fao.org/forestry/site/fire-alerts/en. Accessed 13 Aug 2010
FAO Statistics Division (2008) http://faostat.fao.org/site/573/DesktopDefault.aspx?PageID=573#ancor. Accessed 1 July 2011
Gordon D (2010) Reducing agricultural fire emissions in Russia. Pacific Environment. A report and plan produced for clean air-cool planet. Pacific Environment, San Francisco, 23 pp (unpublished report)
IUCN (2008) 2008 IUCN red list of threatened species. www.redlist.org
Karyakin IV (2008) Saker Falcon in Russia. Raptors Conserv 12:28–47
Korolyuk AYu (2002) Vegetation. In: Gadjiev IM et al (eds) Steppes of Inner Asia. SB RAS Publisher, Novosibirsk, pp 45–94 (in Russian)
Korolyuk AYu, Yegorova AV, Smelyansky IE, Filippova NV (2008) Structure of vegetation of steppe hills in the Altai Piedmonts. Contemp Probl Ecol 1:49–58
Krivenko VG (ed) (1998) Wetlands of Russia. Vol. 1. Wetlands of international importance, AEME publication No. 52. Wetlands International, Moscow
Krivenko VG (ed) (2000) Wetlands of Russia. Vol. 3. Wetlands of the Ramsar shadow list, Wetlands International global series No. 6. Wetlands International, Moscow
Kucheruk VV (1959) Steppe faunistic complex of mammals and its place in Palaearctic fauna. In: Geography of terrestrial animals' population and methods of its investigation. Publishing of AS USSR, Moscow, pp 45–87 (in Russian)
Kurganova IN, Lopes de Gerenyu VO, Myakshina TN, Sapronov DV, Lichko VI and co-authors (2008) Changes in the carbon stocks of former croplands in Russia. Agric Sci 15(4):10–15
Lavrenko EM (1970a) Provincial division of the Pontic-Kazakhstanian Subregion of the Steppe Region of Eurasia. Botanicheskiy Jurnal 55:609–625 (in Russian)
Lavrenko EM (1970b) Provincial division of the Central-Asian Subregion of the Steppe Region of Eurasia. Botanicheskiy Jurnal 55:1734–1747 (in Russian)
Lavrenko EM, Karamysheva ZV, Nikulina RI (1991) Steppes of Eurasia. Nauka, Leningrad, 146 pp (in Russian)
Melnikov VV, Sidorov SV (2009) Saigak in Russia: actual state, conservation, and restoration. Stepnoi Bull 27:42–46 (in Russian)
MNR (2007) State report on the state and protection of the environment in the Russian Federation in 2006. MNR, Moscow
Moon D (2008) The Russian steppe: an environmental history, 1700–1914. Centre for Russian, Central and East European Studies, Working Paper Series, WP2008/07. University of Glasgow, Glasgow
Mordkovich VG (1994) Originality of Siberian steppes, level of their disturbance and safety. Sib Ecol J 1:475–481 (in Russian)

Moshkin AV (2010) Is there any scientific basis for decreasing the conservation status of the Saker Falcon? Raptors Conserv 19:37–74

Namzalov BB (1994) Steppes of South Siberia. Novosibirsk – Ulan-Ude (in Russian)

Neronov V, Lushchekina A, Karimova T, Arylova N (2010) The critically endangered saiga antelope in a changing world. Biological consequences of global change. In: Fourth international symposium of integrative zoology, Kunming, China, 4–6 Dec 2010

Nikolaev VA (1999) Landscapes of Asiatic steppes. Publishing of Moscow State University, Moscow, 288 pp (in Russian)

Nikolsky AA, Rumyantsev VYu (2002) Zonal representativeness of the Zapovedniks system of the Russian Federation. In: Scientific issues of ecological problems of Russia. Proceedings of the national conference ad memoriam A.L. Yashin. 1, Moscow, Nauka, pp 160–165 (in Russian)

Ogureeva GN (ed) (1999) Vegetation zones and types of altitudinal belts systems of Russia and adjacent areas. Map 1:800,000. Publishing of Moscow State University, Moscow, 38 pp (in Russian)

Pettus E (2009) Agricultural fires and arctic climate change: a special CATF report, CATF, May 2009, 33 pp. http://www.catf.us/resources/publications/files/Agricultural_Fires_and_Arctic_Climate_Change.pdf. Accessed 11 July 2011

Prisyazhnyuk VE (ed) (2004) 2003 * Red list of rare and endangered species of animals and plants, which are particularly protected in Russia. Part 1. Vertebrates. All-Russian Institute for Nature Protection, Moscow (in Russian)

Red Data Book of the Russian Federation (2001) Animals. AST, Moscow (in Russian)

Regions (1999) Regions of Russia: statistics compendium, vol 1. Federal State Statistics Comitee, Moscow (in Russian)

Tacker GM, Evans MI (1997) Habitats for birds in Europe: a conservation strategy for the wider environment, Conservation series No.6. BirdLife International, Cambridge

Tkachuk T, Kiriliuk O, Simonov E (2008) Temperate grassland region: Daurian steppe. In: Compendium of regional templates on the status of temperate grasslands conservation and protection. Prepared for the world temperate grasslands conservation initiative workshop 'Life in a working landscape: towards a conservation strategy for the world's temperate grasslands', Hohhot, China, 28–29 June 2008

[illegible]

Chapter 3
Steppes of Kazakhstan: Diversity and Present State

E.I. Rachkovskaya and T.M. Bragina

Abstract The paper summarizes modern data on the phytogeography, ecology and typology of the main plant communities of the steppe areas of Kazakhstan. The patterns of zonal steppe vegetation on the plains and altitudinal vegetation sequences in the mountains are shown. A new map of the phytogeographic regionalization is given. The paper reviews human impacts on the steppe vegetation of Kazakhstan and their degree of usage and transformation, and discusses the conservation status of steppe ecosystems in a network of protected areas, as well as the current status of steppe biodiversity in the Republic of Kazakhstan.

3.1 Introduction

Kazakhstan is located in the heart of Eurasia, in the dry continental climate of the temperate zone. Its area is 2,724,000 km^2. The distance from north to south is 1,600 km, from west to east 3,000 km. Plains in the western and northern parts of the country occupy 60% of the territory, and in its central part (30% of the territory) they alternate with hills ('melkosopochnik') and low mountains. The southern and north-eastern parts of Kazakhstan are occupied by the Altai, Saur, Northern and Western Tien Shan mountain systems.

The vast territory covers several natural zones (forest-steppe, steppe and desert) and includes a variety of regions with a rich flora and fauna.

E.I. Rachkovskaya
Institute of Botany and Phytointroduction, Almaty, Kazakhstan
e-mail: evra17927@mail.ru

T.M. Bragina (✉)
Biology and Geography, Kostanay State Pedagogical Institute, Kostanay, Kazakhstan

Natural Sciences, Pedagogical Institute and Institute of Biology of Southern Federal University, Rostov-on-Don, Russia
e-mail: tm_bragina@mail.ru; naurzum@mail.ru

M.J.A. Werger and M.A. van Staalduinen (eds.), *Eurasian Steppes. Ecological Problems and Livelihoods in a Changing World*, Plant and Vegetation 6,
DOI 10.1007/978-94-007-3886-7_3,

3.2 Diversity of Steppe Ecosystems (Natural-Territorial Complexes)

The differences in natural conditions in Kazakhstan (climate, topography, soils, vegetation) determine the differentiation in natural-territorial complexes or ecosystems in the country. For our purpose an ecosystem, or natural-territorial complex, is characterized by a homogeneous relief, one type or subtype of soil, a set of plant communities with a common structure (life forms) and characteristic species composition, a similar range in fluctuation in species composition and productivity, a similar successional trend, and similar responses to natural and anthropogenic effects (Ogar 1999, 2006). This ecosystem concept, considering elements of topography, soil, vegetation and associated animal populations, we use to assess the potential development of habitats, their potential floral and faunal diversity, and the potential consequences of disturbance.

This concept was first used in classifying and mapping the natural ecosystems of Mongolia (Ecosystems of Mongolia 1995), and later in creating the first small-scale maps of ecosystems of Kazakhstan, taking the peculiarities of the natural conditions of Kazakhstan into account (Ogar and Rachkovskaya 2006). This allows us to determine the differences in the natural ecosystems of the two largest steppe regions of Eurasia: Kazakhstan and Mongolia. On this basis of small-scale maps of ecosystems, made for the National Atlas of the Republic (2006) and using the materials of the UNEP/GEF/WWF project "Development of the Econet for Long-term Conservation of Biodiversity in the Central Asia Ecoregions", we are able to provide (qualitative and quantitative) areal data on the landscape diversity of the main forest-steppe and steppe ecosystems (Table 3.1).

Forest-steppe ecosystems occupy 2.4% of Kazakhstan, and steppes, including those on the foothills and mountains, 42.4%.

In the plain and melkosopochnik part in the north of the country large areas of dry and deserted steppes (11%) are mainly found on the Sub-Ural and Turgai Plateau (Photo 3.1). Dry and deserted steppes are also associated with the elevated deluvial-proluvial plains, mainly among the hills of Central Kazakhstan (6.7%), and are widespread in the lowland plains of the Caspian Lowland (4.8%), while in the lowland plains of the West Siberian Plain, forest-steppe ecosystems and humid temperate-droughty and droughty steppes are widespread (6.3%). Note that the steppe ecosystems on aeolian plains occupy a very small area in Kazakhstan (1.6%). Special for Kazakhstan are the steppes of melkosopochnik (8.6%). Low inselbergs do not occupy large areas (2.4%), but allow the penetration of the northern steppes far to the south, and they comprise many rare types of ecosystems in the region.

The area of the foothills and mountain ecosystems is small (3.4%), but their importance for the conservation of biodiversity and for economic use, especially in desert regions, cannot be overestimated.

This paper will mainly deal with vegetation, as it is the major and most informative biotic component of terrestrial ecosystems.

Table 3.1 Distribution of forest-steppe and steppe ecosystems of Kazakhstan (area in km^2)

Zonal types and subzonal subtypes of ecosystems	Forest-steppes		Steppes		Dry steppes		Deserted steppes		Total	Total
Classes of ecosystems	Area	% of the territory of Kazakhstan	Area	% of the territory of Kazakhstan	Area	% of the territory of Kazakhstan	Area	% of the territory of Kazakhstan	Area	% of the territory of Kazakhstan
On the plains										
Plateau			9,118.3	0.3	192,513.1	7.1	96,500.1	3.6	298,131.5	11.0
High plains			20,171.7	0.8	62,668.9	2.3	96,637.8	3.6	179,478.4	6.7
Plains of lowlands	46,534.7	1.7	125,878.7	4.6	69,553.1	2.5	61,494.3	2.3	303,460.8	11.2
Aeolian plains					35,153.5	1.3	8,644.7	0.3	43,798.2	1.6
Melkosopochnik			27,793.8	1.0	89,159.2	3.3	117,565.3	4.3	234,518.3	8.6
Low mountains	15,681.9	0.6	23,673.9	0.9	20,873.1	0.8	1,032.6	<1	61,261.5	2.4
Total	62,216.6	2.3	206,636.4	7.6	469,920.9	17.3	381,874.8	14.1	1,120,648.7	41.5
On the mountains										
Piedmonts and low mountains			2,189.1	0.1	6,648.0	0.2	22,316.0	0.8	31,153.1	1.1
Medium high mountains	2,180.2	0.08	29,488.8	1.1	31,182.6	1.1			62,851.6	2.3
Total	2,180.2	0.1	31,677.9	1.2	37,830.6	1.3	22,316.0	0.8	94,004.7	3.4
Grand total	64,396.8	2.4	238,314.3	8.8	507,751.5	18.6	404,190.8	14.9	1,214,653.4	44.9

Photo 3.1 Vast areas of deserted steppes have survived in the southern part of the Turgay Plateau; Southern part of Kostanay Oblast (earlier Turgay Oblast); Amangeldy Rajon. 10.06.2005 (Photo Tatyana Bragina)

3.3 Steppe Research in Kazakhstan

The vegetation cover of Kazakhstan is very diverse and comprises steppes, deserts, forests, grasslands, etc. The herbaceous vegetation types include steppes, meadows (lowland and mountain) and savanna-like vegetation (tall herbaceous and short herbaceous). All these major vegetation types have their characteristic physiognomy (structure) and life forms. Accordingly, the steppe, as defined by the steppe expert Lavrenko (1940), includes communities dominated by perennial microthermic (cold-resistant), xerophytic (drought-resistant) herbaceous plants, mainly bunch grasses of the genera *Stipa, Festuca, Koeleria* and *Helictotrichon.*

There are two natural zones with steppes on the plains of Kazakhstan – forest-steppe and steppe itself. In the mountains of Kazakhstan, steppes occur both on low and middle-high mountains, and fragmented also on high mountains.

The steppes of Kazakhstan have been studied for more than a century. During the early twentieth century, extensive soil-botanical investigations were carried out in Northern Kazakhstan, which resulted in the first data collected on steppe vegetation. Important early works on the steppes of Kazakhstan were Krasheninnikov (1925), Baranov (1925), Sobolev (1948), Pavlov (1948), and Rubtsov (1952). The fundamental work of E.M. Lavrenko (1940, 1956) contributed greatly to the knowledge of the vegetation cover of the Eurasian steppes, including

Kazakhstan. The monograph “Steppes of Eurasia” (1991) comprises this early, as well as newer knowledge.

The diversity of the steppes of Kazakhstan has been summarized in numerous reports and is documented in Vegetation Maps and Legends (1960, 1975, 2006). Botanical research, acquired in the period that virgin steppes existed and fallow lands could regenerate without much interference, has resulted in “The Natural Zonation of the Territory of Northern Kazakhstan” (1960), vegetation maps of the region and descriptions of the main types of steppes (Isachenko and Rachkovskaya 1961). In Central Kazakhstan, in the period 1964–1968, a series of detailed vegetation studies were carried out, summarized in Karamysheva and Rachkovskaya (1973) and in the vegetation map of the Kazakh Melkosopochnik (1975). Mountain steppes of Tarbagatai were described in detail by Stepanova (1962), and of the Tien Shan and Dzungarian Alatau by Rubtsov (1948, 1954). Sokolov (1977) published a survey of the vegetation of the Altai, and Chibilyev (1998) and Nikolayev (1999) published accounts of the steppes of Russia and Kazakhstan. More recently overviews were published by Marynich (2005) and Rachkovskaya (2006).

3.3.1 Zonation

The vegetation of Kazakhstan varies considerably from north to south due to differences in climatic conditions (Table 3.2). With respect to the steppes, we deal with two zones – the forest-steppe and the steppe – with seven subzones.

The **Forest-steppe zone** is found only in the northernmost part of Kazakhstan, on the plains of the Western Siberian Lowland. The forest-steppe zone is characterized by the alternation of birch and aspen-birch forests (*Betula pendula, B. pubescens, Populus tremula*) with non-forested land occupied by steppe, meadows with steppe elements, wetland meadows, and marshes. The forest-steppe zone can be divided into two subzones: the southern forest-steppe and steppe with “kolki” (Natural zonation 1960).

In the *southern forest-steppe*, forests cover, in some places, up to 50% of the watershed plains, but the average forest cover is 20–30%. The soils beneath them are grey forest soils. Between the forest patches grass-forb and forb-grass meadow steppes and steppe meadows occur on leached chernozem and meadow-chernozem soils.

The forest-steppe with “kolki” occurs rather widespread on flat lake-alluvial plains and in areas with ridge relief within the West Siberian Lowland. These forests occur on sites with additional moisture. A lower forest cover (sometimes 25–30% of the watershed) characterizes this subzone. Prior to the mass plowing of the flat plains, the Ubagan-Ishim watershed carried such rich forb–feather grass steppes (*Stipa zalesskii*), and east of the River Ishim there was feather grass steppe (*Stipa zalesskii, Peucedanum morisonii*). The soils under the forests are solod, under the steppes ordinary chernozem.

The **Steppe zone** of Kazakhstan occupies 41.4% of the territory of the Republic. It covers the northern part of the Caspian Lowlands, the Torgay Plateau, the Sub-Ural

Table 3.2 Climatic parameters of the natural steppe zones and subzones of Kazakhstan

Natural zones	Subzones	Sum of mean temperatures over 10°C	January, monthly average t°C	July, monthly average t°C	Precipitation mm/year	Humidity coefficient
Forest-Steppe	Southern forest-steppe	2,000–2,100	−16 to −19	19–20	310–350	0.6–0.7
	Forest-steppe with "kolki"	2,200	−19 to −20	19–20	300	0.5–0.6
Steppe	Temperate-droughty steppes	2,200–2,250	−18 to −19	19.5–20.5	300–320	0.5–0.6
	Droughty steppes	2,300–2,400	−18 to −19	19.5–20.5	280–300	0.45–0.5
	Temperate-dry steppes	2,400–2,500	−14 to −16	20–22	240–260	0.4–0.55
	Dry steppes	2,500–2,600	−18	21–23	240–260	0.35–0.4
	Deserted steppes	2,800–2,900	−15 to −17	23–25	150–200	0.2–0.3–0.35

piedmont plateau, the Trans-Ural, the West Siberian Lowland, and the Central Kazakhstan Melkosopochnik (hills) (Sary-Arka). In correspondence with the differences in climatic parameters over this wide zone, the steppe zone is divided into five subzones:

- Temperate-droughty rich forb – feather grass steppes on ordinary chernozem;
- Droughty forb – feather grass steppes on southern chernozem;
- Temperate-dry bunch grass steppes on dark chestnut soils;
- Dry xerophytic forb – bunch grass steppes on chestnut soils;
- Deserted sagebrush – bunch grass steppes on light chestnut soils.

Temperate-droughty rich forb – feather grass steppes on ordinary chernozem are distributed along the southern periphery of the West Siberian Lowland. The basic species of these steppes are *Stipa zalesskii* and *Festuca valesiaca*, mixed with xero-mesophytic, loose bunch grasses (*Poa angustifolia, Bromopsis inermis*) and meso-xerophytic and euryxerophytic steppe forbs (*Hieracium virosum, Artemisia latifolia Veronica spuria, Lathyrus tuberosus, Pulsatilla multifida*).

The indigenous steppe type of this area is a rich forb (*Peucedanum morisonii*) – red feather grass (*Stipa zalesskii*) steppe. It is characterized by a high species richness (45–60 species per 100 m^2) with a cover of 80–90% to 100%.

Droughty forb – feather grass steppes are confined to the flat lake-alluvial plains of the West Siberian Lowland, the denudative plains of the Trans-Ural Plateau and northeast of the Kokchetav Heights. Zonal soil types are the southern chernozem soils and their carbonate variants, as well as saline soils and solonetz with a complex vegetation. Cobbly southern chernozem is common in the Trans-Ural and the periphery of the Central Kazakhstan Melkosopochnik.

In these steppes *Stipa zalesskii* and *Festuca valesiaca* are dominant species while *Koeleria cristata* as a co-dominant. Forbs are less abundant. They are mainly mesoxerophytic and euryxerophytic steppe species (*Salvia stepposa, Seseli ledebourii, Medicago romanica, Galium ruthenicum, Eryngium planum*, etc.), though xeromesophytes still occur (*Thalictrum minus, Xanthoselium alsaticum*, etc.). Total coverage is 70–80%.

On calcareous soils the forb – feather grass steppes are common, with *Stipa lessingiana* as a co-dominant and *Stipa korshinskyi*, an indicator of carbonate. Currently, most of the forb – grass steppe is plowed (about 60% of the area), and it became one of the most important agricultural regions of the country.

Temperate dry bunch grass steppes on dark chestnut soils, often calcareous and loamy, are common on the plains of most of the Turgay Plateau, in the eastern periphery of the West Siberian Lowland, and the plains among the hills of the Central Kazakhstan Melkosopochnik.

The main vegetation features of dry steppes include xerophytic bunch grasses (*Stipa lessingiana, S. capillata, Festuca valesiaca, Koeleria cristata*), loose bunch grasses (*Agropyron pectinatum, Cleistogenes squarrosa* in the east) and rhizome grasses (*Leymus ramosus*), with little occurrence of forbs, though some xerophytic forbs occur (*Dianthus leptopetalus, Phlomoides agraria, Jurinea multiflora, Galatella divaricata*). The proportion of ephemeroid and hemi-ephemeroid forbs (species that are short-lived aboveground) is considerable (*Tulipa schrenkii, T. patens, T. biflora, Ornithogalum fischerianum*). The cover is sparse, usually less than 50–60%. *Stipa lessingiana* steppes, characterized by a small admixture of forbs, are widespread (Photos 3.2 and 3.3).

Dry xerophytic forb – bunch grass steppes on chestnut soils occupy large areas in the Sub-Ural and Turgay Plateau. The main zonal type is a xerophytic forb – fescue – feather grass steppe (*Stipa lessingiana, Festuca valesiaca, Galatella divaricata, G. tatarica, Phlomoides agraria, Tanacetum achilleifolium*) which can be regarded as a southern version of the dry steppes. *Stipa capillata* occurs on the plains among the hills of Sary-Arka.

The total area of temperate dry and dry bunch grass steppe is 53 million ha. This is about 20% of the country, and one-third of this is under cultivation.

Deserted sagebrush – bunch grass steppes on light chestnut soils are the southernmost type of steppes, their southern boundary roughly coinciding with the boundary between steppes and deserts (48° N.L.). This subzone of deserted steppes on light chestnut soils covers the Caspian Lowlands, the Sub-Ural, the Turgai Plateau, Mugodzhary, the Central Kazakhstan Melkosopochnik and the Kulundinskaya Plain. In the northern part of this subzone these are sagebrush – feather grass steppes with *Stipa lessingiana* and species of the genus *Artemisia* and in the southern part with *Stipa sareptana* and *Artemisia* sp. Compared to the dry steppes, in the subzone of deserted steppes the vegetation is more complex.

The main edificators of communities of deserted steppes are *Stipa lessingiana, S. sareptana* and *Festuca valesiaca*. The euxerophytic semi-shrubs *Artemisia lerchiana* (in the west) and *Artemisia semiarida* (in the centre of the subzone), as well as *Artemisia gracilescens* and *A. sublessingiana* (in the centre and east of Kazakhstan), are co-edificators. Forbs are relatively few and typically include xerophytic species. Characteristic are *Kochia prostrata* (although it is absent in the northern steppes) and hemi-ephemeroids (species of the genus *Ferula*), ephemeroids (species of the genus *Tulipa* and *Gagea*) and ephemerals (annuals).

3.3.2 *Vertical Zonation*

Steppes occur in all the mountains of Kazakhstan – from Altai to Karatau. In the mountains, the distribution of the steppes is embedded in the vertical zonation of the vegetation. In Kazakhstan the distribution of mountain steppes is determined by the big mountains – Altai, Saur, Tarbagatai, and Tien Shan.

Vegetation belts in mountains, determined by the vertical gradient in climatic conditions, comprise a wide range of plant communities, depending on landform, water availability, soil characteristics, exposition, etc., but nevertheless, each belt carries a dominant type of vegetation, which is characteristic for that belt (Volkova 1994, p. 14).

Photo 3.2 Feather-grass steppes with *Stipa lessingiana*. Eastern-Turgai Plateau; Kostanay Oblast; Naurzum Rajon. 25.05.2009 (Photo Tatyana Bragina)

Photo 3.3 Feather-grass steppes in spring with *Tulipa schrenkii*. Naurzum Reserve; Central-Eastern part of Kostanay Oblast; Naurzum Rajon. 5.05.2005 (Photo Tatyana Bragina)

At a general level all mountain vegetation of Kazakhstan can be divided into four groups of zonation types:

- Altai (for the mountains located in the dry steppes);
- Saur-Tarbagatai (for the mountains located in the desert steppes);
- Dzungaria-Northern Tien Shan (for the mountains located in the middle [of temperate to cold] deserts);
- Western Tien Shan (for the mountains located in the warm-temperate southern deserts).

3.3.2.1 The Altai-Group of Zonation Types

The dry steppes in Kazakhstan's part of the Altai occur on the piedmonts of the mountains and vary in structure according to the vertical zonation (Sokolov 1977).

The *North-western part of the Altai Mountains* (the mountain ranges Tigiretsky, Ivanovsky, Holzun, Ubinsky, Ulbinsky, Bukhtarminsky, Kalbinsky) is the most humid of the mountain systems in Kazakhstan. These ranges are affected by Atlantic air masses, bringing heavy precipitation. Precipitation is relatively equal from spring to autumn. In winter the Siberian anticyclone is set over the region (Photo 3.4a, b).

The vertical sequence of belts in this area is: mountain steppes – mountain steppes with shrubs – small-leaved forests – dark coniferous forests – subalpine and alpine tall herb and short herb meadows. Steppes occur from 250 m (piedmont plains) to 1,200 m.

In the *Eastern and Southern parts of the Altai Mountains* (the ridges Katun, Southern Altai, Narymsky, Kurchum) the climate is more arid. There the belt sequence is: mountain steppes – mountain steppe with shrubs – light coniferous forests of *Larix* – subalpine-like and alpine-like short herb meadows – mountain tundra. The steppes occur from 600 to 1,800 m (Table 3.3).

3.3.2.2 The Saur-Tarbagatai

The *Saur Mountains* and *Tarbagatai* are located in the deserted steppes zone on light chestnut soils in the mountain foothills.

The sequence of belts in the mountains of *Saur* is: mountain steppes – light coniferous forests of *Larix* – subalpine-like and alpine-like short herb meadows (Table 3.4). Park-like *Larix* forests in conjunction with steppes (expositional forest-steppe) occur at medium heights, while shrub steppes and brushwoods predominate at low heights.

The *Tarbagatai Ridge* is lower than Saur and does not exceed 2,990 m. It has no permanent snow and glaciers, and thus the forest belt is not developed. In the high mountains alpine meadows dominated by species of *Kobresia* are found and subalpine forb – grass and forb meadows dominated by species of *Alchemilla* and cryophytic steppes with *Helictotrichon hookeri, Festuca valesiaca* and *Elytrigia gmelinii* are common and varied. On medium-high mountains typical droughty

Photo 3.4 (**a**) Altai Mountains: Steppe belt (zone). Eastern Kazakhstan; Eastern-Kazakhstan Oblast. 26.06.2008 (Photo Tatyana Bragina). (**b**) Altai Mountains; Steppe belt (zone). Eastern Kazakhstan; Eastern-Kazakhstan Oblast. 26.06.2008 (Photo Tatyana Bragina)

Table 3.3 Distribution (in m) of the steppes along the vertical profile in the Altai Mountains and in the Saur Mountains (by Sokolov 1977)

Altitudinal belt (Zone)	Northern-Western Altai	Ridge Kalba	Southern Altai	Saur
Meadow steppes in the shrub belt	600–1,200	1,000–1,200	1,400–1,800	1,600–2,000 (2,200)
Temperate- droughty and droughty steppes	300–600	400–1,000	1,200–1,400	1,300 (1,600)–1,800 (2,000)
Dry steppes	250–350		800–1,200	800–1,300 (1,600)
Deserted steppes			600–800	500–800

Table 3.4 Vertical zonation (in m) of steppes in the Tarbagatai Ridge (by Stepanova 1962)

Altitudinal belts (zone)	Southern slope (m)	Northern slope (m)
Deserted steppes	500–700	500–900
Steppes	700–1,000 (1,200)	900–2,000
Brushwood	1,000 (1,200)–1,700 (1,800)	
Subalpine with fragments of mountain steppes	1,700 (1,800)–2,400	2,000–2,200
Alpine with fragments of mountain steppes	2,400–3,100	2,200–3,100

steppes and brushwoods, often of the rare endemic species *Amygdalus ledebouriana* and *Calophaca howenii*, are found and on low mountains shrubby dry steppes (Stepanova 1962).

3.3.2.3 The Zhungar-Northern Tien Shan

This group includes the mountain ridges Kyrgyzskiy, Zailiyskiy Ala-Tau, Kungey Ala-Tau, Zhungar Ala-Tau, Ketmen, and the northern face of the Terskey Ala-Tau. The sequence of the zonation types is: steppes (deserted steppes, typical steppes, meadow steppes) – dark coniferous forests, woodlands and meadows – subalpine-like meadows and juniper dwarfshrub – alpine meadows and meadows dominated by species of *Kobresia*.

These mountain ridges have some individual features. On the northern slope of the Zhungar Ala-Tau the steppe zone covers a much greater area than in the Zailiyskiy Ala-Tau. Deserted steppes of ephemeroid sagebrush – feather grass with *Stipa sareptana, S. lessingiana, S. caucasica*, species of *Artemisia*, and *Poa bulbosa* occur in the piedmonts of the northern slope of the Zhungar Ala-Tau between 650 and 1,000 m and mountain dry steppes predominate between 1,000 and 1,200 m, forb and forb – feather grass steppes up to 1,400. Still higher we find only meadow steppes in combination with brushwoods, broad- and small-leaved forests, and meadows. In the Zailiyskiy Ala-Tau the steppe zone consists of three clear subbelts, including deserted steppes and dry steppes, where some spring ephemeroids play a significant role.

The altitudinal profile of the Almaty district is as follows: deserted and dry steppes occur at 650–850 m, forb – grass steppe with shrubs from 850 to 1,200 m, and meadow steppes are found in a belt of small-leaved and fruit forests at 1,200–1,600 m.

Ephemerals and ephemeroids do not occur in the mountain steppes of the Kungey Ala-Tau, Ketmen and Terskey ridges. Here the high-altitude steppe belt typically is broader, especially on southern macro-slopes. The highlands are more arid here and cryophytic steppes are more important, which brings them closer to the high mountains of Central Asia (Volkova 2003).

The Kyrgyz Ridge, which is connected to the ridges of the Western Tien Shan, carries some features of the vegetation of the Central Asian mountains. This is visible in the relatively high proportion of ephemeroid grasses and hemi-ephemeroid tall herbs in the steppe zone of the piedmonts and low mountains .

In general, all steppes of the piedmont plains of the Northern Tien Shan have long ago been converted to agricultural lands.

3.3.2.4 The Western Tien Shan

Here steppes occur as fragments. In the low mountains of the Western Tien Shan savanoid-bunch grass steppes occur, and in the Karatau mountain-xerophytes – bunch grass steppes.

3.4 The Main Types of Steppes of Kazakhstan

Steppe vegetation comprises communities dominated by perennial microthermic, xerophytic grasses, mainly of bunch grasses of the genera *Stipa, Festuca, Koeleria* and *Helictotrichon.* This vegetation covers large areas on the plains, plateaus, hills and mountain slopes. A distinctive feature of all steppe communities of melkosopochnik and mountains of Kazakhstan (Mugodzhary, the Central Kazakhstan Melkosopochnik (Sary-Arka), the Altai Mountains, Tarbagatai and the Tien Shan), is the occurrence of shrub steppes (mainly involving species of the genera *Spiraea* and *Caragana*).

Lavrenko (1940, 1991) divided the steppes into three main subtypes: meadow steppes, steppes, and deserted steppes, which occur successively in a latitudinal zonal alignment. The remaining subtypes of steppes are mainly related to mountain systems: cryophytic steppes in the high mountains, ephemeroid – bunch grass steppes at the periphery of the ridges of the Northern Tien Shan, savanoid – bunch grass and mountain-xerophytes – bunch grasses steppes in the Western Tien Shan and Karatau.

3.4.1 Plain Steppes

Meadow steppes are composed of euxerophytic and mesoxerophytic herbaceous perennial grasses with a constant admixture of mesophytic and xero-mesophytic species. They are typical for the non-forested patches in the forest-steppe zone.

Bunch grasses, particularly *Stipa pennata, Stipa zalesskii* and *Festuca valesiaca*, and rhizoid grasses, such as *Poa angustifolia, Helictotrichon schellianum* and *Elytrigia repens*, are common, and there are many forbs, often including hydrophilic species, e.g. *Lathyrus pisiformis, L. pratensis, Vicia cracca, Artemisia pontica, A. latifolia, Filipendula vulgaris, Fragaria viridis*. Meadow steppes have a tall (60–80 cm on average) and dense herb layer with a total coverage of about 100% (Isachenko and Rachkovskaya 1961).

Real steppes are species-rich forb – feather grass and dry bunch grass steppes dominated by euxerophytic bunch grasses with a constant and significant admixture of drought-resistant forbs (mesoxerophytic and euxerophytic). Communities of this steppe subtype widely occur on the plains and in the mountains, mainly the medium-high mountains. The various mountain ranges support different forb and shrub species.

In the subtype of ***deserted steppes*** euxerophytic bunch grasses predominate and semi-shrubs, mainly of *Artemisia*, are common.

The most widespread steppe formations (i.e. plant communities with one dominant species) of the above-mentioned subtypes of steppes are briefly reviewed.

Feather grass (Stipa zalesskii) steppes are the northernmost, hydrophilic type of plain steppes on chernozem; further south they occur only on mountain slopes. They are characterized by a very rich floral composition. Forbs are especially abundant and diverse. The interesting *Stipa zalesskii* steppes, with an admixture of the tall *Peucedanum morissonii*, previously occurred widely on the plains of northern Kazakhstan but were plowed up and completely destroyed. Only some fragments are preserved in the low mountains, e.g. in the Kokshetau, Erementau, Bayan-Aul, etc.

Forb – sedge – feather grass steppes (Stipa zalesskii, Carex pediformis, Helictotrichon desertorum) are floristically very rich and typical of the melkosopochnik. Colorful steppe forbs are *Pulsatilla patens, Gypsophila altissima, Polygala hybrida, Hieracium virosum, H. echioides, H. umbellatum, Veronica spicata, V. incana, V. spuria*, and many others.

Helictotrichon (Helictotrichon desertorum) steppes are always associated with stony and cobbly soils. In the low mountains of the eastern part of Sary-Arka they exist as part of the vertical zonation.

Shrub-rich *forb – sedge – Helictotrichon* steppe is the wettest type of steppe on low mountains. Petrophilous shrubs, such as *Spiraea crenata, Rosa spinosissima* and *Cotoneaster melanocarpa*, are abundant but also the hydrophilic *Carex pediformis* commonly occurs in the stony steppes.

Among the dry steppes, *petrophytic forb – Helictotrichon* steppes (*Helictotrichon desertorum, Onosma simplicissima, Hedysarum gmelinii, Centaurea sibirica, Goniolimon speciosum* and *Seseli ledebourii*) as well as *sagebrush – Helictotrichon* (*Helictotrichon desertorum, Artemisia frigida*) steppes are the typical communities on slopes of melkosopochnik. These steppes are common in the mountains of the Altai, Tarbagatai and Tien Shan.

Stipa lessingiana steppes widely occur on the dry plains of the Sub-Ural and the Turgay Plateau, as well as on plains among melkosopochniks of Sary-Arka. *Fescue – feather grass* (*Stipa lessingiana – Festuca valesiaca*) *steppes* with xerophytic forbs (*Galatella tatarica, Pyrethrum achillaefolium*) on carbonate soils are typical. Most of them have been plowed up.

Deserted feather grass steppes are mainly characteristic for the northern part of the subzone, where *sagebrush – fescue – feather grass* (*Stipa lessingiana, Festuca valesiaca, Artemisia sublessingiana* or *Artemisia gracilescens*) communities are common. *Sagebrush – feather grass* steppes with abundant *Artemisia compacta* occur in the easternmost part of the subzone.

Stipa capillata steppes on light soils, mainly sandy (Photo 3.5) and cobbly, occupy a huge area on the plains and melkosopochnik. *Fescue – feather grass* steppes with psammophytic forbs (*Artemisia marschalliana, Gypsophila paniculata*) are common, and in the eastern part of Sary-Arka (Karamysheva and Rachkovskaya 1973) and Tarbagatai (Stepanova 1962) *Stipa capillata – Caragana pumila* steppes and *Stipa capillata – Helictotrichon desertorum – Caragana pumila steppes* occur. In their composition and structure these steppes are very similar to the Mongolian dry steppes. Their basic floristic elements are East Kazakhstanian or East Kazakhstan-Mongolian petrophytic species: *Cleistogenes squarrosa, Potentilla acaulis, Veronica pinnata, Convolvulus ammanii.* The East Kazakhstan shrub *Caragana pumila* is common.

Deserted *Stipa capillata* steppes, in particular *Spiraea hypericifolia, Artemisia sublessingiana, Stipa capillata* communities, are closely associated with non-saline, easily destroyed soils on slopes of melkosopochnik.

In the south of the region *Stipa kirghisorum steppes* replace the steppes with *Helictotrichon desertorum* on stony and cobbly southern slopes of mountains and melkosopochnik. *Stipa kirghisorum, Artemisia sublessingiana, Artemisia frigida, Caragana balchaschensis steppe* is common. In the northern part of the Balkhash region, in the contact zone between steppes and deserts, there are large tracts of *Stipa kirghisorum, Festuca valesiaca, Artemisia sublessingiana steppe* on plains underlain by cobbly and pebble-stony deposits.

Stipa korshinskyi steppes are found sporadically on carbonate-rich soils in northern Kazakhstan.

Stipa sareptana steppes are the southernmost type of deserted steppes in this subzone. *Stipa sareptana, Festuca valesiaca, Artemisia gracilescens* or *Artemisia sublessingana*, or *Artemisia semiarida* communities should be mentioned as typical.

Festuca valesiaca steppes occur widely all over the steppe zone, both on plains and slopes of melkosopochnik and mostly on saline soils.

Stipa pennata steppes are restricted to sandy soils. They are common in western Kazakhstan, and on the Turgai and the plains near the Irtysh. Such communities consist of psammophytic species. The most common communities of these steppes are psammophytic forb – feather grass steppes with *Stipa pennata, S. capillata,*

Koeleria glauca, Cleistogenes squarrosa, Potentilla glaucescens, Centaurea sibirica, Scabiosa isetensis, Syrenia saliculosa, Astragalus onobrychis, Gypsophila paniculata, Onosma simplicissima, Artemisia marschalliana, Scorzonera ensifolia, Helichrysum arenarium, Asperula danilewskiana, etc. Here and there *Artemisia marschalliana* is co-dominant in such steppes. In the south of the steppe zone steppes with *Stipa pennata, Agropyron fragile, Cleistogenes squarrosa, Phleum phleoides, Artemisia marschalliana, Jurinea amplexicaule, Scorzonera ensifolia, Medicago romanica, Potentilla glaucescens* are common. *Stipa pennata, Festuca beckeri, Koeleria glauca, Cleistogenes squarrosa, Artemisia marschalliana, Scorzonera ensifolia, Helichryzum arenarium* steppes are common on loose sandy soils in hilly areas, often interspersed with pine forests (*Pinus sylvestris*) (e.g. in the Naurzum pine forests and near the Irtysh).

3.4.2 Mountain Steppes

The ***real mountain bunch grass steppes*** include rich forbs – feather grass steppes and dry bunch grass steppes with a small number of predominantly drought-resistant forb species. *Stipa zalesskyi steppes* and *Stipa capillata steppes* are most common, as well as *Helictotrichon desertorum steppes* and *Stipa kirghisorum steppes* which are associated with more stony substrates. Also *Festuca valesiaca steppes* are widespread. All mountain steppes contain shrubs, but the floristic composition of forbs and shrubs differ in the various mountain ranges.

Salvia stepposa, Medicago falcata, Galium verum, Peucedanum morisonii, Origanum vulgare are reperesentative forb species in the Altai and Tarbagatai. Besides these species, *Bupleurum aureum* and *Adenophora lilifolia* are also found in the steppes of Tarbagatai and on stony substrates the steppes contain *Festuca valesiaca, Crepis tenuifolia, Sedum hybridum, Thymus serpyllum* and *Ziziphora clinopodioides.* The forb flora of the *Stipa capillata steppes* (being part of bunch grass steppes) of Tarbagatai corresponds to that of the Mongolian steppes with *Stipa capillata, Potentilla acaulis, Veronica pinnata, Artemisia frigida.*

The feather grass steppes of the Northern Tien Shan typically contain *Festuca valesiaca, Stipa capillata, S. lessingiana, Bothriochloa ischaemum* and *Ajania fastigiata*. The mountain steppes have been described in detail by Stepanova (1962) for Tarbagatai and Rubtsov (1948, 1954) for the Tien Shan and the Zhungar Alatau.

Cryophytic steppes in the highlands are located on southern slopes. In the Northern Tien Shan *cryophytic fescue steppes* (with *Festuca musbelica, F. olgae, Phlomoides oreophila, Geranium saxatile, Kobresia humilis*) in combination with cryophytic meadows and the prostrate *Juniperus pseudosabina* are found.

In the Western Tien Shan, on southern slopes and the more cobbly parts of northern slopes, special *cryophytic subalpine steppes* are found with *Helictotrichon hookeri, Festuca valesiaca, Poa relaxa, Potentilla hololeuca, Artemisia ashurbajewii, Cerastium cerastoides* and *Oxytropis aulieatensis.*

Ephemeroid – bunch grass steppes are found on low ridges in the periphery of the Northern Tien Shan, such as ephemeroid – sagebrush – feather grass steppes with *Stipa sareptana, S. lessingiana, S. caucasica, Festuca valesiaca, Artemisia* spp., *Kochia prostrata, Poa bulbosa*, and sometimes also containing the shrubs *Spiraea hypericifolia, Cerasus tianschanica*, and species of *Atraphaxis*.

In the Western Tien Shan, on the southern slopes of Dzhabagly, *savannoid bunch grass communities* occur, by many authors considered as "warm steppe" (Karmysheva 1973). The savannoid steppe communities are characterized by dominance of *Festuca valesiaca*. They contain significant numbers of savannoid plants, such as *Ferula tenuisecta, Hordeum bulbosum, Centaurea squarrosa, Schrenkia golickeana* and *Tulipa greigii*. The shrubs *Cerasus tianschanica, Rosa kokanica* and *R. fedtschenkoana* are also important in these steppes. Savannoid – fescue (*Festuca valesiaca, Stipa caucasica, Poa bulbosa, Ferula tenuisecta* and *F. karatavica*) steppes also occur in the Karatau mountains, predominantly on northern slopes with thin, cobbly soils.

Mountain xerophytes – bunch grass steppes are widespread in the steppe zone of the Karatau mountains, especially on southern slopes (Photo 3.6). They contain *Festuca valesiaca, Stipa caucasica, Poa bulbosa, Rhaphidophyton regelii, Pseudolinosyris grimmii, Cousinia albertii, C.karatavica* and *Acantholimon alberti*. *Mountain-xerophytes – fescue* (*Festuca valesiaca, Acantholimon aulieatense*) communities dominate on the flat tops of the Malyi Karatau. In the Central Karatau, shrub – fescue (*Festuca valesiaca, Spiraea hypericifolia, Lonicera nummularifolia, Cotoneaster allochrous*) *steppes* are common. These steppes comprise numerous species that are endemic in the Karatau (*Cousinia mindschelkensis*, species of *Acantholimon*). Feather grass communities with *Stipa orientalis, S. caucasica* or *S. macroglossa* occur locally.

The European-Kazakhstanian feather grass and fescue steppes dominate in the steppe zone of Kazakhstan; however, they are often represented by specific geographic variants with unique floristic compositions.

3.5 Rare, Endemic and Endangered Species of the Forest-Steppe and Steppe Zones

The flora of the forest-steppe and steppe zones of Kazakhstan is comprised of approximately 2,000 species. The floristic endemism of the plains is low in comparison with the mountain-steppe habitats. There are about 30 narrow endemics, which do not occur outside of the steppe zone of Kazakhstan (1.5% of the total) (Rachkovskaya et al. 1999). Most of them are perennial grasses while some are semi -shrubs. Some rare species, associated with rocky surfaces or outcrops of chalky soils, occur in the Mugodzhar mountains (*Megacarpaea mugodsharica, Vincetoxicum mugodsharicum,*) or in some isolated low mountains (Ulytau, Chingiztau) within the Central Kazakhstan Melkosopochnik (e.g. *Clausia kasakorum, Tanacetum ulutavicum, Lepidium eremophyllum, Scorzonera diacanthoides, Potentilla kasachstanica*). A large group

Photo 3.5 Feather-grass steppes with *Stipa capillata.* Central-Eastern part of Kostanay Oblast; Naurzum Reserve; Naurzum Rajon. 10.06.2005 (Photo Tatyana Bragina)

Photo 3.6 Mountain xerophytes – bunch grass steppes with *Stipa* sp. Slopes of Karatau mountains; Southern Kazakhstan; South-Kazakhstan Oblast; Turkestan. 13.05.2010 (Photo Tatyana Bragina)

of endemic species are found in the melkosopochnik of Central and Eastern Kazakhstan (*Euphorbia andrachnoides, Astragalus kasachstanicus, Erysimum kasakhstanicum, Thymus cerebrifolius, Seseli eriocarpum*) or in several parts of the steppe zone of Kazakhstan (*Thymus kirghisorum, Thymus kasakhstanicus, Oxytropis gebleriana*). *Astragalus kustanaicus* (on the Turgay Plateau) and *Linaria dolichocarpa* (in Western and Central Kazakhstan) are associated with sandy substrates, and *Leymus akmolinensis* with meadows. Western Kazakhstan has the following endemic species: *Artemisia lessingiana, Astragalus zingeri, Jurinea mugodsharica, J. fedschenkoana*.

Betula kirghisorum, Berberis karkaralensis (Kent, eastern part of the Central Kazakhstan Melkosopochnik) and *Caragana bongardiana* (its south-western part) are endemic trees and shrubs.

Stipa korshinskyi, Astragalus macropus, Silene suffrutescens, Dianthus acicularis and *Scabiosa isetensis* are typical Trans-Volga-Kazakhstanian endemics.

A number of species have their main distribution area outside the steppe zone, but rarely also occur in the steppe zone. These include a large group of mesophilous species: *Alnus glutinosa* (Mugodzhary, Turgay, eastern part of the Central Kazakhstan Melkosopochnik), *Quercus robur* (Urals), *Convallaria majalis* (Mugodzhary, Turgay), *Drosera rotundifolia, Eriphorum polystachion, Pyrola rotundifolia, Chimaphylla umbellata, Moneses uniflora, Carex angarae, C. cappilaris* (in the Kokshetau Heights or in the low mountains of the eastern part of the Central Kazakhstan Melkosopochnik). Some species (*Soluria geoides, Daphne altaica, Caragana arborescens, Spiraea trilobata, Craniospermum echioide*, etc.), which usually characterize the adjacent mountain areas of Altai, Tarbagatai and Sayan, occur in the eastern part of the melkosopochnik.

The group of rare species which find their northern boundary in the steppe zone is numerous and ecologically diverse: *Ferula pachycarpa, Scaligeria setacea, Pachypterigium multicaule, Soranthus meyeri, Arthrophytum korovinii, Brachanthemum kasakorum, B. fruticulosum, Astragalus temirensis*, etc.

The steppe zone also harbours a group of plant species that require special protection and are listed as endangered in the Red Book of Kazakh SSR (1981). They were also recommended for inclusion in the second edition of the Red Book of Kazakhstan (Ivashchenko 2005). Ornamental species of tulips (*Tulipa schrenkii, T. patens, T.biebersteiniana*) are especially noteworthy among the "red-listed" steppe plants.

3.6 Rare Communities of the Forest-Steppe and Steppe Zones of Kazakhstan

There are four groups of unique plant communities in the steppe zone of Kazakhstan:

1. Rare communities in Eurasia and specific for the territory of Kazakhstan;
2. Rare, mainly relict communities, persisting under special conditions, or composed of endemic species;

3. Rare communities occurring just locally or with a unique combination of species in special habitats;
4. Fragmentary remains which have become rare due to the agricultural development of the steppe zone.

1. *Rare steppe communities only found in Kazakhstan.* This is a particularly interesting botanical and geographical category (Rachkovskaya et al. 1999). These are various communities of the *Helictotrichon* steppe formation, containing a special flora, including the endemics of the stony-cobbly habitats of the Central Kazakhstan Melkosopochnik (Karamysheva and Rachkovskaya 1973).

Rather rare are the *Stipa korshinskyi steppes*, a typical Trans-Volga-Kazakh steppe type, confined to carbonate-rich soils. *Stipa capillata steppes* with *Caragana pumila* are particular to the Eastern Kazakhstan type of steppes. They can be regarded as a phytogeographical analogue of the Mongolian steppes and some species geographically replace one another (*Caragana pumila* → *Caragana pygmaea; Stipa capillata* → *Stipa krylovii*).

The deserted steppes of Central and Eastern Kazakhstan are a very special category of steppes. Some communities are botanically and geographically interesting as they only occur in Kazakhstan, such as the sagebrush – feather grass steppes dominated by *Stipa lessingiana, Stipa sareptana* and *Stipa kirghisorum* and containing several species of *Artemisia* (*A. lerchiana, A. lessingiana, A. subblessingiana, A. gracilescens, A. semiarida*) (Rachkovskaya et al. 1999).

2. *Rare, or relict communities.* This group is quite extensive and includes, first of all, boreal forests and marsh communities. Relict forests of *Alnus glutinosa* and *Betula kirghisorum* (endemic), and *Pinus sylvestris* forests, penetrating into the steppe region in particular habitats (sands, low granite mountains), are most notable and contain many boreal mesophilous species (*Linnaea borealis, Vaccinium myrtilllus, V. vites-idaea, Drosera rotundifolia, Pyrola rotundifolia*) isolated from their main distribution areas. Also *Sphagnum fuscum* bogs with rare, hydrophilic species are present, especially in more northern areas.

We consider some meadow steppes of low mountain tops with subalpine mountain steppe species (*Papaver tenellum, Oxytropis chionobia*) as relict steppes, remnants of a more humid and cold period that existed in the Pleistocene in the Central-Kazakhstan Melkosopochnik (Sary-Arka). The brushwood made up of the ancient, arid relict and narrow endemic xerophytic shrub *Caragana bongardiana* also falls into this group.

3. *Rare local communities.* These are a diverse set of communities. In vast regions of Kazakhstan, mainly in lowlands with low hills, steppes with steppe shrubs are typical. They differ in species composition and structure and have vast distribution areas in Kazakhstan, mainly in the melkosopochnik-low mountain landscape (Bykov and Stepanova 1953; Safronova 1967; Karamysheva and Rachkovskaya 1973). In the western part of the steppe zone we find brushwoods of *Cerasus fruticosus*, and *Amygdalus nana*. In the Sary-Arka *Caragana balchaschensis* is typical of the subzone of deserted steppes and the narrow endemic *C. bongardiana* occurs locally. Shrub steppe communities of *Caragana frutex,* petrophytic steppes with

Spiraea trilobata, steppes of *Helictotrichon desertorum, Stipa capillata* steppes and steppes of *Stipa sareptana* with *Caragana pumila* (analogous to the Mongolian steppes) of the eastern part of melkosopochnik are very special. In Tarbagatai and the southeastern part of the melkosopochnik, steppes and brushwood of *Salophaca soongorica* occur and from the Kokshetau Heights further westward we find the interesting brushwoods of *Cerasus fruticosus* and *Amygdalus nana.*

Unique are also the numerous serial communities of stony habitats dominated by *Artemisia lessingiana, A. glabella, A. frigida, Sedum hybridum, Stipa orientalis* and some rare and endemic petrophilous species, and also chalk deposits and outcrops of Tertiary clay have their specific floristic composition.

4. *Fragmentary remains which became rare due to anthropogenic activities.* This group comprises practically all types of steppes that earlier occurred widespread on the vast plains with soils that are good for agricultural activities (chernozem, dark chestnut and chestnut soils). Cultivated farmland now occupies these soils and the original steppe communities are now preserved in fragments only and thus became rare. These are meadow steppes, rich forb – feather grass steppes, and most of the various *Stipa* steppes mentioned earlier. These preserved natural steppe communities provide sources of seed and soil fauna needed to restore the damaged and neglected areas of virgin steppe.

3.7 Phytogeographic Regionalization of the Vegetation

In west – east direction the steppe zone stretches for some 6,000–7,000 km. Within this territory there are regions that differ in relief, rocks and soils: Common Syrt, the Caspian Lowland, the Sub-Ural piedmont plateau, Mugodzhary, the Turgay Plateau, the Trans-Ural Plateau, the Western Siberian Lowland, and the Central Kazakh Melkosopochnik (Sary-Arka). It is clear that this area requires some botanical and geographical regionalization.

The phytogeographic division used here is based on various criteria, including characteristics of the vegetation cover, ecological requirements of important species, environmental characteristics, and floristic elements. The geographical delineation was established on the basis of a new vegetation map of Kazakhstan (2006), compiled for the National Atlas on the basis of work of several specialists (Lavrenko 1970; Karamysheva and Rachkovskaya 1973; Rachkovskaya et al. 2003). We use the following units: region or zone, subzone, province, and subprovince.

Phytogeographically Kazakhstan lies in the large Eurasian steppe zone, in the Black Sea-Kazakhstan subzone. It is divided into three provinces, namely the West Siberian forest-steppe province, and the Trans-Volga-Kazakh steppe province (with five sub provinces), which are both in the plains; and the Altai mountainous province (with three sub provinces) (Rachkovskaya 2006) (Fig. 3.1).

All provinces and subprovinces are defined along climatological criteria, reflecting their degree of continentality, but also on criteria reflecting the historical

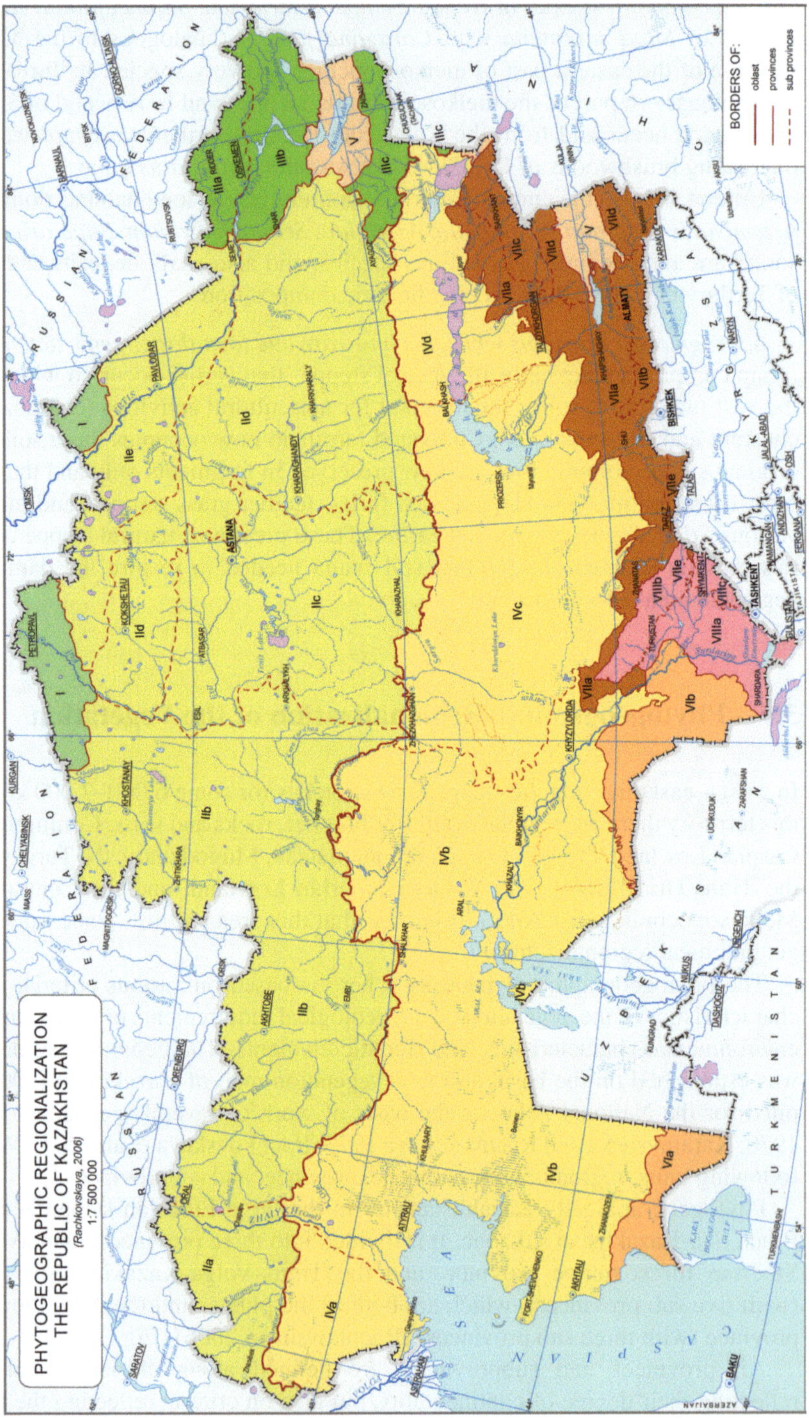
PHYTOGEOGRAPHIC REGIONALIZATION
THE REPUBLIC OF KAZAKHSTAN
(Rachkovskaya, 2006)
1:7 500 000
BORDERS OF:
oblast
provinces
sub provinces

PHYTOGEOGRAPHICAL REGIONS

Oblast	Sub oblast	Provinces	Sub provinces	Plain	Melko-sopochnik	Mountain
Eurasian steppe	Black Sea – Kazakhstanian steppe	West Siberian forest – steppe	North Kazakhstanian forest– steppe	I		
		Trans Volga – Kasakhstanian steppe	West Kazakhstanian	IIa		
			Trans Ural – Torgai	IIb		
			Central Kazakhstanian		IIc	
			East – Central Kazakhstanian		IId	
			Ertiss – Kulunda	IIe		
		South Altai mountain	West Altai			IIIa
			Kalba – South Altai			IIIb
			Sauyr – Tarbagatai			IIIc

Oblast	Sub oblast	Provinces	Sub provinces	Plain	Piedmont	Mountain
Sahara – Gobian desert	Iran – Turanian	North Turanian	Caspian	IVa		
			West – North Turanian	IVb		
			Central – North Turanian	IVc		
			East – North Turanian	IVd		
		Dzungarian	Dzungarian	V		
		South Dzungarian	West – South Turanian	VIa		
			East – South Turanian	VIb		
		Zhetysus – Alatau – North Tien Shanian	Cis-North Tien Shanian		VIIa	
			Trans Ile			VIIb
			North Zhetysus -- Alatau			VIIc
			Kungei – Teriskei – Uzynkarin – South Zhetysus – Alatau			VIId
			Kyrghyzian			VIIe
		Mountain-Middle Asian	Cis-West Tien Shanian		VIIIa	
			Kara Tau			VIIIb
			West Tien Shanian			VIIIc

Fig. 3.1 Phytogeographic regionalization of the Republic of Kazakhstan (From Rachkovskaya 2006)

development of the flora and vegetation, and on orographical criteria. In brief their main floristic and vegetational features are:

The Western Siberian forest-steppe province stretches from the Urals to the Kuznetsk Alatau in the north of Kazakhstan. Unlike the European forest-steppe with the broad-leaved *Quercus* forests, here the small-leaved *Betula* and *Populus-Betula* forests dominate, interspersed with meadow steppes and rich forb – feather grass steppes. Other vegetation types, such as steppe meadows, sedge marshes and willow brushwoods are common in depressions.

The Northern Kazakhstan*ian *forest-steppe subprovince includes a small section of the southern forest-steppe, namely the forest-steppe with "kolki" in the plains and low hills. In the typical forest-steppe the forests are confined to the watershed, and forest-steppe with "kolki" to depressions. Steppes in the southern forest-steppe are represented mainly by meadow steppes, but in the forest-steppe with "kolki" by fragments of rich forb – feather grass steppes and even forb – feather grass steppes (north-east Kazakhstan – Baraba lowland).

The Trans-Volga-Kazakhstanian steppe province is located southward and occupies the area from the lower Volga to the Altai. Throughout the province, the main dominants are bunch feather grasses (Lavrenko 1970): In the northern steppes on the chernozem *Stipa zalesskii* dominates; in the dry steppes *Stipa lessingiana*; but in the deserted steppes *Stipa sareptana* on loamy sands; *Stipa capillata* and *Stipa pennata* on sands, and on stony substrates *Helictotrichon desertorum*. On highly calcareous soils *Stipa korshinskyi* is typical.

Regional differences from west to east allow a division into subprovinces.

The West Kazakhstanian (Ergeninsko-Zavolzhskaya) plain subprovince (Steppes of Eurasia 1991) lies in the northern part of the Volga-Ural interfluve. The zonal range there is: droughty steppes, temperate-dry steppes, dry steppes and deserted steppes. The vegetation cover is complex.

Droughty steppes within the Common Syrt are forb – bunch grass steppes with *Festuca* sp., *Stipa zalesskii, Stipa lessingiana* and *S. capillata.*

In the Caspian lowlands the complexes of bunch grass steppes with *Artemisia lerchiana* and *Artemisia pauciflora* communities dominate. The sagebrush – feather grass (*Stipa sareptana*) complexes are found in the southern part of this subprovince, especially the communities with *Artemisia lerchiana* being characteristic of the deserted steppe sub zone.

The Trans-Ural – Turgay subprovince (Steppes of Eurasia 1991) occupies the southern outskirts of the West Siberian Plain, the Sub-Ural and Turgay Plateau of the Trans-Ural denudation plains and the Mugodzhary. A full range of communities of the steppe zone, from rich – feather grass steppes to deserted steppes, is represented. Before plowing the rich-forb – feather grass steppes and forb – feather grass steppes dominated the droughty steppes of the plains, in the north; the fescue – feather grass (*Helictotrichon desertorum, Stipa zalesskyi*) steppes occur on the slopes of the low melkosopochnik near rivers (Ural). However, most of the subprovince is occupied by the fescue – feather grass (*Stipa lessingiana, Festuca valesiaca*) steppes and xerophytic forb –

feather grass (*Stipa lessingiana, Galatella tatarica, Tanacetum achilleifolium*) steppes, which are currently well under cultivation. Complex feather grass steppes (*Stipa sareptana*) are distributed in the south. The steppes with *Stipa capillata* and *Stipa pennata* are very typical for light soils. Also typical is the occurrence of shrub communities of *Amygdalus nana, Cerasus fruticosa* and *Cythisus ruthenicus*.

At the Sub-Ural piedmont plateau *Stipa lessingiana* steppes in conjunction with communities of *Artemisia pauciflora* are typical on soils with influence of solonetz. On chalky sediments in this area there are rare sagebrush communities of *Artemisia gracilescens* and *Artemisia lessingiana* with *Matthiola fragrans, Zygophyllum pinnatum, Scabiosa isetensis* and *Seseli glabratum*, and in the south with *Anabasis truncata* and *Nanophyton erinaceum*. Petrophytic bunch grass – sagebrush, sagebrush – bunch grass (*Artemisia lessingiana*) steppes and brushwoods are characteristic at the low mountains and melkosopochnik of the Mugodzhary. On the Torgai Plateau previously there were huge homogeneous areas of the fescue – feather grass steppes and xerophytic forb – fescue – feather grass steppes, but these are now plowed.

We also like to point out that in the east there are relic pine forests (*Pinus sylvestris*) on sand with infiltration of boreal elements. The eastern part of the area, impacted by the ancient Turgai run-off valley, is dominated by meadow communities and complex steppes on lake terraces and here, in low-lying depressions, desert vegetation penetrates far to the north.

In general the easterly (Kazakhstanian) nature of the vegetation of this subprovince has some features that are transitional to the western Black Sea province.

The West-Central Kazakhstanian Melkosopochnik subprovince (Karamysheva and Rachkovskaya 1973; Steppes of Eurasia 1991). It occupies a lower-lying part of the Central Kazakhstan Melkosopochnik. Here the zonation is: moderately droughty, droughty, moderately dry, dry and deserted steppes. Remarkably, the width of the subzones is about equal. Petrophytic vegetation of cobbly soils and rocky outcrops of different composition and shrub steppes with *Spiraea hypericifolia* are widespread. In the southern part of the region shrub steppes and brushwoods with *Caragana frutex* and *S. balchaschensis* (*S. bongardiana* in the southwest) are common. In this subprovince, on the plains, the composition of the feather grass steppes is similar to those in the Turgay subprovince.

The East-Central Kazakhstanian low-mountain-melkosopochnik subprovince (Karamysheva and Rachkovskaya 1973). The zonation comprises four subzones: droughty, moderately dry, dry and deserted steppes. The zonation pattern is complicated and considerably varied under the influence of low mountains and hills. Within the subprovince, on the slopes of hills and low mountains, the *Helictotrichon* sp. *and feather grass steppes* (*Festuca* sp., *Stipa kirghisorum, Stipa capillata*) occur. A characteristic feature of the subprovince is the vast expansion of diverse shrub steppes with species of *Spiraea* and *Caragana*, including *Spiraea trilobata* growing on granite. Only in this region, steppes with *Caragana pumila* are widely represented, and precisely these steppes have a lot of species in common with Mongolian steppes. In the deserted steppes *Artemisa compacta* can co-occur with *A. sublessingina* and *A. gracilescens*.

A peculiarity of this subprovince is the occurrence of a vertical zonation of vegetation on slopes of low mountains in which forest-steppe landscapes occur on the Kokchetav Uplands and northern forb – feather grass steppes are found on the slopes in the dry steppe subzone. On low mountains the petrophytic rich-forb – *Helictotricon desertorum* and the special meadow petrophytic-forb – sedge grass steppes (*Carex pediformis*) with high-mountain elements are well represented. Betula and *Populus* forests also occur on the low mountains, and unique pine (*Pinus sylvestris*) forests and woodlands are linked with the granites.

The Irtysh-Kulunda plain subprovince (Rachkovskaya 2006).

This subprovince is represented only in the western part of Kazakhstan. A striking feature here is the unique steppe with the tall umbellate *Peucedanum morisonii*, but unfortunately, it is largely destroyed on the plains. The steppes with *Stipa capillata*, sandy feather-grass (*Stipa pennata*) and fescue (*Festuca beckeri*) steppes dominate the dry steppes on loamy sands and sands along the right bank of the Irtysh River. Also forest belts of *Pinus sylvestris* are typical on the sands.

The Altai mountain-steppe province comprises the western part of the Altai, Tarbagatai and the steppe mountains of Dzungaria. The steppes here are the same bunch grass steppes that are usually found in the Trans Volga-Kazakhstanian province. The zonation includes steppes, brushwoods, forests (mainly of *Larix* sp., and in the northwest of *Abies* sp.), high mountains meadows, and fragmentary tundra.

The West-Altai subprovince. This area is located in the most humid part of the Altai Mountains, within the westerly flow of the atmospheric circulation. In the lower part of the mountains, steppes and brushwoods are characteristic. The forests at the middle part of the mountains consist of *Abies sibirica*, sometimes mixed with *Pinus sibirica*. On the high mountains colourful montane meadows dominate.

The Kalba-South Altai subprovince (Karamysheva and Rachkovskaya 1973) includes territories with the East-Kazakstanian-Western Altai type of zonation (Volkova 1994). The belt of brushwoods is especially typical in just this subprovince of the Altai. They are formed by *Rosa spinosissima, Spiraea crenata, S. hypericifolia, S. trilobata*, and *Amygdalus ledebouriana*, which alternate with feather grass steppes (*Stipa zalesskii*) and fescue – (*Helictotrochon desertorum*) – feather grass steppes and fescue steppes.

The forests are variable though mainly of *Larix sibirica*. In the Kalba Mountains *Pinus sylvestris* occurs on the granites. There are low herbaceous cryophytic meadows, including communities of *Kobresia*. Only in the east of this subprovince the tundra is found in Kazakhstan (moss – lichen tundra as well as herb and shrub tundra with dwarf *Salix* spp. and dwarf *Betula rotundifolia*).

The Saur-Tarbagatai subprovince (Karamysheva and Rachkovskaya 1973). Here steppes are common in the piedmonts and low mountains. *Stipa capillata* and *Stipa lessingiana* steppes with *Calophaca soongorica* and its brushwoods are peculiar to the Tarbagatai (Stepanova 1962). The shrub belt is present, but it is fragmentary (*Spiraea hypericifolia, S. trilobata, S. crenata. Caragana frutex, Cerasus tianschanica, Calophaca soongorica, Amygdalus ledebouriana*). In Saur the belt of

Larix sp. forests is found mainly on north slopes (expositional forest-steppe). Large areas of the high mountains are occupied by subalpine and alpine meadows.

In the deserts, within the Sahara – Gobi phytogeographic region (Lavrenko 1965), steppes are found in the Tien Shan Mountains, within **the Dzungaria-North Tien Shan province**, mainly on the piedmont plains and low mountains, as well as in some areas in the high mountains (Botanical geography of Kazakhstan and Central Asia 2003).

3.8 The Steppe and the Development of Agricultural Production

Table 3.5 shows the areas of land in the various land use categories in the administrative regions ('oblasts') located in the forest-steppe and steppe zones of Kazakhstan (The Republic of Kazakhstan 2006, vol 6). In the steppe zone (North-Kazakhstan, Akmola region) agricultural lands took up 60–65% in 2004.

The areal composition of land used for agricultural purposes is presented in Table 3.6 (The Republic of Kazakhstan 2006, vol 2). In the steppe zone, arable lands are the most valuable type of agricultural lands. The most extensive areas of arable lands are concentrated in the steppe and forest-steppe zones of the provinces ('oblasts') North-Kazakhstan, Kostanay and Akmola, and represent 64% of their land surface. During the economic downturn and the reforms of agricultural enterprises in the 1990s, the area of arable lands was dramatically reduced and large areas were abandoned. But now they mostly have been re-plowed.

The steppe zone is the most important part of the country as regards agricultural production, especially crop production. Its share in the nationwide index of gross agricultural output is about 43%, in crop production it is 40.7% and in livestock production 45.6%. Gross crop production consists mainly of grain.

In 2003 the agricultural production of the steppe zone was realized by 2,862 agricultural enterprises, 31,969 peasant farms and 736,000 private allotments. (Landscape and Biodiversity of the Republic of Kazakhstan 2005).

Crop sector. More than 70% of the total cultivated area of crop land in the country are found in the steppe zone, of which 84% are grain crops consisting mostly of wheat (Photo 3.7a, b). A very small area is set aside for fodder crops, as well as millet and buckwheat, particularly in areas that are favorable for their cultivation. Industrial crops and oilseeds, especially sunflowers, also are grown. The share of oil-bearing crops accounts for more than 3% of the acreage of crops in the steppe zone, and about 1% is occupied by potatoes and vegetables (mainly cucurbitaceous plants). Root crops are common among vegetable crops. Fodder crops take 12% of the cultivated fields. In 2003 the production of cereal crops was 930 kg/ha, sunflower crop 560 kg/ha, potatoes 12,700 kg/ha, vegetables 16,500 kg/ha, i.e. rather below potential yields in this natural zone (Landscape and Biodiversity of the Republic of Kazakhstan 2005).

Table 3.5 Areas of different land use in the administrative regions of the Republic of Kazakhstan, located in the forest-steppe and steppe zones, per 1 November 2004 (in thousands of hectares) (Republic of Kazakhstan 2006, vol. 6)

Provinces ('Oblasts')	Categories of land use							
	Agricultural lands	Lands of rural settlements	Industrial, transport and other non-agricultural lands	Protected lands	Lands of forest resources	Lands of water resources	Reserve lands	Total
Akmola	9558.6	1332.4	98.3	343. 4	527.7	65.1	2795.2	14620.7
West-Kazakhstan	3489.1	1669.4	32.2	0.2	206.1	74.1	8196.0	13667.1
Kostanai	7921.0	1745.1	194.4	103.8	515.5	66.7	9053.6	19600.1
Pavlodar	3124.3	1458.5	129.5	–	454.2	21.0	7283.0	12479.5
North-Kazakhstan	5901.3	926.4	63.0	0.2	681.2	142.4	2089.8	9804.3

Table 3.6 Areas of agricultural land use in the administrative regions of the Republic of Kazakhstan, located in the forest-steppe and steppe zones, on 1 November 2004 (in thousands of hectares) (Republic of Kazakhstan 2006, vol. 2)

Provinces ('Oblasts')	Total area	Agricultural lands	Arable lands	Perennial plantings	Fallows	Hayfields	Pastures
Akmola	9458.6	9259.8	4969.1	4.5	796.6	149.7	1339.7
West-Kazakhstan	3489.1	3445.8	732.3	1.7	442.0	224.2	2044.2
Kostanai	7921.0	7807.4	5096.2	9.5	45.0	44.0	2612.4
Pavlodar	3124.3	3084.2	1081.7	1.7	421.7	116.7	1462.3
North-Kazakhstan	5901.3	5689.3	4031.1	4.2	196.5	16.0	1440.7

Photo 3.7 (**a**) Agriculture lands: Harvesting wheat. Northern Kazakhstan; Kostanay Oblast; Auliekol Rajon. 28.08.2008 (Photo Tatyana Bragina). (**b**) Agriculture lands: wheat fields after harvest. Northern Kazakhstan; Kostanay Oblast; Kamysty Rajon. 29.09.2004 (Photo Tatyana Bragina)

Livestock. The share of the steppe zone in the cattle population of the country is 44.5%, pigs 65.6%, sheep and goats 18.9%, horses 28.2%, camels 41.1%, and birds of all kinds 44.8%.

The share of the steppe zone in the production of livestock and poultry in live weight was 45.8%, cow milk 50.0%, eggs 46.0%, and wool of all kinds 20.1% (Landscapes and Biodiversity of the Republic of Kazakhstan 2005).

3.9 Pasture Lands in the Steppe Zone

Natural fodder lands in the steppe zone are located on land not suitable for plowing. The northern part of the steppe zone is mainly cultivated for arable farming but large areas are occupied by pastures in the southern part of the steppe zone, and these are predominantly agricultural lands.

Productivity of pastures. The productivity of natural pastures in the meadow steppes in the forest-steppe zone is 500–1,000 kg/ha, and in wet years even 1,500 kg/ha. In the northern steppes on the chernozem the productivity of natural pastures in the subzone of forb – feather grass steppes is 500–800 kg/ha dry weight. In the southern part of the subzone of droughty steppes the forb – feather grass natural steppe on carbonate soils the productivity is 300–500 kg/ha. In the dry natural steppes it is 150–350 kg/ha, and on the sandy pastures in this subzone the productivity of natural pastures is 300–500 kg/ha. In the deserted natural steppes 250–550 kg/ha. On saline soils (on solonetz) it is 100–300 kg/ha.

The productivity of mountain pastures of the natural meadow-steppe on medium-high mountains is 500–1,500 kg/ha of dry mass, and on high mountains 200–900 kg/ha (Republic of Kazakhstan 2006, vol 2).

Seasonality of pastures. In the plains of the steppe zone pastures are used mostly in summer (spring-summer-autumn), and pastures of the deserted sagebrush – feather grass steppes are basically used in spring-early summer with partial grazing in autumn. In winter, deep snow requires that cattle is stabled. In the steppe zone of Kazakhstan, a transhumance of cattle to the winter pasturing, mainly to the sandy areas of the southern desert, is practiced, and with return of the spring cattle is led back further north. The pastures of the mountain steppes and meadows (Altai, Tien Shan) are used for transhumance of cattle in summer. The main pastures are located in the steppes of high mountains and to a lesser extent in low and medium-high mountains. The winter pastures for these regions, such as the Alma-Ata region, are the pastures of the sandy desert areas.

Grazing system. Since ancient times people in Kazakhstan have acquired experience on the proper use of pastures, including seasonal changes in pasturing and continuous changing of grazing area during the grazing season, depending on the specific conditions of the area. Historically pastures in Kazakhstan have been used in the following sequence: unsystematic grazing, large-yard pasturing, small-pen pasturing, and fractional pasturing. Unfortunately, even now unsystematic grazing predominates. To graze the pastures sustainably, it is necessary to utilize them in an ecologically acceptable way, in which fodder production capacity and number of grazing livestock are in equilibrium, and this depends on species composition of the forage resources and water availability.

In Kazakhstan, both permanent and temporary pastures are being used. A permanent pasture is an area, which over time is used only for grazing year after year. Temporary pastures are the aftermath of hay meadows, plots of sown perennial herbs after mowing, or stubble of grain crops.

Currently, there is excessive, irregular grazing of pastures near settlements and watering places, which leads to loss of valuable forage grasses, as grassland are getting overgrown by weeds and inedible plants increase. This situation is due to the fact that currently 82% of all cattle is concentrated at small landowners, and only 18% in farms and large farms. Due to economic factors, small landowners are pasturing their livestock within a radius of no more than 5 km from the settlement.

High mountain steppe grazing was very strong until the 1990s. In the last decade, livestock has been greatly reduced in Kazakhstan, and as a result the primary plant communities are regenerating in the high mountain pastures.

Steppe haying. Steppe communities are suitable for haying mainly in wet years, and are best mown early in the season (not later than the end of June). Late mowing causes poor quality of hay that is badly eaten by cattle. Annual haying is impractical because it leads to a decrease of the overall productivity of herbage. In dry years, mowing is not recommended.

Livestock farming. In the forest-steppe in the northern temperate-droughty steppes and droughty steppes the focus of agricultural production is grain, dairy and meat (beef and some pigs). In temperate-dry steppes and dry steppes grain is the main product, but there we also find dairy and meat industry, and sheep and horse herding and breeding. In the deserted steppes the focus of agriculture is on livestock, especially sheep breeding and horse herding. Sheep and horse breeding, dairy and meat production predominate in the Altai Mountains and the Tien Shan.

3.10 Current Status of Kazakhstan's Steppes and the Main Threats

Agriculture and, in recent years, also other anthropogenic factors have had very extensive effects on the vegetation throughout Kazakhstan. Arable farming (cultivation) replaced natural vegetation by agrocoenoses. Cultivation and regulation of watercourses not only transformed the natural ecosystems of the Kazakhstan steppes, but also damaged local water balances, resulting in lowering of groundwater levels and the disappearance of many small water bodies, while the desiccation of large ones have become more frequent and prolonged (Bragina 2007). Grazing also affects the natural vegetation. Anthropogenic impacts, such as roads, oil wells, underground and open mines, pipelines, and powerlines, are local but often disastrous. In many parts of the steppe zone large areas are set aside for mining operations. In addition to losing their steppe vegetation to backfilling quarries and deposition of waste rock, the stability of ecosystems in the surrounding areas are destroyed as a result of the pumping of the ground water to lower levels, and of pollution from dusty dumps and from transport infrastructure. The impact of constructions, roads and pipelines increase the area of destroyed land, of contaminated soil and the spread of weed species.

We will briefly point out the following, main anthropogenic factors affecting the transformation of the flora, vegetation and soils of the steppes:

- Farming (cultivation);
- Fallows;
- Livestock (the impact of grazing on pastures);
- Burns;
- Other anthropogenic impacts;
- Collection of plants.

Farming (cultivation). In the steppe and forest-steppe zones of Kazakhstan gigantic changes in many types of plant communities resulted from very extensive plowing and cultivation of virgin steppe lands in the 1950s and 1960s. Presently, almost all types of steppe on potentially arable soils of the vast plains (chernozem, dark-chestnut and, partially, chestnut soils) are occupied by farmlands and steppes are preserved in fragments only. These plowed up steppes are: meadow steppes (*Festuca valesiaca, Stipa pennata, S. zalesskii, Calamagrostis epigeios, Phleum phleoides, Helictotrichon schellianum, Filipendula vulgaris, Artemisia sericea*), rich forb – feather grass steppes (*Stipa zalesskii, Festuca valesiaca, Phleum phleoides, Calamagrostis epigeios, Filipendula vulgaris, Lathyrus tuberosus, Onobrychis sibirica*), the same steppes with *Peucedanum morisonii*, forb – feather grass steppes (*Stipa zalesskii, Festuca valesiaca, Seseli ledebourii, Salvia stepposa, Phlomis tuberosa*), the same steppes with *Stipa lessingiana* as co-dominant, fescue – feather grass (*Stipa lessingiana, S. korshinskyi, Festuca valesiaca, Galatella divaricata, Salvia stepposa, Phlomoides agraria*), xerophytic-forb – fescue – feather grass (*Stipa lessingiana, Festuca valesiaca, Tanacetum achilleifolium, Galatella tatarica, G. divaricata*) and feather grass (*Stipa korshinskyi*) steppes.

The northern types of steppes on chernozem soils are particularly affected by cultivation. Cultivation on the plains areas in some places reaches 90%, and 30% in the melkosopochnik. Dry steppes on the plains were cultivated at 50–60%, and in the melkosopochnik at 10–15%. Remaining steppe areas in these subzones (stony steppes and complex steppes) were significantly transformed as a result of overgrazing.

The vegetation of piedmont steppes is almost completely cultivated (piedmonts of the Altai, Tien Shan). Mountain steppes were selectively cultivated and grazed. Preserved natural steppe communities are a valuable source of seeds and soil fauna, which are necessary to restore the damaged and neglected areas of disturbed virgin steppes.

Photo 3.8 Feather-grass on long-fallow lands. Kostanay Oblast; Altynsarin Rajon. 02.06.2008 (Photo Tatyana Bragina)

Restoration of long-fallow lands. In Kazakhstan the restoration of the vegetation on long-fallow lands in the dry steppes has been studied by Marynich and Rachkovskaya (2008). The restoration processes were studied in key areas differing in mechanical composition and chemistry of their soils (heavy loam, light loam, loamy sand, sand, carbonate, soils with elements of solonetz and solonetz). The successional dynamics follow this sequence: the first stage of restoration of the vegetation on long-fallow lands (1–4 years) is a weeds stage (annual and biennial plants), followed by a sagebrush (*Artemisia* sp.) stage (3–15 years), particularly well developed in the dry steppes of the continental sector of Eurasia. The third stage is the forb-grass stage (when the age of the long-fallow lands is more than 15 years) (Photo 3.8). Restored secondary virgin-like steppes appear on loamy sand in about 20 years, in loamy carbonate soils much later.

Fallow lands in the dry steppes of the Kostanai region proved to have to be treated very differently depending on their stage of recovery, the prospects of future land use and the potential occurrence of pests. For the main groups of the long-fallow lands the following approaches to rehabilitation, taking into account their economic feasibility, were worked out:

- the fallow lands on dark-chestnut carbonate soils of the plateau do not need an attempt to being recovered, as it is a temporary phenomenon associated with economic difficulties in recent years. Gradually they will be taken into cultivation again;

- fallow lands on dark-chestnut loamy sands are promising for carrying out rehabilitation work aimed at transforming them into pastures;
- fallow lands on dark-chestnut sandy soils should be recovered as fast as possible with further exclusion from plowing. The reclamation of these lands either should be completely forbidden, or to be an object of a strict control because there is intensive aeolian and water erosion;
- fallow lands on saline soils (complexes of soils with elements of solonetz and solonetz proper) abandoned long ago are in the sagebrush (*Artemisia* sp.) recovery stage now. These lands are overgrown by species of sagebrush (*Artemisia nitrosa, A. pauciflora*) and annual saltworts (*Petrosimonia triandra, Bassia sedoides*) with some weeds (*Descurania sophia, Lactuca serriola*). The vitality of plants in these fallow lands is very good, so such lands, being productive pastures, do not need urgent rehabilitation. Also their agricultural development will not feasible in the future.

The restoration of long-fallow lands in the subzone with chestnut soils requires differentiated approaches. For example, the agricultural lands on chestnut carbonate soils are classified as lands of unstable and non-irrigated farming, that demand specialized agro-technical measures aimed, primarily, at the accumulation of soil moisture. In this respect, these agricultural lands should be carefully and selectively be re-engaged for agricultural use.

Livestock farming (*the impact of grazing on pastures*). Modern methods of distant pasturing often leads to the overgrazing in some pasturing areas and inadequate utilization in others. Overgrazing and the total destruction of vegetation have been observed around all settlements, watering points, as well as cattle camps. Pasture degradation is indicated by the reduction of grasses, especially feather grasses (*Stipa* sp). In addition, the vegetation of such pastures is characterized by a low species diversity of forbs, as well as the occurrence of perennial (*Convolvulus arvensis, Dodartia orientalis*) and annual (*Atriplex tatarica, Descurainia sophia, Lappula consanguinea, Lappula squarrosa, Lepidium perfoliatum, Lepidium ruderale*) weeds.

Pastures of the steppe zone are characterized by weediness: there is an abundance of *Artemisia austriaca* in the northern part, and the summer-autumn annual *Ceratocarpus arenarius* in the south. To restore the productivity of the degraded pastures it is necessary to provide respite and rotation in order to establish normal stocking of the pastures, and to maintain schedules for the termination and beginning of grazing.

Burns. Fires are frequent in the steppes, especially in dry years. Fire has a negative impact on the productivity of steppe grasses. As fire kills young and middle-aged plants easier, older plants come to dominate the stand. Semi-shrubs burn almost completely but after a while new plants emerge and the stand rejuvenates and the vegetation is restored.

From an economic point of view, fire is positive for steppe vegetation. Apart from the straight mechanical cleaning of the soil surface of litter and standing dead material, fires enrich the soil by ash elements and promote the regrowth of green phytomass. This positive effect of soil enrichment and renewal of the vegetation has long been known and is practiced to improve pastures. The use of areas affected by natural fires as pastures must be practiced with the caution. It is better to begin grazing only 2–3 years after the fire, when the damaged grass bunches have recovered. Otherwise pastures can easily turn unproductive.

Other anthropogenic impacts (*the effect of road networks*). In all regions of the steppe zone transport has an impact and in recent years the negative impact of vehicles on the vegetation has increased in many areas.

The vegetation of areas with mining operations also is altered or destroyed. Significant transformation of the steppe vegetation occurred at places in the former Semipalatinsk nuclear test site.

Collecting plants. Steppes support many valuable food, medicinal and ornamental plants, and many of them are collected by the locals and tourists. Decorative spring plants, tulips (such as *Tulipa gesneriana, T. patens, T. shrenkii*) and pasque flowers (*Pulsatilla flavescens, P. patens*) are the most susceptible to collection. For example, the particularly attractive tulip, *Tulipa shrenkii*, nowadays is rarely found in the vicinities of settlements.

3.11 Protecting the Steppes

The IUCN World Commission on Protected Areas considered that temperate grasslands are the least protected biomes among all of 15 terrestrial biomes of the world (Henwood 1998). At the same time, the prospects for expanding the network of protected areas in the steppe landscapes of western Eurasia are insignificant. But possibilities exist in eastern Eurasia, including Kazakhstan, where the steppes have significant landscape value (Bragina 2009a, b).

In Kazakhstan there are different kinds of protected areas, varying from reserves, with strict protection, to sites which provide regulated economic use. All protected areas of Kazakhstan are state-owned. The most important for biodiversity conservation are the following: State Nature Reserves (Category 1a, IUCN), State National Nature Parks (Category 2, IUCN) and State Nature Reservats (Category 1b or 2, IUCN) which have administrative offices with their own full-time staff (administration, rangers, science department, department of environmental education or ecotourism). They have the lands transferred to them in perpetuity with the regime of land use in accordance with the laws of the Republic. State Nature Reserves are strictly protected areas with the highest level of protection; State National Nature Parks have several zones with different protection regimes and uses: a zone with reserve status, a zone for environmental stability, a zone for tourism and recreational activities, a zone with limited economic activity. State Nature Rezervats have a core zone (with reserve status), intended for long-term conservation of genetic resources, biological diversity, ecological systems and landscapes, which has a sufficient size to achieve such goals, and a buffer zone that is part of the territory (within the Reservat boundary), and which is used to conduct environmentally-friendly economic activities and sustainable reproduction of biological resources.

The modern environmental legislation of the Republic provides that in the future the strictly protected natural areas will be connected by ecological corridors, surrounded by protection (buffer) zones with limited natural management. That collection of protected cores, ecological corridors and security (buffer) zones will be an ecological network (Econet) to ensure stable conditions of the natural environment. To develop the concept of an ecological network in Central Asia the project "Creating an ecological network for long-term conservation of biological diversity in eco-regions of Central Asia" was supported by UNEP/GEF/WWF in 2003–2006 (Balbakova et al. 2006). The project recommendations led to the inclusion of the concept of an ecological network and its elements in Kazakhstan's environmental legislation. In recent years, special attention has been given to expanding the network of protected areas in the steppe and desert-steppe zones of Kazakhstan (Bragina 2007). To date, within Eurasia, only Kazakhstan, the Asian part of Russia, and Mongolia have vast areas of steppes in their natural state, promising to conserve the ecosystem as a whole, and to protect some of the rare and diminishing steppe species in the flora and fauna.

The steppe nature reserves in Kazakhstan (Naurzum and Korgalzhin) became the first natural objects of the Republic, included in the UNESCO World Heritage List at the 32nd World Heritage Committee at UNESCO (decision № 1102 from 07.07.2008), in the category of "Sary-Arka – Steppe and Lakes of Northern Kazakhstan" (Bragina 2009a, b). At the same time they are the only two steppe reserves in Kazakhstan (Photos 3.9, 3.10, and 3.11). They are located in the subzones of the temperate-dry steppes (Naurzum Reserve) and the dry steppes (Korgalzhin Reserve). Their area in recent years increased more than twofold and is 191,381 and 543,171 ha, respectively (but in Korgalzhin Reserve 200,000 ha are lakes).

The distribution of the major protected steppe areas in the subzones and subprovinces of the natural steppe zone of Kazakhstan is given in Table 3.7. Reserves and national parks are absent in the forest-steppe zone and on the plains in the subzone of the northern temperate-droughty steppes and droughty steppes. This is due to the fact that there much of the steppe ecosystem was destroyed by very large-scale cultivation. In the forest-steppe zone within the Kokshetau Upland there are two national parks in which, along with unique pine forests on granites, moist, forb-rich types of steppe are preserved (see Table 3.7). In the coming years the National Nature Park "Buiratau" will be established in this belt.

In the low mountains of the eastern part of Central Kazakhstan (Table 3.8) there are two very important National Parks (Bayanaul and Karkaraly) in which *Pinus sylvestris* forests and stony steppes are protected. In the planned National Nature Park "Buiratau," located in the Erementau mountains, stony steppes, steppe ecosystems of the subzonal type of temperate-droughty steppes, notably scrub with *Caragana* – feather grass steppes with *Stipa zalesskii*, as well as the relict forests with *Alnus glutinosa* and others with *Betula*, occurring at the southern border of the area, are

Photo 3.9 Feather-grass steppe (with *Stipa pennata*) on the edge of Naurzum pine forest; Central-Eastern part of Kostanay Oblast; Naurzum Rajon. 10.06.2005 (Photo Tatyana Bragina)

Photo 3.10 Feather-grass steppes in spring with *Pulsatilla flavescens*. Naurzum Reserve; Central-Eastern part of Kostanay Oblast; Naurzum Rajon. 29.04.2005 (Photo Tatyana Bragina)

Table 3.7 Distribution of the existing protected areas in the Zavolzhsky-Kazakhstanian province over the subzones and the subprovinces

Subzones/subprovinces	West-Kazakhstanian	Sub-Ural – Turgai		West-Central Kazakhstanian	East-Central Kazakhstanian	Irtysh-Kulunda
Regions		Sub-Ural- Mugodzhary	Turgai			
Forest-steppes					National Parks 1. Burabai 2. Kokshetau	
Temperate-droughty steppes						
Droughty steppes						
Temperate-dry steppes			Nature Reserve Naurzum	National Park Buiratau	National Parks 1. Karkaraly 2. Bayanaulskiy	
Dry steppes				Nature Reserve Korgaldzhyn		Nature Reservats 1. Semei Ormany 2. Ertys Ormany
Deserted steppes			Nature Reservat Altyn Dala[a]			

[a]Planned for protection in the period up to 2014

Table 3.8 The main types of protected steppes and other significant vegetation in the reserves, national parks and nature reservats in the steppe zone of the Republic of Kazakhstan (as of 1 August 2011)

Protected areas	Area (ha)	Main types of protected steppes	Other types of vegetation that are important to protect
Naurzum National Nature Reserve	191,381	Feather grass steppes (*Stipa lessingiana, Tanacetum achilleifolium, Galatella tatarica*) of the slopes of the Turgai Plateau	*Pinus sylvestris* forests on sands
		Psammophyte forb – fescue steppe (*Stipa capillata, Festuca valesiaca, Artemisia marschalliana, Helichrysum arenarium*)	*Betula kirghisorum* forests
		Psammophyte forb – sandy feather grass steppe (*Stipa pennata, Agropyron fragile, Gypsophylla paniculata, Asperula danilewskiana*)	Brushwoods of shrubs (*Amygdalus nana, Cerasus fruticosa*)
		Sandy steppes (feather grass steppe with *Stipa pennata*, feather grass steppe with *Stipa cappillata*, fescue steppes with *Festuca valesiaca*)	Wetlands along the shores of lakes
		Complex steppes on the saline plains	
Korgalzhin National Nature Reserve	543,171	Fescue – feather grass steppes (*Stipa lessingiana, Festuca valesiaca, Galatella tatarica*)	Wetlands along the shores of lakes and deltas
		Sagebrush – fescue – feather grass steppe (*Stipa lessingiana, Festuca sulcata, Artemisia gracilescens*)	
		Stony steppes of melkosopochnik (feather grass steppes with *Stipa capillata*, feather grass steppes with *Stipa sareptana*), fescue grass steppes with *Festuca valesiaca* with petrophytes (*Artemisia sublessingiana, Scabiosa isetensis, Silene suffrutescens*)	
		Complex steppes on the saline plains	
National Nature Park "Burabai"	129,935	Meadow steppes (*Festuca valesiaca, Stipa pennata, S. zalesskii, Calamagrostis epigeios, Phleum phleoides, Helictotrichon schellianum, Filipendula vulgaris, Artemisia sericea*)	*Pinus sylvestris* forests on granites

(continued)

Table 3.8 (continued)

Protected areas	Area (ha)	Main types of protected steppes	Other types of vegetation that are important to protect
		Rich forb – feather grass steppes (*Stipa zalesskii, Festuca valesiaca, Phleum phleoides, Calamagrostis epigeios, Peucedanum morisonii Filependula vulgaris, Lathyrus tuberosus, Onobrychis sibirica*)	Relict *Sphagnum fuscum* bog
Karkaralinsk National Nature Park	112,120	Meadow steppes	*Pinus sylvestris* forests on granites
		Forb – feather grass steppes with *Stipa zalesskii*	Relict *Sphagnum fuscum* bog
		Stony steppes with *Helictotrichon desertorum*	
		Feather grass steppes with *Stipa capillata*	
		Fescue grass steppes	
National Nature Park "Kokshetau"	182,076	Forb – feather grass steppes with *Stipa zalesskii, Peucedanum morisonii, Filipendula hexapetala*	*Pinus sylvestris* forests on granites
		Forb – feather grass steppes with *Stipa stenophyla, Stipa zalesskii, Calamagrostis epigeios*	
		Forb – *Helictotrichon* steppes with *Helictotrichon desertorum*	
Bayanaulskiy National Nature Park	68,453	Fescue – feather grass steppes with *Stipa zalesskii*	*Pinus sylvestris* forests on granites
		Temperate-droughty rich forb – red feather grass (*Stipa zalesskii, Festuca valesiaca, Peucedanum morisonii*) steppes	Forests with *Alnus glutinosa*
		Different stony steppes with *Helictotrichon desertorum* and petrophytic forbs (*Onosma simplicissima, Hedysarum gmelinii, Centaurea sibirica, Goniolimon speciosum, Seseli ledebourii*)	
		Petrophytic steppes with shrubs (*Spiraea crenata, Cotoneaster melanocarpa*)	
National Nature Park "Buiratau"	88,968	Stony steppes with *Helictotrichon desertorum, Gypsophila altissima, Polygala hybrida, Hieracium virosum, H. echioides, H. umbellatum, Veronica spicata, V. incana, V. spuria*	Forests with *Alnus glutinosa*

(continued)

Table 3.8 (continued)

Protected areas	Area (ha)	Main types of protected steppes	Other types of vegetation that are important to protect
		Steppes with *Stipa capillata, Artemisia fridida, Caragana pumila, Veronica pinnata*	
Tarbagatai National Nature Park[a]	144,672[a]	Mountain steppes (*Helictotrichon desertorum, Festuca valesiaca, Salvia stepposa, Medicago falcata, Peucedanum morisonii, Crepis tenuifolia, Sedum hybridum, Thymus serpyllum, Bupleurum aureum, Adenophora lilifolia, Origanum vulgare, Ziziphora clinopodioides*)	Apple–tree (*Malus sieversii*) copses Brushwoods of shrubs (*Calophaca soongorica, Cerasus fruticosa*)
State Nature Reservat "Altyn Dala"[a]	489,774[a]	Feather grass steppes (*Stipa lessingiana, Agropyron pectinatum, Festuca valesiaca, Tanacetum achilaefolium*) Deserted steppes with *Stipa sareptana, Artemisia semiarida, Artemisia gracilescens* Grass steppes (*Koeleria cristata, Festuca valesiaca, Stipa sareptana, Agropyron fragile, Artemisia lercheana*)	Turanga woodlands (*Populus diversifolia*) and willow (*Salix triandra, S. acutifolia, S. dasyclados*) – oleaster (*Elaeagnus*) forests on sands
State Nature Reservat « Ertys Ormany»	277,961	Sandy steppes (sandy feather grass steppes with *Stipa pennata*, feather grass steppes with *Stipa capillata*, fescue steppes with *Festuca valesiaca*)	*Pinus sylvestris* forests on sands (pine belt forests)
State Nature Reservat « Semey Ormany»	662,167	Sandy steppes (sandy feather grass steppes with *Stipa pennata*, feather grass steppes with *Stipa capillata*, fescue steppes with *Festuca beckeri*)	*Pinus sylvestris* forests on sands (pine belt forests)

[a]Planned to become protected in the period up to 2014

protected. In the east the reserves cover the famous belt of forests on sands and sandy steppes.

In the subzone of deserted steppes of Kazakhstan, in the near future, the State Nature Reservat "Altyn Dala" will be established in the Turgai plains with an area of 489,774 ha (Photo 3.12). Only here, in this southern region of the steppe zone, conservation of large areas of the unique deserted steppes of Central Kazakhstan in their original state remain.

In the western part of Kazakhstan there are no large-scale protected areas with IUCN 1 or 2 status.

Photo 3.11 Feather-grass steppes with *Stipa zalesskii.* Naurzum Reserve; Central-Eastern part of Kostanay Oblast; Naurzum Rajon. 10.06.2005 (Photo Tatyana Bragina)

Photo 3.12 Grasslands with *Achnatherum splendens.* South-Eastern part of Kostanay Oblast (earlier Turgay Oblast); Amangeldy Rajon; Planned Atyn Dala State Nature Reservat. 11.07.2007 (Photo Tatyana Bragina)

Mountain forb – feather grass steppes (*Stipa zalesskii, Poa angustifolia, Bupleurum aureum, Centaurea scabiosa, Peucedanum morisonii*) and meadow steppes (*Festuca valesiaca, Carex pediformis, Stipa pennata, Stipa zalesskii, Helictotrichon schellianum, Aster alpinus, Phleum phleoides, Filipendula hexapetala, Fragaria viridis*) are protected in the National Nature Park "Katon-Karagay" (Altay), and dry steppes (*Festuca valesiaca, Stipa capillata, S. lessingiana, Bothriochloa ischaemum, Koeleria cristata, Ajania fastigiata, Salvia stepposa*) and forb – feather grass steppes (*Stipa zalesskii, S. kirghisorum, Alcea nudiflora*) in National Nature Park "Altyn-Emel" (south-facing slopes of the Dzungarian Alatau) and the newly formed Tarbagatai National Nature Park.

Fragments of the forb – feather grass steppes (*Stipa zalesskii, S. kirghisorum, S. capillata, Poa stepposa, Medicago falcata, Origanum vulgare, Lathyrus pratensis, Thalictrum simplex, Iris brevituba*), and the cryophytic steppes (*Festuca musbelica, F. olgae, Phlomoides oreophila, Geranium saxatile, Kobresia humilis*) occur in small areas in the National Nature Park "Ile-Alatau" (Northern Tien Shan) and small areas of the peculiar savannoid – bunch grass steppes(*Festuca valesiaca, Ferula tenuisecta, Hordeum bulbosum, Centaurea squarrosa, Schrenkia golickeana, Tulipa greigii*) and the mountain xerophytes – bunch grasses steppes (*Festuca valesiaca, Stipa caucasica, Poa bulbosa, Rhaphidophyton regelii, Pseudolinosyris grimmii, Cousinia albertii, C. karatavica, Acantholimon alberti*) in the Nature Reserve "Aksu-Dzhabagly" (Western Tien Shan) and the mountain xerophytes – bunchgrasses steppes (*Festuca valesiaca, Stipa caucasica, Poa bulbosa, Ferula tenuisecta, F. karatavica*) in the Karatau Mountains (the Karatau Reserve and the Karatau plot of the planned Turkestan National Nature Park).

In addition, in the plains of Kazakhstan, there are several sanctuaries of national importance: in the forest-steppe (6) and in the steppe zone (21), and their number is expected to increase.

Priorities for the conservation of the steppes of Kazakhstan are as follows:

1. Identification and conservation of the remaining northern areas of the rich forb – feather grass steppes and forb – feather grass steppes (as sanctuaries, natural monuments at national or local level);
2. Inventory of the steppe ecosystems in each phytogeographical subprovince to identify and conserve the key steppe areas and establish the steppe ecological network (the steppe econet);
3. Creation of the trans-boundary steppe ecological network.

Realization of these priorities will ensure the proper conservation of the striking diversity of Kazakhstan's steppes, in terms of its ecosystems as well as in terms of its steppe species. The proper conservation of these natural resources of Kazakhstan's steppes will also promote the proper development of the agricultural productivity of its steppe region.

Acknowledgments We wish to thank all those who assisted in conducting researches in Kazakhstan and editing of this article. Special thanks to Mrs. Heather Triplett for her viewing the English version.

References

Balbakova F, Berkeliev T, Bragina T, Chikin Yu, Mirutenko M, Onufrenya I, Rachkovskaya E, Sadvokasov R, Safarov N, Tsaruk O (2006) Econet – Central Asia. E-Poligraph, Moscow, p 49

Baranov VI (1925) The southern boundary of the chernizem soil steppes in Kustanai province. Tr. Gen. Learn., Kazakhstan, etc. 6, 36 pp

Biodiversity Conservation in Central Azia: an analysis of biodiversity and current threats and initial investment portfolio. WWF, Moscow, 1997, 203 pp

Botanical Geography of Kazakhstan and Middle Asia (desert region) (2003) St. Petersburg, pp 424

Bragina TM (2007) Protected areas of Kazakhstan and the prospects for establishment of ecological network (with a legislative framework for protected areas). Kostanai Printing House, Kostanai, 164 pp

Bragina TM (2009a) Status and prospects for sustaining the steppes of Kazakhstan. In: Steppes of Northern Eurasia. Proceedings of the fifth international symposium, Orenburg, pp 39–43

Bragina TM (2009b) Naurzum Ecological Network (exploration history, current status and long-term conservation of biological diversity in the region of representation of natural World Heritage Site by UNESCO). Kostanaipoligrafiya, Kostanai, p 200

Bykov BA, Stepanova EF (1953) Shrub steppe-like vegetation type. Izv. VGO. T.85.Vyp. 1, pp 50–56

Chibilyev AA (1998) Steppes of Northern Eurasia. (Ecological and geographical outline and bibliography). Urals Branch of RAS, Ekaterinburg, 192 pp

Henwood WD (1998) An overview of protected areas in the temperate grasslands biome. PARKS 8(3):3–8

Isachenko TI, Rachkovskaya EI (1961) The main types of steppes of Northern Kazakhstan. In: Proc Botan of the Statute of the USSR, Series III. Geobotany. Vyp.13, pp 133–397

Ivashchenko AA (2005) Tulips and other bulbous plants of Kazakhstan. Publishing House "Two Capitals", Almaty, 192 pp

Karmysheva NH (1973) Flora and vegetation of the reserve Aksu-Zhabagly. Publishing House "Nauka", Alma-Ata, 178 pp

Karamysheva ZV, Rachkovskaya EI (1973) Botanical geography of the steppe part of Central Kazakhstan. Botanical Institute RAS, St. Petersburg, 278 pp

Krasheninnikov IM (1925) Vegetation cover of the Kyrgyz Republic. In: Proceedings of the Society for the Study of the Kyrgyz territory, Orenburg, vol 5(2), 104 pp

Landscape and biological diversity of the Republic of Kazakhstan (2005) Information-analytical review of the UN Development Program. OO "OST-XXI Century", Almaty, 242 pp

Lavrenko EM (1940) Steppes of the USSR. Vegetation of the USSR, vol II. Publishing House of the USSR Academy of Sciences, Moscow, pp 1–265

Lavrenko EM (1956) Steppe and agricultural lands in their place. Explanatory text to the "Map of the USSR Geobotanical, Scale 1: 4000000", vol II, Ed. AN USSR. Moscow, Leningrad., T.2, pp 595–730

Lavrenko EM (1965) The provincial division of the Central Asian and Irano-Turanian subregions. Afro-Asian desert region. Bot J 50:3–15

Lavrenko EM (1970) The provincial division of the Black Sea-Kazakhstan subregion steppe region of Eurasia. Bot J 55:609–625

Map of ecosystems of Kazakhstan (2006) Scale 1:5 000 000. National Atlas of the Republic of Kazakhstan, vol 1. Natural resources. Republican Cartographic Factory, Almaty, pp 128–133

Map of Mongolia's ecosystems. Scale 1: 1000000. M.: EKOR, 1995, 15 sheets

Marynich OV (2005) Steppe landscapes and biodiversity of the Republic of Kazakhstan. Landscape and biological diversity of the Republic of of the Republic of Kazakhstan. Information-analytical review of the UN Development Program. OO "OST-XXI Century", Almaty, pp 62–67

Marynich OV, Rachkovskaya EI (2008) Dynamics of recovery of vegetation on fallow lands in the dry steppes of Naurzum area of Kustanai region, Kazakhstan. Bot J 93:1083–1100

Natural regionalization of Northern Kazakhstan. Publishing House of USSR Academy of Sciences, Moscow-Leningrad, 1960, 468 pp

Nikolayev VA (1999) Landscapes of the Asian steppes. MGU, Moscow, p 288

Ogar NP (1999) Vegetation of the river valleys of semi-arid and arid regions of mainland Asia. Abstract of doctoral thesis, Almaty, 36 pp

Ogar NP (2006) Principles of separation of ecosystems as territorial units for environmental assessment and mapping. J "Terra", number 1:139–145 (Almaty)

Ogar NP, Rachkovskaya EI (2006) Natural ecosystems. Republic of Kazakhstan. National Atlas of the Republic of Kazakhstan, Natural resources, vol 3. Republican Cartographic Factory, Almaty, pp 447–459

Pavlov NV (1948) Botanical Geography of the USSR. Kazakh Academy of Sciences, Alma-Ata, 704 pp

Rachkovskaya EI (2006) Vegetation. Republic of Kazakhstan. Natural conditions and resources, vol 1, pp 362–393

Rachkovskaya EI, Ogar NP, Marynich OV (1999) Rare plant communities of the steppes of Kazakhstan and their protection. Steppe Bull 3–4:41–46

Rachkovskaya EI, Safronova IN, Volkova EA (2003) Botanical and geographical regionalization. Botanical Geography of Kazakhstan and Middle Asia (desert region), St. Petersburg, pp 192–222

Red Book Kazakh SSR (1981), Part 2, Plants. Publishing House "Nauka" of Kazakh SSR, Alma-Ata, 266 pp

Rubtsov NI (1948) Vegetation cover of Djungarian Alatau. Publishing House of Academy of Sciences of the Kazakh SSR, Alma-Ata, 184 pp

Rubtsov NI (1952) Vegetation cover in Kazakhstan. Essays on the physical geography of Kazakhstan. Publishing House of Academy of Sciences of the Kazakh SSR, Alma-Ata, pp 48–61

Rubtsov NI (1954) The steppes of Northern Tien Shan. J Izv Kazakh SSR Ser Biol 7:3–27

Safronova IN (1967) Biogeography and phytocoenology of steppe shrubs of Central Kazakhstan hills. Bot J 52:844–851

Sobolev LN (1948) Natural fodder lands in Kazakhstan. In: Proceedings of expedition on the study of land funds of the Kazakh SSR, vol 9. Alma-Ata, p 285

Sokolov AA (1977) Common features of soil-forming and soils of eastern Kazakhstan. Publishing House "Nauka", Alma-Ata, 231 pp

Stepanova EF (1962) The vegetation and flora of the ridge Tarbagatai. Publishing House of Academy of Sciences of the Kazakh SSR, Alma-Ata, 434 pp

Tishkov AA (2002) Nature protection and conservation. The physical geography of Northern Eurasia. Oxford regional environments. Oxford University Press, Oxford, pp 227–245

The steppes of Eurasia. Nauka, Leningrad, 1991, 146 pp

The Republic of Kazakhstan (2006) National Atlas of the Republic of Kazakhstan, Socio-economic development, vol 2. Republican Cartographic Factory, Almaty, 243 pp

The vegetation map of the steppe part of Kazakh melkosopochnik (1975) Scale 1:1 500 000. UTS, Moscow, 2 sheets

Vegetation map of Kazakhstan. M: 15000000 (2006) National atlas of the Republic of Kazakhstan, vol 1. Natural resources, p 110

Vegetation map of Kazakhstan. Scale 1:5 000 000 (2006) National Atlas of the Republic of Kazakhstan, Natural resources, vol 1. Republican Cartographic Factory, Almaty

Vegetation map of Northern Kazakhstan. Scale 1:1,500,000 (1960) Natural regionalization of Northern Kazakhstan. Publishing House of USSR Academy of Sciences, Moscow-Leningrad

Volkova EA (1994) Botanical geography of the Mongolian and Gobi Altai. Russian Academy of Sciences, St. Petersburg, 129 pp

Volkova EA (2003) Botanical geography of the mountain provinces of Kazakhstan and Middle Asia (Desert Region). Russian Academy of Sciences, St Petersburg, pp 402–407

Chapter 4
The Central Anatolian Steppe

Harald Kürschner and Gerald Parolly

Abstract The characteristic landscape features of Central Anatolia (Turkey) include large ovası and basins, which are naturally bare of forests and woodlands, but were formerly occupied by steppe vegetation. These steppes evolve under a pronounced continental climate, which is extremely cold in winter and dry and hot during summer. Rainfall is less than 300 mm/year, favouring treeless steppe vegetation dominated by well-adapted dwarf-shrubs, a few herbs, and a larger number of geophytes and annuals. We review the present knowledge on Central Anatolian steppe vegetation (Onobrychido armenae-Thymetalia leucostomi, Astragalo-Brometea) and provide insight into the complex structure and species composition of today's primary and secondary steppes and their replacement communities. In addition, the changes in vegetation due to the long-lasting human impact such as grazing and agricultural activities (ca. one-third of Turkey's grain production concentrates in the former steppe area) are shown, which generally led to a loss of species and a massive decline of the diversity in the area. Finally, we outline some perspectives that may stop the continuing soil erosion and degradation and re-establish a natural equilibrium in the remaining steppe fragments of Central Anatolia.

H. Kürschner (✉)
Institut für Biologie, Systematische Botanik und Pflanzengeographie,
Freie Universität Berlin, Berlin, Germany
e-mail: kuersch@zedat.fu-berlin.de

G. Parolly
ZE Botanischer Garten und Botanisches Museum Berlin-Dahlem,
Freie Universität Berlin, Berlin, Germany
e-mail: g.parolly@bgbm.org

M.J.A. Werger and M.A. van Staalduinen (eds.), *Eurasian Steppes. Ecological Problems and Livelihoods in a Changing World*, Plant and Vegetation 6,
DOI 10.1007/978-94-007-3886-7_4,

4.1 Situation and Topography

4.1.1 Situation and Subdivision

Central Anatolia (Fig. 4.1) occupies approximately 19% of the total area of Turkey and with 151,000 km^2 is the second largest region of Turkey (Turkish News Agency 2000). It forms a plateau-like, semi-arid highland with several basins (ovası) in the centre of Turkey, which can be divided in four natural areas (subregions), (1) the Upper Sakarya, (2) the Konya, (3) the Middle and (4) the Upper Kızılırmak subregion (Fig. 4.1; Erol 1982). To the west, it is bordered by the Aegean mountains (Aegean region), and to the north and south by two zones of folded mountains, namely the Euxine folded mountain chain and the Karadeniz Dağları in the north, respectively the Toroslar Dağları (Taurus Mountains) in the south. Towards the east, the Munzur and Kop Dağları separate the area from the highlands of Eastern Anatolia.

4.1.2 Topography and Vegetation

From west to east, Central Anatolia varies in altitude from 600 to 1,200 m, including several large basins (ovası) of Neogene (Tertiary) origin: the Konya ovası, the Ereğli ovasi, the Aksaray ovası, and the Tuz Gölü basin. They are separated by flat plunges and tectonic lines and surrounded by rolling plateaus, dissected gentle slopes, small hills, and mountainous areas, which connect Central Anatolia with the surrounding folded mountain systems.

In the southern part (Fig. 4.1), some spectacular, high volcanoes overtop the basins and plateaus, such as Karadağ (2,288 m), Karacadağ (2,025 m), Hasan Dağı (3,269 m), Melendiz Dağı (2,963 m), and Erciyes Dağı (3,917 m), with a strong volcanic activity from the Pliocene until historical times. Their lavas and tufas have created a spectacular landscape around Nevşehir (Göreme area, Cappadocia), consisting of soft and porose airfall tufa which forms high pyramids (peri baçaları), cone-shaped rocks, fairy chimneys, columns or mushroom-like structures. All these volcanoes concentrate on a tectonic line, running from south-west towards north-east (Fig. 4.1).

For the most part, the ovası and basins are bare of forests and woodlands and are occupied by fragments of primary and secondary steppes. This steppe vegetation is peripherically delimited by *Quercus* (*Q. cerris*, *Q. ithaburensis*, *Q. pubescens*), *Juniperus* (*J. excelsa*, *J. foetidissima*), and *Pinus* (*P. nigra* var. *caramanica*) woodlands and has been exploited by man and his life stock, especially through intensive grazing and agricultural activities (dry-farming). Today, agriculture and animal husbandry are important sources of income in Central Anatolia and approximately one-third of the grain production of Turkey concentrates in this area.

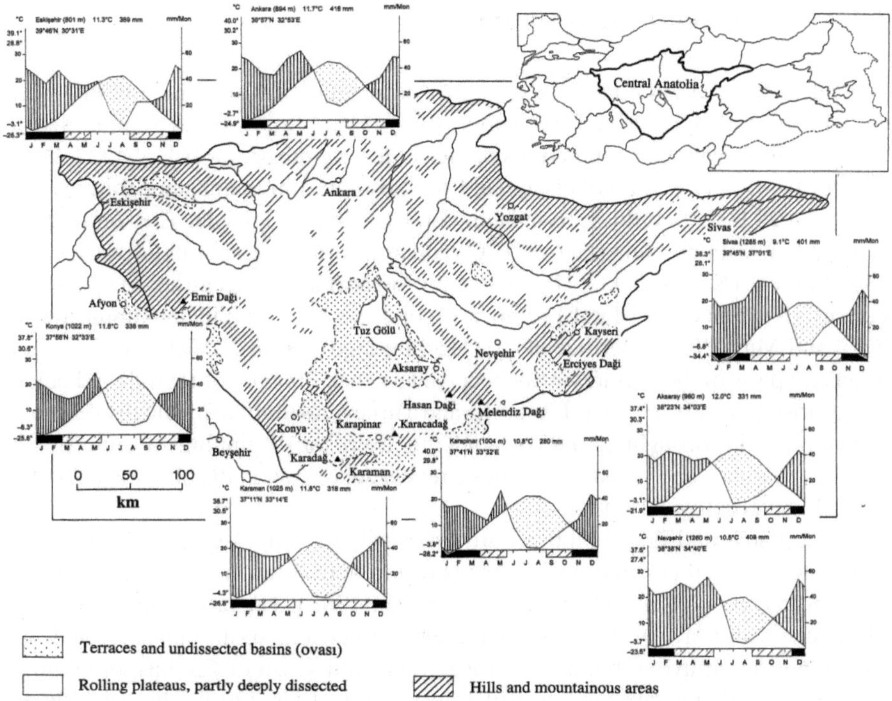

Fig. 4.1 Central Anatolia (delimination after Erol 1982) and its climate (Climatological data after Alex 1985)

4.1.3 Phytogeographical Position

The remnants of the steppes of Central Anatolia form a western outlayer of the vast Eurasian steppe belt, which stretches from Central Asia via Afghanistan and Iran to Turkey. Phytogeographically, this area belongs to the large Irano-Turanian floristic region (Zohary 1973; Takhtajan 1986; Irano-Anatolian subregion), which is very different in terms of climate, flora and vegetation from the other parts of Turkey.

The floristic links to the surrounding phytogeographical regions (Euro-Siberian region, represented by the Euxine subregion in the north; Mediterranean region, represented by the East-Mediterranean subregion in the west and south) are weak and various genera, such as *Acantholimon*, *Astragalus*, *Centaurea* s. l., *Cousinia*, or *Verbascum*, indicate the outstanding character of this area. Most of these genera here have their evolutionary and main centre of diversity. Further genera with a high number of (endemic) species in the Irano-Anatolian area are *Achillea*, *Aethionema*, *Alyssum*, *Consolida*, *Echinops*, *Nepeta*, *Onosma*, *Phlomis*, *Salvia*, *Silene* and *Thymus*.

4.2 Climate and Vegetation Cover

Typical of Central Anatolia is a strongly continental climate, which is extremely cold in winter and marked by heavy, long-lasting snows and regular frosts. The minimum average temperatures of the coldest month are far below 0°C (–2.7°C Ankara to –7.8°C Şereflikoçhisar), and the absolute minimum temperatures vary between –21.0°C (Şereflikoçhisar) and –34.4°C (Sivas, Fig. 4.1). The summers are dry and hot, with maxima up to 40.0°C (Ankara, Konya). Total rainfall varies from 280 mm/year (Karapınar) to 416 mm/year (Ankara) and concentrates in the winter months (November to March), with a second, smaller rainfall maximum during April and May ("Regenperiode der 40 Nachmittage"; Birand 1970). The Tuz Gölü basin and parts of the Aksaray ovası receive less than 300 mm of rainfall and belong to the most arid parts of Central Anatolia and Turkey. As rainfall drastically decreases in summer, the drought period is rendered from the beginning of June to mid October. Garcia Lopez (2001) provides a recent survey about the climatic data of all Turkish meteorological stations.

These harsh and dry conditions are responsible for treeless steppe vegetation, dominated by well adapted dwarf-shrubs (arido-active and arido-passive species), a few herbs, geophytes and annuals (Kürschner 1983). The latter finish the whole life cycle within a few weeks (ephemerals, winter and spring annuals) and thus withstand the unfavourable dry and/or winter-cold season by a diaspore bank in the soil. Woodlands and remnants of forests are restricted to the higher hills and mountains, surrounding the large basins of Central Anatolia.

4.3 Soils and Edaphic Conditions

Most prominent in Central Anatolia are greyish and/or brownish marly soils (sierozomes), which derived under semi-arid to arid climate by weathering from the Neogene limestone. They are poor in humus, have a low water holding capacity and are often characterised by a calcareous crust or hard anhydrite layer just below (30–40 cm) the A-horizon. Around the volcanoes and at the southern periphery, they are mixed with various lavas and tufas consisting of soft and porose airfall tufa. Towards the centre of the basins, they are frequently in close contact to saline soils (solonchak, solonetz), which often additionally show a salt crust by evaporation on the surface. For the most part, however, the soils are alkaline, sometimes neutral and rarely slightly acidic.

4.4 Flora and Vegetation

4.4.1 General Outline

Çetik (1985), in his textbook "Vegetation and Ecology of central Anatolia", provides a comprehensive introduction to all vegetation types of the area and puts a particular focus on steppe vegetation. Floristically, the steppes of Central Anatolia are well known, and first vegetation studies date back to 1915 (e.g., Bilger 1955; Birand 1938, 1947, 1952, 1970; Çetik 1985; Krause 1915; Kürschner 1983, 1984; Kürschner et al. 1997; Louis 1939; Markgraf 1961; Walter 1956; Wenzel 1937).

According to Walter (1956), the climatic data of Central Anatolia indicate a former *Bromus-Stipa* steppe (Photos 4.1 and 4.2), rich in grasses on the dissected plateaus, hilly slopes and foothills surrounding the basins, and a chamaephyte-rich, primary *Artemisia santonicum* steppe (Yavşanlık) in the larger basins (Photo 4.3). Within Central Anatolia, the *Bromus-Stipa* steppe today occurs only sporadically, but can be found in relict areas or areas fenced and protected from grazing. Outstanding from a floristic point of view is the occurrence of several *Stipa* species, such as *S. arabica*, *S. barbata*, *S. hohenackeriana*, *S. holosericea*, *S. lagascae*, *S. lessingiana*, *S. pulcherrima*, *S. syreistchikowii*, and *S. zuvantica*, which contribute much to the typical physiognomy of the *Bromus-Stipa* steppe. Co-dominant are other grasses, such as *Agropyron cristatum*, *A. repens*, *Botriochloa ischaemum*, *Bromopsis cappadocica*, *B. tomentella*, *Chrysopogon grillus*, *Festuca valesiaca*, *Dactylis glomerata*, *Hordeum bulbosum*, *Koeleria macrantha*, *Poa bulbosa* or *P. timoleontis*. These steppes rich in grasses strongly differ from the so-called "feather-grass steppes" of the Pontic-Pannonian territories of south-eastern Europe, which deviate in soil conditions and life forms (Walter 1956).

In contrast, the large basins in Central Anatolia are vegetated by a chamaephyte-rich steppe, dominated by *Artemisia santonicum* (primary *Artemisia* steppe, Yavşanlık). These steppes are more or less greyish green in appearance and made up of chamaephytes (dwarf-shrubs) and hemicryptophytes, with a wealth of annuals and geophytes, at least after the spring rainfall.

Both types of steppe predominate within a semi-arid climate with less than 300 mm of rainfall per year, where the occurrence of woodlands is limited by the dry and arid summers. Today, these natural steppes are rare and mostly have been transformed into a secondary *Artemisia santonicum* steppe or further substitute communities, due to intensive grazing.

The available floristical and vegetation studies clearly indicate that

1. originally, most of the area of Central Anatolia was occupied by a natural *Bromus-Stipa* steppe, including many hemicryptophytes and geophytes;
2. due to intensive grazing through thousands of years, these primary steppes have been degraded and – due to pastoral selection – have transformed into secondary *Artemisia santonicum* steppes. Additionally, a strong increase in therophytes can be observed;
3. primary *Artemisia santonicum* steppes (semi-deserts) were limited to the most arid parts of the basins, characterised by ± saline, flat, greyish and marly steppe soils. They are not dominated by grasses, but intensive grazing of these steppes increases the occurrence of therophytes, mostly unpalatable chamaephytes, and poisonous and/or spiny/thorny hemicryptophytes.

In edaphical respect, gypsum steppe (e.g., Aydoğdu et al. 1994; Ketenoğlu et al. 1983, 2000), salt steppe (e.g., Aydoğdu et al. 2004; Çetik 1985; Kürschner 1983) and sand steppe (e.g., Birand 1970; Çetik 1985; Kürschner et al. 1997; Tatlı 1991) are worth being distinguished. It is presently unclear if the "lowland" steppe stands

Photo 4.1 *Bromus-Stipa* steppe east of Tuz Gölü (Şereflikoçhisar) (Photo by G. Parolly)

Photo 4.2 *Stipa pulcherrima* (Photo by H. Kürschner)

Photo 4.3 Central Anatolian primary steppe (*Artemisia santonicum* steppe; Yavşanlık), Konya ovası (Photo by H. Kürschner)

growing on ultramafic soils deserve a classification into a particular syntaxon in analogy to the situation of the syntaxonomy of the dwarf-shrub and thorn-cushion communities of higher elevations (Parolly 2004).

4.4.2 Adaptations

Life forms, dominating the Central Anatolian steppe, are chamaephytes, hemicryptophytes, therophytes and geophytes, which all represent ecological responses to the prevailing climatic conditions. They show various ecomorphological adaptations to survive the intensive summer drought. Therophytes (e.g., *Alyssum desertorum*, *Arenaria serpyllifolia* agg., *Minuartia hamata*, *M. meyeri*, *Trigonella* spp., *Ziziphora* spp.; often as disturbance indicators) and geophytes (e.g., *Allium* spp., *Colchicum* spp., *Crocus* spp., *Gagea* spp., *Muscari* spp., *Ornithogalum* spp., *Tulipa* spp.) dominate the spring period, and survive the unfavourable dry summer period in an arido-passive state, either by seeds (therophytes) or vegetative diaspores, or bulbs, corms and rhizomes (geophytes) in the ground (soil diaspore bank). By contrast, most of the chamaephytes and hemicryptophytes are arido-active. Most of these species, such as *Artemisia santonicum*, form very small leaves during the summer period, thus reducing their transpiration rate or they shed (most of) their leaves (e.g., *Alhagi camelorum*). Their renewal buds are close to the ground surface and protected in winter by snow against strong frosts. "Normal" leaves are often formed only after the winter and spring rainfalls. Further adaptations of perennial steppe plants, which must be seen in the context of protection against desiccation and reduction of water-loss by transpiration, are small leaves (e.g., *Scabiosa argentea*, *Thymus* spp., *Ziziphora tenuior*), fistulose leaves (e.g., *Festuca* spp., *Stipa* spp.) or xeromorphic leaves with thick cuticles and immersed stomata. Many of the Boraginaceae (*Anchusa* spp., *Alkanna* spp., *Onosma* spp.), Lamiaceae (*Marrubium* spp., *Phlomis* spp., *Salvia* spp., *Sideritis* spp., *Stachys* spp.) or Scrophulariaceae (*Verbascum* spp.) are characterised by a dense, woolly or felty indumentum, which protects the plants against high radiation loads and water-loss.

A further effective adaptation against water stress in summer is an enlarged root system, typical of many of the steppe plants (Bilger 1955; Birand 1952). This phenomenon is well known from desert and steppe plants and responsible for the patchy, mosaic-like above ground distribution of the vegetation cover (Monod 1954). In contrast to the very patchy and diffuse plant scatter above ground, the roots often completely fill the space between the individual plants and are in strong competition for the available space and the sparse water resources. Often the below ground biomass is much higher than the above ground one, and the roots reach depths of up to 50 cm (*Stipa* spp.), 2–3 m (e.g., *Artemisia santonicum*, *Moltkia caerulea*, *Marrubium parviflorum*) or even up to 8 m (e.g., *Alhagi camelorum*; Birand 1970).

The studies of Birand (1938, 1952) demonstrated the correlation of the root length with the annual development of the steppe plants. All species that grow during the drought period (arido-active) show a strong and long main root (tap root), reaching the moist soil layers. Their osmotic value is below 1.5 MPa even during summer-drought, despite a relatively high transpiration (e.g., *Artemisia santonicum*, *Eryngium campestre*, *Euphorbia myrsinites*, *E. tinctoria*, *Peganum harmala*, *Scolymus hispanicum*). New leaves of *Artemisia santonicum*, one of the most typical steppe plants, show osmotic values of 1.2–1.5 MPa, and the dried up but revivable ones the double values. The same holds true for *Peganum harmala*, which characterises nitrophytic substitute communities (Photos 4.4 and 4.5). Fresh leaves vary from 1.6 to 1.9 MPa, whereas these values increase to 2.8 MPa in drying-up leaves. Both taxa have long roots (more than 1.2 m) which can compensate their transpirational water-loss and supply the renewal leaf buds with water also during summer drought.

In contrast, plants, with short roots, as typically found in most of hemicryptophytes, soon die in the early summer period. There osmotic values steadily increase to 3.0 MPa and often show a high daily variation. With increasing water deficit, their maximum osmotic value is reached and the above ground parts dry up. In contrast, the osmotic values of most of the geophytes and therophytes (spring ephemerals), which flower during early spring time or during the rainy season, is low. Their values vary between 0.6 and 0.9 MPa (Birand 1938, 1952), although most of the therophytes root in the upper soil layer (ca. 4–8 [–20] cm).

4.4.3 Species Composition and Phytosociology

In general, the steppe vegetation of Central Anatolia is reasonably well known in terms of vegetation structure, floristic composition, edaphic subtypes and by many local communities. However, and in spite of, or maybe even because of the many

Photo 4.4 Strongly degraded, nitrophytic *Peganum harmala* steppe, Tuz Gölü (Photo by H. Kürschner)

Photo 4.5 *Peganum harmala* (Photo by H. Kürschner)

recent studies devoted to it, the syntaxonomic classification of the Anatolian steppe communities is far from being consolidated. A much-needed, critical syntaxonomic synthesis of the many local studies describing an array of associations is (in contrast to Eastern Anatolia; Hamzaoğlu 2006) not yet available. It is obvious that any phytosociological classification of Central Anatolian steppe vegetation was and is hampered by the long-lasting and severe human impact on the area, which has changed and destroyed most of the natural and more diverse secondary vegetation by overgrazing. Nowadays, the researcher is facing a bewildering mosaic-like pattern of steppe fragments, various degradation stages and replacement communities, which are difficult to understand and to classify.

First insights in the complex structure and species composition of the Central Anatolian steppes along with a preliminary classification were given by Zohary (1973), who proposed the Artemisietea fragrantis anatolica class [nom. nud., cf. art. 2b ICPN (Weber et al. 2000)]. His intuitive classification was more or less based on the dominance principle and is today out of date (for a criticism, see Léonard 1993). The first pioneering and more detailed phytosociological studies, following the Braun-Blanquet (1964) approach, were performed by a team of French and Turkish botanists around Pierre Quézel and Yildirim Akman and their Turkish scholars (e.g., Akman 1974, 1990; Akman et al. 1984, 1985, 1994, 1996; Aydoğdu et al. 1994, 1999, 2004; Hamzaoğlu 2005; Hamzaoğlu et al. 2004; Ketenoğlu et al. 1983, 1996, 2000; Vural 1981; Vural et al. 1995; Yurdakulol and Aydoğdu 1990). According to them, all steppe communities can be classified within the Astragalo-Brometea class. For the Central Anatolian communities, the Onobrychido armenae-Thymetalia leucostomi order was established (e.g., Akman et al. 1985) and later further subdivided into two more or less zonal subunits, classified at subordinal level (Onobrychido armenae-Thymenetalia leucostomi, ca. 800–1,200 [–1,800 m] and Asperulo phrygiae-Thymenetalia chaubardii, ca. [1,300 m] 1,500 to high montane elevations; Akman et al. 1991). The distinction between upland steppe vegetation (Asperulo phrygiae-Thymenetalia chaubardii) and the zonal xerophytic dwarf-shrub communities, grasslands and thorn-cushion communities of the subalpine belt of the Taurus range, grouped into the Astragalo-Brometalia (Quézel 1973), remains still unclear (Parolly 2004).

The present contribution focuses on the "lowland steppe" communities (Onobrychido armenae-Thymenetalia leucostomi) of the Anatolian plateau. The conspectus provided below follows with some slight taxonomic and nomenclatural modifications Kurt et al. (2006) as the most recent compilation of the major units. However, it must be noted that this valuable "synoptic view" (Kurt et al. 2006) is neither a revision nor a critical synopsis, which would tackle the inflation of local "associations" and alliances, continuously published till today. Many of these formally described syntaxa often represent nothing but basal, derivative or fragmentary stands of earlier described and widespread communities. We have made no approach to here resolve the inconsistencies related to the evaluation of the diagnostic species of the order and alliances.

How unsettled the syntaxonomic classification of the Central Anatolian steppe communities really is, is also revealed by the gappy and artificial distribution patterns in the distribution map of the major syntaxa as compiled by Kurt et al. (2006). At present, eight alliances are distinguished for the lower elevations of Central Anatolia. They include many different associations of often only local distribution and value. Their character species are provided in the following survey.

Although a good deal of the syntaxa and their diagnostic species inventory is in urgent need of revision, the list certainly will provide a good and reliable impression of the floristic composition of the Central Anatolian steppe vegetation.

4.4.4 Syntaxonomic Conspectus

Class: Astragalo-Brometea Quézel 1973 em. Parolly 2004
Character species: *Anthemis cretica* subsp. *anatolica, Astragalus angustifolius, A. ornithopodioides, Bromopsis tomentella, Centaurea urvillei, Cruciata taurica, Dianthus zonatus, Erysimum crassipes, Euphorbia macroclada, Festuca valesiaca, Galium incanum* subsp. *elatius, Globularia orientalis, Helianthemum canum* agg., *Hypericum origanifolium, Koeleria macrantha, Leontodon asperrimus, Linum austriacum, L. hirsutum* subsp. *anatolicum, Lomelosia rotata, Minuartia anatolica, Ononis adenotricha, Phlomis armeniaca, Scutellaria orientalis, Stachys lavandulifolia, Stipa holosericea, S. lessingiana, Teucrium chamaedrys, T. polium, Thymus sipyleus, Veronica multifida, Ziziphora clinopodiodes.*

Order: Onobrychido armenae-Thymetalia leucostomi Akman, Ketenoğlu & Quézel 1985
Character species: *Acantholimon acerosum, A. venustum, Achillea wilhelmsii, Allium rotundum, Alyssum pateri, A. sibiricum, Anthemis tinctoria, Artemisia santonicum, Asperula liliaciflora, Astragalus condensatus* (incl. *A. brachypterus*), *A. lydius, A. vulneraria, Bromopsis cappadocica, Bungea trifida, Bupleurum boissieri, Centaurea virgata, Cousinia birandiana, Dianthus anatolicus, D. crinitus, Digitalis lamarckii, Ferulago pauciradiata, Fumana aciphylla, F. paphalogonica, Genista sessilifolia, Hedysarum varium, Helianthemum nummularium, Hypericum aviculariifolium, Inula montbretiana, Jurinea consanguinea, Malabaila secacul, Marrubium parviflorum, Noaea mucronata, Onobrychis armena, O. hypargeia, Onosma cinerea, O. aucheranum, Paronychia kurdica, Phlomis herba-ventii, Polygala pruinosa, Salvia cryptantha, S. hypargeia, Scabiosa argentea, Silene supina* subsp. *pruinosa, Stachys cretica* subsp. *anatolica, Thymelaea passerina, Thymus leucostomus* var. *leucostomus, Ziziphora taurica, Z. tenuior.*

Alliance: Phlomido armeniacae-Astragalion microcephali Akman, Ketenoğlu, Quézel & Demirörs 1984
On various soils derived from serpentine, radiolarite and flysch under cold climatic conditions. Recorded southwest and west of Ankara, the Tuz Gölü basin and the Konya ovası between 750 and 1,300 m altitude.
Character species: *Acantholimon acerosum, Astragalus lycius, A. microcephalus, Helianthemum canum, Marrubium parviflorum* subsp. *oligodon, Phlomis armeniaca, Teucrium chamaedrys.*

Alliance: Convolvulo holosericei-Ajugion salicifoliae Akman, Ketenoğlu, Quézel & Demirörs 1984

On marly and marly-gypsaceous soils under xeric climatic conditions. Recorded west of Ankara, Ayaş, Temelli Polatlı, and Beypazarı between 650 and 1,200 m altitude.

Character species: *Ajuga salicifolia*, *Convolvulus holosericeus*, *Euphorbia macroclada*, *Galium verum*, *Linum flavum* subsp. *scabrinerve*.

Alliance: Salvio tchihatcheffii-Hedysarion variae Akman, Ketenoğlu, Quézel & Demirörs 1984

On steeper, marly slopes under very cold climatic conditions. Widespread southwest of Ankara between 800 and 1,250 m altitude.

Character species: *Asyneuma limonifolium*, *Hedysarum varium*, *Helianthemum nummularium*, *Linum hirsutum* subsp. *anatolicum*, *Salvia tchihatcheffii*.

Alliance: Arenario-Astragalion plumosi Akman 1990

On siliceous (andesite) soils between 1,300 and 1,800 m altitude, mainly north of Ankara (Cubuk, Karagöl, Aydos Mts.).

Character species: *Arenaria ledebouriana*, *Astragalus plumosus* subsp. *plumosus*, *Galium verum* subsp. *glabrescens*, *Sideritis germanicopolitana*, *Stachys iberica* subsp. *stenostachya*.

Alliance: Alysso lepidoto-stellati-Astragalion condensati Aydoğdu, Ketenoğlu & Hamzaoğlu 1999

On various soils derived from limestone, mainly in the Kirşehir area between 1,300 and 1,600 m altitude.

Character species: *Alyssum lepidoto-stellatum*, *Anchonium elichrysifolium* subsp. *canescens*, *Astragalus condensatus*, *A. densifolius*, *A. micropterus*, *Centaurea paphlagonica*, *Euphorbia anacampseros*, *Minuartia anatolica* var. *arachnoidea*, *Salvia blepharochlaena*, *S. modesta*, *Sideritis galatica*, *Verbascum vulcanicum*.

Alliance: Phlomido nissolii-Onobrychidion tournefortii Kurt 2002 (sphalm. "Onobrychion", orthogr. corr. acc. Art. 41a ICPN)

On calcareous, marly soils southwest of Afyon between 900 and 1,000 m altitude.

Character species: *Eryngium bithynicum*, *Hypericum avicularifolium* subsp. *depilatum*, *Onobrychis tournefortii*, *Phlomis nissolii*.

Alliance: Astragalo karamasici-Gypsophilion eriocalycis Ketenoğlu & Quézel 1983

On gypsaceous soils, mainly between 600 and 850 m altitude. Widely distributed northwest of Ankara (Çankiri area), the Tuz Gölü basin (east of Şereflikoçhisar) and in the wider Sivas area (different suballiances)

Character species: *Allium flavum* subsp. *tauricum*, *Astragalus karamasicus*, *Bupleurum boissieri*, *Centaurea patula*, *Gypsophila eriocalyx*, *G. parva*, *Linum mucronatum* subsp. *gypsicola*, *Thymus leucostomus* var. *gypsaceus*.

Alliance: Achilleo wilhelmsii-Artemision santonicum Aydoğdu, Kurt, Hamzaoğlu, Ketenoğlu & Çansaran 2004

On salty soils and saline steppe soils (salt steppe) around the Tuz Gölü, the Seyfe Gölü and in large parts of the Konya ovası between 960 and 1,000 m altitude.

Character species: *Acantholimon halophilum*, *Achillea wilhelmsii*, *Allium pseudoflavum*, *A. scabriflorum*, *Alyssum blepharocarpum*, *Anthemis fumariifolia*, *Artemisia santonicum*, *Krascheninnikova ceratoides*, *Reaumuria alternifolia*, *Verbascum helianthemoides*.

More commonly associated species, widely distributed in the steppes of Central Anatolia, include *Achillea phrygia*, *Aegilops triuncialis*, *Alyssum desertorum*, *Androsace maxima*, *Arnebia decumbens*, *Asperula stricta*, *Astragalus oxytropifolius*, *A. strigillosus*, *Atraphaxis billardierii*, *Callipeltis cucullata*, *Centaurea depressa*, *C. solstitialis*, *Convolvulus lineatus*, *Consolida hellespontica*, *Cousinia halysensis*, *Crepis foetida* subsp. *rhoeadifolia*, *Glaucium corniculatum*, *Globularia trichosantha*, *Gypsophila heteropoda*, *Haplophyllum thesioides*, *Helianthemum salicifolium*, *Helichrysum arenarium* subsp. *aucheri*, *Hesperis angorensis*, *Inula anatolica*, *Koelpinia linearis*, *Lappula barbata*, *Leymus cappadocicus*, *Linaria corifolia*, *Linum mucronatum* subsp. *armenum*, *Marrubium trachyticum*, *Melica cupanii*, *M. transsilvanica*, *M. hamata*, *Moltkia coerulea*, *Muscari longipes*, *Nigella arvensis*, *N. nigellastrum*, *Onobrychis argyrea*, *O. sulphurea* var. *sulphurea*, *Reseda lutea*, *Salvia cyanescens*, *Senecio vernalis*, *Trigonella coerulescens*, *Valerianella coronata*, *Verbascum pycnocephalum*, *Vincetoxicum tmoleum* and *Xeranthemum annuum*.

4.5 Changes in Vegetation Cover Due to Human Impact

Turkey has experienced rapid economic and population growth. These pressures have accelerated the destruction of various ecosystems and vegetation units by many processes that still keep going on. The ongoing anthropozoogenic effects are drastically and in most parts irreversible. This can be seen in the vegetation remnants: natural or semi-natural communities hardly exist in most parts of Central Anatolia, and the remaining units are fragmentary or replaced by substitute communities of various degrees of disturbance (Fig. 4.2). Whereas a number of studies documented the floristic composition and plant cover of Central Anatolia (e.g., Birand 1947, 1970; Çetik 1985; Nalbantli 1964; Bakir 1970; Yilmaz 1977; Kürschner 1983, 1984; Akman et al. 1984) nearly no research has been conducted comparing vegetation changes in grazed and protected areas.

4.5.1 Regressive Succession

Most of the area has been under severe anthropogenic effects since 5000 BP (e.g., Great Konya Basin – Çatalhüyük; Phrygian region) as it is an indigenous land of

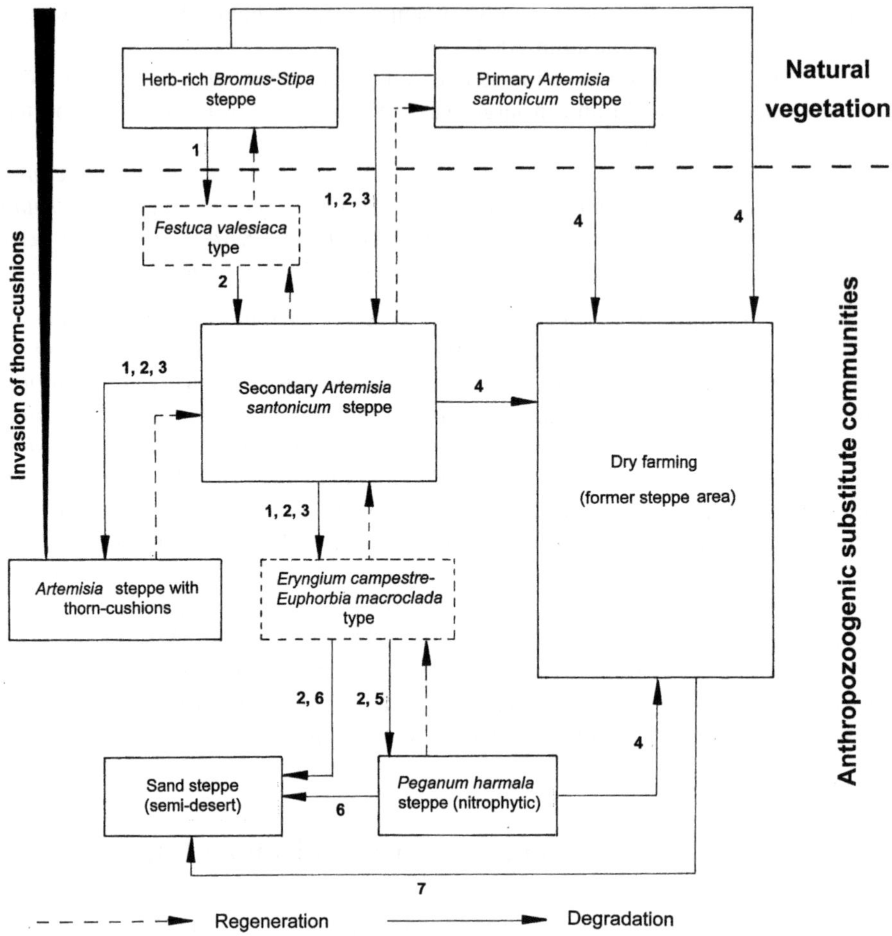

Fig. 4.2 Regressive succesion (primary and secondary steppes) in Central Anatolia. *1–4* direct effects: *1* grazing; *2* overgrazing; *3* collecting of medicinal and forage plants, digging of roots; *4* ploughing; *5–7* indirect effects: *5* settlements, and seasonal settlements (yayla); *6* soil erosion; *7* fallow (uncultivated) fields (Modified after Birand 1970; Kürschner 1983, 1984)

fodder production since the Neolithic (Atalay 2002; Böçük et al. 2009). Great parts of the natural vegetation at all elevations (primary steppes: *Bromus-Stipa* steppe, *Artemisia santonicum* steppe), were destroyed and largely replaced by substitute communities, degraded to various degrees and with a highly reduced floristic diversity. Since historical times, continuous deforestation, overgrazing, and unsuitable tillage and irrigation management have transformed the natural vegetation and caused soil erosion (Kapur et al. 2006).

The transformation and degradation of the vegetation results from numerous direct and indirect effects, which are shown with their consequences in Fig. 4.2 [e.g., 1 grazing; 2 overgrazing; 3 collecting of medicinal and forage plants, digging of roots; 4 ploughing, 5 settlements, and seasonal settlements (yaylas); 6 soil erosion; 7 fallow (uncultivated) fields]. The secondary *Artemisia santonicum* steppes, today typical of the basins and widely distributed in Central Anatolia, derived from a primary formation by overgrazing, collecting of plants and digging of roots. If this process continues, secondary thorn-cushion formations will derive, which are mainly dominated by thorny *Astragalus* species (e.g., *A. condensatus*, *A. microcephalus*, *A. plumosus*) and spiny *Acantholimon* species (*A. acerosum*, *A. venustum*). These taxa migrate from the neighbouring montane-oreal belt into the steppe area, especially when the former forests and woodlands, which originally limited their distribution, were destroyed. If overgrazing continues, a nitrophytic community will develop around the settlements and yaylas, dominated by the unpalatable and poisonous *Peganum harmala* (*Peganum harmala* steppe, hüyük; Photo 4.4). Today, these heavily degraded "steppes" are widespread in the Tuz Gölü and Konya basins, where they replace the former *Artemisia* steppe.

Additionally, overgrazing causes soil erosion and during the frequent summer drought, dust storms blow a fine yellow powder across the plains of Central Anatolia. This locally results in the development of a semi-desert sand steppe, as can be observed in the surroundings of Karapınar (Birand 1970; Tatlı 1991 and references therein). Here regeneration is nearly impossible due to soil erosion and deflation of the fertile soil particles. Tatlı (1991) reported a similar situation from the Doğu Iğdir plain in Eastern Anatolia. Intensive ploughing and the continuing expansion of lands for crop production affect the substitute communities as well as the few remaining sites with primary communities. Agricultural land (dry farming) today covers wide areas of Central Anatolia, where more than one-third of the total crop production of Turkey concentrates on the former steppe area. Mainly hard wheat and barley are grown and, as regards legumes, mostly beans, chickpeas and a small amount of lentils are sown. Important industrial plants in the area are sugar beets.

These fields and fallow fields often show a rich and variously coloured segetal flora, dominated by *Bupleurum* and *Convolvulus* species, or dense and attractive stands of *Acroptilon repens*, *Agrostemma githago*, *Aristolochia maurorum*, *Boreava orientalis*, *Centaurea depressa*, *C. solstitialis*, *Consolida* spp., *Fumaria* spp., *Isatis floribunda*, *Papaver* spp., *Saponaria prostrata*, *Turgenia latifolia*, *Vicia villosa*, *Vaccaria pyramidata* and *Wiedemannia orientalis* (Photo 4.6). Good indicator species for traditional agriculture are the endemic geophytic *Aristolochia* species, which tolerate only shallow ploughing. For a first introduction and more detailed species list of this segetal vegetation, see Çetik (1985). It should be noted that presently the diverse segetal vegetation is increasingly threatened by the advent of modern, industrialized agriculture with intensive use of agrochemistry and deep ploughing.

Today, six, often intergrading types of steppes (anthropozoogenic substitute communities, see Fig. 4.2) dominate Central Anatolia physiognomically:

1. a *Festuca valesiaca* type, which can be considered as an early degeneration stage of the *Bromus-Stipa* steppe after grazing. Due to pastoral selection, most of the palatable species are removed. However, a progressive regeneration is possible, as shown in areas which are fenced and protected from grazing;

Photo 4.6 Segetal flora and wild orchards in Central Anatolia (Photo by H. Kürschner)

2. an *Artemisia santonicum-Thymus leucotrichus* type (secondary *Artemisia santonicum* steppe), today the most common formation, resulting from grazing;
3. an *Artemisia* steppe rich in thorn-cushions, abundant on the hilly slopes and mountains between the basins (ovası);
4. an *Eryngium campestre-Euphorbia macroclada* type, typical of intensively grazed or overgrazed places;
5. a *Peganum harmala* steppe, a very species-poor, nitrophytic substitute community around villages and at lairs;
6. a sand steppe, formed by soil-erosion and heavy dust storms.

The same effects and changes can be observed in the natural vegetation on the hills and mountain sites surrounding the basins in Central Anatolia. Here, most of the formerly widely-distributed forests and woodlands concentrate (e.g., *Quercus cerris* and/or *Q. pubescens* woodlands, *Pinus nigra* var. *caramanica* forests and woodlands, *Juniperus excelsa/foetidissima* forests and woodlands). Most of these natural units are completely destroyed and replaced by degraded woodlands and shrublands. The responsible anthropogenic factor is wood cutting and grazing that have affected the sites for thousands of years. Most of the former woodlands are replaced by very open woodlands, solitary trees or a low shrubland, the latter dominated by *Amygdalus orientalis*, *Atraphaxis billardieri*, *A. spinosa*, or *Rhamnus* spp. Various such stands were described by Çetik (1985). Their understory is dominated by thorn-cushions, which often form pure secondary stands on many hill sites in Central Anatolia, where nearly all of the woody species were cut and

Photo 4.7 Wild orchards in Central Anatolia (Photo by H. Kürschner)

removed by man. Originally, thorn-cushions concentrated in the windswept, treeless subalpine belt of the adjacent mountains (Kürschner 1984, 1986a, b). Today, as in the basins, dry farming already extends into the hill sites, whenever the area is not too rocky and ploughing is possible.

4.5.2 Wild Orchards

Many of the plateaus, hills and foothills of Central Anatolia, which for ages have supported cultivated land for crop production, still show a pattern of scattered fruit trees, the so-called "wild orchards" (*Pyrus elaeagnifolia* woodland; Photo 4.7). Their tree-layer consists mainly of *Celtis tournefortii*, *Crataegus aronia*, *Prunus armeniaca*, *P. divaricata* subsp. *ursina*, *Pyrus amygdaliformis* and *P. elaeagnifolia*. The stands are kept for their fruits, are used as rootstocks for cultivated fruit varieties and serve as shady resting-places in the hot summers. Many authors (e.g., Louis 1939; Kürschner 1984; Mayer and Aksoy 1986; Zohary 1973) regarded these trees as remnants of a former *Quercus* woodland, indicating that these parts of Central Anatolia were forested in former times.

However, vegetation surveys show that oak trees are uncommon in the "wild orchards" and, in contrast, wild fruit trees today are virtually absent in the remnants of *Quercus* woodlands surrounding the steppe area (Woldring and Cappers 2001). Palynological investigations revealed that the *Quercus* woodlands of Central Anatolia reached a maximum expansion between 8000 and 4000 BP, and deforestation started c. 4000 BP, when an abrupt decline of oaks can be observed in palynological cores (Woldring and Cappers 2001). Archaeobotanical research confirms the presence of many of these fruit trees already before the expansion of *Quercus* woodland in Central Anatolia (before 8000 BP). At this time, the competitive capacity of woodland trees probably restricted the occurrence of wild fruit trees to isolated locations. They concentrated mainly on rocky outcrops, a suitable habitat for the light-demanding fruit trees, but unfavourable for *Quercus* woodlands. In response to the large-scale devastating action of man and his livestock (as from c. 4000 BP) they were "... taking advantages of the deforestation and rapidly spread into the newly created habitats ..." (Woldring and Cappers 2001). Today, they occupy areas of former oak and juniper woodlands, and at the same time indicate the potential woodland areas of Central Anatolia.

4.6 Perspectives

Over the past 50 years, almost half of the former steppe area of Central Anatolia has been transformed into cropland (dry farming, interrupted every 2 years by fallow fields) and at the same time there was no adequate reduction in grazing animals (Birand 1970; Firincioğlu et al. 2007; Uslu 1960). Although it is difficult to provide comparable figures, the few data given in Uslu (1960) and others (Anonymous 2001, 2007) at least demonstrated the drastic change in rangeland, agricultural areas and numbers of livestock (Table 4.1, figures only given for goats and sheep). Animal husbandry produces 5% of the gross national product (approx. 122 billion US $; Koç 2000). In total, ca. 12 million animal units graze on Turkish rangelands (Koç 2000). Whereas the proportion of rangeland decreased from 6.2 in 1970 to 3.1 million ha in 2004, the number of grazing goats and sheep increased from nearly 1.7 to 3.2 million. As at the same time large parts of the rangeland area have been transformed into agricultural land (increase from 2.95 to 4.5 million ha, Table 4.1), the grazing pressure on the remaining rangeland vegetation continues to be dramatic. "At present, the rangelands of Turkey are grazed 2–3 times higher than their carrying capacity. As a consequence of bad management, up to 90% of the climax vegetation of Turkey's rangelands has been lost" (Genckan et al. 1990 cited in Koç 2000). This has

Table 4.1 Rangeland, agricultural area and numbers of livestock in Central Anatolia

		1950[a]	1970[b]	2001[b]	2004[c]
Total (Central Anatolia) 9.158.666 ha					
Rangeland (ha)	–	6,200,000	3,300,000	3,117,923	
Agricultural area (ha)	–	2,958,666	–	4,504,276	
Number of livestocks (only sheep and goats)					
		1,671,454	–	–	3,132,095

[a]From Uslu (1960)
[b]From Anonymous (2001)
[c]From Anonymous (2007)

generally led to a massive decline of the plant diversity of vast territories of Central Anatolia and to strong shifts within the species inventory, including furthering the occurrence of non-palatable species (pastoral selection).

The remaining fragments (mostly secondary steppes) underlay a heavy grazing pressure and presently are degraded ecosystems (Evrendilek and Doygun 2000; cf. Fig. 4.2).

The effect of this human impact has been shown in detail by the studies of Birand (1970) near Karapınar and by Firincioğlu et al. (2007) near Ikizce, 45 km southwest of Ankara.

These studies indicate:

1. exclusion of grazing will increase species richness. Firincioğlu et al. (2007) showed that ungrazed plots contain 32 plant species more than grazed ones (in total 113 species);
2. grazing has removed most of the sensitive species;
3. pastoral selection supports unpalatable taxa.

According to Sternberg et al. (2000), grazing causes two kinds of effects:

1. a direct effect that occurs through selective and differential removal of plant tissue or species;
2. an indirect effect on species diversity and composition when selective grazing on dominant species reduces their presence, favouring the spread of less competitive but more tolerant species.

To summarize, all studies clearly show that heavy grazing may cause a loss of rare and endangered species, disrupts regeneration and results in the encroachment of non-palatable plant species, such as *Euphorbia macroclada* (poisonous alkaloids), *Peganum harmala* (harmaline, harmin alkaloids, which affect the central nervous system leading to spasms and paralysis), *Artemisia santonicum* or various Lamiaceae (high content of essential oils), spiny or thorny taxa (*Astragalus* spp., *Acantholimon* spp., *Centaurea* spp., *Cousinia* spp., *Echinops* spp., *Eryngium* spp., *Noaea mucronata*, *Onopordum* spp.) as well as noxious species or plants of little forage value (e.g., *Achillea wilhelmsii*, *Erodium ciconium*, *Erysimum crassipes*, *Marrubium parviflorum*). Thus, plant diversity has been under permanent pressure of uncontrolled land-use and overgrazing, reducing the plant cover that protects the soil and generally resulting in soil erosion. And these disturbing anthropogenic effects still continue.

Until today, nearly half the population of Central Anatolia have lived in rural areas and their main economic resources are pasture, degraded steppes, remnants of woodlands; their agricultural areas (for the production of barley, wheat, lentil and chickpea) were formed by destruction of natural areas and their vegetation. It is very likely that these negative anthropogenic effects will continue in Central Anatolia, leading to even more reduction of plant cover and further soil erosion. Therefore, all remaining vegetation is in a critical state in terms of land degradation and erosion (Böçük et al. 2009). In addition, the demand for higher crop production yields due to the increasing population pressure leads to an excessive use of water resources. This creates salinity in many of the fertile alluvial plains of Central Anatolia, which actually are the gene zones of many crop plant species, particular cereals and legumes and many other useful plants (Boulos et al. 1994; Kapur et al. 2006).

In order to diminish the continuing degradation and to re-establish the natural equilibrium in steppe areas, Böçük et al. (2009) recommended that

1. the heavy grazing still continuing on the remnants of natural pasture areas must be prevented;
2. a rotational grazing plan should be applied which is based on herbage productivity (Kondoh 2003);
3. the cutting of the remaining trees (mostly *Quercus cerris*, *Q. ithaburensis*, *Q. pubescens*) must be prohibited in order to maintain the natural regeneration of the steppe forests;
4. the eroded areas should be, at least, partly reforested and afforested by native tree species.

Clearly, the development of management programmes and their implementation in the areas that are already protected seems most urgent. However, one has to keep in mind that the sustainable use of the natural resources, including the protection of environment, is highly dependent on and connected with the economic and social development of the local population. Therefore, it will be hard to convince local people to take care of the protection of their natural environment without offering any integrative programmes.

Acknowledgements We cordially thank Mrs Gabriela Michaelis (BGBM) for brushing up our manuscript linguistically. Prof. Dr Adnan Erdağ (Adnan Menderes Üniversitesi Aydın, Turkey) kindly provided references.

References

Akman Y (1974) Etude phytosociologique de la region de Beypazarı-Karaşar et Nallihan. Commun Fac Sci Univ Ankara 18:51–113

Akman Y (1990) Etude de la végétation steppique des montagnes d'Aydos située au Nord-Ouest d'Ankara. Ecol Mediterr 16:223–230

Akman Y, Ketenoğlu O, Quézel P, Demirörs M (1984) A syntaxonomic study of steppe vegetation in Central Anatolia. Phytocoenologia 12:563–584

Akman Y, Ketenoğlu O, Quézel P (1985) A new syntaxon from central Anatolia. Ecol Mediterr 11:111–121

Akman Y, Quézel P, Barbéro M, Ketenoğlu O, Aydoğdu M (1991) La végétation des steppes, pelouses écorchées et à xérophytes épineaux de l'Antitaurus dans la partie sud-ouest de l'Anatolie. Phytocoenologia 19:391–428

Akman Y, Quézel P, Aydoğdu M, Ketenoğlu O, Kurt L, Evren H (1994) A phytosociological research on the steppe vegetation of the Yaprakli mountains (Çankiri, Turkey). Ecol Mediterr 20:1–7

Akman Y, Vural M, Quézel P, Kurt L, Ketenoğlu O, Serin M, Barbero M (1996) Etude de la végétation steppique de la région de Karaman et d'Ermenek (sud de l'Anatolie Centrale). Ecol Mediterr 22:1–7

Alex M (1985) Klimadaten ausgewählter Stationen des Vorderen Orients. Beihefte zum Tübinger Atlas des Vorderen Orients, Reihe A. (Naturwissenschaften), No. 14. L. Reichert, Wiesbaden, pp 1–419

Anonymous (2001) The general agriculture census. State Statistics Institution, Ankara
Anonymous (2007) TR7 Central Anatolian agricultural master plan, Ankara (in Turkish)
Atalay I (2002) Ecoregions of Turkey. Publications of the Ministry of Forestry, no. 163, Ankara
Aydoğdu M, Akman Y, Quézel P, Barbero M, Ketenoğlu O, Kurt L (1994) Syntaxonomic analysis of gypsaceous vegetation of the surrounding area between Ayas-Polatli and Beypazari (Ankara, Turkey). Ecol Mediterr 20:9–19
Aydoğdu M, Ketenoğlu O, Hamzaoğlu E (1999) New syntaxa from Cappadocia (Kirşehir, Türkiye). Israel J Plant Sci 47:123–129
Aydoğdu M, Kurt L, Hamzaoğlu E, Ketenoğlu O, Cansaran A (2004) Phytosociological studies on salty steppe communities of Central Anatolia, Turkey. Israel J Plant Sci 52:71–79
Bakir Ö (1970) A range vegetation survey in the range areas of Middle East. Agriculture Faculty Publication 382, Scientific Research Study no. 232, Ankara
Bilger H (1955) Vergleichend-anatomische Untersuchungen über die Wurzeln einiger in der Umgebung von Ankara vorkommender Steppenpflanzen. Commun Fac Sci Univ Ankara Sér C 4:42–70
Birand H (1938) Untersuchungen zur Wasserökologie der Steppenpflanzen bei Ankara. Jahrb Wiss Bot 87:93–172
Birand H (1947) Über die Vegetationsverhältnisse der *Artemisia*-Steppe als Weide. Ann Univ Ankara 1947:197–208
Birand H (1952) Untersuchungen über Wurzelsysteme von Steppenpflanzen. Commun Fac Sci Univ Ankara Sér C 3:219–235
Birand H (1970) Über die Verwüstung der *Artemisia*-Steppe bei Karapinar in Zentralanatolien. Vegetatio 20:20–47
Böçük H, Türe C, Ketenoğlu O (2009) Plant diversity and conservation of the north-east Phrygia region under the impact of land degradation and desertification (Central Anatolia, Turkey). Pak J Bot 41:2305–2321
Boulos L, Miller AG, Mill RR (1994) Regional overview: South West Asia and the Middle East. In: Davis SD et al (eds) Centres of plant diversity. A guide and strategy for their conservation, vol 1. Europe, Africa, South West Asia and the Middle East. WWF/ICUN, Cambridge, pp 293–350
Braun-Blanquet J (1964) Pflanzensoziologie. Grundzüge der Vegetationskunde. 3. Aufl. Springer, Wien/New York
Çetik R (1985) Türkiye Vejetasyonu: I. Iç Anadolu'nun Vejetasyonu ve Ekolojisi. Selçuk Üniversitesi, Konya
Erol O (1982) Türkei. Naturräumliche Gliederung. Karte A VII 2 Tübinger Atlas des Vorderen Orients, Wiesbaden
Evrendilek F, Doygun H (2000) Assessing major ecosystem types and the challenge of sustainability in Turkey. Environ Managem 26:479–489
Firincioğlu HK, Seefeldt SS, Şahin S (2007) The effects of long-term grazing exclosures on range plants in the Central Anatolian region of Turkey. Environ Managem 39:326–337
Garcia Lopez JM (2001) Mediterranean phytoclimates in Turkey. Ecol Mediterr 27:15–32
Hamzaoğlu E (2005) The steppe vegetation of the Dinek Mountain (Kırıkkale). Gazi Üniv Fen Bilim Derg 18:1–15
Hamzaoğlu E (2006) Phytosociogical studies on the steppe communities of East Anatolia. Ekoloji 15(61):29–55
Hamzaoğlu E, Aydoğdu M, Kurt L, Cansaran A (2004) New syntaxa from the west part of Central Anatolia. Pak J Bot 36:235–246
Kapur S, Akca E, Kapur B, Öztürk A (2006) Migration: an irreversible impact of land degradation in Turkey. In: Kepner WG et al (eds) Desertification in the Mediterranean region: a security issue. Springer, Dordrecht, pp 291–301
Ketenoğlu O, Quézel P, Akman Y, Aydoğdu M (1983) New syntaxa on the gypsaceous formations in the Central Anatolia. Ecol Mediterr 9:211–221
Ketenoğlu O, Kurt L, Akman Y, Serin M (1996) A new alliance from Central Anatolia, Minuartion juniperino-pestalozzae. Doğa. Turk J Bot 20:457–464

Ketenoğlu O, Aydoğdu M, Kurt L, Akman Y, Hamzaoğlu E (2000) Syntaxonomic research on the gypsicole vegetation in Cappadocia, Turkey. Israel J Plant Sci 48:121–128
Koç A (2000) Turkish rangelands and shrub culture. Rangelands 22:25–26
Kondoh M (2003) Habitat fragmentation resulting in overgrazing by herbivores. J Theor Biol 225:453–460
Krause K (1915) Über die Vegetationsverhältnisse des westlichen und mittleren Kleinasiens. Beibl Bot Jahrb 116:284–313
Kürschner H (1983) Vegetationsanalytische Untersuchungen an Halophytenfluren Zentralanatoliens (Türkei). Beihefte zum Tübinger Atlas des Vorderer Orients, Reihe A (Naturwissenschaften) Nr. 11, L. Reichert, Wiesbaden, pp 1–89
Kürschner H (1984) Der östliche Orta Toroslar (Mittlerer Taurus) und angrenzende Gebiete. Eine formationskundliche Darstellung der Vegetation Südost-Anatoliens. Beihefte zum Tübinger Atlas des Vorderer Orients, Reihe A (Naturwissenschaften), No. 15, L. Reichert, Wiesbaden, pp 1–146
Kürschner H (1986a) Die syntaxonomische Stellung der subalpinen Dornpolsterformationen am Westrand SW-Asiens. Phytocoenologia 14:381–397
Kürschner H (1986b) The subalpine thorn-cushion formations of western South-West Asia: ecology, structure and zonation. Proc R Soc Edinb 89B:169–179
Kürschner H, Raus T, Venter J (1997) Pflanzen der Türkei. Ägäis – Taurus – Inneranatolien, 2. Aufl. Quelle & Meyer, Wiesbaden
Kurt L, Nilhan TG, Ketenoğlu O (2006) Synoptic view of the steppe vegetation of Central Anatolia (Turkey). Asian J Plant Sci 5:733–739
Léonard J (1993) On the plant associations mentioned in Iran by M. Zohary (1). Bull Jard Bot Natl Belg 62:113–118
Louis H (1939) Das natürliche Pflanzenkleid Anatoliens geographisch gesehen. Geographische Abhandlungen Reihe 3, Heft 12, Stuttgart
Markgraf F (1961) Die jahreszeitliche Entwicklung einer Steppenfläche bei Ankara. Ber Geobot Inst ETH Rübel 32:236–244
Mayer H, Aksoy H (1986) Wälder der Türkei, G. Fischer, Stuttgart
Monod T (1954) Mode 'contracté' et 'diffus' de la végétation saharienne. In: Cloudsley-Thompson JL (ed) Biology of deserts. Institute of Biology, London, pp 35–44
Nalbantli GM (1964) Studies on the plant composition and grazing pressure of community-used rangelands in the Central Anatolia, Turkey. Ph.D. thesis, Ankara University, Ankara
Parolly G (2004) The high mountain vegetation of Turkey – a state of the art report, including a first annotated conspectus of the major syntaxa. Turk J Bot 28:39–63
Quézel P (1973) Contribution à l' étude phytosociologique massif du Taurus. Phytocoenologia 1:131–222
Sternberg M, Gutman M, Perevolotsky A, Ungar ED, Kigel J (2000) Vegetation response to grazing management in a Mediterranean herbaceous community: a functional group approach. J Appl Ecol 37:224–237
Takhtajan A (1986) Floristic regions of the world. University of California Press, Berkley/Los Angeles/London
Tatlı A (1991) Phytosociological investigation of vegetation of wind-eroded areas of Eastern Anatolia. Ecol Mediterr 17:161–168
Turkish News Agency (2000) Turkey, Istanbul
Uslu S (1960) Untersuchungen zum anthropogenen Charakter der zentralanatolischen Steppe. Giessener Abh Agrarforsch Wirtschaftsforsch Europ Osten, vol 12. Kommissionsverlag W. Schmitz, Giessen, pp 1–71
Vural M (1981) Mut, Ermenek, Karaman arası orman-step bölgesinin fitososyolojik ve fitoekolojik yönden araştırılması. Ph.D. thesis, Selçuk Üniversitesi, Konya
Vural M, Duman H, Adıgüzel N, Kol Ü (1995) Göreme Milli Parkı'nin (Nevşehir) vejetasyonu. Turk J Bot 19:389–400
Walter H (1956) Das Problem der Zentralanatolischen Steppe. Die Naturwiss 43(5):97–102
Weber HE, Moravec J, Theurillat J-P (2000) International code of phytosociological nomenclature, 3 rd edition. J Veg Sci 11:739–768

Wenzel H (1937) Forschungen in Inneranatolien II. Die Steppe als Lebensraum. Schrift Geogr Inst Univ Kiel 7:1–27

Woldring H, Cappers R (2001) The origin of the ‚Wild Orchards' of Central Anatolia. Turk J Bot 25:1–9

Yilmaz T (1977) Studies on the botanic composition of some problematic range areas in Konya Province. Publication no. 46, Konya Regional Directorate of Soil-Water Research Institute, Konya

Yurdakulol E, Aydoğdu M (1990) Kırıkkale-Kalecik-Kirşehir arası step vejetasyonunun bitki sosyolojisi yönünden araştırılması. Doğa. Turk J Bot 14:215–234

Zohary M (1973) Geobotanical foundations of the Middle East, 2 vols. Gustav Fischer Verlag, Stuttgart

Chapter 5
Ukrainian Steppes in the Past, at Present and in the Future

Iryna Korotchenko and Mykyta Peregrym

Abstract Steppe is a zonal vegetation type in Ukraine and is most likely of Holocene origin. We briefly review the main steppe communities of the Forest-Steppe and Steppe zones, in terms of their floristic composition and diversity, and their habitats.

Until the seventeenth century Ukrainian steppes were little inhabited and in a near natural state. Then new settlers immigrated and locally founded agrarian settlements which slowly increased in number and area. Grand-scale destruction of steppe ecosystems occurred in the Soviet era and about all plain steppes were tilled, even in a number of reserves. After the collapse of the Soviet Union agriculture in the steppe zone fell in decay and large areas became fallows on which steppe vegetation regenerated. Under full protection litter, organic matter and soil moisture increase in the steppe communities and species diversity declines. Grazing or mowing can prevent this, but may also cause other problems, just as too frequent fires. Fires and flower picking greatly reduce the populations of showy bulb plants.

Conservation had an early start in Ukraine but strongly decreased in the mid-twentieth century. Presently, Ukraine works on an ecological network of steppe reserves, but the program meets with several difficulties, for practical and policy reasons.

I. Korotchenko
M.G. Kholodny Institute of Botany, National Academy of Sciences of Ukraine, Kyiv, Ukraine
e-mail: korotch@bigmir.net

M. Peregrym (✉)
O.V. Fomin Botanical Garden, National Taras Schevchenko University of Kyiv, Kyiv, Ukraine

M.G. Kholodny Institute of Botany, National Academy of Sciences of Ukraine, Kyiv, Ukraine
e-mail: peregrym@ua.fm; mykyta.peregrym@gmail.com

M.J.A. Werger and M.A. van Staalduinen (eds.), *Eurasian Steppes. Ecological Problems and Livelihoods in a Changing World*, Plant and Vegetation 6,
DOI 10.1007/978-94-007-3886-7_5,

5.1 Steppes in Ukraine

Steppes form the youngest zonal type of vegetation in Ukraine and were formed during the Holocene. The development of Ukrainian culture mainly took place in the steppes, even though steppes occupied only about 40% of the modern territory of Ukraine. The boundless space of the steppe was the natural habitat for many animal and plant species, the home and source of food for many people, and inspired them to heroic deeds and even death in defense of their native land. Unfortunately, presently only about 3% of the natural and semi natural steppes of Ukraine remain.

How were the Ukrainian steppes formed? How were these unique territories destroyed and transformed? What is their present condition? What will be the future of the steppes in Ukraine? We will try to answer these questions in this chapter.

5.1.1 The Origin of the Ukrainian Steppes

All reconstructions of the history of the formation of the modern Ukrainian flora and vegetation are based on analyses of fossil plants and pollen, comparisons of modern regional floras, and of the distribution of taxa, taking into account data on geology, climate, etc.

In general, two views can be found in the literature with respect to the development of the flora of Ukraine: an anti-glacialistic and a glacialistic one. The anti-glacialists denied the existence of glaciers on the continental East European land masses in the Pleistocene. The distribution of boulders at medium latitudes in Eastern Europe was, for instance, explained by the existence of a big ocean which furthered their distribution. Also, the wide disjunctions in the distribution of modern plant species were explained by fluctuations in humidity during the Pleistocene (Ruprecht 1866 in Baranov 1954). With the development of phylogeography this view has been discredited, although some leading botanists in Ukraine persisted in this anti-glacialistic view up to only 20 years ago (Pidoplichko 1946–1956; Klokov 1963, 1981; Zaverukha 1985; Udra 1988).

The glacialist view forms the modern paradigm in historical plant biogeography in Ukraine. At present, issues which provoke a very heated debate among Ukrainian scientists are the ages of some relicts (Tertiary or Interglacial), the existence or absence of Tertiary refugia in the modern territory of the Ukraine (especially on the plains) and the migration routes of plants from the Pleistocene onwards.

The first explorations on the biogeography of the Steppe zones in the Russian Empire were published by Ruprecht (1866). He noted that the Chernozem region coincided with the zone of Steppe vegetation. Its northern border was the southern border of the Russian Forest zone and of the distribution of boulders. Ruprecht, an anti-glacialists, explained such a border by the existence of an ancient inland sea which washed the Chernozem mainland.

The next to contribute to the development of historical plant geography in Ukraine was Paczoski (1910). He asserted that the Volyno-Podolian Upland was a refuge of vegetation during the Riss and Wurm Ice Ages, and though this vegetation had changed considerably from the Pliocene onwards, it was the main source for the formation of the modern Black Sea steppes and forests in Polissya (the Forest zone of Ukraine).

Kleopov (1933, 1941), one of the first botanists who tried to explain the process of the formation of the steppe flora in Ukraine, argued that the modern steppes of Ukraine began to form in the beginning of the Holocene from "near glacial" steppes and that they were influenced by immigration of Asian and Mediterranean elements. He also argued that the modern communities of *Carex humilis* Leys. were remnants of "near glacial" steppes that originated from the Tertiary mesophilous flora. Presently, Kleopov's and partially also Paczoski's opinions are accepted by the majority of Ukrainian botanists and were further developed. For example, Didukh (1992), studying the main stages of evolution of the vegetation of the Crimean mountains, ascertained that migrations of some steppe species (*Adonis vernalis* L., *Paeonia tenuifolia* L., *Galium verum* L., etc.) to the Crimea took place after the Riss

Fig. 5.1 Plant geographical classification of Ukraine (according to Didukh 2000). Straight lines are borders of Provinces; crossed lines are borders of Regions:

I. Pannonic: OPAN – Transcarpathian
II. Central European: OKARP – East Carpathian Mountains; RPOD – Roztochia & West Podolia
III. Sarmatic: WLPOD – Volyno-Podolian Upland; POL – Polissya
IV. Subpontic (Forest-Steppe Zone): DPRD – Transdniesteria-Dnieper Upland; OPRD – Dnieper Lowland; SROS – Central Russian Upland
V. Pontic (Steppe Zone): WPONT – West Pontic Steppes; PPONT – East Dnieper Pontic Steppes; SDON – Southern part of the Central Russian Upland; DON – Donetsk Upland; CZAZ – Black Sea Steppes
VI. Mediterranean: KR – Crimea Mountains

Ice Age, and also that anthropogenic influence promoted the development of steppe vegetation in the mountains and plains of the Crimea during the Holocene. Didukh denied the existence of direct land bridges with the Crimea and stated that the Caucasian route was determinative in forming the flora of the Crimea, but this view was criticized by researchers who supposed that a land bridge existed between the Crimea and the Dobrudja region in Romania (Dubovik 2005).

Bezusko et al. (2006), based on integrated spore/pollen, archaeological and radiocarbon studies, established that the steppe vegetation of the left bank of the Dnieper during the Eneolit (about 5,000–6,000 years ago) was more mesophytic than the modern steppe vegetation because the climate was wetter. The modern Steppe zone of Ukraine had a Forest-Steppe nature during that period.

Here we agree with Mosyakin et al. (2005) that "there are almost no phylogeographic studies of the steppe species of the Eurasian flora" and "for overcoming the lack of attention for phylogeographic studies in our countries, collaborative phylogeographic studies of several crucially important groups of vascular plants occurring in Ukraine should be initiated".

Plant geographically the Ukrainian steppes belong to the Subpontic (Forest-Steppe zone) and Pontic Provinces (Steppe zone) (Fig. 5.1).

5.2 Modern State, Distribution and Conservation of Steppe Vegetation in Ukraine

5.2.1 General Description of Steppe Vegetation

Steppe vegetation has developed under a climate of low humidity and with periodic impact of grazing and fires; these factors prevent the development of trees, at least on the plains. In steppes, over 80% of the biomass dies off and rapidly mineralizes every year. These processes set the conditions for the formation of the very fertile black earth (chernozem soil) (Didukh 1998). The structure and differentiation of steppe vegetation depend on climatic and edaphic factors.

Lack of moisture, salinity of the soil, bottom layers and groundwater cause the spread of saline steppes along the marine and Syvash coasts (class *Festuco-Puccinellietea* Soó ex Vicherek 1973). These steppe communities, which are similar to desert steppes and develop on saline chestnut soils, are often discussed in combination with saline lands or solonchaks, and are not further considered here.

Within the Steppe zone and in the south of the Forest-Steppe zone, there are edaphically isolated steppe communities associated with riparian sand areas (class *Festucetea vaginatae* Soó ex Vicherek 1972). In these communities, the moisture deficit is combined with a lack of nutrients (mainly mobile nitrogen compounds) and constrains physical and chemical features of the substrate (acidity, mobility, looseness). These steppe communities are strongly affected by anthropogenic factors and require special protection.

Photo 5.1 Typical steppe landscape with sandstone outcrops of the Donets'k Upland (Lugans'k region) (Photo by M. Peregrym)

Typical steppe plant communities (Photo 5.1) are formed mainly on different types of chernozem and belong to the class *Festuco-Brometea* Br.-Bl. et Tüxen ex Soó 1947. They are characterized by a predominance of species of dry areas while the meadow steppe ecotopes typically have more mesophilic perennial herbs and rootstock grasses mixed with edificators that are mainly xerophilic grasses.

The western regions of Ukraine are dominated by extrazonal steppe communities belonging to Central-European alliance *Cirsio-Brachypodion pinnati* Hadač et Klika ex Klika 1951. They usually occur on rendzina soils and are closely related in their development to the forest edge communities of the class *Trifolio-Geranietea sanguinei* Th. Müller 1962 and shrub communities of the class *Rhamno-Prunetea* Rivas Goday et Borja Carbonell ex Tx. 1962. Within Ukraine it is the most mesophytic of steppe vegetation. They are well represented in Podillya, being the main steppe vegetation there, because of the subhumid climate and carbonate sediments, which are locally common. In Western Europe they are usually replaced by *Fragario viridis-Trifolion montani* communities (Korotchenko and Didukh 1997).

One of the most common associations of Podillya is the *Inuletum ensifoliae* Kozł. 1925. It often occurs on Transnistrian slopes with a high degree of surface washing and soil erosion; sometimes it occupies the entire middle and upper part of slopes. It sporadically occurs in the Bukovinian Carpathians (Abduloeva and Didukh 1999; Korotchenko 2005b, 2009; Korotchenko and Tokaryuk 2005). It is typical on intensely eroded, steep (40–70°), western and south-western slopes with a rubbly substrate on poor sod-carbonate soils. Its total projective canopy cover is low, 30–70% on average; tussocks are not common (5%) as the grass basis is formed by rootstock grasses. Single shrubs of *Chamaecytisus albus* (Hacq.) Rothm usually occur, but there is no evident shrub layer. Its floristic richness is about 20–25 species per 100 m^2.

Large areas with rich sod-carbonate soils in Western Podillya are occupied by the association *Thalictro-Salvietum pratensis* Medw.-Kornas 1959, mainly on slopes of different exposure and steepness (5–35°) and here and there also on flat tops of slopes and at the bottoms of shallow gullies (Didukh and Korotchenko 2003; Korotchenko 2004, 2005a, 2009; Korotchenko and Tokaryuk 2005). Its cover is quite high (80–100%). The grass-sedge base of the herbage is represented by *Carex humilis* Leys., *Briza media* L. and *Festuca pratensis* Huds. Forbs are many and a significant proportion are rootstock species. Its core is formed by xeromesophilic species: *Thalictrum minus* L., *Viola hirta* L., *Astragalus onobrychis* L., *Bupleurum falcatum* L., *Salvia pratensis* L., *Adonis vernalis* L., *Galium verum* L. and others. These communities are floristically rich: 38–55 species per 100 m^2, depending on their developmental stage.

The *Brachypodio pinnati-Seslerietum* (Klika 1929) Toman 1976 is very rare (Kukovytsia et al. 1994; Abduloeva and Didukh 1999; Didukh and Korotchenko 2000; Abduloeva 2002) and only occurs locally in Western Podillya, in the valley of the Dniester and the bottom of the Smotrych river valley, in hollows and mild, north-eastern slopes, or steep (60–70°) western slopes, on turf-carbonate rendzinas up to 40 cm thick, on carbonate rocks. They are rich in species (42–57 species per 100 m^2) and the cover is 60–100%. This is the most mesophytic steppe community in Ukraine.

The alliance *Galio campanulatae-Poion versicoloris* Kukovitsa, Movchart et Shelyag 1994 comprises the Western Podillya meadow steppes on sod-carbonate soils occupying the upper parts of very steep slopes (25–60°). The soil is a washed, poorly developed rendzina on carbonates, which often form outcrops in separate ledges. Total cover is 50–85%; turf content is rather low, 10–15%, and the grass canopy is developed unevenly due to slope steepness. It occurs predominantly on the "shelves" and thus has a mosaic distribution pattern. Species diversity ranges from 17 to 35 species per 100 m^2. The community is well represented in Western Podillya, though it never occupies large areas (Kukovytsia et al. 1994; Didukh and Korotchenko 2000).

Typical meadow steppes with a well-formed grass basis are the communities of the alliance *Fragario viridis-Trifolion montani* Korotchenko et Didukh 1997. They grow on slopes of varying steepness and exposure on typical or leached chernozem soils, where limestone and gypsum lie close to the surface. These plant communities represent the typical meadow-steppe vegetation of the Forest-Steppe zone of Ukraine, occurring rarely in the western part of the Forest-Steppe zone and sporadically in the north of the Steppe zone.

The association *Thymo marschalliani-Caricetum praecocis* Korotchenko et Didukh 1997 is the most mesophytic one. It is found mainly on mid-slopes of varying steepness and exposure, having a high grass cover (95–100%) and, with 33–51 species per 100 m^2, is floristically rich, involving a large number of mesophytic forbs species from the classes *Trifolio-Geranietea sanguinei* and *Molinio-Arrhenatheretea*

R. Tx. 1937. Their soils are medium-developed chernozems. These communities occur sporadically throughout the Forest-Steppe zone and in the northern regions they occupy more elevated areas and steeper slopes. They are confined to the lower slopes with podzolized, leached typical chernozem in southern districts of the Forest-Steppe zone (Korotchenko and Didukh 1997; Korotchenko and Fitsailo 2003). Moderate grazing does not negatively affect the floristic composition of this vegetation.

The most common stands of meadow steppes belong to the association *Salvio pratensis-Poetum angustifoliae* Korotchenko et Didukh 1997. This vegetation has a cover is 90–95% and a species richness of about 50 species per 100 m^2. Characteristic are the great variety of mesophytic and xeromesophytic forbs: *Salvia nutans* L., *Silene nutans* L., *Veronica chamaedrys* L., *Galium ruthenicum* Willd., *Trifolium montanum* L., *Filipendula vulgaris* Moench, *Ranunculus polyphyllus* Waldst. & Kit. ex Willd., *Thymus marschallianus* Willd., *Fragaria viridis* Duchesne and *Hypericum perforatum* L. et al. Rootstock grasses (*Festuca valesiaca* Gaudin and *Koeleria cristata* (L.) Pers.) play a lesser role than the edificator *Poa angustifolia* L. The communities occur on mild slopes (7–20°) of hollow systems with a well-developed chernozem. This is the typical meadow steppe association of the Forest-Steppe zone, growing mainly at the top of the steppe slopes, whereas on the margin with the Steppe zone they occupy the foot slopes, mainly of northern exposure (Korotchenko and Didukh 1997; Honcharenko 2000; Korotchenko and Fitsailo 2003).

The association *Stipetum pennatae* R. Jovanovic 1956 represents typical indigenous feather-grass communities that grow on slopes of different exposures. Stands have a cover of 80–90%, and are floristically rich with up to 55 species per 100 m^2, involving tall and shorter grasses and forbs and a significant number of rare and endangered species. The soil in these stands is matted due to the edificator role of *Stipa pennata* L. These communities occur sporadically in the southern part of the Forest-Steppe zone, and are rare on the boundary with the Steppe zone, mainly in hollow systems, where this natural steppe vegetation is well-preserved (Tkachenko et al. 1987; Korotchenko and Didukh 1997; Korotchenko et al. 2009a).

Shrub steppes of the association *Veronico austriacae-Chamaecytisetum austriaci* Korotchenko et Didukh 1997, with a cover of 90–100% and a high richness of about 50 species per 100 m^2, are mainly confined to northern, north-western and eastern slopes of up to 20°. The shrub layer of *Chamaecytisus austriacus* (L.) Link is 50–70 cm high. The main grasses are *Bromopsis riparia* (Rehmann) Holub, *Festuca valesiaca* Gaudin and *Poa angustifolia* L.. These communities sporadically occur throughout the Forest-Steppe zone (Bayrak 1997; Korotchenko and Didukh 1997; Abduloeva 2002). This association extends rapidly, occupying new areas of protected meadow steppe vegetation which were previously exposed to periodic mowing.

The most typical steppe communities, and the true vegetation core of the Steppe zone, belong to the alliance *Astragalo-Stipion* Knapp 1944 (Photo 5.2). These communities are also found in the southern regions of the Forest-Steppe zone. Compared to the plant communities of the *Fragario viridis-Trifolion montani*, the phytocoenoses of the *Astragalo-Stipion* occupy a much drier habitat, and species of *Stipa*, *Astragalus* and other leguminous plants play a more significant role in the vegetation.

Photo 5.2 Steppe communities of the alliance *Astragalo-Stipion* on a plain part of Crimea Peninsula (Crimea) (Photo by I. Korotchenko)

The association *Stipetum lessingianae* Soó 1948 represents the typical feather grass steppes of Ukraine. The plant communities are formed on different chernozem types of South Ukraine (Bayrak 1997; Korotchenko and Didukh 1997; Krasova and Smetana 1999). Communities are confined to slopes with south-eastern to south-western exposition and of different slope angle (15–40°). The sod content of the soil is rather high, 30–40%, due to the edificator role of *Stipa lessingiana* Trin. & Rupr. and co-domination of *Koeleria cristata* (L.) Pers . The shrub layer is well developed (cover 10–20%), consisting mainly of *Caragana frutex* (L.) K. Koch, which can be a co-dominant on flat hilltops. Cover is 80–90% and species richness varies from 19 to 34 species per 100 m^2.

The association *Thymo marschalliani-Crinitarietum villosae* Korotchenko et Didukh 1997 represents a group of true steppes, occurring on dry southern slopes, rarely on south-eastern slopes, of up to 25°, and sometimes on flat areas, on rich but eroded chernozem soils (Korotchenko and Didukh 1997; Krasova and Smetana 1999; Korotchenko et al. 2009a). Cover varies from 70% to 90% and species richness is high with 40 per 100 m^2. Some patches have a well-developed shrub layer of *Caragana frutex* (L.) K. Koch. The main grass is *Bromopsis inermis* (Leyss.) Holub.

The association *Vinco herbaceae-Caraganetum fruticis* Korotchenko et Didukh 1997 represents the typical shrub steppes of the Steppe zone of Ukraine. It occurs on southern, south-eastern, eastern and western slopes of up to 25°. Cover ranges

Photo 5.3 Steppe communities of the alliance *Festucion valesiacae* in the Biosphere Reserve "Askania-Nova" (Kherson Region) (Photo by I. Korotchenko)

from 70% to 100% and species richness is about 40 per 100 m^2. The shrub layer of *Caragana frutex* (L.) K. Koch and *Amygdalus nana* L. is well developed (cover 10–30%.). *Bromopsis riparia* (Rehmann) Holub. forms the grass basis and xerophytic forbs are well represented, with *Phlomis pungens* Willd., *Phlomis tuberosa* L., *Vinca herbacea* Waldst. & Kit. and others, indicating the southern nature of these steppes. The distribution of this association coincides with the natural habitat of *Caragana frutex*.

The alliance *Festucion valesiacae* Klika 1931 comprises matted grasslands and steppes, degraded due to excessive grazing (Photo 5.3). *Carex humilis* Leys., widespread in Western Europe and very resistant to grazing, along with *Festuca valesiaca* Gaudin, dominate these floristically poor, grazing-resistant communities on turf-carbonate washed rendzinas and chernozem soils.

Most typical of the species-rich communities of the *Festucion valesiacae* is the association *Festuco valesiacae-Stipetum capillatae* Sillinger 1930, which is common on the slopes of hollow systems with different slope exposures and slopes of 15–30° (Korotchenko and Didukh 1997; Abduloeva and Didukh 1999; Kontar 2000; Abduloeva 2002; Kuzemko 2004). This vegetation usually is extensively grazed but its cover remains high (85–95%). It consists of strong vegetation mats (30–40% cover), mainly formed by *Festuca valesiaca* Gaudin, *Koeleria cristata* (L.) Pers. and *Stipa capillata* L. Species richness is 23–48 species per 100 m^2. Stands of this association are among the least disturbed compared to other associations of the alliance. They occur throughout the Forest-Steppe and Steppe zone of Ukraine on hollow slopes with leached chernozem.

The distribution of the association *Festuco rupicolae-Caricetum humilis* Klika 1939 corresponds to the natural range of *Carex humilis* Leys. These are the most common, transformed meadow steppes of Western Podillya and along spurs of the Central Russian Upland, occurring on different slopes and soils under grazing pressure (Korotchenko and Didukh 1997; Abduloeva and Didukh 1999; Didukh and Korotchenko 2000; Abduloeva 2002; Homlya 2005). Cover ranges from 60% to 95% and species richness is 18–40 per 100 m^2. The stands, having well-developed vegetation mats and a high presence of xeromesophytic forbs, occur on sod-chernozems over carbonate limestone rocks (marl, limestone, sandstone), often with outcrops, sometimes on gypsum, or on eroded areas that were previously used in agriculture. The association represents the initial and intermediate stages of disturbance caused by overgrazing.

On eroded loess or chernozem soils, at the top or middle parts of slopes of varying steepness and exposure (5–40°), develop stands of the *Botriochloetum ischaemii* (Krist. 1937) Pop 1977 association. Grass cover ranges from 50% to 100% and the vegetation has 20–36 species per 100 m^2. Mats are well-developed, with rootstock species, such as *Bothriochloa ischaemum* (L.) Keng, *Poa angustifolia* L. and *Elytrigia repens* (L.) Nevski. There is sometimes a shrub layer of *Chamaecytisus austriacus* (L.) Link. Stands of this disturbed, secondary grassland are well represented on hollow slopes with leached chernozem in the Forest-Steppe and Steppe zones of Ukraine (Korotchenko and Didukh 1997; Abduloeva and Didukh 1999; Krasova and Smetana 1999; Kontar 2000; Abduloeva 2002; Kuzemko 2004).

The association *Festucetum valesiacae* (Solodkova et al. 1986) Tkachenko et al. (1987) represents the most anthropogenically altered community of the alliance *Festucion valesiacae* (Photo 5.3). This vegetation develops under constant and extensive grazing pressure and occupies the largest areas in the Forest-Steppe and Steppe zones of Ukraine (Korotchenko and Didukh 1997; Krasova and Smetana 1999; Kontar 2000; Kuzemko 2004; Korotchenko et al. 2009b). It occurs on slopes of different exposure and on plateaus. Soils are poorly developed chernozems on rendzinas, or loesses exposed to erosion processes. Slope steepness ranges from 10° to 45° and species richness (under extensive grazing) is 37 per 100 m^2. Cover depends on grazing intensity and ranges from 65% (under strong grazing) to 95% (under moderate grazing). There is no shrub layer. In this vegetation species with a wide ecological amplitude, able to withstand moderate anthropogenic pressure and disturbance (e.g., *Artemisia austriaca* Jacq., *Verbascum lychnitis* L., *Echium vulgare* L., *Nonea pulla* DC., *Medicago minima* (L.) Bartal., *Achillea millefolium* L., *Centaurea rhenana* Boreau, *Securigera varia* (L.) Lassen et al.) occur.

The alliance *Artemisio-Kochion prostratae* Soo 1964 includes rather xerophytic communities with a poorly developed sod layer and a number of typically southern species in their floristic composition (*Astragalus sulcatus* L., *Artemisia austriaca* Jacq., *Phlomis pungens* Willd. etc.). These communities are abundant in the southern part of the Forest-Steppe and Steppe zones in narrow strips on the well-drained but eroded hilltops, on poor, dry and leached chernozem soils (Bayrak 1997; Korotchenko and Didukh 1997; Korotchenko et al. 2009a). The most widespread are the plant communities belonging to the *Agropyro pectinato-Kochietum prostratae* Zolyomi 1958 corr. Soó 1959 association. They occur on steep slopes (25–45°) of southern, sometimes eastern exposure on well-washed sandy chernozem soils. Vegetation cover is rather diffuse (cover 60–80%), and sods are scarce. The association is dominated by xerophytic species, such as *Agropyron pectinatum* (M. Bieb.) P. Beauv., *Kochia prostrata* (L.) Schrad., *Galatella villosa* (L.) Rchb.f. and others.

The communities of the alliance *Artemisio marschalliani-Elytrigion intermediae* Korotchenko et Didukh 1997 develop under severe erosion and drainage. They occupy steep loess slopes of different exposure (steepness 35–45°), although sometimes they occur on intensely eroded plain areas with leached soil and on the lower parts of steep slopes of southern exposure. Cover is relatively high with 70–90%. A shrub layer is often present and well-developed, consisting of *Chamaecytisus austriacus* (L.) Link., sometimes together with *Cerasus fruticosa* (Pall.) Woronow. A characteristic feature of the association is the dominant position of rootstock grasses (*Elytrigia intermedia* (Host) Nevski, *Bromopsis inermis* (Leyss.) Holub, etc.), poor sod development and a patchy occurrence of forbs. These communities are poor in species and disturbed compared with the typical meadow-steppe. Characteristically, the floristic richness of the associations of this alliance decreases from south to north. They occur sporadically throughout the Forest-Steppe and Steppe zones of Ukraine (Korotchenko and Didukh 1997; Honcharenko 2000; Abduloeva 2002).

The communities of the alliance *Centaureo carbonatae-Koelerion talievii* Romaschenko, Didukh et V.Sl. 1996 are found on the southern spurs of the Central Russian Upland, in areas with chalk outcrops, which form different varieties of carbonate-rich chernozem soils. Communities occur on steep (35°) slopes of different exposure; still steeper chalk outcrops are occupied by plant communities of the class *Helianthemo-Thymetea* Romaschenko, Didukh et V.SI. 1996. Sometimes shrubs of *Caragana frutex* (L.) K. Koch are present. The communities have well-developed sod layers formed by *Carex humilis* Leys., *Stipa capillata* L., *S. pulcherrima* K. Koch and *Festuca valesiaca* Gaudin. Stands are floristically rich, combining typical steppe species (*Salvia nutans* L., *Thalictrum minus* L., *Adonis vernalis* L., *Jurinea arachnoidea* Bunge, *Oxytropis pilosa* (L.) DC., *Polygala sibirica* L.) and calciphylic endemic species (*Androsace koso-poljanskii* Ovcz., *Onosma tanaitica* Klokov, *Centaurea carbonata* Klokov and others). They occur on the banks of the Oskol and Siversky Dinets rivers and in hollow systems of their basins (Romashchenko et al. 1996; Korotchenko and Didukh 1997).

The Crimean Mountains and foothills are marked by a significant diversity of steppe vegetation types. In particular, the alliance *Carici humilis-Androsacion* Didukh 1983 comprises the primary petrophytic meadow steppes of the yailas (mountain pastures) in the western and central parts of the Crimean Mountains at altitudes of 300–1,500 m. On the higher yailas at an altitude of 600–800 m in the eastern part of Mountain Crimea, communities of the *Adonidi-Stipion tirsae* Didukh 1983 alliance occur (Didukh 1983; Didukh and Vakarenko 1984; Vakarenko 1997).

Those of the alliance *Veronici multifidae-Stipion ponticae* Didukh 1983 occur on the rich chernozems or stony eroded soils of piedmont Crimea. Main species are *Stipa ucrainica* P. Smirn., *Stipa poetica* Klokov and *Bromopsis cappadocica* (Boiss. & Balansa) Holub aggr. Also obligate petrophytes related to species of the Mediterranean 'tomillares' are part of the floristic composition of this vegetation.

5.2.2 Anthropogenic Influences on Steppe Vegetation

There are several forms of anthropogenic influence on the steppes in Ukraine, causing destruction and strong changes in the natural vegetation. Foremost is the destruction of the integrity of the soil cover as a result of ploughing, mining operations and excavations. There is no doubt that the two last types of human activities destroy steppes only locally, in places of minerals deposits or at important historical places. Generally, these processes were and are under the control of the governments of the former Soviet Union and Ukraine, but big steppes areas in Crimea, the Black Sea Lowland and the Donets'k Upland illegally were destroyed by "black" archeologists and miners of coal and sandstones, mining by hand, during the last 20 years. In some places, especially in the south of the Lugans'k region, this still occurs at a catastrophic scale.

Ploughing up virgin steppes began at the time when man adapted to settled life and mastered agriculture. But, it long had local effects in the Steppe zone of the modern Ukraine, because the Ukrainian steppes were very little inhabited till the eighteenth century. At that time the Russian Empire began its policy of consolidation of its southern borders, defending them against Mongol-Tatar and Turkish forays. In connection with this a lot of new settlements were founded in the Steppe zone, including the agriculture colonies of Germans, Swiss, Serbs and others. As a result the total area of tilled steppes gradually began to increase.

The Soviet Power saw extensive agriculture as the main vehicle for development of the country, with the result that during the years 1922–1941 and 1945–1960 all plain steppes in the territory of modern Ukraine were tilled completely. Even some reserves did not escape this common lot. N.S. Khrushchev (Head of the Communist Party of the Soviet Union) signed an ukase on ploughing up most of the "Askania-Nova" Reserve and declared that: "To give a Soviet citizen a notion about steppe, 300 ha is enough", a phrase that very good describes Soviet policy at that time.

Agriculture in the Steppe Zone fell into decay after the disintegration of the Soviet Union, and as result many formerly tilled lands gradually became fallow lands. During the first years these areas are occupied by annual weeds, but with lapse of time they are replaced by typical steppe plants (*Festuca* L. and *Stipa* L. species and others) or long-rhizomatous grasses (*Elytrigia repens* L. and others). Whether the fallow lands develop into secondary steppes or mesophytic weed formations depends on the presence or absence of steppe areas with their store of diaspores near these fallow lands. Today we can observe both regenerating steppes and formations dominated by *Elytrigia repens* on formerly tilled steppes.

Other forms of anthropogenic influence on steppes in Ukraine are mowing, grazing and burning. Mowing stimulates the vegetative development of a number of species (*Caragana frutex* (L.) K. Koch, *Chamaecytisus ruthenicus* (Fisch. ex Wol.) Klask.) that have underground stems which allow the formation of dense thickets.

Attempts to stabilize steppe ecosystems demonstrated that only the option of annual mowing prevented the accumulation of organic matter and thus can partially stabilize meadow steppe vegetation. Mowing once in 2 years ensures the rapid increase of the productivity of the vegetation, soon saturates the species richness, and increases the thickness of the layer of organic matter. Mowing 4 years successively and then skip 1 year is not unfavourable for the development of shrub meadow steppes with *Chamaecytisus ruthenicus* (Fisch. ex Wol.) Klask. The best option for these shrub steppe communities is mowing every 1 or 2 years. The density of *Chamaecytisus ruthenicus* is strongly related to the frequency of mowing and burning, as this shrub grows rapidly and profusely after the removal of aboveground biomass (Tkachenko 2004).

It was found that a mowing regime with any regularity makes it impossible to completely stabilize the meadow steppe ecosystem in the Forest-Steppe zone, while in some semi-desert steppes of the Black Sea biosphere reserve it was reported to be quite an adequate measure.

Controlled grazing by ungulates is currently poorly tested on different types of steppe communities due to financial difficulties and organizational problems. However, in some unprotected steppe areas spontaneous periodic cattle grazing can be observed. On the one hand it ensures the reduction of biomass and reduces litter accumulation. On the other hand it contributes (in case of overgrazing) to a reduction in species diversity and the development of depleted steppe communities of the *Festucion valesiacae* alliance with a significant participation of common ruderal species. Therefore the intensity and frequency of grazing must be controlled.

Fires are of great importance for steppe life, especially since the drastic decrease in the number of hoofed steppe animals. But its influence may be both positive and negative: it depends on the frequency and the season of the burn. Burning out the mulch in steppe formations once in 10–12 years in an autumnal burn has very favorable effects on the development of a xerophytic vegetation and prevents the development into a mesophytic type. More frequent and/or spring burning of the steppes reduces the numbers and population area of some species which leads to floristic impoverishment of the plant communities. This is illustrated in the values for the changes in area sizes, density and numbers of individuals of two *Tulipa gesneriana* L. populations in Table 5.1 (Peregrym et al. 2009b) (Photo 5.4).

Unfortunately, steppe burning is not controlled in Ukraine, so that the majority of areas, even protected ones, are on fire every year and sometimes both in spring and autumn. Mostly local people start these fires in order to clean the pastures from last year's herbs or for entertainment. It demonstrates the very low level of ecological understanding of a large part of the Ukrainian population.

It is interesting to note that there are several groups of steppe plants in the Ukrainian natural flora that are very much used by local populations for different aims. First of all these are the spring flowers (*Tulipa* species, *Crocus reticulatus*

Table 5.1 Area sizes, density and numbers of individuals of two *Tulipa gesneriana* L. populations during the years 2004–2009

Population	Time of observation	Area, ha	Average density, individuals/m²	Maximal density, individuals/m²	Numbers, individuals	Comments
Ukraine, Lugans'k region, Antratsyt district, near Malomykolaivka village, steppe slopes (*Festuco-Brometea*)	2004	0.2	0.39±0.25	12	≈800	Spring burning in 2004
	2008	0.24	2.37±1.73	12	≈5,000	Autumnal burning in 2007
	2009	0.24	1.86±1.04	12	≈4,500	Dry and cold spring in 2009
Ukraine, Lugans'k region, Lutugine district, near Rozkishne village, steppe slopes in Ploska Gully (*Festuco-Brometea*)	2004	0.035	0.98±0.75	15	≈300	–
	2008	0.007	2.00	3	35	Spring burning in 2008
	2009	0.004	0.53	1	21	Spring burning in 2009; dry and cold spring in 2009

Photo 5.4 Flowering of *Tulipa gesneriana* L. (*T. schrenkii* Regel.) on Kuyuk-Tuk Island after an autumn fire in the Azov-Syvash National Natural Park (Kherson region) (Photo by M. Peregrym)

Steven ex Adams, *Hyacinthella pallasiana* (Steven) Losinsk., *Adonis vernalis* L., *Pulsatilla pratensis* (L.) Mill. s.l. and others) which are collected in huge quantities for bouquets and sold. Another group concerns plants which are transplanted from the field to domestic front gardens or graves (e.g., *Paeonia tenuifolia* L., *Iris pumilla* L., *Ornithogalum* species and others). A last group concerns medical plants (*Adonis vernalis* L., *A. wolgensis* Steven ex DC., *Thymus* species, *Salvia* species and others). Many of these species, especially those from the first and second groups, are rare and vanishing plants in Ukraine and are included in the Red Data Book of Ukraine. They received this status as a result of the unlimited use of these resources by local people.

5.2.3 Conservation of Steppe Vegetation

5.2.3.1 Changes in Steppe Vegetation as a Result of Conservation

An attempt to effectively conserve steppe vegetation was the introduction of an absolute protection regime. It has both positive and negative aspects. The good side is the preservation of areas with true steppe vegetation from ploughing and terracing. The negative impacts are indirect and more complicated. It lies in the increase of plant litter accumulation and the subsequent mesophytization of the steppe.

Several important mechanisms of development of steppe vegetation were described by Tkachenko (2004). Under absolute protection steppe development goes together with a significant accumulation of organic matter. As a result nitrophilic plants with long rhizomes increase. Usually these species are the precursors of shrub-forest vegetation. The direction and speed of that succession depends on the moisture and thermal regime, and salt and nitrogen content, which are, in turn, affected by the stand structure as it develops and which may accelerate the succession process. Therefore, management of grasslands and steppes should decrease the accumulation of organic matter content and retard the succession process.

Long-term monitoring in the department of the Ukrainian Steppe Natural Reserve "Mykhaylivska Tsilina" demonstrated that absolute protection leads to constant expansion of rootstock grasses in all protected steppe areas and a decrease of the role of bunchgrasses. Bilyk (1957) concluded that the introduction of absolute protection in "Mykhaylivska Tsilina" has contributed to the prairiefication of the steppe ecosystems.

The conservation of steppe biodiversity requires a vast diversity of steppe ecosystems. Steppes need a diversification of use and protection regimes that would not lead to their destruction or degradation. Such an approach does not completely exclude the existence of areas with absolute protection, but their area should not exceed 25% of the reserve. A correct management is needed to maintain steppe vegetation within the Forest-Steppe zone. In the Steppe zone resources are sufficiently limited to prevent succession into scrub and forest and steppe communities can maintain their stability with no significant changes. This is particularly apparent for the grass steppes of the *Astragalo-Stipion* Didukh 1998.

Today, the optimal categories for the protection of steppe areas are preserves or sanctuaries. They make it possible to introduce different modes of land management. Grazing and mowing do not require the exclusion of lands from the protected area, and the preserve itself may cover an area of any size (from several to hundreds of hectares). At the same time, the disadvantage of preserves is that nobody is officially liable for their condition.

5.2.3.2 Steppes in Natural Reserve Network of Ukraine

Ukraine has the long history of steppe conservation which probably began in the Scythian time. In those times the first prototypes of steppe reserves appeared in the Steppe zone. Those were high places (hills, ridges, and mountains) with stone images which had a religious meaning and limited man's activities (Dzybov 1983).

One of the first reserves with the aim of conserving steppe vegetation was created in the South of the Russian Empire (now the Kherson region, Ukraine) in 1898 by the son of a German colonist, Friedrich Falz-Fein, in their estate "Askania-Nova" and had a total area of some 520 ha (Falz-Fein 1997). A little earlier (in 1894) a similar precedent aimed at the creation of steppe reserves had taken place in the territory of modern South-East Ukraine. Count Vorontsov-Dashkov, by consent of the Minister of Agriculture of the Russian Empire, allocated three steppe areas in the modern Lugans'k and Donets'k regions of Ukraine for V.V. Dokuchaev's investigations on steppes; from then on these areas were protected (Melnik 2000). It was one of the first and good examples of steppe conservation in Europe.

The number of state and local reserves in which steppe vegetation was under protection increased during the next decades in Ukraine, but the period from 1940 till 1960 was, for different political and economical reasons, and as a results of the Second World War, very unfavorable for nature conservation in the country. Today steppes in Ukraine are under government protection in 2 Biosphere Reserves, 11 Natural Reserves, 14 National Natural Parks, nearly 20 Regional Landscape Parks, and more than 100 state and local botanical, zoological, geological, or landscape reserves (Table 5.2). Besides, all these protected areas are part of the National Ecological Network of Ukraine, according to the laws on the National Ecological Network and also on the State Program for forming the National Ecological Network of Ukraine in the period 2000–2015. Main aim of this program is creation and development of a network of natural and semi-natural areas which will be connected by ecological corridors for the migration of animals and plants. Undoubtedly, the program promotes improvement of nature conservation, and the creation of new National Natural Parks and Regional Landscapes Parks during the last 10 years is an important gain. But bad financing of the National Program by the Government and problems with the allotment of lands for new reserves and ecological corridors has led to a breach of fulfillment with respect to the timescale for the development of the Network. This considerably diminished the positive effects of the Program.

Unfortunately, these numbers fall short of qualitative characteristics, and Ukrainian steppes still are in danger of vanishing. For example, some National Natural Parks only exist on paper (Dzharylgach National Natural Park, National Natural Park "Tuzlovs'ki Lymany"), while others (National Natural Park "Buz'kyi Gard", National Natural Park "Biloberezhzhia Sviatoslava") received only regional status and as yet have no Government support. Besides, many National Natural Parks don't own the land on which they are situated and there is therefore a great uncertainty of their implementation. Many other problems exist with state and local reserves, and a main problem for the majority of the reserves is that they aren't marked by any borders or marks whatsoever. As result local people often are unaware of the existence of local reserves near villages or cities and continue to exploit them unlimited.

Some valuable natural and semi-natural steppe territories still remain without protection, especially in the Kerch Peninsula in Crimea, and in the Lugans'k and Donets'k regions. These are not plain areas, but these steppes are situated on slopes of gullies, hills and mountains or in former military areas, and their plant cover is rich and natural. Besides, the results of floristic investigations by Moysiyenko and Sudnik-Wójcikowska (2006a, b) show that kurgans, or burial mounds, urgently need protection, because they are peculiar refugia of typical and rare steppe plants among the completely tilled steppes of the Black Sea Lowland. The urgency and importance of conservation of these steppe areas are very high, especially in the context of the coming new law on the sale of land in Ukraine as planned by the Government and Verkhovna Rada of Ukraine for 2012.

Table 5.2 List of Ukrainian Biosphere Reserves, Natural Reserves and National Natural Parks in which steppe vegetation is protected

	Protected territory	Region	Year of foundation	Total area, ha
1.	"Askania-Nova" Biosphere Reserve	Kherson	1898	33,307.6
2.	Chornomors'kyi Biosphere Reserve	Kherson, Mykolaiv	1927	89,129
3.	Dniprovs'ko-Orils'kyi Natural Reserve	Dnipropetrovs'k	1990	3,766
4.	Natural Reserve "Elanets' Steppe"	Mykolaiv	1996	1,675.7
5.	Kazantyp Natural Reserve	Crimea	1998	450.1
6.	Kaniv Natural Reserve	Cherkassy	1923	2,027
7.	Karadag Natural Reserve	Crimea	1979	2,872
8.	Crimea Natural Reserve	Crimea	1923	44,175
9.	Lugans'k Natural Reserve	Lugans'k	1968	2,122
10.	Natural Reserve "Medobory"	Ternopil'	1990	10,521
11.	Opuk Natural Reserve	Crimea	1998	1,592.3
12.	Ukrainian Steppe Natural Reserve	Donets'k, Sumy, Zaporizhzhia	1961	3,335.6
13.	Yalta Mountain-Forest Natural Reserve	Crimea	1973	14,523
14.	Azovo-Syvashs'kyi National Natural Park	Kherson	1993	52,154
15.	National Natural Park "Podils'ki Tovtry"	Khmelnyts'kyi	1996	261,316
16.	National Natural Park "Svyati Hory"	Donets'k	1997	40,609
17.	National Natural Park "Velykyi Luh"	Zaporizhzhia	2006	16,756
18.	Holosiivo National Natural Park	Kyiv	2007	4,525
19.	Slobozhans'kyi National Natural Park	Kharkiv	2009	5,200
20.	Pyriatyn National Natural Park	Poltava	2009	12,000
21.	Dzharylgach National Natural Park	Kherson	2009	10,000
22.	Dvorichna National Natural Park	Kharkiv	2009	?
23.	National Natural Park "Kremenets'ki Hory"	Ternopil'	2009	?
24.	National Natural Park "Buz'kyi Gard"	Mykolaiv	2009	6,138
25.	National Natural Park "Biloberezhzhia Sviatoslava"	Mykolaiv	2009	35,223
26.	National Natural Park "Tuzlovs'ki Lymany"	Odessa	2010	27,865
27.	Pryazovs'kyi National Natural Park	Zaporizhzhia	2010	78,127

5.2.3.3 Conservation of Rare Species and Associations of the Ukrainian Steppes

The history of conservation of rare plants and associations of the Ukrainian steppes is closely associated with the history of the Red and Green Listings in the former Soviet Union and modern Ukraine. The first list of rare and endangered species of the Ukrainian natural flora including steppe plants (*Eremurus spectabilis* M.Bieb., *Paeonia tenuifolia*, *Bulbocodium versicolor* (Ker.-Gawl.) Spreng. and others) was published by Kotov (1962, 1964), though data about the rarity of some species have been reported since the end of the nineteenth century (Taliev 1896, 1899; Lavrenko 1927, and others). Next Chopyk (1963, 1970, 1978) presented three editions of the list of rare and endangered species of the natural flora of Ukraine, in which rare steppe species (*Adonis vernalis* L., *Calophaca wolgarica* (L. f.) DC., *Otites graniticolus* Klokov and others) were analyzed in a separate chapter. These lists formed a good basis for the preparation of the Red Data Book of the Ukrainian SSR (Sytnik 1980) which got official status and was recommended as a tool for the development of effective methods of nature conservation. The Red Data Book covers 151 species of vascular plants including 32 steppe species (21.2%).

The publication of the second edition of the Red Data Book of Ukraine in 1996 (Shelyag-Sosonko 1996) was a turning point in the nature conservation in the country because that Red Data Book of Ukraine had the status of a National Law and imposes a ban on any use of rare plants listed in it. The total number of vascular plants in this edition was 439 species, including 121 steppe species (27.6%). The third edition, published in 2009, contained 611 species of vascular plants of which 145 or 23.7% are steppe species (Didukh 2009b).

An inventory showed that the majority of the rare and endangered steppe species have natural populations in the Ukrainian Reserves and National Natural Parks (Onyshchenko et al. 2002; Popovych 2002), though some (e.g. *Ornithogalum oreoides* Zahar., *Iris furcata* M. Bieb., *Astragalus sareptanus* A. Beck. and others) do not.

Species included in the IUCN Red List, the European Red List and the Annexes of the Bern Convention have special protection status in Ukraine. 41 species of the Ukrainian steppe flora (e.g., *Hyacinthella pallasiana* (Steven) Losinsk., *Çolchicum fominii* Bordz., 5 species of *Stipa* and others) are included in the IUCN Red List, corresponding to 38.7% of all species of the Ukrainian natural flora included in the list. The Annexes of the Bern Convention contain 63 species of the flora Ukraine of which 14 or 22.2% are steppe species (e.g., *Paeonia tenuifolia*, *Achillea glaberrima* Klokov, *Centaurea pseudoleucolepis* Kleopow and others). The European Red List includes 178 Ukrainian species of which 60 or 33.7% are steppe species (e.g., *Cymbochasma borysthenica* (Pall. ex Schlecht.) Klokov & Zoz, *Astragalus dasyanthus* Pall., *Elytrigia stipifolia* (Czern. ex Nevski) Nevski and others).

The first edition of the Green Data Book of the Ukrainian SSR was published in 1987. Main aim of the publication was the recommendation of rare and disappearing plant formations for conservation. It included 26 steppe formations (e.g., *Amygdaleta nanae*, *Elytrigieta stipifoliae*, and 13 formations dominated by *Stipa* species).

This book was worldwide the first that recommended the conservation of plant formations. The second edition of the Green Data Book of Ukraine of 2009 contained 25 herb and bush steppe formations and 8 xerophytic herb and bush formations of outcrops and sands. All these rare steppe formations occur within the Ukrainian Reserved Lands.

5.2.4 The Future of the Steppes

During the last decade agricultural activity has conspicuously decreased throughout the territory of Ukraine. This is seen in the decrease of livestock numbers and, hence, in grazing intensity. For the steppe ecosystems this is a rather positive trend as it leads to their recovery. However, complete cessation of grazing is negative, as it contributes to the accumulation of organic matter and causes mesophytization and promotion of rootstock grasses instead of sod grasses.

A second trend in modern agriculture in Ukraine is the decrease of tilled field area, and the increase of fallow field area. These fallows can gradually develop into steppe vegetation when sufficient steppe plant diaspores are available and soil types are suitable. Such "renewable" areas are important in the development of an ecological network in Ukraine.

Large areas of steppe are currently threatened in yailas in mountain and piedmont Crimea. This is due to the intense construction in this region, massive unauthorized seizure of lands by the Crimean Tatar population and unorganized development of non-sustainable tourism.

Global Climate Change also is expected to have an affect on the ecosystem level. The expected drying of the climate will promote the decomposition of humus in the Ukrainian steppes and as a result mesophytic and more nitrophilous species (meadow and bush plants) may increase and some steppes may change to forest. Besides, leaching will decrease the quality of humus in the soils, and in fact this has been observed in Ukrainian steppe reserves since 1985 (Didukh 2009a). As a result of this, steppes may be transformed to meadow and forest communities when strictly protected, but anthropogenic influences may re-direct these processes and lead to invasions of foreign species. Global Climate Change also may affect the distributional ranges of species, e.g. with species with broad ecological amplitudes increasing and those with very narrow ecological amplitudes decreasing.

5.2.5 Governmental Policy on Steppes in Ukraine

Solving several problems at Government level, together with the continued development of the National Ecological Network and the creation of new protected territories, will give opportunities for recultivation and provide solid, effective and long-term conservation of the steppes of Ukraine.

Problem 1. Steppes do not constitute a category on their own in the National Cadastre of Lands of Ukraine but belong to the category "Pastures and eroded lands". This makes it impossible to accurately calculate the territorial extent of the steppe area. As a consequence steppes are not taken into account during the preparation of new state laws, programs and also during allotment of lands for private or state needs. Therefore, inclusion of the new category "Steppes" in the National Cadastre of Lands of Ukraine, followed by an accurate inventory of all steppe areas, is an important target.

Problem 2. The State Program "Forests of Ukraine" was begun by the Ukrainian Government in 2002. Main aim of the program is to increase forests cover to 19–20% of the country and reforming Ukrainian Forestry according to European standards by 2015. Ex facte, it is a very good program, but unexpectedly its basic realization was begun in the Steppe zone. Foresters resolved to create new pine forests on steppic slopes of gullies, hills and mountains. As a consequence, many steppes areas were ploughed up for new forest plantations, but the new forests didn't grow because the young pines perished under the steppe conditions which are atypical for the pines. Now several public organizations and initiative groups of scientists try to reach out to the Ukrainian Government calling for changing this situation (Peregrym et al. 2009a). Results so far are none, because practice has shown that the Program "Forests of Ukraine" is a good vehicle for corruption at different state levels. Thus, urgent changes in the State Program "Forests of Ukraine" on a legislative level are a necessary condition for the conservation of the Ukrainian steppes, as the Program presently has disastrous effects.

Problem 3. As yet, there is no purposeful state program on the sustainable use, conservation and recultivation of steppes whatsoever, in contrast to the program for the other major zonal type of vegetation in Ukraine (forests). This demonstrates that steppe areas have no priority in the National Policy at present. Undoubtedly, the development and approval of a state steppe program will promote the more effective use and conservation of these unique areas in the country.

References

Abduloeva OS (2002) Towards a syntaxonomy of xerophytic herb vegetation of the western Forest-Steppe (Prydniprovs'ke Podillia, Tovtrovyi Kryazh, Kremenets' Mountains). Ukr Phytosociol Collect Ser A 1:124–144 (in Ukrainian)

Abduloeva OS, Didukh YaP (1999) Meadow-Steppe vegetation of eroded slopes of the Prydnistrov'e (National Natural Parks "Podolian Tovtras") in aspect of its conservation. Ukr Phytosociol Collect Ser A 3:10–36 (in Ukrainian)

Baranov VI (1954) Stages of development of flora and vegetation of the USSR in the Tertiary period. Part 3. Results of studying fossil Tertiary floras and problems of relicts in modern vegetation of the USSR. Scientific notes of the V.I. Ulianov-Lenin Kazan' State University, Ser. Botany, vol 114, Book 4. Kazan, 364 pp (in Russian).

Bayrak OM (1997) Rare plant communities of the Left-bank Dnieper and state of its protection. Ukr Phytosociol Collect Ser A 2:19–27 (in Ukrainian)

Bezusko LG, Bezusko AG, Mosyakin SL, Kotova NS (2006) Palynological characteristics of the eneolithic deposits from the Razdol'ne multilayer settlement site (Donetsk region, Ukraine). Ukr Bot J 63:783–793 (in Ukrainian)
Bilyk GI (1957) Vegetation of "Mykhailivs'ka Tsilyna" Reserve and its transformation under influence of human activity. Ukr Bot J 14(4):26–39 (in Ukrainain)
Chopyk VI (1963) Rare plants of the Ukrainian SSR and its conservation. Naukova dumka, Kyiv, 45 pp (in Ukrainian)
Chopyk VI (1970) Rare plants of Ukraine. Naukova dumka, Kyiv, 188 pp (in Ukrainian)
Chopyk VI (1978) Rare and endangered plants of Ukraine. Naukova dumka, Kyiv, 212 pp (in Russian)
Didukh YaP (1983) Experience of classification of xerophytic suffrutescent and herbaceous vegetation of the Mountain Crimea. Bot J 68:1456–1466 (in Russian)
Didukh YaP (1992) Plant cover of the Crimea Mountains (structure, dynamics, evolution and conservation). Naukova dumka, Kyiv, 256 pp (in Russian)
Didukh YaP (ed) (1998) Comparative analysis of plant diversity of reserved steppe ecosystems of Ukraine with aim of optimization regimes of its conservation. M.G. Kholodny Institute of Botany of NAS of Ukraine, Kyiv, 75 pp (in Ukrainain)
Didukh YaP (ed) (2000) Ecoflora of Ukraine, vol 1. Phytosociocentre, Kyiv, 284 pp (in Ukrainian)
Didukh YaP (2009a) Ecological aspects of Climate Global Change: causes, results, actions. Herald NAS Ukr 2:34–44 (in Ukrainian)
Didukh YaP (ed) (2009b) The red data book of Ukraine. Plant kingdom. Globalconsulting Press, Kyiv, 912 pp (in Ukrainian)
Didukh YaP, Korotchenko IA (2000) Classification of steppe vegetation of the Pokuttia. Ukr Phytosociol Collect Ser A 1:3–15 (in Ukrainian)
Didukh YaP, Korotchenko IA (2003) Xeroterm vegetation of the north-west Podolian Upland. Herald L'viv Univ 34:82–91 (in Ukrainian)
Didukh YaP, Vakarenko LP (1984) Comparative analysis of syntaxons of floristic classification of steppes and outcrops of the Mountain Crimea. Ukr Bot J 41(3):11–20 (in Ukrainian)
Dubovik ON (ed) (2005) Florogenesis of the Crimean-Neorossicus Province. "Fyton" Press, Kyiv, 180 pp (in Russian)
Dzybov DS (1983) About strategy of conservation of rare species of plants and plant vegetation in districts of developed agriculture and animal husbandry. Protection of genofond of natural flora. Nauka Press, Novosibirsk, pp 28–36 (in Russian)
Falz-Fein V (1997) Askania-Nova. Agrarna Nauka, Kyiv, 350 pp (in Russian)
Homlya LM (2005) Vegetation of the Khorol River valley. Ukr Phytosociol Collect Ser A 1, 187 pp (in Ukrainian)
Honcharenko IV (2000) Coenotic diversity of herb type of vegetation of Sumy geobotanical region. Ukr Phytosociol Collect Ser A 1:117–131 (in Ukrainian)
Kleopov YuD (1933) Plant cover of the south-west part of the Donets'k Upland (former Stalin region). Herald Kyiv Bot Gard 15:9–162 (in Ukrainian)
Kleopov YuD (1941) "Near glacial" steppes in the European part of the USSR. 1. The relict section Caricion humilis. Scientific notes of the Kharkiv University, vol 22, pp 167–183 (in Russian)
Klokov MV (1963) Basic stages of development of the plain flora of the European part of the USSR. Materials about the history of the flora and vegetation of the USSR, vol 4. Academy of Sciences of the USSR Press, Moscow-Leningrad, pp 376–406 (in Russian)
Klokov MV (1981) Psammophyte floristic complexes in the territory of the Ukrainian SSR. News of taxonomy of vascular plants in 1979. Naukova dumka, Kyiv, pp 90–150 (in Russian)
Kontar IS (2000) Syntaxonomy of vegetation of crystal rocks outcrops of Forest-Steppe of Ukraine. II. Classes Festuco-Brometea and Sedo-Scleranthetea. Ukr Phytosociol Collect Ser A 1:16–28 (in Ukrainian)
Korotchenko IA (2004) Steppe vegetation of the south part of National Natural Park "Podolian Tovtras", Scientific Herald of Chernivtsi University, vol 223, Biology. Ruta Press, Chernivtsi, pp 197–221 (in Ukrainian)

Korotchenko IA (2005a) Peculiarities of distribution of communities of alliance Cirsio-Brachypodion pinnati Hadač et Klika 1944 em Krausch 1961 in the territory of Ukraine, Collection of scientific publications "Advances in Botany and Ecology", vol 1. Phytosociocenter, Kyiv, pp 133–138 (in Ukrainian)

Korotchenko IA (2005b) Steppe vegetation of Chornyi Potik karst-speleological reserve (Chernivtsi region) and its conservation. In: Materials of conference "Academician M.M. Grishko's scientific heritage". Hlukhiv State Pedagogical University Press, Hlukhiv, pp 80–82 (in Ukrainian)

Korotchenko IA (2009) Steppe vegetation of the Right-Bank Forest-Steppe of Ukraine. In: Proceedings of International scientific conference "Vegetation of East Europe: classification, ecology and protection". Ladomir Press, Briansk, pp 110–113 (in Russian)

Korotchenko IA, Didukh YaP (1997) Steppe vegetation of the south part of the Laft-Bank Forest-Steppe of Ukraine. II. Class Festuco-Brometea. Ukr Phytosociol Collect Ser A 1:20–39 (in Ukrainian)

Korotchenko IA, Fitsailo TV (2003) Steppe vegetation of the Kyiv plateau, Scientific notes. Natural sciences, 21. "KM Academia" Press, Kyiv, pp 20–35 (in Ukrainian)

Korotchenko IA, Tokaryuk AI (2005) Floristic and ecological peculiarities of steppes of the Bukovyns'ke Prykarpattia. Conserv Ukr 11(2):1–9 (in Ukrainian)

Korotchenko IA, Mala YuI, Fitsailo TV (2009a) Syntaxonomy of steppe vegetation of the south of the Right-Bank Forest-Steppe of Ukraine. Sci Notes. Biol & Ecol 93:54–69 (in Ukrainian)

Korotchenko IA, Mala YuI, Fitsailo TV (2009b) Syntaxonomy of steppe vegetation of the north of the Right-Bank Forest-Steppe of Ukraine. Scientific Herald of Chernivtsi University. Biology (Biological systems). Chernivtsi National University Press, Chernivtsi. 1:73–84 (in Ukrainian)

Kotov MI (1962) Rare, endemic and endangered species of plants of the Ukrainian SSR and necessity of its conservation. Protect Conserv Business USSR 7:50–53 (in Russian)

Kotov MI (1964) Rare and endemic plants of the flora of Ukraine and necessity of its conservation, To protect the native nature, 3. Urozhai Press, Kyiv, pp 5–12 (in Ukrainian)

Krasova OO, Smetana MG (1999) Steppe vegetation of Kobylna gully. Ukr Phytosociol Collect Ser A 1–2:21–30 (in Ukrainian)

Kukovytsia GS, Movchan YaI, Solomakha VA, Shelyag-Sosonko YuR (1994) Syntaxonomy of meadow steppes of the Western Podolian Upland. Ukr Bot J 51(2–3):35–48 (in Ukrainian)

Kuzemko AA (2004) Steppe vegetation of the Ros' River valley, Collection of scientific publication "Steppe and halophilic ecosystems of Ukraine". Institute of Botany of NAS of Ukraine, Kyiv, pp 197–217 (in Ukrainian)

Lavrenko EM (1927) List of rare plants which were found in borders of Taganrog and Mariupol' districts between Mius and Kalmius Rivers. News Kharkiv Agric Instit 8/9:41–48 (in Ukrainian)

Melnik VI (2000) Rare species of flora of plain forests of Ukraine. Phytosociocenter, Kyiv, 212 pp (in Ukrainian)

Mosyakin SL, Bezusko LG, Mosyakin AS (2005) Relicts, refugia, and migration routes of plants in Europe in the Pleistocene – Holocene: a brief overview of phylogeographic evidence. Ukr Bot J 62:777–789 (in Ukrainian)

Moysiyenko II, Sudnik-Wójcikowska B (2006a) The flora of Kurgans in the Desert Steppe Zone of Southern Ukraine. Chornomors'kiy Bot J 2(1):5–35 (in English)

Moysiyenko I, Sudnik-Wójcikowska B (2006b) The flora of Kurgans in the Steppe Zone of Southern Ukraine – Phytogeographic and ecological aspects. Pol Bot Stud 22:387–398 (in English)

Onyshchenko VA, Andrienko TL, Ostapko VM et al (2002) Representative of rare species of vascular plants in National Natural Parks of Ukraine. Ukr Bot J 59:476–485 (in Ukrainian)

Paczoski JK (1910) Main features of development of the South-West Russia flora. Kherson, 430 pp (in Russian)

Peregrym M, Moysiyenko I, Kolomiychuk V, Zagorodnyuk I (2009a) Is it possible that the President of Ukraine will help steppes? Steppe Bull 27:30–34 (in Russian)

Peregrym M, Moysiyenko I, Peregrym Yu, Melnyk V (2009b) Tulipa gesneriana L. (Liliaceae) in Ukraine. Kyiv University Press, Kyiv, 144 pp (in Ukrainian)
Pidoplichko IG (1946–1956) About the ice ages, vol 1–4. Kyiv University Press and Academy of Sciences of the Ukrainian SSR Press, Kyiv, in Russian
Popovych SYu (eds) (2002) Catalog of rare biodiversity of Reserves and National Natural Parks of Ukraine. Phytosociocenter, Kyiv, 276 pp (in Ukrainian)
Romashchenko KYu, Didukh YaP, Solomakha VA (1996) Syntaxonomy of Class Helianthemo-Thymetea cl. nov. of chalky outcrops vegetation of north-eastern Ukraine. Ukr Phytosociol Collect Ser A 1:49–62 (in Ukrainian)
Ruprecht FI (1866) Geobotanical researches about chernozem with map of distribution of chernozem in the European Russia, Attachment to vol 10 of the Notes of Academy of Science, 6. Academy of Science Press, St. Petersburg (in Russian)
Shelyag-Sosonko YuR (ed) (1996) The Red data book of Ukraine. Plant Kingdom. Ukrains'ka entsyklopedia, Kyiv, 608 pp (in Ukrainian)
Sytnik KM (ed) (1980) The Red data book of the Ukrainian SSR. Naukova Dumka, Kyiv, 504 pp (in Ukrainian)
Taliev VI (1896) Chalky pine forests of Donetsk and Volga Basins. Kharkiv, 58 pp (in Russian)
Taliev VI (1899) Materials for botanical-geographic description of the Donetsk Upland. In: Proceedings of Society of Naturalists of Kharkiv University 34:167–278 (in Russian)
Tkachenko VS (2004) Phytocenotic monitoring of reservation successions in the Ukrainian Steppe Natural Reserve. Phytosociocenter, Kyiv, 184 pp (in Ukrainian)
Tkachenko VS, Movchan YaI, Solomakha VA (1987) Analysis of syntaxonomical changes in meadow steppes of "Mykhailivs'ka Tsilyna" Reserve. Ukr Bot J 44(2):65–73 (in Ukrainian)
Udra IF (1988) Distribution of plants and questions in paleo- and biogeography. Naukova dumka, Kyiv, 196 pp (in Russian)
Vakarenko LP (1997) Steppe and outcrop communities near of the Mountain Crimea. Ukr Phytosociol Collect Ser A 1:101–109 (in Ukrainian)
Zaverukha BV (1985) The Volyno-Podolian flora and its genesis. Naukova dumka, Kyiv, 192 pp (in Russian)

Chapter 6
Cessation of Traditional Management Reduces the Diversity of Steppe-Like Grasslands in Romania Through Litter Accumulation

Eszter Ruprecht

Abstract Romania still has extensive dry grasslands that are outstanding in diversity and conservation status compared to European standards. Land-use change is one of the greatest threats to biodiversity and conservation of these grasslands. Litter accumulation was proved to have a prime role in governing community processes in dry grassland following abandonment by affecting the regeneration from seed of constituent species. Accumulated litter reduces bare soil surface and lowers light availability below adequate levels for seedling emergence and thus decreases microsite quantity and quality. Besides these mainly negative physical effects, shown in experiments, it was found that there can be a chemical pressure on germinating seeds as well, since plant leaves of one of the most common dominant species in abandoned sites (*Stipa pulcherrima*) contains allelopathic substances with documented inhibiting effect on different processes related to regeneration from seeds of co-occurring grassland species. Because of the mainly negative effects of litter on seed germination, re-introduction of a management regime which comprises litter removal, e.g. mowing or grazing, can restore the plant diversity and open vegetation structure of the dry steppe-like grasslands in Romania, and probably elsewhere.

6.1 Continental Steppe-Like Grassland in Romania

Species-rich dry grassland communities still occur in low-intensity farming systems in many European countries. Their characteristic species are adapted to human land use, and grassland diversity is strongly related to past and current management practices (Poschlod and WallisDeVries 2002).

In Romania, a large variety of grassland types exist (Sanda et al. 2008; Doniţă et al. 1992), and their occurrences are determined mostly by geographic position,

E. Ruprecht (✉)
Hungarian Department of Biology and Ecology, Babeş-Bolyai University, Cluj Napoca, Romania
e-mail: eszter.ruprecht@ubbcluj.ro

M.J.A. Werger and M.A. van Staalduinen (eds.), *Eurasian Steppes. Ecological Problems and Livelihoods in a Changing World*, Plant and Vegetation 6,
DOI 10.1007/978-94-007-3886-7_6,

geomorphology, soil type, and land use. In some regions, e.g. in the Transylvanian Basin (central Romania), extended dry grasslands still exist that are outstanding in diversity and conservation status compared to European standards. An international team sampled Romanian dry grassland stands dominated by *Stipa capillata*, *S. lessingiana*, *S. pulcherrima*, *S. tirsa*, *Bothriochloa ischaemum*, *Brachypodium pinnatum*, *Briza media*, *Bromus erectus*, *Festuca rupicola*, *F. pallens*, *Helictotrichon decorum*, *Sesleria heuflerana*, as well as *Carex humilis* and *C. tomentosa*, and analysed their scale-dependent diversity patterns. In the Romanian grasslands very high species richness values were found at all spatial scales (0.0001–100 m^2), being similar to values previously recorded from semi-dry grasslands in the White Carpathians or from alvar grasslands in the hemiboreal zone. The highest richness values occurred in semi-dry basiphilous grasslands dominated by *Brachypodium pinnatum*, and it appears that the maximum diversity values at 0.1 m^2 (43 vascular plant species) and at 10 m^2 (98) are possibly the highest ever recorded in any plant community worldwide (Wilson et al. in press; Todorova et al. in preparation).

Steppe-like grasslands in Romania are of a special interest because they are considered to be relic vegetation types from the Late Glacial that escaped Holocene woodland invasions (Illyés and Bölöni 2007; Kuneš et al. 2008). For this reason, they harbour many plant species of eastern and southern origin as well as several endemic taxa. Such vegetation types are typical for steep, south-facing slopes with eroded carbonate chernozemic soils on clayish or marly substrate in the Transylvanian Basin, where the climate is temperate continental, with an annual precipitation of 520–690 mm with a summer peak (June, July, August), and a mean annual temperature of 8.4–9.7°C (Kun et al. 2004). This region represents the westernmost outpost of the distribution for many species of Siberian or Pontic origin (e.g. *Nepeta ucranica*, *Cephalaria uralensis*, *Centaurea ruthenica*, *Peucedanum tauricum*; see Soó 1942). These grasslands represent an important link between eastern European grasslands dominated by feather grass species (*Stipa* spp.) and central European dry and semi-dry grasslands (Photos 6.1 and 6.2). In addition, due to their high species richness and the occurrence of rare and endangered species, continental dry steppe-like grassland belong to the priority habitat type 62C0 (Ponto-Sarmatic steppes) of the Habitats Directive of the European Union (European Commission 2007; Gafta and Mountford 2008).

This conservation interest revealed a serious lack of data regarding the actual state and vegetation dynamics caused by changes in management of these xeric grassland types in the Transylvanian Basin (Cremene et al. 2005; Enyedi et al. 2008; Ruprecht et al. 2009).

6.2 Land-Use Changes and Their Effects on Steppe-Like Grassland

Traditional management of steppe-like grasslands in the Transylvanian Basin was grazing by sheep or cattle from spring to autumn. Through the last 50 years, land-use has changed in this region, greatly influencing several dry grassland sites

Photo 6.1 Species-rich steppe-like grassland dominated by feather grass species (*Stipa* spp.) in the Transylvanian Basin (Photo by Anna Szabó)

Photo 6.2 On shallow soils *Salvia nutans* colours the species-rich steppe-like grasslands (Photo by Anna Szabó)

(Enyedi et al. 2008; Ruprecht et al. 2009). Livestock numbers have decreased due to the low profitability, which generated a large-scale abandonment of grassland areas. Because these grassland types have the lowest productivity (600 $g\,m^{-2}$ $year^{-1}$ dry matter) among the region's grasslands, they were the first to be abandoned. Likewise, some stands have been converted to plantations of non-indigenous species, such as *Pinus nigra* or less often *Robinia pseudoacacia* for both profit and erosion control, others are periodically burned to control for litter accumulation and shrub encroachment, while many of the stands are long-term unused and abandoned from any management.

At a global scale, land-use change is one of the greatest threats to biodiversity and ecosystem functioning, and restoration interventions often fail, both because of abiotic, e.g. atmospheric nutrient deposition and acidification, and biotic constraints, e.g. under-representation of target species in the seed bank and dispersal drawbacks in a fragmented landscape (Bakker and Berendse 1999).

Both with agricultural abandonment and with intensification of agricultural use, plant diversity in semi-natural grassland communities dramatically changes (Luoto et al. 2003; Wellstein et al. 2007a). For example, abandonment of a pasture adapted to grazing can have significant effects on the composition of the vegetation. The often experienced decrease in species numbers is mainly due to (1) an increasing cover and biomass of a few highly productive species, which exert strong competitive effects on subordinate species, and (2) the accumulation of dead plant remains, i.e. litter, which interfere with seed germination and seedling establishment (Bosy and Reader 1995; Kahmen et al. 2002).

In line with the above mentioned general trends, in a study on steppe-like grasslands in the Transylvanian Basin, Enyedi et al. (2008) found that grazed and long-term abandoned sites differ in their vegetation structure, species composition, and diversity. Grazed stands have a more open structure (with 20% bare soil surface) and harbour more species than abandoned stands. Especially, rosette-forming and hemi-rosette species (e.g. *Hieracium praealtum* subsp. *bauhinii*, *Sanguisorba minor*, *Leontodon crispus*, *Astragalus monspessulanus*, *Stipa lessingiana*, *Potentilla cinerea*, *Koeleria macrantha*, *Festuca rupicola* and *Dichanthium ischaemum*) are more abundant in grazed stands, while abandoned stands are characterised by a strong dominance of *Stipa pulcherrima*, a closed structure (3% bare soil surface), higher portion of erosulate species (e.g. *Teucrium chamaedrys*, *Vinca herbacea*, *Galium verum* and *Coronilla varia*), litter accumulation (30% litter cover compared to no litter cover in case of grazed stands) and lower species richness.

6.3 Community Processes in Dry Grassland Following Abandonment: The Prime Role of Litter in Affecting Seed Germination

The regeneration of plants from seeds is a process that can be broken down into several conditional components, such as reproductive fertility and seed production, seed viability, germination, seed resource-based initial seedling growth (radicle elongation, leaf expansion), autotrophic seedling growth and establishment. As such, it is a particularly vulnerable stage in the life cycle of many plant species, and the suppression of seedling recruitment can have serious consequences for population viability and species diversity (e.g. Tilman 1993; Zobel et al. 2000). It is known that litter affects seed germination and seedling establishment and thus may determine the fate of species in a community or govern community processes (reviewed by Facelli and Pickett 1991; Xiong and Nilsson 1999). Litter effects are complex, i.e. physical (e.g. affecting light availability, soil humidity, temperature and its amplitude, protecting the seeds from predators, acting as a mechanical barrier for establishing seedlings) and chemical (e.g. affecting soil nutrient content, releasing secondary metabolic or decay compounds with potential toxic effect upon species), and their outcome depends on a delicate balance between facilitative and inhibitory effects, determined by environmental conditions and life history characteristics of species.

Litter accumulation may reduce the availability of suitable microsites for recruitment (safe sites) and thus regeneration possibilities for most plant species (Bergelson 1990). Expectedly, dry grasslands are more prone to negative effects of litter accumulation due to their low productivity and open structure. Typical plant species of these habitats are adapted to a high proportion of bare soil, high levels of irradiation as well as high temperatures at the soil surface, and are thus expected to be especially sensitive to changes in micro-environmental conditions caused by the accumulation of plant litter.

Several studies have found that reintroducing disturbance is favourable for seedling establishment through the increased availability of safe sites for recruitment (e.g. Hofmann and Isselstein 2004; Eskelinen and Virtanen 2005). In a 2-year field experiment we assessed the effect of litter removal alone and in combination with vegetation cutting on natural seedling emergence and seedling survival in two, long-term abandoned steppe-like grassland stands in the Transylvanian Basin (Ruprecht et al. 2010a). Since Foster and Gross (1998) have found that litter and living above-ground biomass have additive effects on seedling emergence, we expected a higher impact of a combination of litter removal and vegetation cutting than of litter removal alone. Treatments were applied in 1 m × 1 m main plots with eight replications and seedling establishment was observed in two 25 cm × 25 cm subplots designated within each main plot.

Our treatments increased the percentage of bare soil and the availability of light on the soil surface compared to the control. In accordance with altered physical conditions, the treatments significantly affected the emergence of dry grassland species, and the effect was consistent across sites and during the two study years, however, the number of seedlings appearing in the first year (3,383 seedlings in 2006) was twice as high as in the second (1,560 seedlings in 2007). Litter removal alone or in combination with vegetation cutting significantly increased the cumulative seedling number compared to the control in the first year of the study, and this effect was close to significance in the second year (Fig. 6.1). This difference may be attributed to differences in weather conditions between the 2 years, since 2006 was a rainy and 2007 a dryer year, or traced back to the high germination values from the seed bank already in the first study year (Ruprecht et al. 2010a).

Litter removal alone had no effect on seedling survival compared to the control in any of the two sites and 2 years, and in turn, seedling survival was significantly enhanced by the combination of the treatments (litter removal and vegetation cutting). Contrary to our expectations that under harsh environmental conditions, as

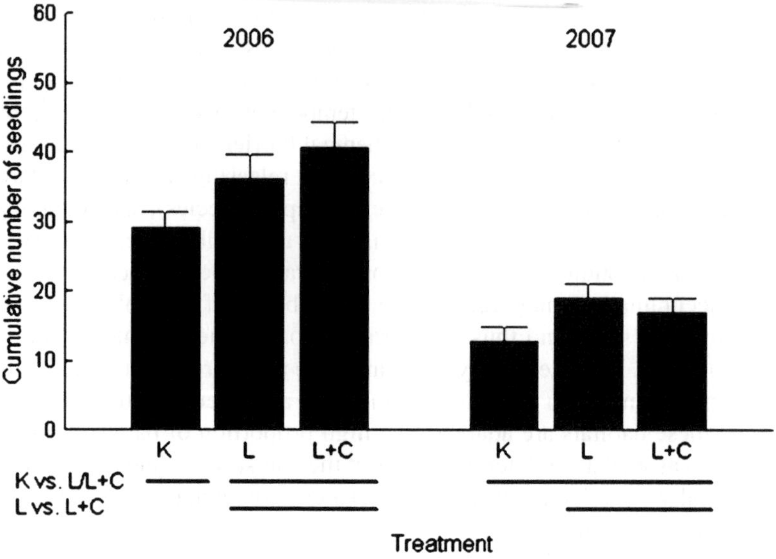

Fig. 6.1 Results of a 2-year (2006, 2007) field experiment conducted in two, long-term abandoned steppe-like grassland sites (Suatu, Puini) in the Transylvanian Basin, Romania. Effect of applied treatments (K = control, L = litter removal, L + C = litter removal and vegetation cutting with clippings removed) on the cumulative number of seedlings (mean ± SE) appeared in the two subplots (25 cm × 25 cm) of each experimental main plots (n = 8). Results of a factorial ANOVA. Lines below the bars indicate significance as revealed by orthogonal contrasts between the treatments: K vs. L and L + C, L vs. L + C; the interruption in the line of K vs. L/L+C for 2006 indicates a significant difference at $p < 0.05$; for 2007 this difference was close to statistical significance (Based on Ruprecht et al. 2010a)

in case of the studied dry grassland system, litter may exert facilitative effects and increase seedling survival through increase of water availability (e.g. Eckstein and Donath 2005; Ruprecht et al. 2010b) and by attenuating extremes in soil temperature (e.g. Weltzin et al. 2005), our results indicate that denser living and dead biomass had an additive negative effect on seedling survival.

As shown by monthly observations over these 2 years (Ruprecht et al. 2010a), field germination of dry grassland species occurred mostly in wet periods during the year, with a highly synchronised germination peak in early spring (March or April). Seedlings counted and determined during the 2 years belonged to 72 grassland species, and of those general grassland species, such as *Teucrium chamaedrys*, *Dorycnium pentaphyllum* subsp. *herbaceum*, *Asperula cynanchica*, as well as typical dry-grassland species of conservation interest, such as *Veronica spicata* subsp. *orchidea*, *Campanula sibirica*, and *Dichanthium ischaemum*, were the most abundant. Germination of the majority of species (74%) was favoured by one (22%) or both (52%) treatment types at least at one of the sites. Additionally, there were five species, three of them typical or rare dry grassland species with high conservation value (*Euphorbia seguierana*, *Veronica prostrata*, and *Cephalaria uralensis*) and formerly not present in the established vegetation, which germinated exclusively in managed plots. These species germinated very probably from the seed bank since most of them are known to build up a persistent seed bank (Csontos 2001), and their germination was likely stimulated by increased light availability through litter removal and vegetation cutting. Contrary to the findings of former studies that challenged the re-establishment of dry grassland species from the seed bank after apparent disappearance from the vegetation (e.g. Halassy 2001; Wellstein et al. 2007b), these results reveal that even after long-term abandonment of dry grassland for 40 years, litter removal alone or in combination with vegetation cutting may enhance the re-emergence of rare and even locally apparently extinct plant species (Ruprecht et al. 2010a).

Our findings demonstrate that in abandoned grassland, accumulated litter reduces bare soil surface and lowers light availability below adequate levels for seedling emergence and thus decreases microsite quantity and quality. Re-introduction of a management regime which comprises litter removal, e.g. mowing (cutting combined with biomass and litter removal by raking) or grazing (biomass removal and trampling-induced litter erosion), can promote re-emergence of species-rich grasslands from long-term abandoned, species-poor sites.

6.4 Allelopathy as an Important Driver of Community Assembly and Succession in Steppe-Like Grassland Through Its Negative Effects on Recruitment

Allelopathy, i.e. the leaching or volatilization of phytotoxins from plant tissues, is considered as one possible mechanism acting on seed germination, but also on seedling establishment (Bosy and Reader 1995). Chemicals involved are generally secondary metabolites, mainly simply structured, low molecular weight compounds, such as coumarins, terpenoids, phenolics or tannins. Within a community, especially the dominant plant species may exert a strong influence on the performance (e.g. germination) of co-occurring species (Quested and Eriksson 2006), but also on ecosystem processes. Hence, to understand these interactions and the potential influence of dominant species on ecosystem processes, it is necessary to identify plant species with strong allelopathic effects on other species, but equally important is to reveal the differential susceptibility of community constituents to these toxins. It is important to mention that not only living plant parts but also decaying material (senescent plant parts, litter) can release phytotoxins, thus exert allolopathic effects (Bonanomi et al. 2006).

We investigated possible chemical effects of the most common dominant species in abandoned steppe-like grassland, *Stipa pulcherrima*, on the germination of co-occurring dry grassland species by an experimental series: (1) a laboratory experiment in climate chambers; (2) a pot experiment under controlled watering (intermittent and frequent); and (3) a field seed sowing experiment in a grassland stand.

1. In the laboratory experiment seeds of eight species were put to germinate in Petri-dishes under two constant (10°C and 20°C) and two fluctuating (5/15°C and 10/25°C) temperature regimes, with 12 h of light and 12 h of darkness. One group of samples was watered with the aqueous extract of *S. pulcherrima* leaves and the other with distilled water. We found that the leaf extract significantly and strongly reduced percentage germination (on average by between 41% and 95%) of the seven co-occurring grassland species, but not of *S. pulcherrima* seeds, compared to controls (watering with distilled water) (Fig. 6.2). This outcome appeared to be prevalent across all different temperature regimes, the effect of the extract being significantly lower only at 10°C constant temperature than in the temperature regimes with higher daytime temperatures (Ruprecht et al. 2008). In addition, *Stipa* leaf extract significantly delayed germination by about 1.5 weeks on average, and reduced the radicle length of seedlings, which, under natural conditions, can have serious consequences for the establishment success and survival of seedlings under the harsh field conditions, with only a relatively narrow time window favourable for germination (see above). Missing the favourable period for germination and/or slow seedling growth as a consequence of reduced radicle protrusion in the presence of leaf extract may be fatal for the recruitment and survival of species in this dry grassland system (Ruprecht et al. 2008).
2. In the controlled pot experiment we tested the effect of different litter types on the germination of eight dry grassland species, using three natural litter types differing in decaying state and composition (*S. pulcherrima* fresh leaves, partly decomposed leaves, mixed partly decomposed plant material) and an artificial litter (cut cloth of synthetic fibres), with two levels of water addition. By using the artificial litter as well we intended to separate physical and chemical litter effects, since the plastic litter was expected not to release nutrients and inhibitory chemical substances. Surprisingly, we found almost no differences in the effect of artificial and natural litters on the germination of grassland species. Conversely, we found a difference between the three natural litter types in their effect on seed germination. Under frequent watering, probably because of more intensive leaching of chemical substances, *Stipa* fresh leaves exerted a significant negative effect on the germination of two species, *Dichanthium ischaemum* and *Jurinea mollis*, and a similar, close to significant trend was observed on the germination response of *Viola hirta*, compared to the other natural litter types (Fig. 6.2, Ruprecht et al. 2010b). This result, although substantially weaker, is in line with the findings of the laboratory experiment, where the aqueous extract of *Stipa* leaves also inhibited seed germination.
3. In the field experiment we conducted seed sowing of six dry grassland species in small plots (25 cm × 25 cm) designated within a long-term abandoned grassland site while applying litter removal or litter removal and artificial litter application as treatments. We used the same artificial litter as in the pot experiment, applied in the same quantity as natural litter (30 g $plot^{-1}$ which corresponds to 470 $g\,m^{-2}$). We followed the germination of the six sowed species during 13 month and compared cumulative germination between treatments and the control (natural litter left intact). We found that artificial litter represented a

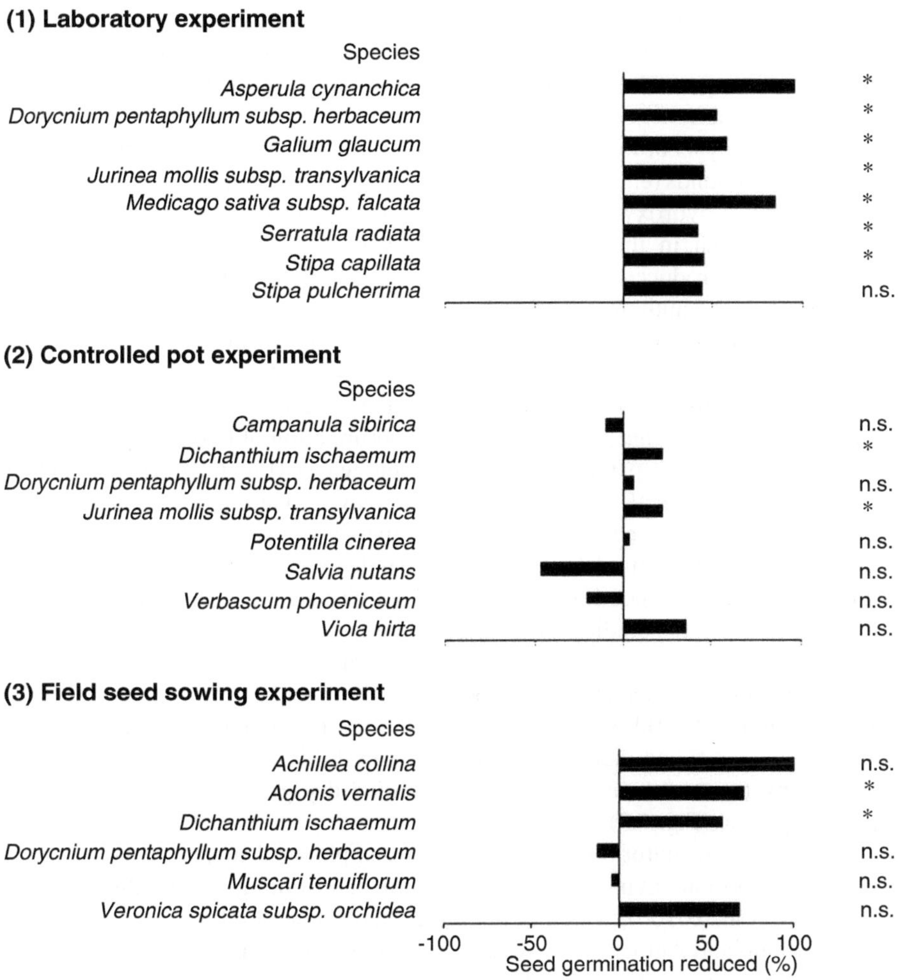

Fig. 6.2 Results of an experimental series on chemical effects of *Stipa pulcherrima* leaf extract (*1*) or *S. pulcherrima* fresh and senesced leaves (*2* and *3*) on the germination of in total 18 plant species, typical constituents of steppe-like grasslands. One species was used in all the three experiments, two species in two experiments and 15 in only one of the experiments. Asterisks denote significant differences at $p<0.05$, *n.s.* non significant effect (Based on Ruprecht et al. 2008, 2010b)

more suitable environment compared to natural litter for seed germination of two out of the six species sowed in field plots. This can be explained by the negative chemical effects of natural litter, since the two litter types had similar physical properties (Fig. 6.2).

The results of the three experiments are slightly different regarding the allelopathic potential of *S. pulcherrima* fresh or senesced leaves (representing the largest quantity in the natural litter). Based on the findings of the laboratory experiment it is unambiguous that *S. pulcherrima* is exerting strong negative effects on different processes related to regeneration by seeds of co-occurring grassland species, but such effects appeared to be weaker and more species-specific in the pot and field experiments.

GC-MS analysis of the aqueous extract of *S. pulcherrima* leaves detected more phenolic compounds (e.g. o-hydroxy-benzoic acid, trans-cinnamic acid, 4-hydroxy-benzoic acid, 4-hydroxy-3-methoxy-benzoic acid) with known allelopathic effect in the fresh than in the senescent leaves, very probably because allelopathic substances are produced more intensely in the first part of the vegetation period, while towards the autumn plant synthesis could switch to storage, as we observed the appearance of glycerine or arabinitol in the autumn-leaves of *S. pulcherrima* (Ruprecht et al. 2010b; and see also Datta and Sinha-Roy 1975; Bokhari 1978). This explains why we have found significant differences between the inhibiting effects of *S. pulcherrima* fresh leaves on the germination of two test species in the pot experiment as compared to that of senescent leaves (Fig. 6.2). Also in the field situation, the chemical impact of senesced leaves (natural litter) affected the germination performance of only two out of the six test species, and this suggests that under close-to-natural (pot) or natural (field) conditions allelopathic effects can be alleviated, e.g. possibly by microbial activity degrading the allelochemicals (Lankau 2010). Other studies, even on the same target species, often found varying results concerning the strength and importance of allelopathic inhibition, suggesting that the process may depend on the specific environmental context (e.g. Bais et al. 2003; Blair et al. 2006). In addition, as a very important aspect of chemical interactions, we found that the degree of susceptibility to allelopathic substances released by the dominant species in the studied dry grassland system is highly species-specific (Fig. 6.2).

Under close-to-natural or natural conditions physical litter effects proved to be more important than chemical effects in determining germination success (Ruprecht et al. 2010b), which is in concert with other studies on dry grassland (e.g. Rotundo and Aguiar 2005; Amatangelo et al. 2008). However, we emphasize that chemical litter effects should not be underestimated, since species-specific chemical interactions can be important drivers of community processes in steppe-like grassland.

Our experiments in the field and under controlled conditions clearly have demonstrated the negative effects of the accumulation of litter on the regeneration of dry grassland species. But our experiments also showed that appropriate management regimes, involving the removal of litter or living plant material from abandoned grasslands, that have decreased in apparent species numbers, can restore the plant diversity and open vegetation structure of the dry steppe-like grasslands in Romania, and probably elsewhere.

Acknowledgement Participation at the conference "Eurasian Steppes: Status, Threats and Adaptation to Climate Change", organised in the Hustai National Park, Mongolia, brought new ideas and conceptions in my way of understanding dry grassland and especially the steppe vegetation. The author was supported by the Romanian Ministry of Education, Research and Innovation (CNCSIS-UEFISCSU, project PN II-RU TE 296/2010), while writing this chapter.

References

Amatangelo KL, Dukes JS, Field CB (2008) Responses of a California annual grassland to litter manipulation. J Veg Sci 19:605–612

Bais HP, Vepachedu R, Gilroy S, Callaway RM, Vivanco JM (2003) Allelopathy and exotic plant invasion: from molecules and genes to species interactions. Science 301:1377–1380

Bakker JP, Berendse F (1999) Constraints in the restoration of ecological diversity in grassland and heathland communities. Trends Ecol Evol 14:63–68

Bergelson J (1990) Life after death: site pre-emption by the remains of *Poa annua*. Ecology 71:2157–2165

Blair AC, Nissen SJ, Brunk GR, Hufbauer RA (2006) A lack of evidence for an ecological role of the putative allelochemical catechin in spotted knapweed invasion success. J Chem Ecol 32:2327–2331

Bokhari UG (1978) Allelopathy among prairie grasses and its possible ecological significance. Ann Bot 42:127–136

Bonanomi G, Sicurezza MG, Caporaso S, Esposito A, Mazzoleni S (2006) Phytotoxicity dynamics of decaying plant materials. New Phytol 169:571–578

Bosy JL, Reader RJ (1995) Mechanisms underlying the suppression of forb seedlings by grass (*Poa pratensis*) litter. Funct Ecol 9:635–639

Cremene C, Groza G, Rakosy L, Schileyko AA, Baur A, Erhardt A, Baur B (2005) Alterations of steppe-like grasslands in Eastern Europe: a threat to regional biodiversity hotspots. Conserv Biol 19:1606–1618

Csontos P (2001) A természetes magbank kutatásának módszerei (Methods for studying the natural seed bank), Synbiologica Hungarica 4. Scientia Kiadó, Budapest

Datta SC, Sinha-Roy SP (1975) Phytotoxic effects of *Croton bonplandianum* Baill. on weedy associates. Vegetatio 30:157–163

Doniţă N, Ivan D, Coldea G, Sanda V, Popescu A, Chifu T, Paucă-Comeănescu M, Mititelu D, Boşcaiu N (1992) Vegetaţia României (Vegetation types of Romania) [in Romanian]. Editura Tehnică Agricolă, Bucureşti, p 407

Eckstein RL, Donath TW (2005) Interactions between litter and water availability affect seedling emergence in four familial pairs of floodplain species. J Ecol 93:807–816

Enyedi ZM, Ruprecht E, Deák M (2008) Long-term effects of the abandonment of grazing on steppe-like grasslands. Appl Veg Sci 11:53–60

Eskelinen A, Virtanen R (2005) Local and regional processes in low-productive mountain plant communities: the roles of seed and microsite limitation in relation to grazing. Oikos 110:360–368

European Commission (2007) Interpretation manual of European Union habitats – EUR27. European Commission, DG Environment, Brussels, p 144

Facelli JM, Pickett STA (1991) Plant litter: its dynamics and effects on plant community structure. Bot Rev 57:2–31

Foster BL, Gross KL (1998) Species richness in a successional grassland: effects of nitrogen enrichment and plant litter. Ecology 79:2593–2602

Gafta D, Mountford JO (eds) (2008) Manual de interpretare a habitatelor Natura 2000 din România (Manual for interpreting Natura 2000 habitats from Romania) [in Romanian]. Risoprint, Cluj, p 101

Halassy M (2001) Possible role of the seed bank in the restoration of open sand grassland in old fields. Commun Ecol 2:101–108

Hofmann M, Isselstein J (2004) Seedling recruitment on agriculturally improved mesic grassland: the influence of disturbance and management schemes. Appl Veg Sci 7:193–200

Illyés E, Bölöni J (eds) (2007) Lejtősztyepek, löszgyepek és erdőssztyeprétek Magyarországon (Slope steppes, loess steppes and forest-steppe meadows in Hungary) [in Hungarian, with English summary]. Illyés Eszter, Budapest, p 236

Kahmen S, Poschlod P, Schreiber KF (2002) Conservation management of calcareous grasslands. Changes in plant species composition and response of functional traits during 25 years. Biol Conserv 104:319–328

Kun A, Ruprecht E, Szabó A (2004) Az Erdélyi-medence bioklimatológiai jellemzése (The bioclimatological characteristics of the Transylvanian Basin (Romania)) [in Hungarian, with English summary]. Múz Füz 13:63–81

Kuneš P, Pelánková B, Chytrý M, Jankovská V, Pokorný P, Petr L (2008) Interpretation of the last-glacial vegetation of eastern-central Europe using modern analogues from southern Siberia. J Biogeogr 35:2223–2236

Lankau R (2010) Soil microbial communities alter allelopathic competition between *Alliaria petiolata* and a native species. Biol Invas 12:2059–2068

Luoto M, Pykälä J, Kuussaari M (2003) Decline of landscape-scale habitat and species diversity after the end of cattle grazing. J Nat Conserv 11:171–178

Poschlod P, WallisDeVries M (2002) The historical and socioeconomic perspective of calcareous grasslands – lessons from the distant and recent past. Biol Conserv 104:361–376

Quested H, Eriksson O (2006) Litter species composition influences the performance of seedlings of grassland herbs. Funct Ecol 20:522–532

Rotundo JL, Aguiar MR (2005) Litter effects on plant regeneration in arid lands: a complex balance between seed retention, seed longevity and soil-seed contact. J Ecol 93:829–838

Ruprecht E, Donath TW, Otte A, Eckstein RL (2008) Chemical effects of a dominant grass on seed germination of four familial pairs of dry grassland species. Seed Sci Res 18:239–248

Ruprecht E, Szabó A, Enyedi MZ, Dengler J (2009) Steppe-like grasslands in Transylvania (Romania): characterization and influence of management on species diversity and composition. Tuexenia 29:353–368

Ruprecht E, Enyedi MZ, Eckstein RL, Donath TW (2010a) Restorative removal of plant litter and vegetation 40 years after abandonment enhances re-emergence of steppe grassland vegetation. Biol Conserv 143:449–456

Ruprecht E, Józsa J, Ölvedi TB, Simon J (2010b) Differential effects of several "litter" types on the germination of dry grassland species. J Veg Sci 21:1069–1081

Sanda V, Popescu A, Doltu MI, Öllerer K, Burescu P (2008) Fitocenozele din România. Sintaxonomie, structură, dinamică şi evoluţie (Associations of Romania. Syntaxonomy, structure, dynamics and evolution) [in Romanian]. Ars Docendi, Bucureşti, p 570

Soó R (1942) Az Erdélyi Medence endémikus és reliktum növényfajai (Die Endemismen und Reliktarten des Siebenbürgischen Beckens) [in Hungarian]. Acta Geobot Hung 5:139–183

Tilman D (1993) Species richness of experimental productivity gradients: how important is colonisation limitation? Ecology 74:2179–2191

Todorova S, Becker T, Boch S, Dolnik C, Jeschke M, Kuzemko A, Löbel S, Otýpková Z, Pedashenko H, Ruprecht E, Szabó A, Vassilev K, Kiehl K, Chytrý M, Dengler J (in preparation) Species-area relationships in European dry grasslands – a comparative analysis across regions, taxa, and scales

Weltzin JF, Keller JK, Bridgham SD, Pastor J, Allen PB, Chen J (2005) Litter controls plant community composition in a northern fen. Oikos 110:537–546

Wellstein C, Otte A, Waldhardt R (2007a) Impact of site and management on the diversity of Central European mesic grassland. Agric Ecosyst Environ 122:203–210

Wellstein C, Otte A, Waldhardt R (2007b) Seed bank diversity in mesic grasslands in relation to vegetation type, management and site conditions. J Veg Sci 18:153–162

Wilson JB, Peet RK, Dengler J, Pärtel M (in press) Plant species richness: the world records. J Veg Sci

Xiong S, Nilsson C (1999) The effects of plant litter on vegetation: a meta-analysis. J Ecol 87:984–994

Zobel M, Otsus M, Liira J, Moora M, Mols T (2000) Is small-scale species richness limited by seed availability or microsite availability? Ecology 81:3274–3282

Chapter 7
Past Trends, Present State and Future Prospects of Hungarian Forest-Steppes

Zs. Molnár, M. Biró, S. Bartha, and G. Fekete

Abstract In Hungary a countrywide vegetation mapping project carried out between 2003 and 2006 provided immense, detailed data on the current status of the forest-steppe vegetation (MÉTA database). In addition, two fundamentally important historical sources from the late eighteenth century have been analyzed recently. Using these sources we reconstruct and evaluate the past history and current status, and forecast the expected future of the vegetation types within the forest-steppe zone.

We show that by the end of the eighteenth century most forest-steppe habitats had undergone considerable change. While steppe woodlands had largely disappeared, large areas of sand steppes and closed steppes on chernozem remained and were used for grazing, some having been degraded into dune areas with windblown sand.

As a result of man's activities, including mechanized land-use, the forest-steppe vegetation underwent great changes during the past 200 years. We review these changes. In the past decades cessation of mowing and grazing is problematic. Presently, approximately 251,000 ha (6.8%) of the total of 3,700,000 ha of forest-steppe vegetation have survived in Hungary, of which only 5.5% of the stands may be considered natural, 38% semi-natural, 46% moderately degraded, and 10% strongly degraded.

We predict future trends in the forest-steppe vegetation by evaluating (1) past trends, (2) current threats and regeneration potential, and (3) expected climate change. Important threats are (1) spread of invasive species; (2) abandonment of traditional land-use; (3) drop of the groundwater table due to regulation and draining, (4) plowing; (5) overgrazing; (6) excessive wild game populations; (7) afforestation; and (8) forest management practices. They will lead to further decline and fragmentation of most dry vegetation types, and we predict that steppe woodlands on loess and sand will almost fully disappear in the coming decades, though lack of grazing may lead to the extension of Juniperus-Populus scrub in sand dune areas.

Zs. Molnár (✉) • M. Biró • S. Bartha • G. Fekete
Department of Functional Ecology, Centre for Ecology, Hungarian Academy of Sciences, Vácrátót, Hungary
e-mail: molnar.zsolt@okologia.mta.hu; biro.marianna@okologia.mta.hu; bartha.sandor@okologia.mta.hu; h6868fek@helka.iif.hu

M.J.A. Werger and M.A. van Staalduinen (eds.), *Eurasian Steppes. Ecological Problems and Livelihoods in a Changing World*, Plant and Vegetation 6, DOI 10.1007/978-94-007-3886-7_7, © Springer Science+Business Media B.V. 2012

7.1 Introduction

Information on past changes of the vegetation in any studied landscape may greatly help to interpret the current situation and predict future changes (Rackham 1980; Pickett 1989). This is particularly true in the case of the forest-steppe vegetation in Hungary. Compared to deciduous forests and grassland habitats in the mountain ranges, this vegetation has been transformed by humans to such a great extent that its current pattern and dynamics may be hardly understood by direct contemporary studies (Biró 2006; Molnár 2007).

The forest-steppe is a distinct vegetation zone in the transitional belt between the climatic zones supporting closed forests and steppes. It incorporates a large number of different vegetation types, including those that are much more broadly distributed than the zone itself, and are thus not exclusively linked to this vegetation zone (for edaphic factors, for example). Characteristics of the zone are determined by specific vegetation types that are primarily restricted to this zone (climatically zonal and intra-zonal communities) (Berg 1958).

The characteristics of the East-European forest-steppe, including physiognomy, structure, biogeographical and ecological patterns, vegetation dynamics and post-glacial history, have already been described in detail (Aliokhin 1950; Artushenko 1970; Berg 1958; Lavrenko 1956, 1981; Lavrenko et al. 1991; Lavrenko and Karamysheva 1993; Tarasov et al. 1998, 2000). However, botanical knowledge on the past changes, current status and expected changes of this forest-steppe vegetation is much less, and has not yet been reviewed. For Hungary, until recently such a review was not even possible, because of lack of quantitative data, and available data on past changes concerned only few locations. However, the past 10–15 years have brought great progress in this field.

A country-wide vegetation mapping project carried out between 2003 and 2006 provided data on the current status of the forest-steppe vegetation, which were degrees of magnitude more detailed than ever before (the MÉTA-program, see Molnár et al. 2007b; Horváth et al. 2008). The overall goal of the project was to collect a great variety of information on the natural vegetation as part of the natural heritage of Hungary. 199 persons completed the project in about 7,000 field days.

Recent studies on vegetation history have also yielded new results. New locations (e.g., the drier, interior part of the Hungarian forest-steppe zone) have been researched and dating methods became more refined. In addition, two fundamentally important historical sources from the eighteenth century have been botanically analyzed (the outstandingly detailed diary of the botanist Pál Kitaibel with data from thousands of locations in Hungary, and the 1:28,800 scale maps and supplementary country description of the First Military Survey with a detailed legend and full coverage of the territory of Hungary) (Biró 2006; Molnár 2007). These sources are particularly important, because they allow the reconstruction of the environment from before the beginning of major landscape transformations (river regulation, development of dispersed farmsteads, plowing up the steppe, adoption of intensive agricultural technologies).

The new methods and data gave us the unique opportunity within the Eurasian forest-steppe zone to reconstruct and evaluate the past history, describe the current status, and forecast the expected future of each vegetation type in detail. Here we outline the Holocene history of the 13(15) most significant vegetation types of the forest-steppe zone in Hungary, and describe their status at the end of the eighteenth century with the changes since then (Table 7.1). We also provide information on their current extent and condition along with their regeneration potential and threatening factors.

The following vegetation types are discussed (codes of the Hungarian Habitat Classification System are in parentheses, Bölöni et al. 2007): Closed steppe on chernozem soil (H5a) , *Artemisia* steppe on loess and clay cliffs (I2), Steppe woodland of *Quercus* on loess (M2), Continental deciduous steppe thicket (M6), Open sand steppe (G1), Closed sand steppe (H5b), *Juniperus*, *Populus* steppe woodland and scrub (M5), Steppe woodland of *Quercus* on sand (M4), *Artemisia* steppe on solonetz soil (F1a), *Achillea* steppe on meadow solonetz (F1b), Saline meadow (F2), Tall herb meadow steppe on solonetz soil (F3), Annual vegetation of salt lakes and mud-flats, and vegetation of saline flats (F5), *Puccinellia* meadow (F4), and Steppe woodland of *Quercus* on saline soil (M3).

Forest-steppe and forest-steppe-like vegetation also occur on the foothills and even at higher regions in the mountain ranges. These are, however, either extra-zonal communities, such as relic stands of *Spiraea* shrubberies, now found only locally in the mountains, or are secondary or edaphic in origin as the *Brachypodium pinnatum*-dominated vegetation developing after forest clearing, and grasslands on calcareous and siliceous bedrock, respectively. These are not treated here (as salt marshes are also left out of the analysis). Because of their small area, we do not discuss the isolated stands of closed *Acer* and *Tilia* forests, which represent the cool continental forest-steppe vegetation (Fekete 1965), and the forest-steppe-like *Pinus sylvestris* forests (Pócs 1966). Because the literature on the forest-steppe vegetation in Hungary has been published mostly in Hungarian, our paper also serves as an introduction to this literature for the international scientific community.

7.2 The Hungarian Forest-Steppe – Biogeographical Background

By far the largest part of the Great Hungarian Plain has been considered by most Hungarian botanists as part of the forest-steppe biome (Soó 1926, 1929, 1931, 1960; Borhidi 1961; Zólyomi 1957, 1989; Varga 1989; Zólyomi and Fekete 1994). There, the annual precipitation ranges from 450 to 600 mm (250–350 in the vegetation period and 200–250 from November to April), and the mean annual temperature is 10–11°C (17–18 in the vegetation period and 3–4 from November to April) (HMS 2000). The forests are composed typically of oaks (mostly *Quercus robur*), while the steppes are dominated by *Festuca* and *Stipa* species. It is still debated, however,

Table 7.1 Historical change of some attributes of the main forest-steppe habitats in Hungary in the last 200 years based mainly on Biró (2006) and Molnár (2007)

Code	Short name	(1) Likelihood of historical continuity	(2) Main period(s) of extensive change (century)	(3) Change in naturalness in the past 200 years	(4) Change in intensity of grazing	(5) Degree of degradation due to afforestation	(6) Degree of degradation due to amelioration	(7) Current proportion with neglected management	(8) Change in the groundwater table	(9) Leaching (–) salinization (+)	(10) Change in the extent of bare ground	(11) Change in patch size	(12) Change in total area	(13) Amount of invasive species
H5a	Closed steppe on chernozem soil	•••	18, 19	– – –	–	•	•	••				– – –	– – –	+
I2	Vegetation of loess/clay cliffs	•••	20	– –	–	•••		•			–	–	–	+++
M2, M6	Steppe thicket and woodland on loess	•••	<18	– –	– –	•		••				– –	– –	+
G1	Open sand steppe	•••	19, 20	+	– –	•••		•			– –	– –	– –	++
H5b	Closed sand steppe	•	19, 20	– – –	– –	••	•	•			–	– – –	– – –	++
M5	Juniperus-Populus scrub	•	19, 20	++	– –	•••		••	– –		–	++	+++	++
M4	Steppe woodland on sand	•••	<18, 19, 20	– – –	– – –	•		•	– – –			– –	– – –	+++
F1a	Artemisia steppe on solonetz soil	•••	20	–	– –		••	••	–	–		–	–	0
F1b	Achillea steppe on asolonetz soil	•	19–20		– –	•	•••	••	–	–		++	+++	0
F2	Saline meadow	••	20/2	– – (–)	– – –		••	••	– (– –)	– (– –)	–	– (– –)	– –	0
F3	Saline meadow steppe	•••	19, 20	– – –	+	••	••	•	– –	+		– –	– – –	+
F4, F5	Puccinellia meadow and vegetation of saline- and mud-flats	•••	20/2	– (– –)	– –		•	•	– (– –)	– (– –)	– –	– –	– –	0
M3	Steppe woodland on saline soil	•••	<18, 19, 20	– –	– –	•		••	– –			–	– –	+

whether the driest innermost parts of the plain still meet the criteria of the forest-steppe biome, or should be regarded as representatives of the steppe zone. Although the severity of summer droughts in this area does not reach the aridity level experienced in the zone of typical *Stipa* steppes, the length of the arid period is longer and water deficit well exceeds that observed in the area of the East European forest-steppe zone. About 40% of the years are so-called steppe years with weather characteristics typical of the steppe zone (Borhidi 1961; Kun 2001).

The forest-steppe vegetation in Hungary is the westernmost representative of the Eurasian forest-steppe (Soó 1926, 1960; Zólyomi 1989; Niklfeld 1973–1974) with some small outposts in Austria and the Czech Republic. The largest continuous steppe of Central Europe, the Hortobágy, also is found in Hungary (most of it is preserved now in a national park).

There are two types of forest-steppe distinguished within the European forest-steppe with broad-leaved forests. These are the continental and the sub-Mediterranean forest-steppes (Borhidi 1961; Niklfeld 1973–1974). The continental forest-steppe is characterized by closed and rather mesic *Quercus* forests and adjacent lush, forb-rich meadow steppes. This continental forest-steppe extends from the east as far as Podolia, and also occurs in isolated patches in the eastern part of the Carpathian Basin (Transylvania), and in the Gödöllő Hills in north-central Hungary (Fekete 1965; Niklfeld 1973–1974). The sub-Mediterranean forest-steppe is spread out in a long strip to the south and southwest of the continental forest-steppe and appears in several disjunct patches at the foothills of Cis Caucasia, the Crimean Peninsula, in Bessarabia and southern Moldova, and along the lower Danube in Romania and Bulgaria (Lavrenko 1970; Niklfeld 1973–1974). Most of the forest-steppe in the Carpathian Basin also belongs to this type especially in the western and central parts of Hungary. The main and typical communities of this forest-steppe type are dry and open woodlands and parklands of *Quercus* spp. developed under the partial influence of the submediterranean climate, and with a variable number of submediterranean floristic elements (one of the commonest is *Quercus pubescens*). Forest and steppe plants exhibit similar distributional pattern: species numbers of the two groups show decreasing gradients towards the driest and warmest central part of the plain (Fekete et al. 2010). The lack of chorological symmetry may presumably indicate the lack of the true steppe biome in the Carpathian Basin.

Characteristic of the Hungarian forest-steppe is its mosaic structure. The zonal arrangement of climate, soil types and vegetation so typical of Eastern Europe is disrupted here, and replaced by a mosaic landscape (Kádár 1975; Varga 1992). Owing partly to relief and pedological features, the forest-steppe zone is discontinuous: loess tables and plateaus and sand dune areas are fringed by river valleys with saline vegetation at the margins and in depressions. Zonal forest-steppe vegetation has developed on loess and humus-rich sand, whereas the vegetation on saline soils and wind-blown sand is intra-zonal. Extra-zonal forest-steppe vegetation occurs on the southern slopes and rocky outcrops in the surrounding hills and mountains (Zólyomi 1958, 1989; Jakucs 1961; Soó 1964; Borhidi 2003; Illyés and Bölöni 2007).

The Hungarian forest-steppe exhibits the characteristics of the more eastern forest-steppe, and also displays some unique features, which are generally attributed to submediterranean climatic influences and their isolated geographical situation (being surrounded by the high mountains of the Carpathians). For these reasons, it is regarded a distinct biogeographical unit, the Pannonian Biogeographical Region, first mapped out by Soó (1933), and recently by Varga (Varga 2003; European Environmental Agency 2006). The Pannonian Region is the western neighbour of the Pontic Region.

Presently we consider that most of the Great Plain (with exception of its marginal areas) was part of the forest-steppe zone throughout the entire Holocene (Magyari et al. 2009). It is likely that most of the floodplains and parts of the sand dunes and fens were covered with forests, whereas the loess tables and plateaus were probably wooded only to a small extent. The saline areas, however, may have been almost completely devoid of woody vegetation. Plowing up the forest-steppe habitats started as early as the Neolithic Age. The plowed areas became truely extensive probably by the Bronze Age. By the eleventh century, most of the area suitable for agriculture had been plowed. The process of forest loss, however, is not known in detail, but it seems that the decreasing trend has been continuing over millennia. The forest cover reached its minimum by the end of the eighteenth century (Zólyomi 1936, 1952; Járai-Komlódi 1966, 1987; Fairbairn 1992, 1993; Sümegi 1998; Gyulai 2001; Sümegi et al. 2000, 2006; Jakab et al. 2004; Magyari et al. 2009; Molnár 2009) (Photos 7.1, 7.2, 7.3, 7.4, 7.5, 7.6, 7.7 and 7.8).

7.3 The Data Bases

In this paper, we use the units of the Hungarian Habitat Classification System (Bölöni et al. 2007), which is the most frequently used and thoroughly tested system of vegetation classification in Hungary. In this system, a unit includes all plant communities that occur in similar habitats, and have similar species composition and physiognomy (Bölöni et al. 2011). For a detailed phytosociological description of forest-steppe plant communities see Borhidi (2003, 2012). Species names follow Király (2009).

Our reconstruction of the Holocene history of the vegetation is based on available relevant literature data. Since we were primarily interested in the development of the current vegetation, we mostly reviewed the data since the Boreal Period (and sometimes the Late Glacial Period). Unfortunately, for several reasons significantly less information is available on the Holocene history of the Hungarian forest-steppe than needed. First, only a few sites have remained that are suitable for paleobotanical research (few bogs, swamps and mires, extensive saline areas, and dry habitats). Second, the location of the pollen source (whether moor or the adjacent steppe habitat) is uncertain in the case of several taxa (i.e. *Quercus* spp., Poaceae). Finally, the pollen of several characteristic species or species groups are indistinguishable from that of other species or even sometimes other genera. As a consequence, it is not possible to determine, e.g., the proportion of *Quercus robur* versus *Q. pubescens* in steppe woodlands, or the exact habitat (whether dry steppe on chernozem soil or saline soil, or even arable land) of *Artemisia* and Chenopodiaceae species, and not even to identify the major grasses of the steppes to at least the generic level (*Stipa*, *Festuca*, *Bromus*, etc.). Nevertheless, the more recent research (e.g. Sümegi 1998, 2004, 2005a, d; Sümegi et al. 2006; Jakab et al. 2004; Magyari 2002, Magyari et al. 2009) has modified the results of earlier works (e.g. Rapaics 1918; Zólyomi 1936, 1952; Járai-Komlódi 1966, 1987; Győrffy and Zólyomi 1994; Medzihradszky and Járainé Komlódi 1995) in many details.

For the reconstruction of the vegetation at the end of the eighteenth century (Biró 2006; Molnár 2007) two main sources were used: (1) the field notes of Pál Kitaibel (Gombocz 1945; Lőkös 2001), and (2) maps of the First Military Survey of Hungary (1783–1785) and the supplementary written description. Kitaibel was a traveling field botanist being way ahead of his time in many respects. As a botanist, he described many plant species new to science, and at the same time he had a very broad interest in other fields (balneology, geology, ethnography, etc.). His diary, which was continually written during his journeys, is very detailed. Of the more than 1,500 pages of the diary, about 500 pages touch on areas of forest-steppe vegetation. While traveling on a wagon, Kitaibel usually took only short notes at any given locale. Nonetheless, his several thousand data outline many features of the vegetation of the Great Plain in his days. By reading his diary, we were able to follow his path on the sheets of the First Military Survey. We identified old scientific plant names mostly with the help of the work of Sándor Jávorka (1926–1945). The data were sorted by habitat types.

Features of the landscape are shown on the maps of the First Military Survey (1:28,800, 1783–1785) in such great detail that the characteristics of the plains at the end of the eighteenth century may be much better determined than from any other maps of that time. The so-called Country Description, a supplement to the maps, contains short notes on the quality of the roads, the condition of the forests, the quality of meadows and grazing fields, and topographical features for each map sheet and each settlement (Borbély and Nagy 1932). These data were also sorted by habitat types.

We were able to recover such a large amount of data for saline habitats from the field notes of Kitaibel that we could analyze them by the COCKTAIL algorithm (Bruelheide 2000) and the JUICE software (Tichy 2002). The JUICE software groups individual species that more frequently co-occur than expected by chance. Accidental species only infrequently occurring in saline habitats and/or species typical of other habitats were grouped into "pseudo-species" to simplify analysis (such as species of dry steppe on chernozem soils, meadow species, arable weeds). We listed all significant species groups and individual species, and also provided the first few non-significant species or species groups in parentheses, because the program was designed for analyzing phytosociological samples, whereas we worked with fragmentary species lists.

Most of the information on the present state of forest-steppe habitats was derived from the MÉTA database (Molnár et al. 2007b; Horváth et al. 2008). Data were collected from a survey using the so-called MÉTA method (Molnár et al. 2007b). This is a vegetation mapping method, supported by satellite images, which applies a grid of hexagonal cells in which vegetation is mapped directly in the field.

Photo 7.1 Species-rich meadow steppe on a hillside (*Festuca rupicola, Brachypodium pinnatum, Salvia pratensis, Inula hirta, Chamaecytisus austriacus, Euphorbia glareosa*) (Photo by Zs. Molnár)

Photo 7.2 The flat landscape of a loess tableland with intensive agricultural areas and degraded remnants of grasslands along intermittent watercourses and roadsides

Photo 7.3 A sand dune area with *Festuca vaginata* and *Stipa borysthenica* grasslands and *Juniperus communis-Populus alba/canescens* woodlands and scrub. Invasive species (*Robinia pseudacacia, Ailanthus altissima, Pinus nigra, Asclepias syriaca*) are present (Photo by Á. Molnár)

Photo 7.4 Aerial view of a section of the forest-steppe landscape with sand dunes, pine plantations and abandoned arable fields of varying age with *Robinia* invasion

The survey was based on cells with an area of 35 ha each (267,813 cells cover the entire territory of Hungary). Habitat types were listed per hexagon, then the area, habitat condition based on naturalness values, spatial pattern, connectedness, and threatening factors were estimated and recorded for each habitat type. For standardization of mapping, three different pre-printed data sheets and two detailed guides (Mapping Guide and Habitat Guide) were prepared, and field trainings were organized (Molnár et al. 2007b; Bölöni et al. 2007).

One hundred contiguous hexagons formed a MÉTA quadrat (35 km^2), the basal unit of distribution maps. The total number of quadrats was 2,813. For the preparation of distribution maps, SQL queries (Horváth and Polgár 2008) were carried out per

Photo 7.5 One of the last remnants of the open steppe woodlands on sand with closed sand steppe in the openings (*Quercus robur*, *Pyrus pyraster*, *Crataegus monogyna*, *Festuca wagneri*, *Iris variegata*) (Photo by Á. Molnár)

Photo 7.6 A well-structured forb-rich saline meadow steppe developed on an ancient floodplain (and with some salt in deeper soil layers) with *Peucedanum officinale*, *Aster punctatus*, *A. linosyris*, *Artemisia pontica*, *Limonium gmelini* and *Iris spuria*. This is the dominant vegetation of the glades in steppe woodlands on saline soil (Photo by T. Baranyai)

Photo 7.7 A fine-scale vegetation mosaic in an area with salt-affected soil. The various vegetation types follow a particular pattern from the highest to the lowest terrain: *Achillea* and *Artemisia* steppes with *Festuca pseudovina* and *Artemisia santonicum*, vegetation of saline flats with *Camphorosma annua* and *Puccinellia limosa*, saline meadows with *Alopecurus pratensis*, and salt marshes and lakes with *Bolboschoenus maritimus* (Photo by Zs. Molnár)

Photo 7.8 A saline landscape with special geomorphological and vegetation patterns. This is the most widespread steppe type in Hungary. On the loess tablelands among saline steppes, arable fields are typical

habitat and per MÉTA quadrat. The so-obtained data were quality-checked, which included the deletion of apparently erroneous data and the estimation of missing values (ca. 6% of the total).

We used a naturalness-based attribute in the MÉTA-database (Table 7.2) to characterize overall quality of vegetation stands in the field. This attribute quantifies

Table 7.2 Percentage of stand area of each habitat type per degree of naturalness based on the MÉTA database

		Naturalness value				
Code	Short name	2	3	4	5	Total area (ha)
H5a	Closed steppe on chernozem soil	25	62	12	0	25,000
I2	Vegetation of loess/clay cliffs	59	18	23	0	95
M6	Steppe thickets	0	38	30	32	3
M2	Steppe woodland on loess	8	62	30	0	100
G1	Open sand steppe	22	44	31	3	10,700
H5b	Closed sand steppe	20	55	23	1	28,000
M5	Juniperus-Populus woodland and scrub	6	33	48	12	3,000
M4	Steppe woodland on sand	10	41	42	7	290
F1a	Artemisia steppe on solonetz soil	2	34	54	10	33,800
F1b	Achillea steppe on solonetzi soil	12	58	29	1	46,000
F2	Saline meadow	5	41	48	6	93,000
F3	Saline meadow steppe	28	41	29	3	1,120
F4	Puccinellia meadow	1	13	52	34	7,000
F5	Vegetation of saline- and mud-flats	1	19	43	37	2,500
M3	Steppe woodland on saline soil	21	57	22	0	130
	All habitats together	10.3	45.9	38.2	5.5	250,738

2: heavily degraded, 3: moderately degraded, 4: semi-natural, 5: natural (for details see Bölöni et al. 2008a, b)

habitat quality according to the following scale: 1: completely degraded, 2: heavily degraded, 3: moderately degraded, 4: semi-natural, 5: natural (see details in Németh and Seregélyes 1989; Molnár et al. 2007b). Naturalness values integrate variables related to structural properties and species richness into a single number. Though the estimation of naturalness values includes subjective elements, we took great care to standardize the method during the MÉTA project (Molnár et al. 2007b; Bölöni et al. 2007).

For the assessment of endangerment, we listed a number of potential threatening factors and evaluated their effect on each habitat unit. Then we estimated the cumulative area of the affected habitat unit for each factor (Table 7.3). The effective threatening factors (the ones that are most likely to affect the survivorship of the habitat type in a MÉTA hexagon in the next 10–15 years, Molnár et al. 2007b, Molnár et al. 2008c) were selected from a list of 28 factors. Severity of threat was not recorded. We developed 12 synthetic indicators from the threatening factors by thematic grouping, which we then used for the evaluation of general endangerment of each habitat unit (interpretation of the 12 indicators is explained in detail in Molnár et al. 2008c). We ordered the indicator values across habitats and within factors according to their rank, and then averaged these ranks across factors (Molnár et al. 2008c).

Regeneration potential of a focal vegetation unit was assessed at the spatial scale of the MÉTA quadrats (i.e. with regard to all other units detected in the quadrat, Table 7.4). We evaluated three regeneration scenarios (for more details see Seregélyes et al. 2008): (1) regeneration success of an existing stand following moderate

Table 7.3 Factors threatening forest-steppe habitats (% of the total habitat area threatened) in Hungary, based on the MÉTA habitat database

Code	Habitat name	Draining	Undergrazing	Overgrazing	Lack of grazing	Lack of mowing	Amelioration	Shrub encroachment	Burning	Improper management of forests	Premature clearing	Tree plantation	Wild game	Invasive species	Plowing	Construction	Mining
H5a	Closed steppe on chernozem soil		2	11	10	12	2	39	2			6	3	31	22	11	2
I2	Vegetation of loess/clay cliffs			6	6			14					3	59	10	12	
M6	Steppe thickets							23	4				16	13	7	5	4
M2	Steppe woodland on loess				4	11		26		10	35	12	27	42		1	
G1	Open sand steppe			13	2			48				7	0	69	3	7	
H5b	Closed sand steppe		2	15	5	7	3	32				11	1	47	34	12	5
M5	Juniperus-Populus woodland and scrub							2		19	5	24	1	49			
M4	Steppe woodland on sand		4		2	2		9		20	20	26	18	74			
F1a	Artemisia steppe on solonetz soil	25	5	14	8	6							0	4	13	4	
F1b	Achillea steppe on solonetz soil	35	11	10	10	10	3						0	12	15	6	
F2	Saline meadow	41	5	9	10	12	2	5					0	16	24	6	3
F3	Saline meadow steppe	24			3	11	3	11				4	0	20	27	14	
F4	Puccinellia meadow	35		8		6							0		9	6	7
F5	Vegetation of saline - and mud-flats	21	4	9	7								0		4	4	
M3	Steppe woodland on saline soil	8			14	18		19	5	12	16	8	12	45	9		

Values below 2% are not shown (for details see Molnár et al. 2008c)

Table 7.4 Percentage of stand area of each habitat type with at least medium regeneration potential, based on the MÉTA database

Code	Short name	(1)	(2)	(3)
H5a	Closed steppe on chernozem soil	67	37	27
I2	Vegetation of loess/clay cliffs	63	19	18
M6	Steppe thickets	69	59	7
M2	Steppe woodland on loess	53	30	2
G1	Open sand steppe	76	45	34
H5b	Closed sand steppe	84	30	20
M5	Juniperus-Populus woodland and scrub	96	90	65
M4	Steppe woodland on sand	23	13	0
F1a	Artemisia steppe on solonetz soil	99	78	64
F1b	Achillea steppe on solonetz soil	100	93	78
F2	Saline meadow	99	78	62
F3	Saline meadow steppe	76	59	44
F4	Puccinellia meadow	100	75	32
F5	Vegetation of saline - and mud -flats	95	58	12
M3	Steppe woodland on saline soil	46	44	0

For details see Seregélyes et al. 2008
Three types of regeneration potential were assessed: *1*: on site following disturbance, *2*: in other adjacent habitat units, and *3*: on old-fields separately. Note that the habitat types totals can add up to more than 100%

disturbance or degradation; (2) successful development in a different habitat type following environmental change (e.g. development of steppe vegetation in a dried-out meadow); (3) successful development in vegetation-free areas, i.e. the ability to colonize abandoned old-fields. Although the evaluation of regeneration potential is necessarily subjective, we standardized the evaluation method as much as possible. In the Habitat Guide we (1) specified the factors that determine regeneration potential of a certain habitat type, and (2) for each vegetation type we provided a detailed list of examples for each regeneration scenario (we described about 720 examples altogether, Bölöni et al. 2003, 2007; Molnár et al. 2007b).

7.4 The Vegetation of the Hungarian Forest-Steppe

7.4.1 Closed Steppe on Chernozem Soil (Habitat Code: H5a)

These are closed steppes on humus-rich chernozem soils typically developed over loess, and dominated usually by *Festuca rupicola*. Most of the surviving stands occur on slopes in the lowland or on foothills (Virágh and Fekete 1984; Zólyomi and Fekete 1994; Horváth 2002; Illyés and Bölöni 2007; Bartha 2007a).

Characteristic, dominant and constant species are: *Bothriochloa ischaemum*, *Brachypodium pinnatum*, *Bromus inermis*, *Carex humilis*, *Chrysopogon gryllus*,

Elymus hispidus, Festuca rupicola, Koeleria cristata, Poa angustifolia, Stipa capillata, Stipa joannis (*S. pennata*), *S. pulcherrima, Adonis vernalis, Ajuga laxmannii, Allium paniculatum, A. rotundum, Astragalus austriacus, A. dasyanthus, A. onobrychis,Chamaecytisus austriacus agg., Crambe tataria, Dorycnium* spp., *Euphorbia pannonica* (*E. glareosa*), *Fragaria viridis, Galium glaucum, Hypericum elegans, Inula germanica, Linaria biebersteinii, Nepeta parviflora, Phlomis tuberosa, Plantago media, Ranunculus polyanthemos, Rapistrum perenne, Salvia austriaca, S. nemorosa, S. nutans, Taraxacum serotinum, Teucrium chamaedrys, Thalictrum minus, Viola ambigua.*

The cold-continental steppes (including meadow steppes and semi-desert-like vegetation) of most of the Great Hungarian Plain were transformed into warm continental steppes during the early Holocene (Nyilas and Sümegi 1991; Sümegi et al. 2000; Sümegi 2005b). These steppes persisted essentially uninterrupted till their transformation into arable lands, since forest development was prevented by climatic conditions, grazing, fires and human influences (see Medzihradszky et al. 2000; Magyari et al. 2009). The late glacial steppes were dominated by grasses, *Artemisia* and Chenopodiaceae species, although other species, such as *Helianthemum, Scabiosa, Knautia arvensis, Trifolium, Sedum, Achillea, Leontodon, Taraxacum, Centaurea scabiosa* agg. were also recorded (Járai-Komlódi 1966; Sümegi 2005b; Sümegi et al. 1999, 2006; Magyari 2002). These genera reveal very little about the actual species composition and degree of dominance in the steppe vegetation (perhaps they suggest the presence of a kind of meadow steppe). It is not known what species of grasses formed the matrix of these steppes, how many and what kind of forbs were intermixed, and to what degree the vegetation was closed.

The Holocene steppe vegetation of present-day Hungary is also little known. It is often assumed that the loess tablelands were covered with forest-steppe with scattered *Quercus* woodlands, although certain parts of the plains may have been continuously treeless (Zólyomi 1936, 1952; Járai-Komlódi 1966; Sümegi et al. 1999; Magyari 2002; Magyari et al. 2009; Sümegi 2005a, b). Plowing up the steppes started in the Neolithic age, and continued with only small interruptions (Fairbairn 1992, 1993; Sümegi 1998; Gyulai 2001). The first major period of destruction of steppes on chernozem soil may have been the eleventh to fourteenth centuries, when the country was characterized by a large number of small settlements and intensive land use (see Győrffy 1966; Blazovich 1985). Changes in the settlement structure, such as desertion of small villages and development of rural towns, a process starting in the second half of the thirteenth century and increasing in intensity throughout the fourteenth century (Blazovich 1985), may have resulted in at least the partial regeneration of the steppe vegetation over large areas.

Referring to this vegetation type, Kitaibel wrote: the grassland stands are "*essentially oldfields.*" He documented rather species-rich stands only at a few places, which suggests that along the roads he traveled such steppes did not occur. Many of the specialist, characteristic steppe species were found on the unplowed strips between arable lands and on earthworks marking the boundary between settlements. Kitaibel reported the occurrence of *Crambe tataria* at several locations. This species was probably quite widespread and occurred adjacent to and even on arable land. At times he also observed *Salvia nutans* in unbelievable amounts ("*everywhere across the grazing field from right to left, as if it had been sown*"). It is apparent from the comparison of maps of the First Military Survey and current soil maps that a substantial part of the loess tables east of the Tisza river was covered with arable land in addition to extensive steppes.

Steppes of the Great Plain were almost completely transformed to arable land by the end of the twentieth century. Sizeable steppe areas have survived in gullies west of the Danube, on the foothills of mountain ranges, and on elevated terrain embedded in saline steppes. Part of the stands situated on slopes is in good condition and appears stable over time (Virágh and Fekete 1984; Horváth 2002; Bartha 2007a). The stands on level ground, however, are degraded almost without exception (Tóth 1988; Molnár 2007). Steppe remnants on kurgans, earthen fortifications and on unplowed sections among arable fields are sometimes the only survivors of this vegetation type (Zólyomi 1969a; Csathó 2010). Invasive species, though spreading, are not yet abundant. The formerly grazed and currently abandoned stands, particularly on foothills, at places have been invaded by shrubs, and typically show an accumulation of dead organic matter (Bartha 2007a). Mainly stands next to arable land are threatened by plowing (Csathó 2010). Their capacity for regeneration is limited. Good regeneration on oldfields has been observed only on foothills (in abandoned vineyards). This type of steppe vegetation covers about 25,000 ha (of which ca. 6,000 ha are on foothills and slopes). Hardly 10% of the stands are in good condition (naturalness 4 and 5). All stands next to arable lands and those in good conditions are greatly threatened (Fig. 7.1).

7.4.2 Artemisia Steppe on Loess and Clay Cliffs (Habitat Code: I2)

This is the characteristic open vegetation of loess/clay cliffs and steep slopes with *Kochia prostrata* and *Agropyron pectiniforme* as dominant species. It also occurs on kurgans. Too steep loess cliffs are not suitable for the development of this vegetation just like cliffs invaded by *Robinia pseudacacia* or *Lycium barbarum*. The southern slopes of kurgans host usually species-poor stands. Fragmented stands can survive even on the most disturbed kurgans.

Characteristic, dominant and constant species are: *Agropyron pectiniforme* (*A. pectinatum*), *Bothriochloa ischaemum*, *Bromus hordeaceus*, *B. tectorum*, *Poa bulbosa*, *Stipa capillata*, *Allium sphaerocephalon*, *A. flavum*, *Anthemis tinctoria*, *Artemisia austriaca*, *A. campestris*, *A. pontica*, *Bassia sedoides*, *Brassica elongata*, *Ephedra distachya*, *Iris pumila*, *Kochia prostrata*, *Lappula patula*, *Linaria genistifolia*, *Linum austriacum*, *Sedum* (*Hylotelephium*) *maximum*, *Xeranthemum annuum*.

The lower parts of loess cliffs often turn into vertical walls, where the microclimate is rather extreme with strong insolation and only 100–200 mm annual precipitation.

As a consequence, these habitats are inhabited by cryptogamous communities, mainly mosses. They are extra-zonal desert communities exhibiting floristic similarities to the vegetation of gypsum deserts in southern Spain and the cryptogamous desert vegetation of the climatic loess deserts in the Dead Sea area. Based on the current distribution and ecological characteristics of many of the moss species in this vegetation it is presumed that these communities are very old. Several species may have been part of the xeric flora of the late Tertiary, and survived through the glacial period when they had large areas of distribution (Pócs 1999). The cold-continental dry steppes in the Carpathian Basin were transformed into warm-continental steppes during the Holocene. Characteristic species of the former (i.e. *Krascheninnikovia ceratoides*, *Bassia prostrata*) found refuge in extreme habitats, such as loess cliffs (Zólyomi 1936, 1952; Járai-Komlódi 1966; Zólyomi and Fekete 1994). There is hardly any palynological evidence of the presence of this vegetation type during the Holocene (but see Jakab et al. 2004). Since the Neolithic Age, and even more since the Bronze Age, it may have developed also in secondary habitats on the southern slopes of abandoned earthworks and kurgans.

Kitaibel observed this habitat type (*Bassia prostrata*) at several places. He also recorded repeatedly *Agropyron pectiniforme*, but mainly in closed steppes (the species also is listed among plants growing on kurgans).

The present-day stands of this vegetation type seem to be stable, although invasive species are spreading (*Robinia pseudacacia*, *Lycium barbarum*, *Prunus institia*, *Ailanthus altissima*). Certain habitat patches are afforested with *Robinia pseudacacia*. The stands directly adjacent to arable fields are often plowed. The total area of this vegetation type is about 95 ha. It was found at 152 locations in Hungary. The condition of only hardly one quarter of all the stands is rather natural. They are threatened and have limited regeneration potential (Fig. 7.1).

7.4.3 Steppe Woodland of *Quercus* on Loess (Habitat Code: M2) and Continental Deciduous Steppe Thicket on Loess (Habitat Code: M6)

These are low or moderately tall mixed oak forests (*Quercus robur*, *Q. pubescens*) on deep, humus-rich soils developed over loess with usually a dense shrub layer and forest and steppe species in the herb layer (M2), as well as thickets of *Amygdalus nana* (*Prunus tenella*), *Cerasus* (*Prunus*) *fruticosa* and *Rosa* species (*R. gallica*, *R. pimpinellifolia*) forming small patches in grasslands and fringes of xerothermophilous forests (M6) (Zólyomi 1957, 1958).

Characteristic, dominant and constant species are: *Acer tataricum*, *A. campestre*, *Cotinus coggygria*, *Crataegus monogyna*, *Fraxinus ornus*, *Ligustrum vulgare*, *Prunus spinosa*, *Pyrus pyraster*,*Quercus pubescens*, *Q. robur*, *Q. petraea s.l.*, *Q. cerris*, *Ulmus minor*, *Viburnum lantana*, *Adonis vernalis*, *Ajuga laxmannii*, *Buglossoides purpurocaerulea*, *Carex humilis*, *Dictamnus albus*, *Festuca rupicola*,

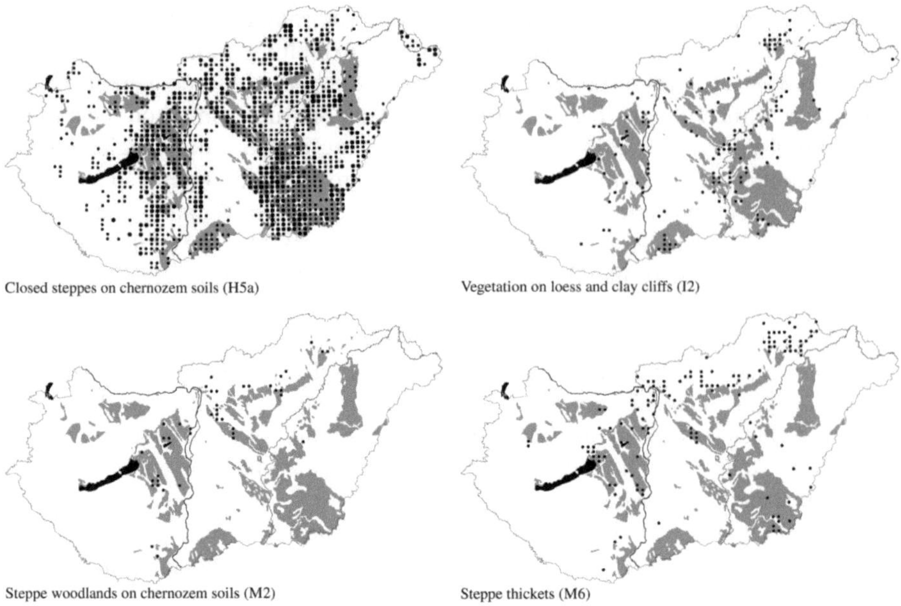

Fig. 7.1 Present distribution of forest-steppe habitats on loess in Hungary. The background *gray color* indicates potential distribution of dry vegetation on loess (Based on Zólyomi 1989)

Nepeta pannonica (*N. nuda*), *Peucedanum cervaria*, *P. alsaticum*, *Phlomis tuberosa*, *Polygonatum odoratum*, *Pulmonaria mollis*, *Stipa capillata*, *Tanacetum corymbosum*, *Thalictrum minus* in woodlands, and *Amygdalus nana*, *Cerasus fruticosa*, *Colutea arborescens*, *Crataegus monogyna*, *Prunus spinosa*, *Rosa spinosissima*, *R. gallica*, *Adonis vernalis*, *Brachypodium pinnatum*, *Dictamnus albus*, *Festuca rupicola*, *Fragaria viridis*, *Geranium sanguineum*, *Inula ensifolia*, *I. hirta*, *Iris variegata*, *Peucedanum cervaria*, *Phlomis tuberosa*, *Stipa* spp., *Vinca herbacea*, in steppe thickets.

The extent and characteristics of steppe woodlands and thickets growing on the Great Plain during the Holocene are not known. Based on the estimates of several authors (Zólyomi 1952; Járai-Komlódi 1966; Jakab et al. 2004; Sümegi 2005a, c; Sümegi et al. 2006), large areas were treeless on the loess tables and plateaus (at most 10–20 [–30]% woodland cover). This is directly or indirectly supported by the richness of the steppe flora, the occurrence of steppe animals, the behavior of nomadic peoples, and the occurrence of chernozem soils without signs of forest presence (see Medzihradszky et al. 2000). The regeneration potential of steppe woodlands on level loess tables with a dry climate and often very homogeneous habitats without refugia, and where the effect of human land use is more intense, is likely to be lower than on the topographically more diverse foothills and sand dune areas with a variety of habitats.

Kitaibel observed steppe woodlands on loess only on few occasions, most of them at the margins of the plains. According to his field notes, there were nearly no trees, except for some *Salix* and in most areas even shrubberies were rare. At one location, however, the now extinct *Spiraea crenata* was abundant. On the sheets of the First Military Survey, the lack of forests on the dry loess tables is particularly noticeable.

Today, rather natural steppe woodlands are found only at the margins of the plains and on loess-covered slopes west of the Danube. Its total area is less than 100 ha (for their current conditions see Lendvai and Kevey 2008). Only about a dozen stands of steppe thickets have remained on the Great Plain. They are small, usually degraded, and thus greatly endangered (Fig. 7.1).

7.4.4 Open Sand Steppe (Habitat Code: G1)

These are drought-tolerant shortgrass steppes with at most 75% vegetation cover, occupying loose, humus-poor sand in the Great Plain and more infrequently in hilly regions and foothills. The dominant species are drought-tolerant tussock grasses (*Festuca vaginata*, *Stipa borysthenica*). Formerly, their stands formed mosaics partly with steppe woodlands of *Quercus* or *Juniperus* and *Populus* scrub, as well as with *Molinia* meadows located in depressions. It is an endemic community.

Characteristic, dominant and constant species are: *Bothriochloa ischaemum*, *Bromus tectorum*, *B. squarrosus*, *Carex liparicarpos*, *Festuca vaginata*, *F. wagneri*, *Koeleria glauca*, *Poa bulbosa*, *Secale sylvestre*, *Stipa capillata*, *S. borysthenica*, *Achillea ochroleuca*, *Alkanna tinctoria*, *Alyssum tortuosum*, *A. montanum* subsp. *gmelini*, *Arenaria serpyllifolia*, *Artemisia campestris*, *Centaurea arenaria*, *Colchicum arenarium*, *Corispermum nitidum*, *C. canescens*, *Dianthus serotinus*, *D. diutinus*, *Ephedra distachya*, *Euphorbia seguierana*, *Fumana procumbens*, *Gypsophila fastigiata* subsp. *arenaria*, *Iris humilis*, *Kochia laniflora*, *Linum hirsutum* subsp. *glabrescens*, *Medicago minima*, *Onosma arenaria*, *Polygonum arenarium*, *Potentilla arenaria*, *Salsola kali*, *Silene borysthenica*, *S. conica*, *Syrenia cana* (*Erysimum canum*), *Teucrium chamaedrys*.

The sand dune areas of the Great Plain were formed during the ice age, and were naturally mobilized several times until recently (Lóki et al. 1995; Jakab et al. 2004; Sipos and Kiss 2006). Areas of shifting sand were also generated by overgrazing. The vegetation of shifting sand is one of the most dynamic communities of Hungary (regarding changes in forest cover and in the dominance-structure of species) (Fekete 1992; Biró and Molnár 1998; Bartha 2007b; Biró et al. 2007, 2008; Bartha et al. 2008a, b, c). There are no direct data on the forest cover of the dunes, but it is likely that they were not forested completely at any time throughout the Holocene (Jakab et al. 2004; Sümegi et al. 2005). Thus, at least parts of the grassland vegetation are likely to be continuous from the late glacial and the early postglacial (Boreal) periods. Also, it is not known since when *Festuca* and *Stipa* species have been the dominant grasses of these open steppes.

In the description of Kitaibel, the current flora of shifting sand may be recognized, but in a much more open and treeless landscape. The maps of the First Military

Survey show extensive treeless sand dune areas with very few arable lands, tree plantations, orchards and vineyards (Biró and Molnár 1998). At that time, the open sand steppes formed an extensive, partly contiguous system of habitat patches in the sandhill area between the Danube and Tisza Rivers (Biró 2003). The primary species stabilizing sand dunes in the nineteenth century was *Festuca vaginata* (Fekete 1992; Biró and Molnár 1998).

With the diminishing effect of wind and declining frequency of overgrazing since the nineteenth century, natural closure of the vegetation has been taking place and shrubs have been spreading. Today, *Juniperus communis*, *Populus alba*, *P. nigra*, *P. canescens*, and *Crataegus monogyna* are common, sometimes abundant, in most sand dune areas. Shifting sand dunes disappeared, and vegetation cover increased strongest in communities dominated by *Festuca wagneri* (Fekete et al. 2002). Afforestation of shifting sand and the simultaneous establishment of orchards and vineyards with the development of a network of distant farms commenced in the nineteenth and were completed in the twentieth century. As a result, the total area of open sand steppes was reduced to 6–8% of their original size (Biró 2008).

Invasive species (*Asclepias syriaca*, *Robinia pseudacacia*, *Ailanthus altissima*, for example) have been spreading rapidly, although mainly in the secondary or more degraded stands, and in oldfields and plantations adjacent to them (Botta-Dukát 2008; Czúcz et al. 2011). Today, these steppes are rarely plowed, but their afforestation is still in progress. Their total area is 10,700 ha, of which only one third is rather natural. It is because many tracts of arable land have been abandoned recently in sandy habitats, which undergo relatively rapid regeneration (Csecserits and Rédei 2001; Bartha 2007c; Bartha et al. 2008b, c), and thus are classified in this habitat category. The vegetation type is a bit endangered (Fig. 7.2).

7.4.5 Closed Sand Steppe (Habitat Code: H5b)

Closed dry and semi-dry steppes on humus rich sandy soils. The minimum cover of the herb layer is 50%. They occur in almost every sandy region of the country.

Characteristic, dominant and constant species are: *Bothriochloa ischaemum*, *Carex humilis*, *Chrysopogon gryllus*, *Festuca wagneri*, *F. rupicola*, *Holoschoenus romanus* (*Scirpoides holoschoenus*), *Koeleria cristata*, *Phleum phleoides*, *Poa angustifolia*, *Stipa capillata*, *S. borysthenica*, *Achillea ochroleuca*, *Adonis vernalis*, *Anemone sylvestris*, *Anthericum liliago*, *A. ramosum*, *Asperula cynanchica*, *Aster linosyris*, *Astragalus austriacus*, *A. dasyanthus*, *A. exscapus*, *A. asper*, *Chamaecytisus ratisbonensis*, *Colchicum arenarium*, *Dianthus giganteiformis* subsp. *pontederae*, *Filipendula vulgaris*, *Geranium sanguineum*, *Inula salicina*, *Iris aphylla* subsp. *hungarica*, *Iris humilis*, *I. variegata*, *Peucedanum arenarium*, *Plantago media*, *Potentilla arenaria*, *Pulsatilla pratensis* subsp. *hungarica*, *P. patens*, *Salix repens* subsp. *rosmarinifolia*, *Salvia pratensis*, *Scabiosa canescens*, *Senecio* (*Tephroseris*) *integrifolius*, *Silene borysthenica*, *Teucrium chamaedrys*, *Trifolium alpestre*, *Veronica pallens* (*Pseudolysimachion incanum*), *V. spicata* (*Pseudolysimachion spicatum*).

The habitat of closed sand steppes is completely suitable for forests (Magyar 1961). Despite this, the available, but not yet decisive historical data suggest that it was only partially forested during the last 10,000 years (Jakab et al. 2004; Sümegi et al. 2005). In the first half of the Holocene, there might have been more extensive steppe woodlands in sandy habitats, the disappearance of which probably took place during the past 5–6,000 years (see Sümegi et al. 2005). From the late Neolithic Age to the eighteenth century, most of the closed steppe stands may have been plowed up (many of them even several times), or were at least grazed for several thousand years. It is unknown how many sand steppe species had disappeared by the eighteenth century. The decline in the number of steppe species from the mountain ranges to the center of the sandhill area is very evident today (Fekete et al. 1999, 2010). Our field experience and earlier floristic data suggest that similarly to salt steppe woodlands, the steppe flora is richer in the vicinity of sand steppe woodlands than at sites having been treeless for long. This phenomenon may be explained by less intense land use, a great diversity of microclimatic sites, a small-scale variation in habitat conditions, and, as a consequence of all these, by a great variety of microrefugia (Molnár 1998; Molnár et al. 2008d).

Judged from his species lists, Kitaibel probably observed mostly disturbed and species-poor, weed-infested, supposedly secondary stands (many of them may have been oldfields). Floristic data, mainly from the nineteenth century, indicate an altogether rich steppe flora (see Molnár et al. 2008d). According to the sheets of the First Military Survey, the plowing of closed sand steppes already began at the end of the eighteenth century, although most of the sandy habitats with humus-rich soil were still covered with almost completely treeless sand steppe.

During the nineteenth century, almost 100% of the closed sand steppes were plowed up. Apart from some military grounds, this vegetation survived only in refugia (fringes and glades of sand steppe woodlands, rather humus-rich depressions among sand dunes, dune tops in marshy and saline environments). Owing to draining of wetlands in the second half of the twentieth century, many secondary and species-poor, atypical stands have developed from former calcareous fens and *Molinia* meadows (Molnár et al. 2008d). Certain stands are invaded by shrubs, others by invasive species (e.g., *Robinia pseudacacia*). Litter accumulation is pronounced in stands recently freed from grazing. It undergoes slow regeneration, often in small areas, because oldfields in this habitat occur mostly at the edges of sand dunes and as enclosures in wetlands. Sand steppes developed from drained meadows are often plowed up. The total area of closed sand steppes is 28,000 ha, of which only a few percent is found in their original habitat. Only one quarter of the stands are considered natural (naturalness 4 and 5). The more species-rich stands are greatly endangered (Fig. 7.2).

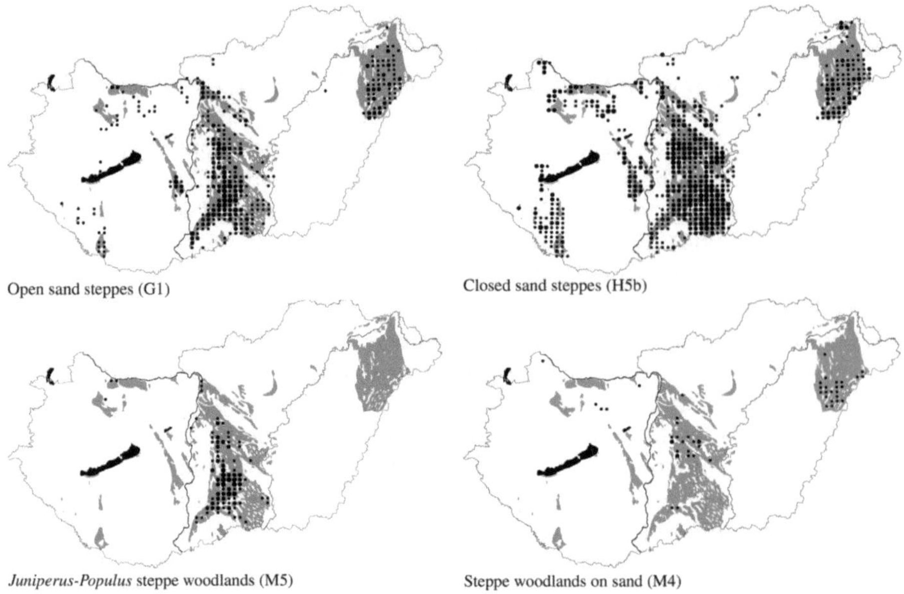

Fig. 7.2 Present distribution of forest-steppe habitats on sand in Hungary. The background *gray color* indicates the potential distribution of sand vegetation (Based on Zólyomi 1989)

7.4.6 *Juniperus-Populus* Steppe Woodland and Scrub (Habitat Code: M5)

These are open woodlands or scrub dominated by *Juniperus communis* and/or *Populus alba* and *P. canescens* on sandy soils with low to moderate humus content. This species poor, open vegetation type is typically poor in forest species, and forms a mosaic with open sand steppes (G1). It is most widespread on the Great Plain, in the sandy region between the Danube and Tisza rivers.

Characteristic, dominant and constant species are: *Berberis vulgaris*, *Crataegus monogyna*, *Juniperus communis*, *Ligustrum vulgare*, *Populus alba*, *P. canescens*, *P. nigra*, *Rhamnus catharticus*, *Brachypodium sylvaticum*, *Convallaria majalis*, *Geum urbanum*, *Hieracium umbellatum*, *Lithospermum officinale*, *Polygonatum odoratum*, *P. latifolium*, *Vincetoxicum hirundinaria*, *Viola hirta*, *V. odorata*.

Although *Juniperus communis* is a widespread species on the sand dunes today, very few data on its occurrence are available from the past 10,000 years (Járainé Komlódi 1985). It may have appeared in the landscape 2,000 years ago simultaneously with the decline of *Quercus* followed by *Carpinus* and *Fagus*.

According to the hypothesis of Gábor Fekete (Fekete 1992), the surfaces of sand dunes well above the groundwater table are settled, in the absence of habitat specific woody species, by the generalist *Juniperus communis* and/or *Populus alba* and *P. canescens* 'borrowed' from floodplain forests. Thus, this scrub is the edaphic climax community in the successional series of sand dunes. With regard to the origin of

Juniperus-Populus scrub, other hypotheses have also been put forward. Soó (1960) does not exclude the possibility that *Juniperus-Populus* scrub may develop from closed *Quercus* forests via degradation, though he also accepts the view that this vegetation may be part of the natural successional series of primary succession on sand dunes. Simon (1985) argues that the direct successional vegetation development from open sand steppe via *Juniperus-Populus* scrub to *Quercus* woodlands is possible. These opposing views stem partly from the fact that stands of oak trees are still present in certain sand dune areas, whereas the forest flora is completely absent in others.

Phytosociological material documenting the species composition of *Juniperus-Populus* scrub is very scant. The only published phytosociological tables are from Szodfridt (1969). Based on these data, he explicitly rejects the idea that *Juniperus-Populus* scrub may develop from *Quercus* woodlands via degradation. Based on the position of this community on dunes, two types were described. The first is found on top of the dunes, in which species typical of sand steppes are still dominant, whereas in the other type, growing in depressions, already some species of open forests and closed steppes with broad ecological tolerances (mostly shrubs) appear. There are only very few exclusive forest species (while mesophilous forest plants are completely absent), even though oak woodlands and forests are not absent in the studied landscape.

In the sand dune areas of Hungary, several types of *Juniperus-Populus* scrub may develop depending on the geomorphology of the dunes. They primarily differ in the dominance relations of the two edifying species, which is determined by their dynamics (for instance, *Populus* species grow larger and form denser stands in concave, more favorable sites, from which *Juniperus* is gradually excluded owing to strong shading). The position of these types relative to one another and the occurrence of these types as a series is characteristic of the landscape (Babos 1955).

As of today, there are no relevant data available from the Holocene to test the two hypotheses (data from landscape analyses currently support both). The total absence of extensive *Juniperus* scrub in the sandhill area between the Danube and Tisza rivers during much of the Holocene cannot be excluded either. In evaluating the hypotheses, it is important to consider that there are fully grown *Quercus robur* and *Q. pubescens* trees in the dry sand dunes, and it is also possible that *Quercus* woodlands could not develop in certain sand dune areas during the last millennia due to continual forest clearing, fires and heavy grazing.

At the time of the First Military Survey, most sand dune areas in the Kiskunság were treeless (Biró 2003). Often, even open *Populus* woodlands and thickets were missing there. Even Kitaibel recorded only loose groups of *Populus* trees on sand dunes with typical species of the sand steppe flora in between.

During the nineteenth century, the sparse woodlands of the dunes extended in area (Biró 2008; Biró and Molnár 2009), which was partly a spontaneous process, and partly the result of afforestation to stabilize shifting sand. It is possible that *Populus nigra* has become a typical species of the sand dune flora this way (Molnár 2003). One reason for their depauperate forest flora may be just their secondary origin (besides their open canopy and dry habitat conditions). Their regeneration potential is very good, particularly in sand steppes, but also in oldfields. The presence of invasive species in the stands is typical (*Robinia pseudacacia*, *Ailanthus altissima*, more recently *Prunus serotina*, Juhász and Bagi 2008; Juhász et al. 2009; Botta-Dukát et al. 2008). Certain stands are under intensive forest management (Babos 1955). Grazing in this vegetation has almost stopped by now. Their total area is about 3,000 ha. Almost two thirds of the stands are rather natural (naturalness 4 and 5). It is not endangered (Fig. 7.2).

7.4.7 Steppe Woodland of Quercus on Sand (Habitat Code: M4)

These are steppe woodlands developed on humus-rich sandy soils and dominated by *Quercus robur*. They are represented both by rather small groups of trees or more extensive stands. In the landscape, they typically form a mosaic with sand steppes. Their shrub layer is generally tall and rather dense. *Festuca rupicola* and *Poa angustifolia*, together with the so-called forest-steppe species are common in the herb layer (Hargitai 1940; Soó 1943).

Characteristic, dominant and constant species are: *Berberis vulgaris*, *Betula pendula*, *Chamaecytisus ratisbonensis*, *Cornus sanguinea*, *Corylus avellana*, *Crataegus monogyna*, *Euonymus europaeus*, *Fraxinus angustifolia* subsp. *pannonica*, *Ligustrum vulgare*, *Malus sylvestris*, *Populus alba*, *P. tremula*, *Pyrus pyraster*, *Quercus robur*, *Q. pubescens*, *Rhamnus cathartica*, *Rosa elliptica*, *R. gallica*, *Salix rosmarinifolia*, *Viburnum lantana*, *Anemone sylvestris*, *Brachypodium pinnatum*, *B. sylvaticum*, *Campanula bononiensis*, *Clinopodium vulgare*, *Convallaria majalis*, *Cucubalus baccifer*, *Dictamnus albus*, *Gentiana cruciata*, *Geranium sanguineum*, *Hieracium umbellatum*, *Iris variegata*, *Jurinea mollis*, *Lithospermum officinale*, *Lychnis coronaria*, *Melampyrum cristatum*, *Origanum vulgare*, *Peucedanum cervaria*, *P. oreoselinum*, *Polygonatum odoratum*, *Silene conica*, *Tanacetum corymbosum*, *Thalictrum minus*, *Trifolium alpestre*.

It is assumed that the extent of forests in areas of dry sand was greater in the first half of the Holocene, and much smaller afterwards. Based on indirect evidence, these woodlands were perhaps composed of *Tilia* and *Ulmus* in addition to *Quercus* (see Járai-Komlódi 1966, 1985; Jakab et al. 2004; Sümegi et al. 2005). In the Middle Ages, much more extensive woodlands existed in certain areas, but they were far in between (Hargitai 1940).

It may be inferred from the *Descriptio* of the First Military Survey that most of the wooded areas were *Quercus* woodlands, most of them with a 15-year clearing cycle (Molnár 1998). Data from Kitaibel suggested the heavy use of these forests along with their protection (for example, dense *Salix* hedgerows protecting them from grazing).

The stands have shrunk to a fraction of their original size during the past 200 years. They were transformed to arable land to a lesser, and tree plantations, first *Robinia* then *Pinus*, to a large extent. The formerly short clearing cycle was gradually increased to 25, then 40 years, and grazing in the forest was abandoned. Owing to water regulation carried out in the region in the twentieth century, the groundwater table has dropped 2–3 m, and the oaks started to die (Molnár 1998). The forests have been degraded first by the invasion of *Robinia pseudacacia*, more recently by *Prunus serotina*, and at places by wild game populations (wild boar, fallow deer) well exceeding carrying capacities. Most of the stands are surrounded by *Robinia* plantations. Several of the original species went locally extinct (for example *Majanthemum bifolium*, *Dracocephalum austriacum*). Establishment of oak plantations by currently adopted forestry methods is hardly successful (Molnár 1998). The total area of this vegetation type is ca. 290 ha. This is one of the most endangered vegetation types in Hungary. Half of the stands are degraded, and their regeneration potential is the lowest among all. Despite their legal protection, stands are cleared even today by forestry companies (Fig. 7.2).

7.4.8 Artemisia Steppe on Solonetz Soil (Habitat Code: F1a)

These steppes occur in the Carpathian Basin on saline, mainly solonetz soils that are periodically (but rarely and then only slightly) moist for a short time. They are shortgrass steppes dominated by *Festuca pseudovina* and *Artemisia santonicum* as the most frequent co-dominant species. Large stands (the so-called *puszta*) are typical. This vegetation type is usually rich in halophytic species, whereas species of wet meadows and dry steppes on chernozem soils are rare. It also hosts a number of endemic taxa.

Characteristic, dominant and constant species are: *Festuca pseudovina*, *Hordeum hystrix*, *Poa bulbosa*, *Allium vineale*, *Artemisia santonicum*, *Bassia prostrata*, *Bupleurum tenuissimum*, *Cerastium dubium*, *C. pumilum*, *Erophila verna*, *Gypsophila muralis*, *Limonium gmelini*, *Lotus tenuis* (*L. glaber*), *Myosotis stricta*, *Ornithogalum tenuifolium*, *Plantago schwarzenbergiana*, *Podospermum canum*, *Ranunculus pedatus*, *Scilla autumnalis*, *Trifolium retusum*, *T. angulatum*, *T. parviflorum* (*retusum*), *Veronica arvensis*.

Sümegi et al. (2006) consider the Hortobágy steppe as an almost completely treeless saline landscape (with graminoids, *Plantago maritima*, *Artemisia santonicum*, many Chenopodiaceae, among them supposedly *Atriplex tatarica*, *Suaeda* spp.) at the end of the Glacial Period and through the Holocene. In the interior of the roughly 100,000-ha-large Hortobágy, the pollen of *Fagus*, *Carpinus*, *Fraxinus* and *Tilia* were absent throughout the entire Holocene (as opposed to the nearby floodplains). These data suggest that the current stands of *Artemisia* steppe of the Great Plain may have directly descended from former salt steppes in earlier parts of the Holocene (Somogyi 1965; Sümegi et al. 2000, 2006). It is not known, however, how much the species pool has changed since the late Glacial Period, and for how long *Festuca pseudovina* and *Artemisia santonicum* have been the dominant species.

We identified the following species group in the 92 detailed lists of salt steppe species of Kitaibel, which were recorded during his journeys through the Great Plain around the turn of the eighteenth to nineteenth century. *Artemisia santonicum*-group: *Artemisia santonicum*, *Limonium gmelini*, *Bromus hordeaceus*, *Achillea collina*,

Plantago maritima, (*Festuca pseudovina*), (*Podospermum canum*), (*Puccinellia limosa*), (*Atriplex hastata*). This list recalls the *Artemisia* salt steppe for contemporary botanists. The written and often very apt descriptions of Kitaibel, which were based on great field experience, also indicate that he saw the saline steppes east of the Tisza River the same as we see them today. Species frequencies in the 92 species lists also correspond well to the current situation. On the sheets of the First Military Survey prepared between 1783 and 1785, the saline steppe cannot be recognized, but it is likely that the grassy fields mentioned in the description as "*moist from rain*" were not part of the floodplain, but of saline steppe. Robert Townson, a British explorer, traveling through the Hortobágy almost the same time (1799) as did Kitaibel, describes the Hortobágy unambiguously as a desolate land and depicts it as an immense barren land devoid of trees (apud Nyilas 1999).

The extension of *Artemisia* steppes has been reduced by plowing, amelioration and establishment of rice paddies during the past 200 years, although substantially less compared to other forest-steppe habitats. The degree of weed infestation is limited (*Bromus hordeaceus*, *Hordeum hystrix*), and the condition of many stands has improved in response to decreasing overgrazing in the past 30–40 years. They are almost completely free of invasive species. A long-term threat is leaching of the upper soil layer prompted by dropping groundwater levels, which results in the replacement of the halophytic species by generalist ones. On sufficiently saline soils, the *Artemisia* steppe has high regeneration potential, and can develop even in oldfields. It is one of the least endangered vegetation types in the forest-steppe zone in Hungary. Its total area currently is 33,800 ha, two thirds of which are in rather natural condition (naturalness 4 and 5) (Fig. 7.3).

7.4.9 Achillea Steppe on Solonetz Soil (Habitat Code: F1b)

This is a secondary, species-poor and usually shortgrass community growing typically on meadow solonetz soil. It is dominated generally by *Festuca pseudovina* and *Achillea setacea* and/or *A. collina*, which are associated with other pseudo-halophytic species, dry grassland species and generalist meadow species (it is usually poor in *Achillea asplenifolia* and stenohalophytic species). It occurs on drained or naturally dry former floodplains and may also develop from *Artemisia* steppes following leaching of the soil.

Characteristic, dominant and constant species are: *Carex stenophylla*, *Cynodon dactylon*, *Festuca pseudovina*, *Koeleria cristata*, *Lolium perenne*, *Poa angustifolia*, *Achillea setacea*, *A. collina*, *Bupleurum tenuissimum*, *Cardaria* (*Lepidium*) *draba*, *Euphorbia cyparissias*, *Inula britannica*, *Limonium gmelini*, *Plantago lanceolata*, *Podospermum canum*, *Ranunculus pedatus*, *Scleranthus annuus*, *Trifolium fragiferum*.

Data on its Holocene history are not available. Because of the regular relocation of riverbeds during the Holocene (Somogyi 1965; Sümegi et al. 2000), and the presumed periodical leaching of saline soils (Somogyi 1964, 1965), its occurrence may have been rather frequent in the past. There is no reference to this vegetation type in the notes of Kitaibel.

In the past 200 years, its area could have multiplied. It is the dominant habitat type in unplowed, but drained floodplains. It may develop in place of riparian forests after their clearing (Molnár and Borhidi 2003). It is often weed-infested, particularly when heavily grazed or when the soil is drained or leached, but invasive species are rare. It may develop in oldfields or from ameliorated *Artemisia* steppes and drained saline meadows. It easily regenerates, even in oldfields. Its total area is 46,000 ha, but only less than one third of the stands are in rather "natural" conditions (naturalness 4: low cover of weeds, dominance of *Festuca* and *Achillea*). It is not an endangered vegetation type (Fig. 7.3).

7.4.10 Saline Meadow (Habitat Code: F2)

These are seasonally inundated tallgrass meadows growing on solonetz or solonchak meadow soils. Their habitat typically is covered with water from October till May or June, and the soil remains moist throughout much of the year. They are less diverse on solonetz soils, where characteristic tussocks on small mounds of mud are formed, and this contributes to increased naturalness and species diversity.

Characteristic, dominant and constant species are: on solonetz: *Alopecurus pratensis*, *Beckmannia eruciformis*, *Glyceria fluitans* subsp. *poiformis*, *Oenanthe silaifolia*, *Plantago schwarzenbergiana; Ranunculus lateriflorus*, *R. sardous*, *Rorippa sylvestris* subsp. *kerneri*, on solonchak: *Carex distans*, *Festuca arundinacea*, *Juncus gerardii*, *Puccinellia limosa*, *Achillea asplenifolia*, *Linum perenne*, *Orchis laxiflora* subsp. *palustris* (*Anacamptis palustris*), *Rhinanthus angustifolius* subsp. *serotinus*, *Scorzonera parviflora*, *Taraxacum bessarabicum*, *Triglochin maritimum.*

It is unknown whether or not these meadows were originally forested, though the former existence of forests in these habitats is very unlikely not only for the presence of salts in the soil, but also for the widely fluctuating water regime and the heavy soil (Debreczy in Molnár and Kun 2000). Sümegi (2005b) demonstrated the presence of regularly desiccating, but not flooded meadows in the Hortobágy during the Holocene, which could have been saline meadows.

We identified three species groups in the species lists of Kitaibel: (1) *Beckmannia* group: *Lythrum virgatum*, *Beckmannia eruciformis*, *Trifolium fragiferum*, *Glyceria poiformis* subsp. *fluitans*, meadow species, (*Festuca pratensis-arundinacea*), (marsh species), (*Alopecurus geniculatus*) and (*Alopecurus pratensis*); (2) *Ranunculus lateriflorus* group: *Ranunculus lateriflorus*, *Elatine alsinastrum*, *Rorippa sylvestris* subsp. *kerneri*, *Eleocharis palustris*, *Glyceria poiformis* subsp. *fluitans*, marsh species, *Alopecurus geniculatus*, (*Plantago tenuiflora*) and (meadow species); (3) *Achillea asplenifolia-Agrostis stolonifera* group: *Festuca arundinacea-pratensis*, *Achillea asplenifolia*, *Agrostis stolonifera*, (fen species), (*Carex distans*), (*Juncus compressus*).

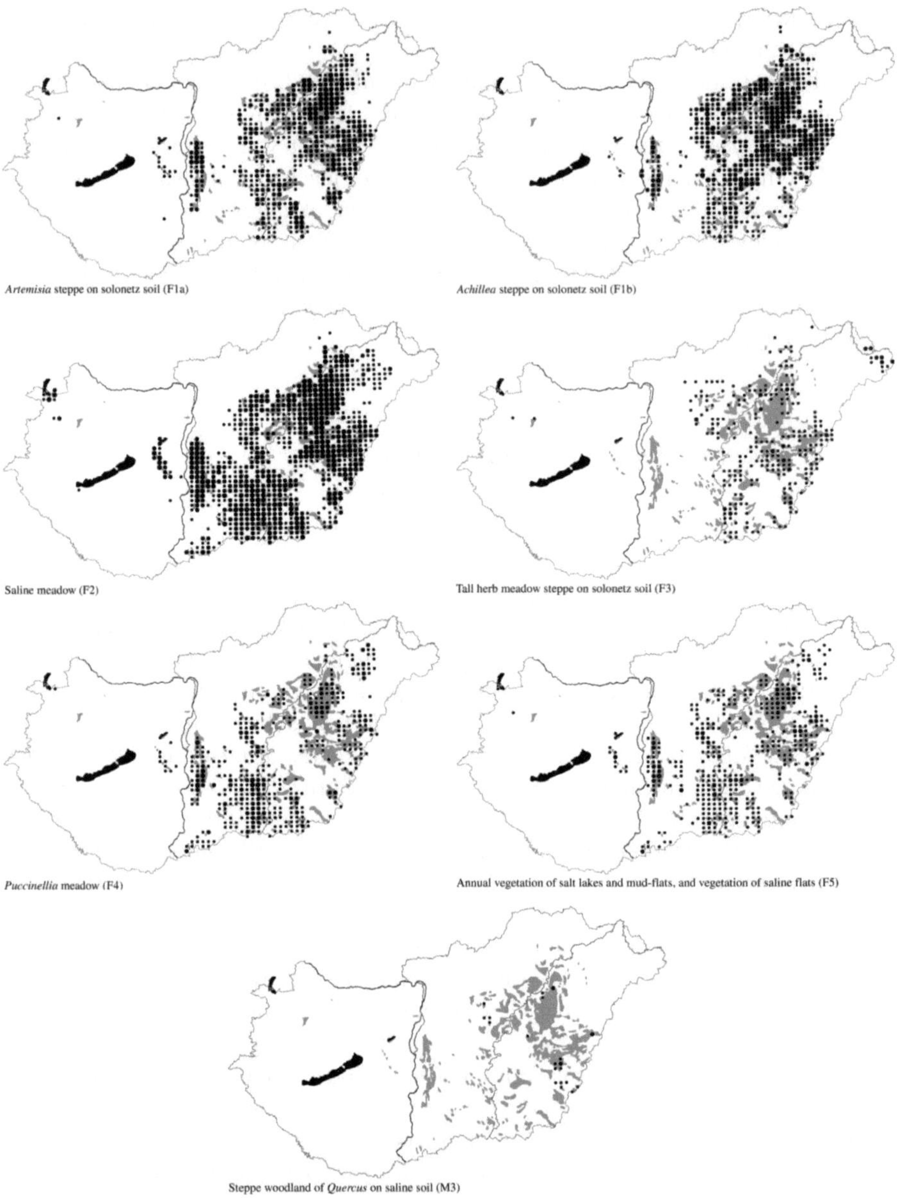

Fig. 7.3 Present distribution of forest-steppe habitats on saline soils in Hungary. The background *gray color* indicates the potential distribution of saline vegetation in general, and that of saline steppe woodlands in particular (Based on Zólyomi 1989)

The *Beckmannia* group suggests the presence of characteristic, wet saline meadows with ample supply of water, whereas the *Ranunculus* and *Achillea-Agrostis* groups indicate muddy habitats with heavily grazed (possibly by cattle) vegetation and solonchak meadows, respectively. Maps of the First Military Survey depict the non-flooded (possibly saline) meadows only irregularly, and thus the reconstruction of their original extent is not reliable. According to Kitaibel, *Alopecurus pratensis* was a common and typical species of saline meadows also at his time.

Saline meadows on solonetz soils have become substantially drier in the past 200 years, and their extension also decreased. Nevertheless, a very large number of stands have survived due to their heavy soils and the topographical features (many small local depressions that are very costly to drain) of their habitats. The extension of saline meadows on solonchak soils, however, has been drastically reduced. The groundwater table has dropped, sometimes as much as 2–3 m (Pálfai 1994), and the soil has been leached. Invasive species are rare. Plowing threatens mainly the stands on sand. They regenerate well, even in oldfields and after amelioration of the habitat. When left unmanaged, leaf litter accumulates rapidly, and diversity drops. Its total area is 93,000 ha, making it the most common habitat type in Hungary. More than half of its stands are in conditions close to natural (naturalness 4 and 5). It is rather threatened on solonchak soils, but not on solonetz (Fig. 7.3).

7.4.11 Saline Tall Herb Meadow Steppe (Habitat Code: F3)

These saline meadow steppes are dominated by meadow steppe and dry steppe species as well as halophytes; primarily Apiaceae and other tall herbs determine their structure. They occur mainly in the eastern part of Hungary on solonetz soils. Their habitat is typically wet in the spring and dry in the summer. The most frequent characteristic species are *Aster punctatus*, *Artemisia pontica* and *Peucedanum officinale*. The more mesic variant resembles tallgrass meadows and is rich in characteristic species and species of meadow steppes, whereas the drier variant is rather short and represents a transitional stage towards *Achillea* steppes. The species-richest stands are linked to steppe woodland of oaks on saline soils or hardwood (*Fraxinus*, *Ulmus*, *Quercus*) gallery forests, but most of the stands occur now in non-wooded environments. Similar habitats occur as far to the East as Mongolia (Varga 1989).

Characteristic, dominant and constant species are: *Alopecurus pratensis*, *Festuca pratensis*, *F. rupicola*, *F. pseudovina*, *Phragmites australis*, *Artemisia pontica*, *Aster linosyris*, *A. punctatus*, *Clematis integrifolia*, *Dianthus pontederae* (*D. giganteiformis* subsp.), *Filipendula vulgaris*, *Fragaria viridis*, *Iris spuria*, *Limonium gmelini*, *Lotus angustissimus*, *Lychnis flos-cuculi*, *Peucedanum officinale*, *Plantago schwarzenbergiana*, *Rumex pseudonatronatus*, *Serratula tinctoria*, *Seseli varium*, *Veronica spicata* (*Pseudolysimachion spicatum*), *Viscaria vulgaris* (*Lychnis viscaria*).

These saline meadow steppes may have been continuously present in the eastern part of the Great Plain since at least the Late Glacial Period, which is suggested by fossil pollen of *Thalictrum* sp., *Peucedanum* sp., *Filipendula ulmaria*, and *Sanguisorba officinalis* found at the margin of the Tisza floodplain (Magyari 2002; Sümegi 2004). More detailed data, however, are not available.

Kitaibel described stands of *Peucedanum officinale* at several locations. The co-occurring species there were essentially the same as those today, but the number of stands linked more strongly to floodplains may have been higher. Analysis using the JUICE software resulted in the following species group: *Peucedanum officinale-Aster punctatus* group: *Peucedanum officinale*, *Aster punctatus*, *Artemisia pontica*, dry steppe species, *Clematis integrifolia*, *Eryngium planum*, *Peucedanum alsaticum*. This list completely concurs with the flora of characteristic contemporary stands. Contrary to our expectation, Kitaibel recorded *Peucedanum officinale* only infrequently suggesting that this kind of meadow steppe was uncommon on floodplains and their margins even 200 years ago. The First Military Survey does not provide information.

During the past 200 years, many of its stands have been plowed or afforested. Others dried up, which has lead to the development of vegetation of shorter stature with an atypical species composition. They are sometimes infested by weeds. Stands adjacent to forests may be invaded by shrubs. They have limited regeneration potential, although some of the typical species readily enter oldfields where they may become abundant (i.e. *Aster punctatus*). The total area of this community is barely 1,120 ha, one third of which is in good condition (naturalness 4 and 5). It is endangered (particularly the species-rich stands) (Fig. 7.3).

7.4.12 *Puccinellia* Meadow, Annual Halophytic Vegetation of Salt Lakes and Vegetation of Saline Flats (Habitat Code: F4 and F5)

In the most saline habitats of Hungary, three main community types are distinguished: (1) annual halophytic vegetation of salt lakes and mud-flats (F5); (2) vegetation of saline flats embedded in dry saline steppes as small patches (F5); and (3) closed *Puccinellia* meadows (F4). These communities are rich in Pannonian endemics, and exhibit pronounced continental characteristics. They develop on strongly saline soils that are under water early in the vegetation period, but then dry out completely. On solonchak soils, precipitation of crystallized salts on the ground surface is common.

Characteristic, dominant and constant species are: *Crypsis aculeata*, *C. alopecuroides*, *C. schoenoides*, *Cyperus pannonicus*, *Pholiurus pannonicus*, *Puccinellia limosa*, *P. festuciformis* subsp. *intermedia*, *Aster tripolium* subsp. *pannonicus*, *Atriplex littoralis*, *Bassia sedoides*, *Camphorosma annua*, *Chenopodium chenopodioides*, *C. glaucum*, *Lepidium crassifolium*, *Matricaria chamomilla* var. *salina*, *Myosurus minimus*, *Plantago maritima*, *P. tenuiflora*, *Plantago schwarzenbergiana*, *Rorippa sylvestris* subsp. *kerneri*, *Salicornia prostrata*, *Salsola soda*, *Spergularia media*, *S. salina*, *Suaeda pannonica*, *S. maritima*, *S. salinaria*.

We assume that stands of considerable size may have existed continuously in Hungary since the Pleistocene. However, direct evidence is very scarce, because the survival chance of pollen grains is greatly reduced by soil salinity and soil cracking. Furthermore, it is very difficult to distinguish among Chenopodiaceae pollen (but see fossil evidence for the occurrence of *Suaeda* and *Atriplex tatarica* in the Hortobágy steppe, Sümegi et al. 2006). Before the draining of wetlands, several hundred salt lakes of variable ages occurred on the Great Plain (Molnár 1979; Boros and Biró 1999). Surprisingly, *Polygonum aviculare*, which is considered an indicator species of human disturbance (trampling by humans and livestock), has been a regular companion in the communities on saline mud-flats in the Hortobágy throughout the Glacial and Postglacial periods (Sümegi et al. 2006) suggesting that it is a natural component of these communities.

Kitaibel observed similar habitats at many places. He often recorded salt harvests by sweaping, and even the collection of soda (sodium carbonate). Once he describes the vegetation zones of a salt lake: open water in the center with *Bolboschoenus maritimus*, then *Puccinellia limosa* outward, followed by other species (*Plantago maritima*, *Podospermum canum*, *Lepidium crassifolium*, *Camphorosma annua*). This perfectly corresponds to the vegetation zones of current salt lakes with white water. We identified the following six species groups by analyzing the 92 species lists of Kitaibel with the JUICE software: (1) *Pholiurus* group: *Trifolium angulatum*, *Ventenata dubia*, *Pholiurus pannonicus*, arable weeds, *Bromus hordeaceus* and *Myosurus minimus*; (2) *Hordeum hystrix-Lepidium ruderale* group: *Matricaria chamomilla* var. *salina*, *Polygonum aviculare*, *Hordeum hystrix*, *Lepidium ruderale*, *Bromus hordeaceus*, *Pholiurus pannonicus*, (*Artemisia santonicum*) and (*Plantago tenuiflora*); (3) *Matricaria chamomilla* group: *Bromus hordeaceus*, *Matricaria chamomilla* var. *salina*, arable weeds, *Lepidium perfoliatum*, *Festuca pseudovina*, (*Myosurus minimus*), (*Podospermum canum*) and (*Lepidium ruderale*). (4) *Suaeda maritima* group: *Suaeda maritima*, *Salicornia prostrata*, *Suaeda pannonica*, *Scorzonera parviflora*, *Aster tripolium*, *Atriplex hastata*, *Crypsis aculeata*, (*Spergularia maritima*); (5) *Lepidium crassifolium* group: *Lepidium crassifolium*, *Camphorosma annua*, (*Plantago maritima*), (*Aster tripolium*); (6) *Puccinellia limosa-Lepidium crassifolium-Camphorosma annua* group: *Camphorosma annua*, *Lepidium crassifolium*, *Puccinellia limosa*, *Limonium gmelini*, *Plantago maritima*, (*Lepidium ruderale*), (*Atriplex hastata*).

The first three groups indicate disturbed stands on solonetz soils. The cause of disturbance may be regular trampling, grazing, and partly the inclusion of secondary stands at the margin of arable fields into the species lists. The fourth and fifth group represent species characteristic of salt lakes and saline flats on solonchak soils, respectively, whereas the sixth group includes the generalist species of *Puccinellia* meadows and saline flats. In the latter cases the species lists also correspond to the current situations.

In the First Military Survey, the Austrian military officers sometimes attached some texts to salt lakes, such as "*Himmelteich*" (a lake from the sky), and "*dries up completely by summer*". These data most likely refer to temporal intermittent lakes without concentrated inflow. The tiny salt lakes and muddy surfaces in the salt steppes are not depicted on the maps, however.

In the past 200 years, these habitats dried up considerably, or their soil often became leached. Certain stands were plowed up or ameliorated. Most of the salt lakes in sandy areas have completely dried up in the last third of the twentieth century, whereas salt lakes on heavy solonetz soils suffered less from drought (there the spread of *Bolboschoenus* causes some problems). Their vegetation rapidly regenerates in suitable habitats, but regeneration ability sharply drops on leached soils. Soil leaching is followed by colonization of meadow, marsh or steppe species, and the development of a closed vegetation cover. Salt accumulation on the surface has ended at many places. Invasive species are missing. Only heavy overgrazing may cause moderate weed infestation. The total area is approximately 9,500 ha. The condition of around 80% of the stands is close to natural (as the degraded ones are often classified into other habitat types). They are not threatened on the short run (except for salt lakes, which are endangered), but on the long run leaching may result in the transformation of many stands (Fig. 7.3).

7.4.13 Steppe Woodland of Quercus on Saline Soils (Habitat Code: M3)

These are open woodlands on saline soils dominated by *Quercus robur*. The stands rarely reach a height of 15 m, and form a habitat mosaic with saline tall herb meadow steppes, halophytic communities, dry steppes and reed beds. Forest elements are mixed with steppe and halophytic species. This habitat occurs almost exclusively in the eastern part of Hungary.

Characteristic, dominant and constant species are: *Acer tataricum*, *Crataegus monogyna*, *Fraxinus angustifolia* subsp. *pannonica*, *Ligustrum vulgare*, *Malus sylvestris*, *Prunus spinosa*, *Pyrus pyraster*, *Quercus robur*, *Ulmus minor*, *Agropyron caninum*, *Alopecurus pratensis*, *Arum orientale*, *Aster punctatus*, *Betonica officinalis*, *Brachypodium sylvaticum*, *Carduus crispus*, *Carex melanostachya*, *Corydalis cava*, *Cucubalus baccifer*, *Doronicum hungaricum*, *Lathyrus niger*, *Melampyrum cristatum*, *Melica altissima*, *Peucedanum officinale*, *Poa nemoralis*, *Polygonatum latifolium*, *Pulmonaria mollis*, *P. officinalis*, *Ranunculus ficaria*, *Scilla vindobonensis*, *Serratula tinctoria*, *Viola cyanea-odorata*.

A substantial portion of the current stands of steppe woodlands on saline soil may have developed from hardwood gallery forests by means of water loss in their habitat, which has been witnessed in the past 150 years, and as a consequence, canopy gaps developed. We propose this hypothesis on the basis of indirect historical evidence, literature data (Máthé 1933; Soó 1960; Zólyomi and Tallós 1967; Zólyomi 1969b, c; Molnár 1989), and the examination of current stands. This hypothesis was put forward first by Máthé (1933). Later it was accepted by Soó (for example Soó 1960) as opposed to Zólyomi (Zólyomi and Tallós 1967; Zólyomi 1969b, c). Another group of saline steppe woodlands may represent stands that developed before draining and water regulation. They may have occurred in floodplains where the soil had a deep saline soil horizon, and which gradually dried out as a consequence of changes in the course of rivers (see Somogyi 1965). This origin is suggested by the geographical situation of certain stands that are now bordered by ancient riverbeds and oxbows, which were naturally cut-off from the river many centuries ago. Habitat desiccation and salinization thus may have taken place before water regulation, which represented only the second step in their development.

On the Great Plain, Kitaibel observed 'forests on saline soils' at several locations, which were rather similar to those today, but exhibited more pronounced gallery forest characteristics. He also described forests growing in floodplain and/or saline environments with large amounts of *Peucedanum officinale* and *Aster punctatus.* During the First Military Survey, data making this vegetation type reliably recognizable were not collected (only the continuous existence of the present forests on saline soil may be demonstrated, but not their saline character 200 years ago).

During the past 200 years, the size of the few remaining stands further decreased, and some even disappeared. The canopy has been closing, and the glades have been invaded by shrubs owing to the cessation of grazing and cutting. The forest interior tended to become atypical (species-poor) rather than weedy. Spread of invasive species has started. Some of the stands are still intensively managed, while most of them are under legal protection. Several saline areas have been afforested with *Quercus robur.* A portion of these plantations have opened up, and the rather old stands gradually attain steppe woodland physiognomy. The regeneration ability of these steppe woodlands is generally good in glades, poor in plantations, and zero in oldfields. Their total area is barely 130 ha, although with the directly adjacent closed forest patches – that are classified in a different habitat type: closed lowland *Quercus* forests (L5) – it is twice as large. Barely one fifth of the stands are in good condition (naturalness 4 and 5). It is a strongly endangered vegetation type (Fig. 7.3).

7.5 Past Trends in the Hungarian Forest-Steppe Habitats

It was at the end of the eighteenth century that the spatial and temporal density, reliability and detail of vegetation data suddenly increased in Hungary. At that time the first detailed data on the extension, species composition and land use history of almost all forest-steppe habitats were collected (Kitaibel's diary, First Military Survey, Biró 2006; Molnár 2007). Luckily, this was before river regulation and afforestation programs were launched in Hungary (Tóth 1997). The data indicate that most forest-steppe habitats already had undergone considerable transformation by this time. Steppe woodlands on loess had almost completely disappeared (in fact we do not even now their extent during the entire Holocene). Steppe woodlands on sand were already missing in many sand areas and those that remained became thin and patchy. The steppe woodlands on saline soils were also small in size. Large sections of the forest-steppe area were treeless and even shrubless. Roughly 20% of all the forests may have been steppe woodland (mostly on sand), 67% were wet or mesic forest on floodplains and along rivers and rarely swamps, and 13% were individual trees or groups of trees and shrubberies. Around 63% of the forests were dominated by *Quercus robur* (Biró and Molnár 2009). According to both Kitaibel and the First Military Survey, the most common woody vegetation types were gallery forests, steppe woodlands on sand and shrubberies.

The forests were cleared usually in short cycles (<20 years) and/or were heavily grazed. For these reasons the species number and abundance of forest herbs were small (Molnár 1998).

According to the data of Kitaibel and the First Military Survey (Biró 2006; Molnár 2007), there were large areas covered with sand steppes and closed steppes on chernozem soil, most of which were used for grazing (although part of them had already been plowed up). The most typical farming method – besides the three-field rotation – was switching between crop cultivation and grazing in several-year cycles. Marshes, as well as saline and sand dune areas were, however, plowed up only very sporadically. The "struggle" between grazing land and arable land was apparent in the landscape. Dune areas with wind-blown sand were extensive and occupied an area larger than expected under natural conditions because of excessive grazing. They were mostly treeless (*Populus* stands were rare, *Juniperus* was almost fully absent), and many of them were shifting dunes. The grassy vegetation (dominated by *Festuca* and *Stipa* spp.) was open and heavily grazed. The area of closed steppes on chernozem soil, which were extensive during the entire Holocene, further expanded following a minimum in the Middle Ages. As the several-hundred-year-old regenerating oldfields were very widespread, species-rich stands may have been rather sporadic. Most of the characteristic steppe species even then occurred on roadsides and in the unplowed strips between arable fields.

According to the descriptions of Kitaibel, both the solonetz and solonchak areas were very similar to those of today (species composition, zonal arrangement of communities, land use). Their extension may have reached its maximum then (Kitaibel saw only few arable fields on saline soil). His data suggest heavy grazing. Kitaibel's data also show that the saline vegetation did not undergo significant changes during the past 200 years (see Molnár 1996b, 2003; Molnár and Borhidi 2003), and paleobiological evidence suggests the continuity of this vegetation since the end of the Last Glacial Period (Sümegi et al. 2000, 2006). We emphasize this, because the extensive (thousands of hectares) steppes on solonetz soil were considered secondary and no more than 150 years old until recently in the Hungarian botanical and pedological literature (see Szabolcs 1961; Jakucs 1976; Varga and Sipos 1993).

The forest-steppe vegetation experienced great changes during the past 200 years (Table 7.1). Former trends continued (degradation, plowing) and new processes started due to motorized technologies in agri- and sylviculture. Amelioration and plowing of saline soils along with transformation into rice paddies started (though the latter attempts have been given up). Indigenous *Quercus* forests have been replaced by *Robinia* and *Pinus* plantations (still ongoing), and sand dunes have been afforested (now slowing down). Draining of wetlands has lead to soil leaching, and colonization of non-salt-tolerant species has resulted in the development of atypical vegetation. In the second half of the twentieth century, the spread of invasive species has intensified and eventually has become a widespread and harmful process. Another threatening factor arising in the past decades is the cessation of mowing and grazing.

Based on a comparison of the latest vegetation map showing the potential natural vegetation (Zólyomi 1989) and the actual data of the MÉTA database, approximately 251,000 ha (6.8%) of the total of 3,700,000 ha of forest-steppe vegetation have survived until today (Figs. 7.1, 7.2 and 7.3). Habitat types that survived most extensively and in the largest patches are those on saline soil, which are unsuited for crop and intensive hay production (184,000 ha, which is at least 27% of the original area). Fragments with sizes of several thousands or even several tens of thousands of hectares are not infrequent. Although stands of the open sand steppe with sizes of several hundreds hectares still remain, of the overall 1,300,000 ha of forest-steppe on sand, only 42,000 ha remain (3.3%). Most stands of the closed steppe, however, are secondary, developed from drained wet meadows. The greatest decline in the extent of forest-steppe vegetation occurred on loess. Of its former 1,730,000 ha, barely 1.5% are left. A substantial portion of them is fragmented and restricted only to roadsides in certain parts of the country (Zólyomi 1969a; Csathó 2010). Presently, they remain to a large extent as enclosures in saline steppes instead of on large loess tablelands, and as dry derivatives of meadows. The woody component of the Hungarian forest-steppe has almost completely disappeared from the loess tablelands and saline soils and has been drastically reduced on sand.

7.6 The Present Status of the Hungarian Forest-Steppe Habitats

The condition of only 5.5% of the stands may be considered natural, 38% semi-natural, 46% moderately degraded, and 10% strongly degraded (or barely to moderately regenerated), based on the combined data of all forest-steppe habitat units in the MÉTA database. The rest of the stands (2.5%) has been degraded to such a degree that they can no longer be identified. The degree of degradation of the stands seems to greatly depend on the suitability of the habitat for crop production, soil condition, and the secondary origin of stands. The proportion of rather natural stands exceeds 80% only in the case of *Puccinellia* meadows and the annual vegetation on saline and mudflats. This number is around 60% in *Artemisia* steppe and *Juniperus-Populus* scrub, and 12–25% in closed steppe on chernozem soil and on sand, in the vegetation of loess and clay cliffs, and in steppe woodland on saline soil and on loess tablelands.

7.7 Future Prospects of the Hungarian Forest-Steppe Habitats

We considered three approaches for predicting future trends in forest-steppe vegetation in Hungary: (1) extrapolation of past trends, (2) current threats and regeneration potential, (3) expected climate change and expected changes in land use.

According to the MÉTA database, the most influential factors threatening the forest-steppe vegetation are (1) spread of invasive species; (2) abandonment of traditional land-use (and, as a consequence, spread of shrubs and accumulation of leaf litter); (3) drop of the groundwater table due to regulation and draining,

(4) plowing; (5) overgrazing; (6) excessive wild game populations; (7) afforestation of grassland habitats; (8) forest management practices (Molnár et al. 2008c) (Table 7.3).

The most successful and common invasive species are *Robinia pseudacacia*, *Asclepias syriaca*, *Ailanthus altissima*, *Elaeagnus angustifolia*, and more recently *Prunus serotina*. In view of predicted climate change and changes in land use we expect new invasive species to appear (Walther et al. 2009).

After a peak in the 1980s, the number of domestic animals kept on grazing fields has been declining in Hungary for 25 years. Today mostly sheep graze on the steppe, while only a few percent of the cattle herds graze in the field. The vegetation of abandoned or undergrazed fields gradually changes: leaf litter accumulates, certain species (for instance, *Alopeurus pratensis*, *Elymus repens*, *Bothriochloa ischaemum*) become overdominant, the abundance of short plants decreases (such as *Thymus* spp., *Potentilla arenaria*, annuals), while shrubs and invasive species start spreading.

It may seem surprising that groundwater directly influences around 55% of the forest-steppe vegetation in Hungary. These are primarily forest-steppe habitats on saline soils and steppe woodlands on sand. The explanation for the latter is that these woodlands are dominated by *Quercus robur*, a rather mesic tree of the lowlands. Under current climatic conditions, this oak can be an edifying species in the forest-steppe on sand only if its roots reach down to the groundwater. In areas covered with sand, however, the groundwater table has dropped drastically (3–5 m, Pálfai 1994; Szilágyi and Vorosmarty 1997) in the past decades, and therefore these forests can barely regenerate themselves, and even older trees die off (Molnár 1998).

The regeneration potential of different forest-steppe habitat units is greatly different (Seregélyes et al. 2008). It depends on the dispersal abilities of species making up the local vegetation, which tends to be good in species of open sand steppe and saline mud-flats (Bartha et al. 2008a, b, c; Csecserits and Rédei 2001; Czúcz et al. 2011; Biró 2006; Molnár 2007), and on the speed of regeneration, which is rather low for forests and closed steppes on chernozem soil (Bartha 2007a; Molnár and Botta-Dukát 1998; Molnár 1996a, 1998). It also depends on the spatial characteristics of the neighbouring landscape, such as the presence of suitable sources of propagules (Molnár and Botta-Dukát 1998; Czúcz et al. 2011) and availability of habitat patches for colonization. Unfortunately, steppe woodlands, which are among the rarest vegetation types of Hungary, are also the ones with the lowest regeneration potential among all Hungarian habitat units. This may be the consequence of (1) their strong fragmentation, (2) the spread of invasive species, (3) the reduced survival and reproductive success of oaks (due to the increasing frequency of arid years and a decrease in the groundwater table, Molnár 1998), and (4) the inherently slower regeneration of dry woodlands. We anticipate further decline in the regeneration potential of almost all forest-steppe habitat units because of the expected harmful changes in land use, the spread of invasive species, and the predicted increase in aridity.

We may provide only crude estimates on the effects of expected climate change on forest-steppe habitats (Czúcz et al. 2009, 2010). It is likely that increasing aridity will facilitate transformation of habitats, spread of invasive species, changes in land use, and will affect the level of under- or over-utilization (Fekete and Varga 2006). The strength of these effects, however, may greatly depend on the forecasted energy crisis (Czúcz et al. 2010). Forecasts based on statistical distribution models (SDM) are, however, not sufficiently reliable because of the many artifacts these models produce.

As we have shown, most of the forest-steppe vegetation in Hungary has been exterminated and replaced by arable land, plantations and settlements. Although ca. two-thirds of the remnants are under legal protection (Horváth et al. 2003, unpublished), the efficiency of protection measures is variable. In the past decades, conservationists have abandoned the approach of focusing exclusively on reserves, and conservation strategies have been developed for entire regions with varied landscapes (Natura 2000 network, EU agri-environmental schemes). We believe, however, that the official bodies of conservation are not yet prepared for the expected changes in land use and climate. They have to face two issues simultaneously: (1) how to maintain disappearing traditional land use, which fundamentally contributes to the survival of the natural heritage (see Molnár et al. 2008a), and (2) how to maintain the ability of natural habitats to regenerate and to adapt to the forecasted climate and land-use change (Bartha 2007a; Seregélyes et al. 2008; Czúcz et al. 2010).

Considering the past changes and current situation of natural habitats, we think that, of all forest-steppe habitat units in Hungary, steppe woodlands on loess and sand will almost fully disappear in the coming decades owing to the concerted effects of improper forest management and spread of invasive species. We further expect significant degradation and reduction in area of tall herb saline meadow steppes due to increased aridity and plowing, and of closed steppes on chernozem soil, steppe thickets and closed steppes on sand due to spread of shrubs and other invasive species, as well as plowing. Degradation of the vegetation on loess and clay cliffs and open sand steppe also is expected owing to the spread of invasive species in both, and afforestation in the latter. The area of *Puccinellia* meadows, vegetation of saline and mud-flats, saline meadows and steppe woodlands on saline soils will further decrease due to increasing aridity and leaching. The extent of the secondary *Achillea* steppe will probably increase (although it may be partly plowed up) as a consequence of leaching of the soil of saline *Artemisia* steppes and drying of saline meadows. Saline *Artemisia* steppes seem to be relatively stable, although soil leaching, and thus their slow transformation, are ongoing processes. As a long-term outcome of lack of grazing, the extension of *Juniperus-Populus* scrub in sand dune areas may increase (though this may be hindered by invasive species) despite the increasingly arid climate.

In sum, we predict further transformation, invasion of weeds, and fragmentation of the forest-steppe habitats in Hungary – although at different speeds and to different degrees depending on habitat types – and, as a consequence, the decline of their resilience and regeneration potential.

References

Aliokhin VV (1950) Karta rastitelnosti SSSR. Masshtab 1:18,000,000. (The map of vegetation of the USSR. Scale 1: 18,000,000.) In: Aliokhin VV (ed), Geografiya rasteniy. (The plant geography.) Uchpedgiz, Moscow

Artushenko AT (1970) Rastitel'nost' lesostepi i stepi Ukrainy v chetvertichnom periode. Naukova Dumka, Kiev

Babos I (1955) A nyárfások homokbuckán előforduló megjelenési formái. (Physiognomical types of poplar groves growing on sand dunes.) Erdészeti kutatások 4:31–86

Bartha S (2007a) Kompozíció, differenciálódás és dinamika az erdőssztyep biom gyepjeiben. (Composition, differentiation and dynamics in the grassland of the forest-steppe biome.) (Slope steppes, loess steppes and forest steppe meadows in Hungary.) In: Illyés, Bölöni (eds) Lejtősztyeppek, löszgyepek és erdősztyepprétek Magyarországon. MTA ÖBKI, Budapest, pp 72–100

Bartha S (2007b) Homoki gyepek spontán dinamikai folyamatai. In: Horváth A, Szitár K (eds) Agrártájak monitorozása. A hatás-monitorozás elméleti alapjai és gyakorlati lehetőségei. (Spontaneous dínamical processes in sand grassland. In: Horváth A, Szitár K (eds) Monitoring agricultural landscapes. Theoretical foundations and practical prospects of effect monitoring.) MTA ÖBKI, Vácrátót, pp 183–188

Bartha S (2007c) Másodlagos szukcesszió felhagyott homoki szántókon. In: Horváth A, Szitár K (eds) Agrártájak monitorozása. A hatás-monitorozás elméleti alapjai és gyakorlati lehetőségei. (Secondary succession in sandy oldfields. In: Horváth A, Szitár K (eds) Monitoring agricultural landscapes. Theoretical foundations and practical prospects of effect-monitoring.) MTA ÖBKI, Vácrátót, pp 202–207

Bartha S, Campetella G, Ruprecht E, Kun A, Házi J, Horváth A, Virágh K, Molnár Zs (2008a) Will inter-annual variability in sand grassland communities increase with climate change? Commun Ecol 9(Suppl):13–21

Bartha S, Molnár Zs, Fekete G (2008b) Patch dynamics in sand grasslands: connecting primary and secondary succession. In: Kovács-Láng E, Molnár E, Kröel-Dulay Gy, Barabás S (eds) The KISKUN LTER, long-term ecological research in the Kiskunság, Hungary. Institute of Ecology and Botany, H.A.S, Vácrátót, pp 37–40

Bartha S, Molnár Zs, Horváth A, Virágh K (2008c) Időjárási szélsőségek – társulásszintű adaptív válaszok. Évelő nyílt homokpusztagyepek 12 éves monitorozásának tapasztalatai. In: Korsós Z, Gyenis Gy, Penksza K (eds) Előadások Összefoglalói. (Meteorological extremities – adaptive community responses. Results of the monitoring of open sand steppes over 12 years.) Abstract of presentations. MBT XXVII, Vándorgyűlés 2008, Szeptember 25–26, pp 1–6

Berg LS (1958) Die Geographischen Zonen der Sowjetunion I. Teubner Verlag, Leipzig

Biró M (2003) A Duna-Tisza köze vegetációja a 18. században. Rekonstruált vegetáció térkép, méretarány 1:100 000. (Vegetation of the area between the Danube and Tisza Rivers in the 18th century. Reconstructed vegetation map. Scale 1:100,000.) In: Molnár Zs (ed) A Kiskunság száraz homoki növényzete. (Dry sand vegetation of Kiskunság, Hungary.) Természet Búvár Alapítvány Kiadó, Budapest, p 30

Biró M (2006) A történeti térképekre alapuló vegetációrekonstrukció és alkalmazásai a Duna-Tisza közén. Doktori értekezés. (The method of vegetation reconstruction based on historical maps and its application to the area between the Danube and Tisza Rivers.) Ph.D. dissertation, Pécs

Biró M (2008) A Duna-Tisza köze fásszárú vegetációjának átalakulása a 18. század óta, különös tekintettel a száraz homokterületekre. (Development of woody vegetation of the Danube Tisza Interfluve since the 18th century.) In: Kröel-Dulay Gy, Kalapos T, Mojzes A (eds): Talaj-vegetáció-klíma kölcsönhatások. (Soil-vegetation-climate interactions.) MTA Ökológiai és Botanikai Kutatóintézete, Vácrátót, pp 23–38

Biró M, Molnár Zs (1998) A Duna-Tisza köze homokbuckásainak tájtípusai, azok kiterjedése, növényzete és tájtörténete a 18. századtól. (Landscape types and their extent, vegetation and history of the sand dune areas between the Danube and Tisza rivers from the 18th century.) Történeti Földrajzi Füzetek 5:1–34

Biró M, Molnár Zs (2009) Az Alföld erdei a folyószabályozások és az alföldfásítás előtti évszázadban. (Forests of the Great Plain before river regulation and afforestation.) In: Kázmér M (ed) Környezettörténet 2006. A környezet változásai történeti és természettudományi források tükrében. Hantken Kiadó, Budapest, pp 169–206

Biró M, Révész A, Molnár Zs, Horváth F, Czúcz B (2007) Regional habitat pattern of the Duna-Tisza köze in Hungary II. – The sand, the steppe and the riverine vegetation; degraded and ruined habitats. Acta Bot Hung 50(Suppl):21–62

Biró M, Horváth F, Papp O, Molnár Zs (2008) Historical landscape changes near Fülöpháza in the Kiskunság. In: Kovács-Láng E, Molnár E, Kröel-Dulay Gy, Barabás S (eds) The KISKUN LTER: Long-term ecological research in the Kiskunság, Hungary. Institute of Ecology and Botany, Vácrátót, pp 11–12

Blazovich L (1985) A Körös-Tisza-Maros köz középkori településrendje. (Settlement structure in the area between the Körös and Maros Rivers.) Dél-alföldi Századok, Békéscsaba, Szeged

Bölöni J, Kun A, Molnár Zs (eds) (2003) Élőhely-ismereti útmutató 2.0. (Habitat guide 2.0.) Manuscript, MTA ÖBKI, Vácrátót. http://www.novenyzetiterkep.hu/eiu/index.html

Bölöni J, Molnár Zs, Illyés E, Kun A (2007) A new habitat classification and manual for standardized habitat mapping. Annali di Botanica nouva serie 7:105–126

Bölöni J, Molnár Zs, Biró M, Horváth F (2008a) Distribution of the Hungarian (semi-)natural habitats II. Woodlands and shrublands. Acta Bot Hung 50(Suppl):107–148

Bölöni J, Molnár Zs, Horváth F (2008b) Naturalness-based habitat quality of the Hungarian (semi-) natural habitats. Acta Bot Hung 50(Suppl):149–159

Bölöni J, Molnár Zs, Kun A (2011) Magyarország élőhelyei. A hazai vegetációtípusok leírása és határozója. (Habitats in Hungary. Description and identification guide of Hungarian vegetation types.) MTA ÖBKI, p 341

Borbély A, Nagy J (1932) Magyarország I. Katonai Felvétele II. József korában. (The first military survey of Hungary under Joseph the 2nd.) Térképészeti Közl 2:35–85

Borhidi A (1961) Klimadiagramme und klimazonale Karte Ungarns. Ann Univ Budapest Ser Biol 4:21–50

Borhidi A (2003) Magyaroszág növénytársulásai. (Plant associations of Hungary.) Akadémiai Kiadó, Budapest

Borhidi A (2012) Plant associations of Hungary. Akadémiai Kiadó, Budapest (in print)

Boros E, Biró Cs (1999) A Duna-Tisza közi szikes tavak ökológiai állapotváltozásai. (Changes in ecological conditions of the salt lakes in the area between the Danube and Tisza Rivers.) Acta Biol Debr Oecol Hung 9:81–105

Botta-Dukát Z (2008) Invasion of alien species to Hungarian (semi-)natural habitats. Acta Botanica Hungarica 50(Suppl):219–227

Botta-Dukát Z, Molnár E, Szitár K (2008) Invasion of alien plants in the sand vegetation of Kiskunság. In: Kovács-Láng E, Molnár E, Kröel-Dulay Gy, Barabás S (eds) The KISKUN LTER: long-term ecological research in the Kiskunság, Hungary. Vácrátót, pp 35–36

Bruelheide H (2000) A new measure of fidelity and its application to defining species groups. J Veg Sci 11:167–178

Csathó AI (2010) Fragments of steppe vegetation in roadsides and railway verges in the Great Hungarian Plain. In: Programme and book of abstracts. 2010 IENE international conference on ecology and transportation: improving connections in a changing environment, pp 85–86

Csecserits A, Rédei T (2001) Secondary succession on sandy old-fields in Hungary. Appl Veg Sci 4:63–74

Czúcz B, Torda G, Molnár Zs, Horváth F, Botta-Dukát Z, Gy K-D (2009) A spatially explicit, indicator-based methodology for quantifying the vulnerability and adaptability of natural ecosystems. In: Filho WL, Mannke F (eds) Interdisciplinary aspects of climate change. Peter Lang Scientific Publishers, Frankfurt, pp 209–227

Czúcz B, Gathman JP, McPherson GR (2010) The impending peak and decline of petroleum production: an underestimated challenge for conservation of ecological integrity. Conserv Biol 24(4):948–956

Czúcz B, Csecserits A, Botta-Dukát Z, Kröel-Dulay Gy, Szabó R, Horváth F, Molnár Zs (2011) An indicator framework for the climatic adaptive capacity of natural ecosystems. J Veg Sci 22:711–725

European Environmental Agency (2006) EEA Report No1/2002. Europe's biodiversity — biogeographical regions and seas. Biogeographical regions in Europe. The Pannonian region. http://www.eea.europa.eu/publications/report_2002_0524_154909

Fairbairn AS (1992) Archaeobotanical investigations at Csőszhalom: a Late Neolithic tell site in north-east Hungary. M.Sc. thesis, University College London, Institute of Archeology, p 67

Fairbairn AS (1993) Plant husbandry at the prehistoric Hungarian tell sites of Csőszhalom and Kenderfüöld. Final report for the British Academy Applied Sciences in Archaeology Fund

Fekete G (1965) Die Waldvegetation im Gödöllőer Hügelland. Akadémiai Kiadó, Budapest, 223 pp

Fekete G (1992) The holistic view of succession reconsidered. Coenoses 7:21–29

Fekete G, Kun A, Molnár Zs (1999) Floristic characteristics of the forest-steppe in the Duna-Tisza Interfluve. In: Kovács-Láng E (ed) Long-term ecological research in the Kiskunság, Hungary. MTA ÖBKI, Vácrátót, pp 13–14

Fekete G, Molnár Zs, Kun A, Virágh K, Botta-Dukát Z (2002) Záródó homokpusztagyep a Duna-Tisza-közén: a Festuca wagneri gyepjei. (Closed sand steppe in the area between the Danube and Tisza Rivers: stands of Festuca wagneri.) In: Salamon-Albert É (ed) Magyar botanikai kutatások az ezredfordulón. Tanulmányok Bohidi Attila 70. születésnapja tiszteletére. (Botanical Research in Hungary at the turn of the millennium. Studies in honor of the 70th birthday of Attila Borhidi.) PTE TTK, Pécs, pp 381–414

Fekete G, Varga Z (eds) (2006) Magyarország tájainak növényzete és állatvilága. (Vegetation and wildlife of Hungarian landscape.) MTA Társadalomkutató Központ, Budapest, pp 460

Fekete G, Somodi I, Molnár Zs, Fekete G, Somodi I, Molnár Zs (2010) Is chorological symmetry observable within the forest-steppe biome in Hungary? A demonstrative analysis of floristic data. Commun Ecol 11:140–147

First Military Survey (1783–1785) Scale 1:28,000. Map collection of the National Museum of Military History, Budapest

Gombocz E (ed) (1945) Diaria Itinerum Pauli Kitaibelii I-II. Verlag des Ungarischen Naturwissenschaftlichen Museums, Budapest, 1083 pp

Győrffy Gy (1966) Az Árpád-kori Magyarország történeti földrajza. (Historical geography of Hungary in the age of the Árpád dinasty.) Bp

Győrffy Gy, Zólyomi B (1994) A Kárpát-medence és Etelköz képe egy évezred előtt. (The Carpathian Basin and Etelköz a thousand years ago.) In: Kovács L (ed) A honfoglalásról sok szemmel I. Honfoglalás és régészet. (The Hungarian conquest from different aspects I. The conquest period and archaeology.) Balassi Kiadó, Budapest, pp 13–37

Gyulai F (2001) Archaeobotanika. A kultúrnövények története a Kárpát-medencében a régészeti növénytani vizsgálatok alapján. (Archeobotany. The history of cultivated plants in the Carpathian Basin on the basis of botanical research.) Jószöveg Kiadó, Bp., 221 pp

Hargitai Z (1940) Nagykőrös növényvilága. II. A homoki növényszövetkezetek. (The flora and vegetation of Nagykőrös. II. Plant associations on sand.) Bot Közlem 37:205–240

HMS (2000) Climate atlas of Hungary. Hungarian Meteorological Service, Budapest

Horváth A (2002) A mezőföldi löszvegetáció términtázati szerveződése. (The spatial organization of the vegetation on the loess plateaux in the Mezőföld.) Scientia Kiadó, Budapest

Horváth F, Polgár L (2008) MÉTA SQL expert interface and access service. Acta Bot Hung 50(Suppl):35–45

Horváth F, Kovács-Láng E, Báldi A, Gergely E, Demeter A (2003) Európai jelentőségű természeti területeink felmérése és értékelése. (Survey and evaluation of natural areas with European significance.) MTA Ökológiai és Botanikai Kutatóintézete, Vácrátót

Horváth F, Molnár Zs, Bölöni J, Pataki Zs, Polgár L, Révész A, Krasser D, Illyés E (2008) Fact sheet of the MÉTA Database 1.2. Acta Bot Hung 50(Suppl):11–34
Illyés E, Bölöni J (eds) (2007) Slope steppes, loess steppes and forest-steppe meadows in Hungary. [Lejtősztyepek, löszgyepek és erdőssztyeprétek Magyarországon.] MTA ÖBKI, Budapest, pp 236
Jakab G, Sümegi P, Magyari E (2004) A new paleobotanical method for the description of Late Quaternary organic sediments (Mire-development and paleoclimatic records from S Hungary.) Acta Geol Hung 47:1–37
Jakucs P (1961) Die phytozönologischen Verhältnisse der Flaumeichen-Buschwälder Südostmitteleuropas. Akadémiai Kiadó, Budapest
Jakucs P (1976) A Hortobágy növényvilága. Flora and vegetation of the Hortobágy. In: Kovács G, Salamon F (eds) Hortobágy. A nomád pusztától a Nemzeti Parkig. (Hortobágy. From the nomadic puszta to the national park.) Natura Kiadó, Budapest, pp 38–56
Járai-Komlódi M (1966) Palinológiai vizsgálatok a Magyar Alföldön a Würm glaciális és a holocén klíma- és vegetációtörténetére vonatkozóan. Kandidátusi Értekezés. (Palinological studies on the Hungarian Plain with regard to the vegetation history and climate of the Würm and the Holocene.) Dissertation for the Degree of Candidate, ELTE, Bp., 280 pp
Járai-Komlódi M (1987) Postglacial climate and vegetation in Hungary. In: Pécsi M, Kordos L (eds) Holocene environment in Hungary. MTA Földrajzi Kutatóintézete, Budapest, pp 37–47
Járainé Komlódi M (1985) Pollenanalitikai vizsgálatok a Kolon-tónál. (Palinological studies at the Lake Kolon.) In: Tóth K (ed) Tudományos kutatások a Kiskunsági Nemzeti Parkban (Scientific research in the Kiskunság National Park 1975–1984) Hungexpo, Budapest, pp 152–155
Jávorka S (1926, 1929, 1934, 1935, 1936, 1945) Kitaibel herbáriuma I-V. és Pteridophyta. (Herbaria and pteridophytes of Kitaibel.) Ann Musei Nat Hung 24:428–585; 26:97–210; 28:147–196; 29:55–102; 30:7–118; 38:85–97
Juhász K, Bagi I (2008) A Prunus serotina EHRH. élőhely preferenciái az invázió diszperziós szakaszában homoki területeken. (Habitat preference of Prunus serotina EHRH. in the dispersal stage of invasion on sand.) Bot Közlem 94(1–2):1–16
Juhász M, Bagi I, Csintalan Zs (2009) The critical effect of drought stress on the invading features of bird cherry (Prunus serotina Ehrh.) during dispersion stage on a sand dune area in Hungary. Cereal Res Commun 37:553–556
Kádár Z (1975) Landscapes, zones and their regional energy. Acta Geogr Debr 13:35–80
Király G (ed) (2009) Új magyar füvészkönyv. Magyarország hajtásos növényei. Határozókulcsok. New Hungarian herbal. Vascular plants of Hungary. Identification keys. Aggteleki Nemzeti Park Igazgatóság, Jósvafő
Kun A (2001) Analysis of precipitation year-types and their regional frequency distributions in the Danube-Tisza Mid region, Hungary. Acta Bot Hung 43:175–187
Lavrenko EM (1956) Stepi i sel'skohoziastvennye zemli na meste stepei. (Steppes and agricultural areas in the places of former steppes) In: Lavrenko EM, Sochava VB (eds) Rastitel'nyi pokrov SSSR, V kn. (The vegetation cover of the USSR, vol. V.) Akademii Nauk SSSR, Moscow
Lavrenko EM (1970) The provincial Division of the Central Asian Steppe Subregion of Eurasia. Botanicheski Zhournal 55:1734–1747
Lavrenko EM (1981) Vegetation of Pleistocene periglacial steppes of the USSR. Bot Zhurn 66:313–327
Lavrenko EM, Karamysheva ZV, Nikulina RI (1991) Stepi Evrazii (Steppes of Eurasia), Leningrad, Nauka
Lavrenko EM, Karamysheva Z (1993) The steppes of the former Soviet Union and Mongolia. In: Coupland RT (ed) Ecosystems of the world, vol 8b. Elsevier, Amsterdam
Lendvai G, Kevey B (2008) Aceri tatarico – Quercetum roboris. In: Kevey B (ed) Magyarország erdőtársulásai. (Forest Associations of Hungary.) Tilia 14:415–417
Lóki J, Sümegi P, Félegyházi E, Hertelendi E (1995) A Kolon-tó fenékszintjébe mélyített fúrás rétegsorának szedimentológiai, pollenanalitikai és malakológiai elemzése. (Sedimentological, palinological and malakological analysis of the core from the bed of Kolon Lake.) Acta Geogr Debr 33:93–115

Lőkös L (ed) (2001) Diaria Itinerum Pauli Kitaibelii III., Természettudományi Múzeum, Bp., 459 pp

Magyar P (1961) Alföldfásítás I-II. (Afforestation in the Great Plain I-II.) Akadémiai Kiadó, Bp., 466+512 pp

Magyari E (2002) Climatic versus human modification of the Late Quaternary vegetation in Eastern Hungary. Unpublished Ph.D. Thesis, University of Debrecen, Hungary, 150 pp

Magyari EK, Chapman JC, Passmore DG, Allen JRM, Huntley JP, Huntley B (2009) Holocene persistence of wooded steppe in the northern Great Hungarian Plain. J Biogeogr 37:915–935

Máthé I (1933) A hortobágyi Ohat-erdő vegetációja. (Vegetation of the hat Forest in the Hortobágy). Bot Közlem 28:163–184

Medzihradszky Zs, Járainé Komlódi M (1995) Az ember természetformáló tevékenysége a holocén folyamán a Kárpát-medencében. Emlékkötet Andreánszky Gábor születésének 100. évfordulójára. (Human impact on the environment in the Carpathian Basin during the Holocene.) Magyar Természettudományi Múzeum, Budapest

Medzihradszky Zs, Bölöni J, Molnár Zs, Kertész M, Varga Z, Debreczy Zs, Molnár A (2000) Mit tudunk a múltról? (What do we know about the past?) In: Molnár Zs, Kun A (eds) Alföldi erdőssztyeppmaradványok Magyarországon. (Forest-steppe remnants in Hungary.) WWF Füzetek 15, Budapest, pp 20–25

Molnár B (1979) A Duna-Tisza köze kialakulása és földtani felépitése. (Geology and development of the area between the Danube and Tisza rivers.) In: Tóth K (ed) Nemzeti Park a Kiskunságban. (National Park in the Kiskunság.) Natura Kiadó, Budapest, pp 64–73

Molnár A (1989) A bélmegyeri Fás-puszta növényzete. (Vegetation of the Fás puszta at Bélmegyer.) Bot Közlem 76:65–82

Molnár Zs (1996a) Ártéri vegetáció Tiszadob és Kesznyéten környékén II: A keményfaligetek (Fraxino pannonicae – Ulmetum) története és mai állapota. (Vegetation of the floodplain near Tiszadob and Kesznyéteny. II. The history and current status of hardwood gallery forests (Fraxino pannonicae – Ulmetum).) Bot Közlem 83:51–79

Molnár Zs (1996b) A Pitvarosi-pusztá és környékük vegetáció- és tájtörténete a Középkortól napjainkig. (Vegetation and landscape history of Pitvaros puszta from the Middle Ages.) Natura Bekesiensis 2:65–97

Molnár Zs (1998) Interpreting present vegetation features by landscape historical data: an example from a woodland-grassland mosaic landscape (Nagykőrös-wood, Kiskunság, Hungary). In: Kirby KS, Watkins C (eds) The ecological history of European forests. CAB International, Wallingford, pp 241–263

Molnár Zs (ed) (2003) A Kiskunság száraz homoki növényzete. (Dry sand vegetation of Kiskunság, Hungary.) TermészetBÚVÁR Alapítvány Kiadó, 159 pp

Molnár Zs (2007) Történeti tájökológiai kutatások az Alföldön. Doktori értekezés, Pécsi Egyetem. Studies in historical landscape ecology in the Great Plain. Ph.D. dissertation, University of Pécs, Pécs

Molnár Zs (2008a) A Duna-Tisza köze és a Tiszántúl növényzete a 18–19. század fordulóján I. módszertan, erdők, árterek és lápok. (Vegetation of the Great Plain east of the Danube River at the turn of the 18th and 19th centuries. I. Methodology, forests, floodplains and fens.) Bot Közlem 95:11–38

Molnár Zs (2008b) A Duna-Tisza köze és a Tiszántúl növényzete a 18–19. század fordulóján II. szikesek, lösz- és homokvidékek, legelők, sáncok, szántók és parlagok. (Vegetation of the Great Plain east of the Danube River at the turn of the 18th and 19th centuries. II. saline areas, loess tables, sand areas, grazing fields, ancient earthworks, arable land and oldfields.) Bot Közlem 95:39–63

Molnár Zs (2009) A Duna-Tisza köze és a Tiszántúl fontosabb vegetációtípusainak holocén kori története: irodalmi értékelés egy vegetációkutató szemszögéből. (Holocene history of the common vegetation types in the areas east of the Danube River, and between the Danube and Tisza Rivers: literature review from the standpoint of a vegetation scientist.) Kanitzia 16:93–118

Molnár Zs, Borhidi A (2003) Continental alkali vegetation in Hungary: syntaxonomy, landscape history, vegetation dynamics, and conservation. Phytocoenologia 21:235–245

Molnár Zs, Botta-Dukát Z (1998) Improved space-for-time substitution for hypothesis generation: secondary grasslands with documented site history in SE-Hungary. Phytocoenologia 28:1–29

Molnár Zs, Kun A (eds) (2000) Alföldi erdőssztyeppmaradványok Magyarországon. (Lowland forest-steppe remnants in Hungary.) WWF Füzetek 15., Budapest, 56 pp

Molnár Zs, Bartha S et al (2007b) A grid-based, satellite-image supported, multi-attributed vegetation mapping method (MÉTA). Folia Geobot 42:225–247

Molnár Zs, Bartha S, Babai D (2008a) Traditional ecological knowledge as a concept and data source for historical ecology, vegetation science and conservation biology: A Hungarian perspective. In: Szabó P, Hedl R (eds) Human nature. Studies in historical ecology and environmental history. Institute of Botany of the ASCR, Brno, pp 14–27

Molnár Zs, Bölöni J, Biró M, Horváth F (2008b) Distribution of the Hungarian (semi-)natural habitats I. Marshes and grasslands. Acta Bot Hung 50(Suppl):59–105

Molnár Zs, Bölöni J, Horváth F (2008c) Threatening factors encountered: Actual endangerment of the Hungarian (semi-)natural habitats. Acta Bot Hung 50(Suppl):199–217

Molnár Zs, Fekete G, Biró M, Kun A (2008d) A Duna-Tisza közi homoki sztyepprétek történeti tájökológiai jellemzése. In: Talaj-vegetáció-klíma kölcsönhatások (eds Kröel-Dulay Gy, Kapalos T, Mojzes A) (Characterization of closed sand steppes between the Danube and Tisza Rivers on the basis of their history and landscape ecology. In: Kröel-Dulay Gy, Kapalos T, Mojzes A (eds) Soil, climate, vegetation interrelations.) MTA ÖBKI, Vácrátót, pp 39–55

Natura (2000) Network, European Comission. http://ec.europa.eu/environment/nature/natura2000/index_en.htm

Németh F, Seregélyes T (1989) Természetvédelmi információs rendszer: Adatlap kitöltési útmutató. (Information system of nature conservation: Guide for completing the data sheets.) Manuscript, Környezetgazdálkodási Intézet (Institute of Environmental Management), Budapest.

Niklfeld H (1974) Natürliche Vegetation Karte, 1:200 000. In: Niklfeld H (ed) Atlas der Donauländer. Deuticke Vlg. Wien, Karte, 171

Nyilas I (1999) Az angol Robert Townson leírása a Hortobágyról. (Description of the Hortobágy by the British Robert Townson, 1793.) In: Rózsa P (ed) Robert Townson magyarországi utazásai. (The travels of Robert Townson in Hungary.) Kossuth Egyetemi Kiadó, Debrecen, pp 143–146

Nyilas I, Sümegi P (1991) The mollusc fauna of Hortobágy at the end of the Pleistocene (Würm 3) and in the Holocene. In: Proceedings of tenth international malacological congress (Tübingen 1989), pp 481–486

Pálfai I (1994) Összefoglaló tanulmány a Duna-Tisza közi talajvízszint-süllyedés okairól és a vízhiányos helyzet javításának lehetőségeiről. In: Pálfai I (eds.): A Duna-Tisza közi Hátság vízgazdálkodási problémái. A synthetic study on the causes of decreasing groundwater table between the Danube and Tisza Rivers, and the options for improving the current situation. In: Pálfai I (ed) Problems in water management between the Danube and Tisza Rivers. Nagyalföld Alapítvány, Budapest, pp 111–123

Pickett STA (1989) Space-for-time substitution as an alternative to long-term studies. In: Likens GE (ed) Long-term studies in ecology: Approaches and alternatives. Springer, Berlin, pp 110–135

Pócs T (1966) A magyarországi tűlevelű erdők cönológiai és ökológiai viszonyai. Kandidátusi értekezés tézisei. (Phytosociological and ecological characteristics of coniferous forests in Hungary.) Theses for the dissertation for the candidate degree, Budapest

Pócs T (1999) A löszfalak virágtalan növényzete I. Orografikus sivatag a Kárpát-medencében. (Non-flowering vegetation of loess cliffs I. Orographic desert in the Carpathian Basin.) Kitaibelia 4:143–156

Rackham O (1980) Ancient woodland. Arnold, London

Rapaics R (1918) Az Alföld növényföldrajzi jelleme. (The pytogeographical characteristic of the Great Plain.) Erdészeti Kísérletek 21:1–164

Seregélyes T, Molnár Zs, Csomós Á, Bölöni J (2008) Regeneration potential of the Hungarian (semi-)natural habitats. I. Concepts and basic data of the MÉTA database. Acta Botanica Hungarica 50(Suppl):229–248

Simon T (1985) Az ELTE Növényrendszertani és Ökológiai Tanszéke komplex ökológiai kutatásai. (Complex ecological research of the Department of Ecology and Plant Systematics of the

ELTE.) In: Tóth K (ed) Tudományos kutatások a Kiskunsági Nemzeti Parkban. (Scientific research in the Kiskunság National Park.) KNP, Kecskemét, pp 140–173

Sipos Gy, Kiss T (2006) OSL mérés lehetőségei: hagyományos módszerekkel nem datálható, kvarctartalmú üledékek korának meghatározása. (OSL measurements: a method for dating quartz-containing sediments that cannot be dated by traditional methods.) In: Kázmér M (ed) A Környezettörténet 2006 konferencia előadásainak összefoglalói. (Abstracts of the conference entitled History of the Environment 2006.) Hantken Kiadó, Budapest, pp 43–45

Somogyi S (1964) A szikes talajok képződésének földrajzi tényezői Magyarországon. (Geographical factors in the development of saline soils in Hungary.) Földr Közlem 3:219–244

Somogyi S (1965) A szikesek elterjedésének időbeli változásai Magyarországon. (Temporal changes in the extent of saline soils in Hungary.) Földr Közlem 4:41–55

Soó R (1926) Die Entstehung der ungarischen Puszta. Ungarische Jahrbücher 6:258–276

Soó R (1929) Die Vegetation und die Entstehung der ungarischen Puszta. J Ecol 17:329–350

Soó R (1931) A magyar puszta fejlődéstörténetének problémája. (Problems in the developmental history of the Hungarian puszta.) Földr Közlem 59:1–17

Soó R (1933) Floren- und Vegetationskarte des historischen Ungarns. A Tisza István Társaság Honismertető Bizottság Kiadványai 30:1–35

Soó R (1943) A nyírségi erdők a növényszövetkezetek rendszerében. (The forests of the Nyírség in the phytosociological system of classification.) Acta Geobot Hung 5:315–352

Soó R (1960) Az Alföld erdői. (Forests of the Great Plain.) In: Magyar P (ed) Alföldfásítás I. (Afforestation on the Great Plain I.) Akadémiai Kiadó. Budapest, pp 419–478

Soó R (1964) Magyarország növénytársulásainak részletes (kritikai) áttekintése. (Critical review of plant associations of Hungary.) In: Soó R (ed) A magyar flóra és vegetáció rendszertani-növényföldrajzi kézikönyve I. (Phytosociological and phytogeographical handbook of the flora and vegetation of Hungary I.) Akadémiai Kiadó, Budapest, pp 130–289

Sümegi P (1998) Az utolsó 15000 év környezeti változásai és hatásuk az emberi kultúrákra Magyarországon. (Environmental changes and their effect on human cultures in Hungary over the past 15,000 years.) In: Ilon G (ed) A régésztechnikusok kézikönyve. (Handbook for archeology technicians.) Savaria Kiadványa, Szombathely

Sümegi P, Magyari E, Daniel P, Hertelendi E, Rudner E (1999) Reconstruction of the Quaternary evolution of the Fehér-lake at Karoskút (Hungary). Földtani Közlöny 129:479–519

Sümegi P (2004) Findings of geoarcheological and environmental historical investigations at the Körös site of Tiszapüspöki-Karancspart Háromága. Anthaeus 27:307–342

Sümegi P (2005a) A comparative geoarcheological report and environmental history of the Bronze Age tell of Polgár, Kenderföld. In: Poroszlai I et al (eds) Bronze Age in Hungary: Koszider Periode. Százhalombatta Archaeological Expedition sorozat, III. kötet, Matrica Múzeum, Százhalombatta, pp 185–234

Sümegi P (2005b) Loess and upper paleolithic environment in Hungary. An introduction to the environmental history of Hungary. AUREA, Nagykovácsi, p 312

Sümegi P (2005c) The environmental history of the Jászság. In: Gál E, Juhász I, Sümegi P (eds) Environmental archaeology in north-eastern Hungary. Varia Archaeologica Hungarica sorozat, XIX. kötet, MTA Régészeti Intézet, Bp., pp 103–110

Sümegi P, Molnár A, Szilágyi G (2000) Szikesedés a Hortobágyon. (Salinization in the Hortobágy.) Természet Világa 131:213–216

Sümegi P, Mucsi M, Fényes J, Gulyás S (2005) First radiocarbon dates from the freshwater carbonates of the Danube-Tisza interfluve. In: Hum L, Gulyás S, Sümegi P (eds) Environmental historical studies from the late tertiary and quaternary of Hungary. University of Szeged, Szeged, pp 103–118

Sümegi P, Bodor E, Törőcsik T (2006) A hortobágyi szikesedés eredete. (The origin of salt affected soils in the Hortobágy.) In: Kiss A, Mezősi G, Sümeghy Z (eds) Táj, környezet és társadalom. Ünnepi tanulmányok Keveiné Bárány Ilona professzor asszony tiszteletére. (Landscape, environment, society. Studies dedicated to the honor of Prof. Keveiné Bárány Ilona.) SZTE Kiadványa, Szeged, pp 633–641

Szabolcs I (1961) A vízrendezések és öntözések hatása a tiszántúli talajképződési folyamatokra. (The effect of water regulation and irrigation on soil formation processes east of the Tisza.) Akadémiai Kiadó, Budapest, 355 pp

Szilágyi J, Vorosmarty Ch (1997) Modelling unconfined aquifer level reductions in the area between the Danube and Tisza rivers in Hungary. J Hydrol Hydromech 45:328–347

Szodfridt I (1969) Borókás nyárasok Bugac környékén. (Juniper-poplar scrub near Bugac). Bot Közlem 56:159–166

Tarasov PE, Webb T et al (1998) Present-day and mid-Holocene biomes reconstructed from pollen and plant macrofossil data from the former Soviet Union and Mongolia. J Biogeogr 25:1029–1053

Tarasov PE, Volkova VS et al (2000) Last glacial maximum biomes reconstructed from pollen and plant macrofossil data from northern Eurasia. J Biogeogr 27:609–620

Tichy L (2002) JUICE, software for vegetation classification. J Veg Sci 13:451–453

Tóth A (1988) Degradálódó hortobágyi löszgyepek reliktum foltjainak synökológiai viszonyai. (Synecological relations in remnant stands of degraded steppe on loess in the Hortobágy.) In: Tóth A (ed) Tudományos kutatások a Hortobágyi Nemzeti Parkban. (Scientific research in the Hortobágy National Park.) Budapest, pp 11–83

Tóth A (1997) A közép-tiszavidéki táj gyökeres megváltozása a folyószabályozási munkálatok nyomán. (Changes in the Middle Tisza Region as a result of river regulation.) In: Füleky Gy (ed) A táj változásai a honfoglalás óta Kárpát-medencében. (Changes of the landscape in the Carpathian Basin since the time of the Conquest.) GATE MSZKI, Gödöllő, pp 347–353

Varga Z (1989) Die Waldsteppen des pannonischen Raumes aus biogeographischer Sicht. - Düsseldorf. geobot. Koll 6:36–50

Varga Z (1992) Biogeographical and ecological background with special reference to lowland habitats. In: Fésüs I et al (eds) Interaction between agriculture and environment in Hungary. IUCN, Budapest, pp 20–52

Varga Z (2003) A Kárpát-medence állatföldrajza. A Kárpát-medence állatföldrajzi helyzete. (Zoogeography of the Carpathian Basin. Place of the Carpathian Basin in the phytogerographical system.) In: Láng I, Bedő Z, Csete L (eds) Növény, állat, élőhely. (Plant, animal, habitat.) MTA Társadalomkutató Központ – Kossuth Kiadó, Budapest, pp 89–128

Varga Z, Sipos J (1993) Hortobágyi Krónika (Hortobágy Chronicle). Alföldi Nyomda Debrecen, 96 pp

Virágh K, Fekete G (1984) Degradation stages in a xeroseries: composition, similarity, grouping, coordination. Acta Bot Hung 30:427–459

Walther G-R, Roques A et al (2009) Alien species in a warmer world: risks and opportunities. Trends Ecol Evol 24:686–693

Zólyomi B (1936) Tízezer év története virágporszemekben. (The history of ten thousand years in pollen grains.) Term Tud Közlöny 68:504–516

Zólyomi B (1952) Magyarország növénytakarójának fejlődéstörténete az utolsó jégkorszaktól. (History of the vegetation in Hungary since the last glacial period.) MTA Biol Oszt Közlem 1:491–530

Zólyomi B (1957) Der Tatarenahorn-Eichen Lösswald der zonalen Waldsteppe. Acta Bot Hung 3:401–424

Zólyomi B (1958) Budapest és környékének természetes növénytakarója. (The natural vegetation of Budapest and its neighborhood.) In: Pécsi M (ed) Budapest természeti képe. (Natural heritage of Budapest.) Akadémiai Kiadó, Budapest, pp 511–642

Zólyomi B (1969a) Földvárak, sáncok, határmezsgyék és a természetvédelem. (Earthworks, earthen fortifications, unplowed land strips and nature conservation.) Természet Világa 100:550–553

Zólyomi B (1969b) Közép-Tiszavidék: Természetes növényzet. (Middle Tisza region: natural vegetation) In: Pécsi M (ed) Magyarország Tájföldrajza. II. A tiszai Alföld. (Regional geography of Hungary. II. The Great Plain along the Tisza.) Akadémiai Kiadó, Budapest, pp 124–131

Zólyomi B (1969c) Körös-Maros közi síkság: Természetes növényzet. (Körös-Maros plain: natural vegetation.) In: Pécsi M (ed) Magyarország Tájföldrajza. II. A tiszai Alföld. (Regional geography of Hungary. II. The great plain along the Tisza.) Akadémiai Kiadó, Budapest, pp 317–319

Zólyomi B (1989) Magyarország természetes növénytakarója. (Map of the natural vegetation of Hungary.) In: Pécsi M (ed) Nemzeti Atlasz. (National Atlas.) Kartográfia Vállalat, Budapest, p 89

Zólyomi B, Fekete G (1994) The Pannonian loess steppe: differentiation in space and time. Abst Bot 18:29–41

Zólyomi B, Tallós P (1967) Galatello-Quercetum roboris. In: Zólyomi B (ed) Guide der Excursionen des Internationalen Geobotanischen Symposiums. Ungarn, Eger-Vácrátót, pp 55–61

Chapter 8
The Dry Grasslands in Slovakia: History, Classification and Management

Daniela Dúbravková and Mária Hajnalová

Abstract Dry steppe-like grasslands are of the most endangered habitats in Slovakia since the natural distribution of suitable sites supporting this vegetation is limited. These sites are refuges for many rare thermophilous species of plants and invertebrates, and significantly contribute to the biodiversity of the European landscapes. The land use of dry grasslands in Slovakia experienced some dramatic changes in the last decades. The most crucial factor negatively influencing the biodiversity of grassland habitats is cessation of traditional extensive management activities, abandonment, afforestation, ploughing and building that resulted in area reduction, fragmentation, and degradation of dry grasslands. We summarise the actual results of rather intense dry grassland research in eastern Central Europe from the perspectives of (I) their establishment history, (II) variability and classification, and (III) conservation and management. Summarizing the actual archaeological and palynological knowledge we polemicize about the potential existence of original dry grassland sites in the contact zone of the Western Carpathians and the Pannonian Basin where the steppe-like vegetation occurred continually since the Holocene. An overview of phytosociological and habitat classification of the steppe-like grasslands according to results of the newest surveys and expert perspectives is given including a call for need of revision of the classification of some European natural habitat types of community interest but with an unclear delimitation. Practical management recommendations, based on the ecological requirements of dry grassland habitats, are discussed.

D. Dúbravková (✉)
Institute of Botany, Slovak Academy of Sciences, Bratislava, Slovakia

Homeland Museum in Považská Bystrica, Považská Bystrica, Slovakia
e-mail: daniela.dubravkova@savba.sk

M. Hajnalová
Department of Archaeology, Faculty of Arts, Constantine the Philosopher University, Nitra, Slovakia
e-mail: mhajnalova@ukf.sk

M.J.A. Werger and M.A. van Staalduinen (eds.), *Eurasian Steppes. Ecological Problems and Livelihoods in a Changing World*, Plant and Vegetation 6,
DOI 10.1007/978-94-007-3886-7_8,

Abbreviations

AT Austria
CZ the Czech Republic

8.1 Introduction

Slovakia is located in the eastern part of Central Europe. It includes the biogeographical regions of the Western Carpathian Mountains and the adjacent northern part of the Pannonian (= Carpathian) Basin. There, the dry grasslands occur in the colline belt of the peri-Carpathian mountains (e.g. Považský Inovec Mts, Tríbeč Mts, Krupinská planina Mts, Slovenský kras Mts) and the lowland landscapes of the Pannonia (Fig. 8.1). The dry grassland vegetation in Slovakia is mostly restricted to smaller extra-zonal stands, although in some attributes of structure and species composition it is similar to the zonal forest-steppe (lesosteppe) in central Eurasia. Central European dry grasslands are western outposts of the vast steppes of Russia and the Ukraine (Walter 1974; Bohn and Neuhäusl 2000–2003). The forest-steppe zone enters Central Europe along the Danube River (Horvat et al. 1974; Illyés and Bölöni 2007). The precise location of the zonal forest-steppes in Central Europe differs in various sources although the main area is considered to be located in the Great Hungarian Plain (Bohn and Neuhäusl 2000–2003; Illyés and Bölöni 2007) (Fig. 8.2).

8.2 History of Dry Grassland Sites

In Central Europe, dry grasslands have existed since the Pleistocene. They occurred in vast areas in lowlands and hilly landscapes with a dry continental climate, harsh winters and short summers (Frenzel et al. 1992; Chytrý et al. 2007; Kuneš et al. 2008). In the full and late glacial the mountain areas of the Carpathians were covered by taiga and hemiboreal forest while towards to the west and southwest the forests tended to be increasingly open or patchy gradually passing into the generally treeless tundra and steppe landscapes (Kuneš et al. 2008). The occurrence of thermophilous species was conditioned by repeated series of their retreat and expansion due to the climatic oscillations in the Quaternary ice ages and interglacial stages (Hewitt 1999). In the interglacials the distribution areas of plant and animal species expanded back to their previous ranges from the cold-stage refugia located in southern Europe (Iberian, Apennine and Balkan peninsulas), Caucasus and Caspian Sea area through various migratory routes (Willis and Whittaker 2000; Ložek 2009). However, many topographically sheltered refugia with favourable microclimates were located also in the present temperate zones (Hewitt 2000). The recent occurrence of thermophilous species of continental, Pontic-Pannonian and sub-Mediterranean distribution in the Pannonian Basin is explained by the survival of some species with wide ecological

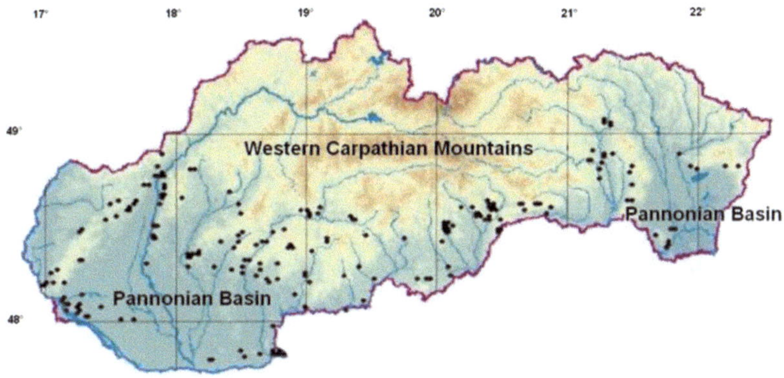

Fig. 8.1 Location of the dry grassland sites in Slovakia. They occur on warm slopes in the colline belt of the peri-Carpathian mountain ranges and in lowland areas of the Pannonian Basin

Fig. 8.2 The distribution of the forest-steppe and steppe zones in Eurasia. The detail shows the approximate location of the forest-steppe zone in the Pannonian Basin (Source: http://wwf.org/, modified by D. Dúbravková)

amplitudes in micro-refugia located directly in this area, while some other species re-colonised Pannonia before and during the early Holocene along the lower Danube (Magyari et al. 2010). Since there are only a limited number of pollen analyses in the Eurasian steppe zone and also only few fossil and phylogeographic studies available it is difficult to appraise the distribution of rare and endemic thermophilous species in the Pannonian Basin and other areas (Hewitt 2004).

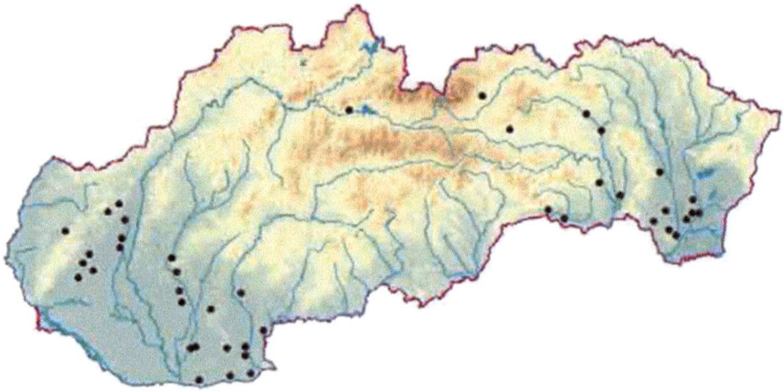

Fig. 8.3 Distribution of Neolithic and Aeneolitic archaeobotanical sites in Slovakia (Hajnalová 2007)

With climate amelioration and forest expansion in the Holocene dry grasslands became fragmented and restricted to the driest landscapes (south facing slopes, dry plateaus) while the forest occupied areas with a humid and cold mesoclimate such as north-facing slopes and valleys (Ložek 1971; Chytrý et al. 2007). In Central Europe the dry grasslands may have existed even before man settled down, although these might not necessarily have been identical with the recent communities (Poschlod and Wallis De Vries 2002). Their spatial extent, however, might have been influenced by human activities (Pott 1996; Bieniek and Pokorný 2005). Humans lived in the lowland and hilly landscapes of the Pannonian Basin since the Palaeolithic (about 2 million years BC). As the climate turned warmer their numbers gradually increased. Agricultural activities in this area started about 7000–6000 BC. The colline areas in Slovakia, particularly the south-western and south-eastern margins of the Western Carpathians that are in direct contact with the Pannonian Basin, were colonised by groups of early farmers in the early Neolithic (5500 BC). They entered current Slovakia from the south along the Danube and the Tisza Rivers and their tributaries (Hajnalová 2007). In these two main corridors, via which farming spread to new areas, archaeobotanical sites contain the oldest macro-remains of cultivated plants excavated in Slovakia (Fig. 8.3). Farming in the montane basins of the Central Western Carpathians appeared later.

Present hypotheses on the character of the vegetation in Central Europe at the time of arrival of the Neolithic farmers differ considerably. Some authors assume that the landscape was covered by deep mixed deciduous forests (Küster 1995) while others claim that Europe's original vegetation were not the closed forests, but rather a more open, park-like landscape of sparse deciduous forests with forest pastures and small grassland stands maintained by the grazing of large primeval herbivores (Vera 2000; Sádlo et al. 2005). As there are very few reliable palynological data from the territory of Slovakia, the following text is based on literature from the adjacent regions and archaeological interpretations on site density. 'The deep forests theory' induces that forest clearing by the slash-and-burn management started near human settlements in the loess areas of the Pannonian Basin about 6000 BC and in the northern areas of the Carpathians some 1,000–2,000 years later. Regarding the sparse population density at those times, the forest clearings might not have left any considerable marks in the landscape and the process of forest re-colonisation in the cleared stands might have been rather quick (Krippel 1986). It is assumed that this is the reason why pollen spectra and archaeological findings dating from those times do not correlate (Dresslerová and Pokorný 2004). Deforestation started to be of a considerable intensity only in the early and full Bronze Age (about 1000 BC) due to the spread of mining and processing of metal, improving the farming techniques and allowing population increase (Lang et al. 2000–2003; Illyés and Bölöni 2007). Large-scale forest clearing as a consequence of the expansion of farming (cultivating, pasturing and hay making) even in the less fertile areas continued with some intermissions until the Modern Time. In the montane areas, e.g. in the Považský Inovec Mts, forests were decimated during the Turkish wars in the seventeenth century. People made charcoal and created pastures for their herds, and flocks sheltered in the mountains (Kňazovický 1962). In Central Europe the area covered by forests was the smallest at about 1700 (Sádlo et al. 2005).

'The open woodland with treeless formations theory' (Vera 2000) assumed that at the time of the arrival of the Neolithic farmers the landscape was covered by open forest with steppe-like enclaves (Sádlo et al. 2005). Much has since been written against, as well as in support of Vera's theory, and several specialists now agree that it can be relevant, although probably not everywhere and not in the exact way as it was presented (Szabó 2009). The theory is also supported by a substantial archaeobotanical analysis (Bogaard 2004) which has presented new facts on the establishment of Neolithic farming in the loess areas of Central Europe. The Neolithic farmers most probably situated their fields at the treeless stands naturally occurring in the landscape. Analyses have dismissed the theory of cyclic slash-and-burn cultivation and, instead, demonstrated the intensive cultivation of permanent plots (Bogaard 2004). Diverse archaeological studies from Slovakia, the Czech Republic and Poland (Hajnalová 1989; Opravil 1999; Bieniek 2002; Bieniek and Pokorný 2005) document archaeobotanical findings of macro-remains of seeds and awns of feather grass (*Stipa pennata* s.l.) in thick layers dated as Neolithic. This potentially edible plant typical for steppe vegetation most probably played a role in the economy of the Neolithic settlers as a material for insulation or for making mattresses. Local gathering, rather than distant transport, best explains the presence in considerable quantities of this xerothermic grass in the remains of Neolithic settlements (Bieniek 2002). The natural occurrence of small steppe stands with xerothermic grassland vegetation in the landscape of Central Europe of those times has now repeatedly been proved by pollen analyses (Nalepka et al. 1998; Bieniek and Pokorný 2005). Some authors suppose that due to its utility value feather grass might have been transported and planted in new areas outside its original range (Behre 1988; Bakels 1992), but considering the ecological requirements of the species, and the difficulties to get it established, this theory is most probably improper. Summing up, it is likely that steppe-like sites with feather grass naturally

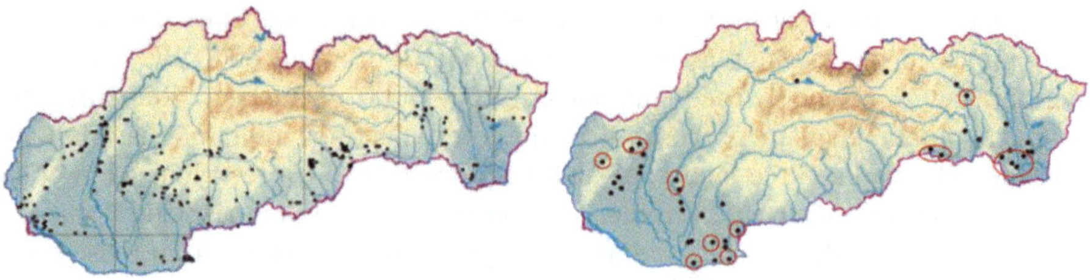

Fig. 8.4 Comparison of the recent distribution of dry grassland sites (*left*) and the Neolithic archaeobotanical excavation sites in Slovakia (*right*). The *red marks* show the co-occurrence of archaeological and dry grassland sites

occurred in south-western Slovakia at the times of establishment of farming in this area (about 5000–3000 BC).

The subsequent climate changes since the Bronze Age (since about 2500 BC) induced in most areas of Europe spontaneous afforestation. However, during this period humans maintained the treeless stands in the landscape of Central Europe and increased the distribution of dry grasslands by forest clearing and extensive livestock grazing (Bieniek and Pokorný 2005). The anthropogenic influences changed the natural boundaries of vegetation belts, induced secondary migrations of some species, and thus caused the creation of the secondary plant communities replacing the original (natural) vegetation. Xerophilous grassland communities developed gradually at sites where *Quercus, Quercus-Carpinus* and eventually *Fagus* forests have been cleared on slopes of warm expositions at low altitudes and on former fields. Xerophytes and steppe species entered these sites from patches of natural steppes located nearby and as seeds transported in the fur of grazing animals.

The problem of a primary or a secondary origin of recent dry grassland sites in the contact zone of the Western Carpathians and the Pannonian Basin is difficult to solve without local palynological studies. The origin of dry grasslands in the more oceanic areas of Western Europe is different. There the first farming communities appeared later (at about 4000 BC) and established themselves in a different climatic zone and another natural vegetation type than the humans dwelling in the Carpatho-Pannonian region. Based on pollen studies and archaeology, it seems that deforestation and maintenance of the treeless sites in Western Europe is a result of the activities of the Neolithic farmers (Pott 1996). Their activities caused the establishment of steppe-like vegetation outside its zonal ranges and isolated glacial steppe refugia (Behre 2000). Thus the dry grassland sites in Western Europe are considered to be of secondary origin. As we showed earlier, this is not valid for the Carpatho-Pannonian region. A very characteristic feature of eastern Central Europe is the co-occurrence of recent dry grassland sites and the areas inhabited and cultivated by humans since the prehistory (Ložek 1999; Bieniek and Pokorný 2005) (Fig. 8.4). We may interpret this fact in two ways: (I) the Neolithic settlements were established in areas of natural distribution of historical (and also recent) steppe-like vegetation, or (II) the dry grasslands subsequently developed at the sites of former

Neolithic settlements and farmland. Whatever the case, it is a fact that recent species-rich dry grasslands occur near archaeological sites that document the agricultural activities of Neolithic farmers (e.g. the hilly landscapes on the south-western edges of the Western Carpathians: Dolní Věstonice-Děvín – CZ, Lančár-Chríb, Nitra, Levice, Kamenín, Bíňa, Čenkov, Štúrovo, and on the north-eastern edge of the Pannonian Basin: Zádiel-Včeláre, Košice; Fig. 8.4), in the Bronze and Iron Ages as well as in the Great Moravian era since about the ninth century (e.g. Ducové, Dražovce, Lupka, Zobor, Bíňa), on the medieval castle hills (e.g. Čachtice, Devín, Turňa nad Bodvou, Krásna Hôrka, Sirotčí hrádek – CZ, Dívčí hrady – CZ, Falkenstein – AT) and at religious sites (e.g. a belfry hill Lančárska zvonica, calvaries in Nitra and Beckov). The dry grassland vegetation at those sites has repeatedly been studied (Maglocký 1979; Vozárová 1986; Unar 2004; David 2009; Dúbravková et al. 2010a). Those prehistoric settlements were set up at naturally sheltered sites with a favourable meso-climate and fertile soil, located mainly on loess terraces and slopes along the river valleys, e.g. along the Morava, Dudváh, Hron, Žitava and Nitra Rivers (Ložek 1999; Bogaard 2004; Hajnalová 2007). Due to the long-term agricultural utilization of the land (fields, pastures, and meadows), soil erosion and occasional fires prevented later establishment of forest at these sites. The hypothesis on the existence of original dry grassland sites, where steppe-like vegetation occurred continuously since the Holocene in the landscape of the Carpatho-Pannonian region, is supported by some palynological studies, analyses of fossil mollusc communities, plant macro-remains as well as by isolated relict occurrences of taxa typical for continental steppes, such as *Helictotrichon desertorum* subsp. *basalticum* (Podpěra 1904; Chytrý et al. 2007).

8.3 Variability and Classification of Dry Grasslands in Slovakia

8.3.1 Phytosociological Classification

In the phytosociological classification system the Central European dry grasslands belong to the class *Festuco-Brometea* Br.-Bl. et Tüxen ex Soó 1947 which includes dry and semi-dry grasslands and Euro-Siberian zonal steppes and forest-steppes (Mucina 1997). The class comprises several orders. The order *Festucetalia valesiacae* Br.-Bl. et R. Tx. ex Br.-Bl. 1949 includes the dry grasslands and steppes on deep to relatively shallow soils dominated by narrow-leaved tussock-forming grasses mostly fescues and feather grasses (e.g. *Festuca pseudodalmatica, F. rupicola, F. valesiaca, Stipa capillata, S. joannis, S. pulcherrima, S. tirsa*, etc.). The species composition is specific due to the occurrence of rare species of continental (e.g. *Astragalus austriacus, A. danicus, A. exscapus, Carex supina, Crambe tataria, Festuca valesiaca, Chamaecytisus ratisbonensis, Iris pumila, Oxytropis pilosa, Potentilla arenaria, Stipa capillata, S. dasyphylla, S. tirsa*), sub-Mediterranean

Photo 8.1 Dry grassland of the *Festuco valesiacae-Stipetum capillatae* Sillinger 1930 association in the Malé Karpaty Mts, Devínska Kobyla Nature Reserve. In the background there is the paleontological sandstone site Sandberg, a Tertiary seabed. The stand is dominated by *Stipa capillata, Festuca valesiaca* and *Koeleria macrantha*. The species in flower are *Jurinea mollis, Lotus borbasii, Dorycnium pentaphyllum* and *Dianthus pontederae* (Photo by D. Dúbravková, May 15, 2006)

(e.g. *Cleistogenes serotina, Fumana procumbens, Chrysopogon gryllus, Melica transsilvanica, Stipa eriocaulis*), and Pontic-Pannonian distribution (e.g. *Cruciata pedemontana, Gypsophila paniculata, Inula ensifolia, Linum hirsutum, Ranunculus illyricus, Tithymalus glareosus*). Based first of all on different geological substrates, and with some corresponding differences in floristic composition, the Slovak dry grasslands of the order *Festucetalia valesiacae* are divided into two alliances. The alliance *Festucion valesiacae* Klika 1931 includes narrow-leaved continental steppe-like communities on alkaline to neutral soils developed over loess, fluvial sediments, carbonate and volcanic rocks (Michálková 2007a; Dúbravková et al. 2010b) (Photo 8.1). The alliance *Koelerio-Phleion phleoidis* Korneck 1974 is restricted to the acidic soils poor in minerals, such as granite, gneiss and quartzite (Photo 8.2). The vegetation includes some acidophilous species (e.g. *Acetosella*

Photo 8.2 A fragment of acidophilous dry grassland in a horse pasture on quartzite (Biele Karpaty Mts, Skalický vrch). The dominant grasses are *Agrostis capillaris, Anthoxanthum odoratum* and *Festuca rupicola; Steris viscaria* is in flower (Photo by J. Košťál, May 13, 2009)

vulgaris, Agrostis vinealis, Armeria vulgaris, Jasione montana) although the overall species composition is similar to the alliance *Festucion valesiacae*. This vegetation occurs in Slovakia only at a few localities, since the distribution of suitable sites is limited (Michálková 2007b).

Besides the closed to semi-closed steppe-like dry grasslands there occur open Pannonian, rocky, dry grasslands on calcareous substrates with *Carex humilis* and *Festuca pallens* (*Bromo pannonici-Festucion pallentis* Zólyomi 1966) and dealpine *Sesleria albicans*-dominated grasslands (*Diantho lumnitherii-Seslerion* (Soó 1971) Chytrý et Mucina in Mucina et al. 1993) in Slovakia (Janišová and Dúbravková 2010) as well as the sub-xerophilous vegetation of the alliances *Cirsio-Brachypodion pinnati* Hadač et Klika ex Klika 1951 and *Bromion erecti* Koch 1926 (Škodová 2007a, b). These, however, are not a topic of the current paper.

8.3.2 Classification at the Habitat Level

The dry grassland communities of the alliance *Festucion valesiacae* present in Slovakia could be classified within three different natural habitat types: 6210 Semi-natural dry grasslands and scrubland facies on calcareous substrates, 6240* Sub-Pannonic steppic grasslands, and 6250* Pannonic loess steppic grasslands

(Stanová and Valachovič 2002; European Commission 2007). The contents of these habitat types overlap a great deal, and this makes the classification extremely complicated. Due to the absence of internationally standardized information, the European Community has accepted these vague units as Natural Habitat Types of Community Interest according to Annex I of the Habitats Directive (92/43/EEC), although such types have unclear delimitations and are not supported by data (Dúbravková et al. 2010b). They were designated using an inappropriate combination of two criteria: (I) occurrence of phytogeographically important (sub-)continental and Pannonian species, and (II) type of geological substrate. Applying the results of recent international comparative studies of grassland habitat diversity (Botta-Dukát et al. 2005; Illyés et al. 2007; Dúbravková et al. 2010b) might help to re-evaluate the delimitation of habitat types and establish better criteria for international habitat classification and subsequent effective monitoring of endangered habitats.

One of the most important environmental factors affecting the variability of (sub-) xerophilous vegetation of the class *Festuco-Brometea* is soil moisture (Chytrý et al. 2007). For this reason we suggest to classify the biotope types according to this criterion into the following two proposed categories. The first unit would include more xerophilous vegetation dominated by narrow-leaved grasses (xerophilous *Festuca* and *Stipa* species). It could be called 'narrow-leaved dry grasslands' and include both the recent 6240* and 6250* habitats. Such a unit would comprise the alliance *Festucion valesiacae* (dry grasslands on loess, calcareous and volcanic substrata) as well as the *Koelerio-Phleion phleoidis* (dry grasslands on acidic substrata) and it might be subsequently divided into a few sub-units according to the type of substrate. On the other hand, the 'broad-leaved semi-dry grasslands' would represent the more humid habitat type dominated by *Bromus erectus* and *Brachypodium pinnatum*. From the phytosociological point of view it would include vegetation of the alliances *Bromion erecti* and *Cirsio-Brachypodion pinnati*. In the current habitat classification it partly represents the 6210 habitat. A similar habitat classification is used in practise in the habitat classification scheme of the Czech Republic (Chytrý 2001). It has repeatedly been pointed out that the Slovak habitat classification published by Stanová and Valachovič (2002) needs revision (Michálková et al. 2006; Dúbravková and Janák 2011).

8.4 Conservation and Management of Dry Grasslands in Slovakia

8.4.1 Socio-economic Perspectives on Land-Use Change

The peculiar species-rich composition of the Central European dry grasslands is a result of long-term agricultural activities. The xerophilous grasslands were used for ages as low-productivity pastures which could be occasionally moved in more humid years (Poschlod and Wallis De Vries 2002). The sites in Slovakia are often located at difficult terrain situations, mostly on steep and rocky slopes, and thus were usually used for grazing of smaller animals such as sheep and goats. However, destitute farmers used to graze there all their livestock, including a few cows, horses, sheep, and goats (Barańska et al. 2010). The dry grasslands were managed in a traditional way until about the end of the 1960s (Buček et al. 2006). The main reason for cessation of the low-intensity grazing practice was the intensification of agricultural production during the period of communism in former Czechoslovakia (1948–1989). Private land was collectivised, the majority of former farmers started to work in industry, the number of livestock kept privately in the countryside decreased rapidly, and thus the area that was grazed decreased. The state farming companies tended to enormously intensify the production, and accordingly the animals of low profitability (sheep and goats) were replaced by a higher number of cows. The dry grasslands could not compete in fodder productivity with the less dry grasslands and meadows, and therefore some substitute 'progressive' utilisation of the dry habitats was applied. Many steppe-like grassland sites were thus destroyed by afforestation (Zlinská 2000), ploughing (Deván et al. 2006), and mining, or they were abandoned which led to degradation of their extraordinary species composition. Similar unsuitable and even detrimental management was applied in other post-communistic countries (Meshinev et al. 2005; Buček et al. 2006; Illyés and Bölöni 2007; Molnár et al. 2008; Ruprecht et al. 2009; Barańska et al. 2010) and in the Western Europe as well (Wallis De Vries et al. 2002).

Although the variability of steppe-like grasslands in Slovakia is considerable, their actual status as regards land use and conservation is not optimistic. The majority of the dry grasslands maintained until today are located in nature protection areas, such as national parks and reserves, and also there the traditional low-intensity grazing was ceased a long time ago and the sites became abandoned. Other dry grasslands occur near villages on fragments of formerly communal pastures which used to be grazed at low intensity, even during the communist days, by a few animals privately kept in the countryside (Buček et al. 2006). After 1989 the land was returned to the private owners. But the modern land holders have only limited experience and few facilities for an active management of the sites, and thus many dry grasslands remain abandoned. Farming seems not to be profitable in modern society, and even though agro-environmental payments are provided, very few dry grassland sites have been managed in an effective and habitat-friendly way. Extensive sheep and goat grazing, which is the most suitable management practice for preserving the biodiversity of the steppe-like habitats, is done only at a few sites located in areas of nature conservation (Rajcová 2010).

We also mention the negative impact of the past activities of the official State Conservation Agency. Due to lack of knowledge on the ecology of grasslands and the linkage between species richness and management, since the 1960s conservationists avoided any disturbance activities, including grazing and mowing, in the grasslands located in nature conservation sites (Buček et al. 2006; Mládek et al. 2006; Barańska et al. 2010). This was detrimental for dry grasslands in the protected areas, and what is more, it raised mistrust by locals towards the conservationists when they recently encouraged them to graze and mow the stands, where such activities were prohibited in the past (Šuvada pers. comm.). Present-day nature conservation strategy emphasizes using the same management methods as were applied traditionally by our farming ancestors. The problem is, however, that only a few people remember in detail the past routines performed regularly at particular sites. This knowledge on e.g. the right timing of pasturing, ways of open-air sheep penning, flock and shepherd migration, as well as rotational pasturing, is gradually disappearing as elder persons pass away. But this precious information results from age-old human experience and should be protected as a part of the cultural heritage of Slovakia.

8.4.2 Environmental Perspectives on Abandonment

The majority of xerophilous plant species naturally occurring in the dry grassland habitats are week competitors that can only establish and persist in nutrient-poor stands with a low intensity of competition (Bylebyl 2007). Such grasslands are either naturally open (such as rocky dry grasslands of the *Bromo pannonici-Festucion pallentis*) or managed (grasslands of the *Festucion valesiacae* and *Koelerio-Phleion phleoidis*). Management is therefore crucial for maintenance of the species composition of xerothermophilous steppe-like grasslands (Virágh and Bartha 1996; Enyedi et al. 2008). Cessation of the traditional management (sheep and goat grazing) as a result of a low economical profitability and changes in lifestyle cause rather quick successional changes in these valuable habitats. If not managed, litter accumulates and changes the micro-climatic conditions at the sites. This results in an unfavourable alteration of the temperature and humidity at soil surface as well as shading and the formation of a mechanical barrier preventing germination and establishment. Therefore, species with short life cycles (annuals, biennials) first disappear from the abandoned stands (Černý et al. 2010). But the germination of the perennials, e.g. *Eryngium campestre*, is also negatively affected by shading and strong competition (Bylebyl 2007).

During the first years of abandonment, the sites pass a ‘ruderal successional stage’ with a higher abundance of species such as *Echium vulgare, Anthemis tinctoria, Bromus sterilis*, etc. (Bylebyl 2007). This stage is usually followed by a ‘grass stage’ with an increased dominance of sub-xerophilous (e.g. *Brachypodium pinnatum, Bromus erectus*) and mesophilous grasses (e.g. *Arrhenatherum elatius, Avenula pubescens*). Due to the accumulated litter more water is retained and, by its decomposition, higher amounts of nutrients (P, N) cycle in the ecosystem. Nitrogen, originating from the industry and traffic, also is augmented (by about 5 g/m^2/year) through aerial deposition (Willems et al. 1993; Bobbink et al. 1998; Wallis De Vries et al. 2002) and is a serious problem in today’s nature conservation. An increase in nutrients, particularly of phosphorus, leads to the decline of oligotrophic species and of species richness (Willems et al. 1993; Chytrý et al. 2009). Since the nutrients persist in the soil for a long time, the vegetation changes induced may remain constant for decades. Abandoned stands with nutrient additions are often invaded by expansive (*Calamagrostis epigejos*) and alien species (e.g. *Ambrosia artemisifolia, Eleagnus angustifolia, Lycium barbarum, Robinia pseudoacacia, Stenactis annua*) (Domán et al. 1997; Buček et al. 2006; David and Záchenská 2010). In the advanced stage of succession polycorm-forming shrubs (e.g. *Crataegus* spp., *Prunus spinosa, Rubus* spp.) and other woody species (e.g.. *Cerasus fruticosa, C. mahaleb, Spiraea media, Quercus* spp., *Fagus sylvatica, Pinus sylvestris*) intrude the sites and develop layers of high cover. Such spontaneously overgrown sites rapidly decrease in species richness.

8.4.3 Management Recommendations

To maintain the species-rich dry grasslands it is necessary to suppress the uncontrolled spread of shrubs and trees at the abandoned sites by manual or mechanical clearing. This is an important precondition for maintaining the light-demanding xerophilous species (Bylebyl 2007). Absence of the woody species, however, does not guarantee the sustainability of dry grasslands. Another important step toward long-term persistence of rare steppe species is regular biomass removal (by grazing, eventually mowing and occasionally a fire) and topsoil disturbance (by grazing, trampling). The disturbance provides creation of gaps with bare soil within the dense grass and litter sod and this allows the seeds to germinate. Germination of small-sized seeds is more disturbance-dependent than that of larger seeds (Burke and Grime 1996; Kupferschmid et al. 2000). While practising any management action it is important not to perform it simultaneously in the entire area of the site. A section of the site (each time a different one) should remain untouched so that the variability in vegetation structure is maintained. Variability of habitats and presence of different successional stages in the landscape supports the diversity of invertebrates (Wallis De Vries et al. 2002; Cremene et al. 2005) and the occurrence of rare and endangered plant species (Enyedi et al. 2008; Ruprecht et al. 2009).

The most convenient management of dry grasslands is grazing by sheep and goats in low intensity (Photo 8.3). These light-weight animals are suitable for grazing sites at slopes with a shallow soil (Háková et al. 2004). The proper mix of animals is three sheep to one goat. A sheep is a selectively grazing animal which grazes the stand to a short height (about 3 cm), and thus the biomass removal is sufficient. Goats, however, graze the stand to a taller-height, preferring grasses in flower, bark and leaves of woody species and thus reduce the height of shrubs (Mládek et al. 2006). In low-productivity stands the native breeds of grazing animals are the most suitable ones because they are best adapted to the nutrient-poor fodder, the climate and the terrain situation (Calaciura and Spinelli 2008). The animals prefer spring pasture when the fodder is fresh. At the end of April, when the grazing usually starts, the stand should not be too wet and muddy and the height of the grass must reach at least 5 cm. Strong trampling in spring causes damage to the vegetation sod. Therefore 'one-shot walk-through grazing' is positive for the vegetation but permanent grazing could harm some species. Later in the season (in July, August) the grass turns dry and less tasty. The grazing becomes more selective and many

Photo 8.3 Extensive grazing management of steppe-like grasslands with *Festuca valesiaca* in the Hainburger Berge Mts in NE Austria (Hundsheimer Berg Nature Reserve). The pasture was divided into a few sections by solar-powered electric wire fences and the flock rotated through the sections according to the stage of grass growth (Photo by K. Hegedüšová, June 27, 2007)

ungrazed patches are created. However, grazing later in the season does not endanger the sensitive species and the majority of plants may produce ripe seeds that can be transported in the animal fur to other sites.

Spontaneous seed dispersal rarely exceeds a distance of 25 m in dry grasslands. For effective site regeneration and maintenance of the biodiversity seed transport from the species-rich sites located nearby might be necessary (Stampfli and Zeiter 1999; Barbaro et al. 2001). This can best be achieved through herding sheep, as seeds show a high attach potential to sheep fur (Bylebyl 2007), and rotating pasturing with flocks grazing a stand for 3–4 weeks (Poschlod et al. 1998; Calaciura and Spinelli 2008; Háková et al. 2004; Barańska et al. 2010a). The recommended animal density is 5 animals per hectare (Barańska et al. 2010). Annual grazing is preferred, but stands in good condition could be grazed once in several years (Buček et al. 2006).

Other types of management action such as mowing, mulching and burning may also be applied, although only under specific circumstances and regulations. Mowing showed to be an effective way to eliminate expansive grasses e.g. *Brachypodium pinnatum, Bromus erectus* and *Calamagrostis epigeios* (Willems et al. 1993; Klimeš et al. 2008; Házi et al. 2010). It has to be done twice a year and preferably late in the growing season (August, September).

Where there are few farming animals available, mulching recently became a popular and relatively cheap management action. But mulching does not cause seed transport and does not remove biomass; it disrupts the solid litter layer and even long-term mulching does not cause a negative change in species composition (Kahmen et al. 2002).

Burning is another cheep management action, but it is not recommended because, if applied regularly, the species composition shifts to a variant that very much resembles the herbaceous layer of a fallow (Kahmen et al. 2002). If applied, the stands must be burnt in a mosaic pattern and only once in 15–20 years (Háková et al. 2004). While a combination of grazing and infrequent burning increases the species richness of dry habitats, burning without grazing leads to a loss of species (Noy-Meir 1995).

An effective method of restoring degraded dry grasslands and former fields is the import of fresh hay from species-rich sites. Compared to seed sowing, a thin layer of hay serves not only as a source of seeds but also creates proper conditions for germination and establishment (Kiehl and Pfadenhauer 2007; Házi et al. 2010).

To further the proper management of dry grasslands in Slovakia, we prepared guidelines (a management model) for land owners, farmers and the State's nature conservancy which are collectively responsible for management of the steppe-like sites (Dúbravková and Janák 2011). The models and the level of their implementation will also be used to design the rules concerning agro-environmental payments. This may positively motivate land holders and set a high standard for habitat-friendly management activities. The management models for all types of grassland habitats in Slovakia are published at www.daphne.sk/mm/manazmentove-modely.

Acknowledgements This study was supported by the grant 'Management models for grassland habitats' (SK0115) supported through the EEA Financial Mechanism, the Norwegian Financial Mechanism and the state budget of the Slovak Republic and grant VEGA 2/0181/09 financed by the Science Grant Agency of the Ministry of Education of the Slovak Republic and Slovak Academy of Sciences.

References

Bakels CC (1992) Fruits and seeds from the Linearbandkeramik settlement at Meindling, Germany, with special reference to *Papaver somniferum*. Anal Praehist Leid 25:55–68

Barańska K, Chmielewski P, Cwener A, Pluciński P (2010) Conservation and restoration of xerothermic grasslands in Poland – theory and practice. Wydawnictwo Klubu Przyrodników, Świebodzin, 52 pp

Barbaro L, Dutoit T, Cozic P (2001) A six-year experimental restoration of biodiversity by shrub-clearing and grazing in calcareous grasslands of the French Prealps. Biodivers Conserv 10:119–135

Behre K-E (1988) The role of man in European vegetation history. In: Huntley B, Webb T (eds) Vegetation history. Handbook of vegetation science. Kluwer Academic Publications, Dordrecht, pp 633–672

Behre K-E (2000) Der Mensch öffnet die Wälder – zur Entstehung der Heiden und anderer Öffenlandschaften. Rundgespräche der Komission für Ökologie 18:103–116

Bieniek A (2002) Archaeobotanical analysis of some early Neolithic settlements in the Kujawy region, central Poland, with potential plant gathering activities emphasised. Veg Hist Archaeobot 11:33–40

Bieniek A, Pokorný P (2005) A new find of macrofossils of feather grass (*Stipa*) in an Early Bronze Age storage pit at Vliněves, Czech Republic: local implications and possible interpretation in a Central European context. Veg Hist Archaeobot 14:295–302

Bobbink R, Hornung M, Roelofs JG (1998) The effect of air-borne nitrogen pollutant on species diversity in natural and semi-natural European vegetation. J Ecol 86:717–738

Bogaard A (2004) Neolithic farming in central Europe. An archaeobotanical study of crop husbandry practices. Routledge, London/New York, 209 pp

Bohn U, Neuhäusl R (eds) (2000–2003) Karte der natürlichen Vegetation Europas. Bundesamt für Naturschutz, Bonn

Botta-Dukát Z, Chytrý M, Hájková P, Havlová M (2005) Vegetation of lowland wet meadows along a climatic continentality gradient in Central Europe. Preslia 77:89–111

Buček A, Lacina J, Laštůvka Z (2006) Panonské stepní trávníky na Moravě. Veronica – časopis pro ochranu přírody a krajiny 20/17:1–58

Burke MJW, Grime JP (1996) An experimental study of plant community invasibility. Ecology 77:776–790

Bylebyl K (2007) Central European dry grasslands: processes of their development and possibilities for their maintenance. Diss Bot 406:1–142

Calaciura B, Spinelli O (2008) Management of Natura 2000 habitats. 6210 Semi-natural dry grasslands and scrubland facies on calcareous substrates (*Festuco-Brometalia*) (*important orchid sites). In: European Commission (ed) Technical report 2008 12/24, pp 1–38

Černý T, Petřík P, Boublík K (2010) How did changes in landscape management influence dry grassland composition after the Velvet Revolution? In: Janišová M, Budzáková M, Petrášová M (eds) Seventh European dry grassland meeting. Succession, management and restoration of dry grasslands. Abstracts & excursion guides, 27–31 May 2010, Smolenice, Slovakia, p 17

Chytrý M (2001) Suché trávníky. In: Chytrý M, Kučera T, Kočí M (eds) Katalog biotopů České republiky. Agentura ochrany přírody a krajiny, CR, Praha, pp 129–140

Chytrý M, Hoffmann A, Novák J (2007) Suché trávníky (*Festuco-Brometea*). In: Chytrý M (ed) Vegetace České republiky. 1. Travinná a keříčková vegetace Academia, Praha, pp 371–497

Chytrý M, Hejcman M, Hennekens SM, Schellberg J (2009) Changes in vegetation types and Ellenberg indicator values after 65 years of fertilizer application in the Rengen Grassland Experiment, Germany. Appl Veg Sci 12:167–176

Cremene C, Groza G, Rakosy L, Schileyko AA, Baur A, Erhardt A, Baur B (2005) Alterations of steppe-like grasslands in Eastern Europe: a threat to regional biodiversity hotspots. Conserv Biol 19:1606–1618

David S (2009) Stepní flóra a vegetace zemních valů hradiska v obci Bíňa (okr. Nové Zámky). Bull Slov Bot Spoločn 31:77–85

David S, Záchenská M (2010) Species richness and threatened vascular plant species of the sub-pannonic steppic grasslands in the southern part of Krupinská planina Mts. In: Janišová M, Budzáková M, Petrášová M (eds) Seventh European dry grassland meeting. Succession, management and restoration of dry grasslands. Abstracts & excursion guides, 27–31 May 2010, Smolenice, Slovakia, p 19

Deván P, Devánová K, Mereďa P, Smatanová J, Rajcová K (2006) Charakteristika flóry záujmového územia floristického kurzu v Pruskom. Bull Slov Bot Spoločn 28(suppl 1):13–21

Domán E, Kalivoda B, Kapocsi J (1997) Management of loess grasslands in the Körös-Maros region. In: Tóth E, Horváth R (eds) Proceedings of the "Research, conservation, management" conference. Aggtelek, 1–5 May 1996. Aggtelek National Park Directorate, Aggtelek, pp 199–204

Dresslerová D, Pokorný P (2004) Vývoj osídlení a struktury pravěké krajiny na středním Labi. Pokus o přímé srovnání archelogické a pyloanalytické evidence. Archeologické rozhledy 56:739–762

Dúbravková D, Janák M (2011) Manažmentový model pre xerotermné travinno-bylinné spoločenstvá. MSc, DAPHNE – Inštitút aplikovanej ekológie a Botanický ústav SAV, Bratislava, p 36. URL: http://www.daphne.sk/sites/daphne.sk/files/uploads/MM03_xerotermy.pdf
Dúbravková D, Hegedüšová K, Janišová M, Škodová I (2010a) New vegetation data of dry grasslands in the Western Carpathians and northern Pannonian Basin. Tuexenia 30:357–374
Dúbravková D, Chytrý M, Willner W, Illyés E, Janišová M, Kállayné Szerényi J (2010b) Dry grasslands in the Western Carpathians and the northern Pannonian Basin: a numerical classification. Preslia 82:165–221
Enyedi ZM, Ruprecht E, Deák M (2008) Long-term effects of the abandonment of grazing on steppe-like grasslands. Appl Veg Sci 11:55–62
European Commission (2007) Interpretation manual of European Union habitats. EUR 27:1–142
Frenzel B, Pécsi M, Velichko AA (1992) Atlas of paleoclimates and paleoenvironments of the Northern Hemisphere. Geographical Institute, Gustav Fischer Verlag, Budapest/Stuttgart
Hajnalová E (1989) Katalóg zvyškov semien a plodov v archeologických nálezoch na Slovensku. In: Hajnalová E (ed) Súčasné poznatky z archeobotaniky na Slovensku. Acta Interdisciplinaria Archaelogica, Nitra, pp 3–192
Hajnalová M (2007) Early farming in Slovakia: an archaeobotanical perspective. In: Colledge S, Conolly J (eds) The origins and spread of domestic plants in southwest Asia and Europe. Institute of Archaeology/Left Coast Press, Walnut Creek, pp 295–312
Háková A, Klaudisová A, Sádlo J (eds) (2004) Zásady péče o nelesní biotopy v rámci soustavy Natura 2000. PLANETA XII, Ministerstvo životního prostředí, Praha, 157 pp
Házi J, Bártha S, Szentes S, Penksza K (2010) Seminatural dry grassland management by mowing of *Calamagrostis epigeios* (L.) Roth in Hungary. In: Janišová M, Budzáková M, Petrášová M (eds) Seventh European dry grassland meeting. Succession, management and restoration of dry grasslands. Abstracts & excursion guides, 27–31 May 2010, Smolenice, Slovakia, p 31
Hewitt GM (1999) Postglacial recolonization of European Biota. Biol J Linn Soc 68:87–112
Hewitt GM (2000) The genetic legacy of the quaternary ice ages. Nature 405:907–913
Hewitt GM (2004) Genetic consequences of climatic oscillations in the Quaternary. Philos Trans R Soc Lond B 359:183–195
Horvat I, Glavač V, Ellenberg H (1974) Vegetation Südosteuropas. Gustav Fischer Verlag, Jena
Illyés E, Bölöni J (eds) (2007) Lejtősztyepek, löszgyepek és erdőssztyeprétek Magyarországon. MTA ÖBKI, Budapest, 236 pp
Illyés E, Chytrý M, Botta-Dukát Z, Jandt U, Škodová I, Janišová M, Willner W, Hájek O (2007) Semi-dry grasslands along a climatic gradient across Central Europe: vegetation classification with validation. J Veg Sci 18:835–846
Janišová M, Dúbravková D (2010) Formalized classification of rocky Pannonian grasslands and dealpine Sesleria-dominated grasslands in Slovakia using a hierarchical expert system. Phytocoenologia 40:267–291
Kahmen S, Poschold P, Schreiber K-F (2002) Conservation management of calcareous grasslands. Changes in plant species composition and response of functional traits during 25 years. Biol Conserv 104:319–328
Kiehl K, Pfadenhauer J (2007) Establishment and persistence of target species in newly created calcareous grasslands on former arable fields. Plant Ecol 189:31–48
Klimeš L, Hájek M, Hájková P, Rozbrojová Z (2008) Vliv kosení na flóru a vegetaci. In: Jongepierová I (ed) Louky Bílých Karpat. ZO ČSOP Bílé Karpaty, Veselí nad Moravou, pp 325–345
Kňazovický J (1962) História dolného Považia. Sráva VÚLH. VÚLH, Zvolen
Krippel E (1986) Postglaciálny vývoj vegetácie Slovenska. Veda, Bratislava, 307 pp
Kuneš P, Pelánková B, Chytrý M, Jankovská V, Pokorný P, Petr L (2008) Interpretation of the last-glacial vegetation of eastern-central Europe using modern analogues from southern Siberia. J Biogeogr 35:2223–2236
Kupferschmid AD, Stampfli A, Newbery DM (2000) Dispersal and microsite limitation in an abandoned calcareous grassland of the Southern Prealps. Folia Geobot 35:125–141

Küster H (1995) Geschichte der Landschaft in Mitteleuropa. Verlag C. H. Beck, München, 424 pp

Lang G, Rybníček K, Rybníčková E (2000–2003) Spätglaziale und holozäne Vegetationsgeschichte Europas. In: Bohn U, Neuhäusl R (eds) Karte der natürlichen Vegetation Europas. Bundesamt für Naturschutz, Bonn, pp 87–104

Ložek V (1971) K otázce stepí ve střední Evropě. Zprávy Českoslov Bot Společn 6:226–232

Ložek V (1999) Ochranářské otázky ve světle vývoje přírody. 8. Zemědělská kolonizace a její dopad. Ochr Přír 54/8:227–233

Ložek V (2009) Refugia, migrace a brány I. Ohlédnutí za starými problémy. Živa 4:146–149

Maglocký Š (1979) Xerotermná vegetácia v Považskom Inovci. Biol Pr SAV 25(3):1–129

Magyari EK, Chapman JC, Passmore DG, Allen JRM, Huntley JP, Huntley B (2010) Holocene persistence of wooded steppe in the Great Hungarian Plain. J Biogeogr 37:915–935

Meshinev T, Apostolova I, Georgiev V, Dimitrov V, Petrova A, Veen P (2005) Grasslands of Bulgaria. Final report on the National Grasslands Inventory Project (PINMATRA/2001/020). Dragon Publishers, Sofia, 104 pp

Michálková D (2007a) *Festucion valesiacae* Klika 1931. In: Janišová M (ed) Travinnobylinná vegetácia Slovenska – elektronický expertný systém na identifikáciu syntaxónov. Botanický ústav SAV, Bratislava, pp 33–49

Michálková D (2007b) *Koelerio-Phleion phleoidis* Korneck 1974. In: Janišová M (ed) Travinnobylinná vegetácia Slovenska – elektronický expertný systém na identifikáciu syntaxónov. Botanický ústav SAV, Bratislava, pp 86–89

Michálková D, Škodová I, Mertanová S (2006) Príspevok k fytocenológii xerotermných rastlinných spoločenstiev v Považskom Inovci. In: Rajcová K (ed) Zborník výsledkov inventarizačného výskumu územia európskeho významu Tematínske vrchy. KOZA a Pre Prírodu, Trenčín, pp 35–44

Mládek J, Pavlů V, Hejcman M, Gaisler J (eds) (2006) Pastva jako prosředek údržby trvalých travních porostů v chráněných územích. VÚRV, Praha, 104 pp

Molnár Z, Bölöni J, Horváth F (2008) Threatening factors encountered: actual endangerment of the Hungarian (semi-)natural habitats. Acta Bot Hung Suppl 50:195–210

Mucina L (1997) Conspectus of classes of European vegetation. Folia Geobot Phytotax 32:117–172

Nalepka D, Wasilykowa K, Tomczyńska Z, Bieniek A (1998) Szata roślinna Pojezierza Kujawskiego i uzitkowanie roślin w okresie osadnictwa kultury lendzielskiej: wstępne doniesienie. Prace Mater Muz Archeol Etnogr w Łodzi Ser Archeol 39:139–174

Noy-Meir I (1995) Interactive effects of fire and grazing on structure and diversity of Mediterranean grasslands. J Veg Sci 6:701–710

Opravil E (1999) Archeologické nálezy kavylu na Moravě. Pravěk 9:153–157

Podpěra J (1904) O vlivu periody glaciální na vývoj květeny zemí českých. Věst Kl Přír Prostějov 6(1903):127–141

Poschlod P, Kiefer S, Tränkle U, Fischer S, Bonn S (1998) Plant species richness in calcareous grasslands as affected by dispersibility in space and time. Appl Veg Sci 1:75–91

Poschlod P, Wallis De Vries MF (2002) The historical and socioeconomic perspective of calcareous grasslands – lessons from the distant and recent past. Biol Conserv 104:361–376

Pott R (1996) Die Entwicklungsgeschichte und Verbreitung xerothermer Vegetationseinheiten in Mitteleuropa unter dem Einfluβ des Menschen. Tuexenia 16:337–369

Rajcová K (2010) Starostlivosť o prírodnú rezerváciu Beckovské Skalice. In: Májsky J, Rajcová K (eds) Beskovské Skalice. Zborník inventarizačných výskumov prírodnej rezervávie Beckovské Skalice. Pre Prírodu, Trenčín, pp 62–65

Ruprecht E, Szabó A, Enyedi ZM, Dengler J (2009) Steppe-like grasslands in Transylvania (Romania): characterisation and influence of management on species diversity and composition. Tuexenia 29:353–368

Sádlo J, Pokorný P, Hájek P, Dreslerová D, Cílek V (2005) Krajina a revoluce. Významné přelomy ve vývoji kulturní krajiny českých zemí. Malá Skála, Praha, 248 pp

Škodová I (2007a) *Bromion erecti* Koch 1926. In: Janišová M (ed) Travinnobylinná vegetácia Slovenska – elektronický expertný systém na identifikáciu syntaxónov. Botanický ústav SAV, Bratislava, pp 71–78

Škodová I (2007b) *Cirsio-Brachypodion pinnati* Hadač et Klika ex Klika 1951. In: Janišová M (ed) Travinnobylinná vegetácia Slovenska – elektronický expertný systém na identifikáciu syntaxónov. Botanický ústav SAV, Bratislava, pp 78–86

Stampfli A, Zeiter M (1999) Plant species decline due to abandonment of meadows cannot easily be reversed by mowing. A case study from the southern Alps. J Veg Sci 10:151–164

Stanová V, Valachovič M (eds) (2002) Katalóg biotopov Slovenska. DAPHNE – Inštitút aplikovanej ekológie, Bratislava, 225 pp

Szabó P (2009) Open woodland in Europe in the Mesolithic and in the Middle Ages: can there be a connection? Forest Ecol Manag 257:2327–2330

Unar J (2004) Xerotermní vegetace Pavlovských vrchů. Sborn Přír Klubu Uh Hradiště Suppl 11:1–140

Vera FWM (2000) Grazing ecology and forest history. CABI Publishing, Wallingford, 506 pp

Virágh K, Bartha S (1996) The effect of current dynamical state of a loess steppe community on its responses to disturbances. Tiscia 30:3–13

Vozárová M (1986) Xerotermné travinno-bylinné spoločenstvá Zoborskej skupiny Tríbeča. Zborn Slov Nár Múz 32:3–31

Wallis De Vries MF, Poschlod P, Willems JH (2002) Challenges for the conservation of calcareous grasslands in northwestern Europe: integrating the requirements of flora and fauna. Biol Conserv 104:265–273

Walter H (1974) Die Vegetation Osteuropas, Nord- und Zentralasiens. Gustav Fischer Verlag, Stuttgart

Willems JH, Peet RK, Bik L (1993) Changes in chalk-grassland structure and species richness resulting from selective nutrient addition. J Veg Sci 4:203–212

Willis KJ, Whittaker RJ (2000) The refugial debate. Science 287:1406–1407

Zlinská J (2000) Vegetácia Holubyho lesostepi pri Vinosadoch v Malých Karpatoch. Acta Environ Univ Comen 10:139–152

Chapter 9
Iberian Steppes

Helios Sainz Ollero and Marja A. van Staalduinen

Abstract In Spain, the extensive, largely treeless areas in semi-arid environments are considered as steppes, dominated by small shrubs, with forbs and grasses (dwarf shrub steppes). The main factor for their occurrence seems to be climatic continentality and drought.

The Iberian steppe vegetation contains plant species of South Mediterranean Irano-Turanian and of Mediterranean Saharo-Sindian affinities. The Mediterranean-Irano-Turanian element consists of eastern steppe species that spread through the Mediterranean basin, with many vicariances at generic level. The Mediterranean-Saharo-Sindian element of the Iberian steppes consists of plants derived from the Tertiary subtropical flora of the open African spaces. Plant endemism reaches 42% in the Iberian steppes.

The steppes harbour steppe bird communities with highly endangered species. Steppe birds in general depend very much on the structure of the vegetation. Many bird species are seriously at risk, while processes, like agricultural intensification, abandonment of grazing, changes of land use, and construction activities, cause degeneration and destruction of many steppe areas.

9.1 Iberian Steppes in Perspective

One of the fascinating features of the Iberian vegetation concerns the presence of steppe areas. Although the Iberian peninsula is one of the European areas with the largest forest areas, – about a quarter is covered by trees (26% according to WWF-Spain 2009) –, it is far from the mythically referred "squirrel traveller",

H. Sainz Ollero • M.A. van Staalduinen (✉)
Department of Ecology, Universidad Autónoma de Madrid, Cantoblanco, Madrid, Spain
e-mail: helios.sainz@uam.es; marjavs@xs4all.nl

M.J.A. Werger and M.A. van Staalduinen (eds.), *Eurasian Steppes. Ecological Problems and Livelihoods in a Changing World*, Plant and Vegetation 6,
DOI 10.1007/978-94-007-3886-7_9,

attributed to Pliny and Strabo. Accordingly, in ancient times the continuity of the Iberian forests was such that a squirrel could travel the peninsula from coast to coast through the treetops, without descending to the ground (Costa et al. 1997). This story is highly questionable in the light of paleogeographic and floristic data that point to the persistence, throughout the Holocene, of steppe areas related to extensive cold steppes that developed during the glacial periods (González-Sampériz 2004).

In Spain the largely treeless areas in semi-arid environments are considered as steppes. Most of these open formations are dominated by small shrubs, mixed with forbs and grasses, and thus are dwarf shrub steppes, differing considerably in structure from the grassy Russian steppes. The main factor seems to be the climatic continentality and drought: the Spanish steppes mostly occur on upland plains and plateaus surrounded by mountain ranges causing rain shadows.

The Mediterranean region is located in an ecotone position between the Holarctic and Paleotropical realms and carries plants with a cold, continental desert affinity (biome VII) and others with a thermophilic, arid or semi-arid affinity and with a subtropical root (Biome III) (Walter 1977, 1984). Accordingly, typical of the Iberian steppe vegetation is the co-occurrence of plant species of South Mediterranean Irano-Turanian and of Mediterranean Saharo-Sindian affinities. Additionally, the Iberian steppes harbour endemic, Ibero-North African taxa, with small distribution areas in the more arid parts of the Iberian peninsula.

The **Mediterranean-Irano-Turanian element** of the Iberian steppes consists of eastern steppe species that spread through the Mediterranean basin in the latter half of the Tertiary as the Mediterranean Sea desiccated and the Alpine orogeny took place (Jäger 1971; Suárez et al. 1992). The Tethys Sea was gradually reduced, in the late Miocene (Messinian period), and the drying-up Mediterranean basin was converted into a series of brackish lagoons, surrounded by extensive halophytic steppes. These complex events facilitated the emergence of a steppe environment and stimulated the evolution of the thorny-cushion ('tragacanthic') life form in several shrubby species in the Mediterranean high mountains, above the level of cloud condensation, as well as the establishment of 'species of dry affinity' lower down (Sainz Ollero and Dominguez Lozano 2011). Many floristic vicariances at generic level (*Salsola*, *Suaeda*, *Astragalus*, *Artemisia*, *Stipa*, *Microcnemum*, *Gypsophila*, *Thymelaea*, *Juniperus* sect. *Sabina*, etc.), and some remarkable disjunct occurrences (*Pistacia atlantica*, *Krascheninnikovia ceratoides*, *Microcnemum coralloide*) document these processes.

The **Mediterranean-Saharo-Sindian element** of the Iberian steppes consists of plants derived from the Tertiary subtropical flora of the open African spaces, which had savanna-like landscapes (Quézel 1978) and gave rise to landscapes related to hot deserts. The esparto grass (*Stipa tenacissima*) is probably the best example of this floristic element which is so prominent in the Maghreb countries. The formations in which esparto dominates precisely delineate the Mediterranean thermophilic steppes in Spain as well as in northern Africa. Other characteristic species of this element are *Periploca laevigata*, *Maytenus senegalensis*, *Ziziphus lotus*, *Whitania frutescens* and, *Launaea arborescens*.

Endemism is considerable in the Iberian steppe areas. Already the classic work of Willkomm (1852) highlights the existence of 126 Iberian endemics (42%) in a total of 302 steppe species. The floristic peculiarity of the Iberian steppes is further augmented by the abundance of Ibero-Mauretanian species with small distribution areas.

Probably the most significant endemics in this steppe flora are the three endemic woody crucifers which are ancient and evolutionary isolated taxa (paleoendemics) (Gómez Campo 1981), whose distribution areas are fragmented and small, and clearly reflect their relict character. These are *Vella pseudocytisus* (basin of the Tagus, Hoya de Baza steppe near Teruel), *Boleum asperum* (Monegros, valley of the Ebro) and *Euzomodendron bourgaeanum* (Tabernas, Almería).

In the gypsum steppes endemics are very abundant. Willkomm (1852) already mentioned 32 endemic species of the gypsum steppe, and Mota et al. (2009) listed 69 gypsophilous species in a total of 140 species which occur in these habitats.

9.2 Landscape Diversity in Iberian Steppes

Given the amplitude of the concept of 'steppe', particularly as used in Spain, it is difficult to precisely delineate their distribution in the Iberian peninsula. While in the Iberian steppes the major differences have to do with substrates (Sainz Ollero 2003; Sainz Ollero et al. 2010; García Antón et al. 2002), all the steppes have in common an aridity above the mean value which limits the presence of trees. In some cases the aridity is related to the general climate, as in the Murcia-Almeria sector, but in many others it is the local phenomenon of a rain shadow caused by surrounding ridges, as in the Ebro basin and interior plateaus. Often continentality exacerbates the local climatic conditions favouring stable steppe formations, as happens in the 'parameras'. These terrains are very characteristic of the Iberian peninsula and are located at altitudes of 1,000–1,200 m along the periphery of the Iberian System. They constitute the primary habitat of forest-steppe, open, treeless vegetation in a mosaic with *Juniperus* ('sabinares' and 'enebrales') and *Pinus* ('pinares') stands, and of thorn scrub ('aulagares') and dwarfshrubs ('tomillares'). Soil-related constraints are complementary factors of great importance that often determine the presence of steppe. Here we mention the nutrient-poor lithosols, as in La Serena and other parts of the basement in the western peninsula, and the evaporite substrates from Miocene ancient lakes that are rich in marl-gypsum materials and salt. Endorheic depressions are frequent in the inland basins (Ebro, Tagus, Duero, Hoya de Baza and Guadix) (Fig. 9.1c).

The first maps of Willkomm (1896) and Reyes Prosper (1915) (Fig. 9.1a, b) show a series of nuclei located in the interior depressions on Tertiary substrates (Oligocene and Miocene) in the basins of the Duero, the Tagus (steppes of La Mancha) and the Ebro, and in the semi-arid area of Murcia-Almeria, the driest part of the Iberian peninsula. Reyes Prosper delineated larger spaces, but they essentially coincide with those of Willkomm. Ornithologists (De Juana et al. 1988) suggested larger areas, determined by the distribution of steppe birds. These extensions would also

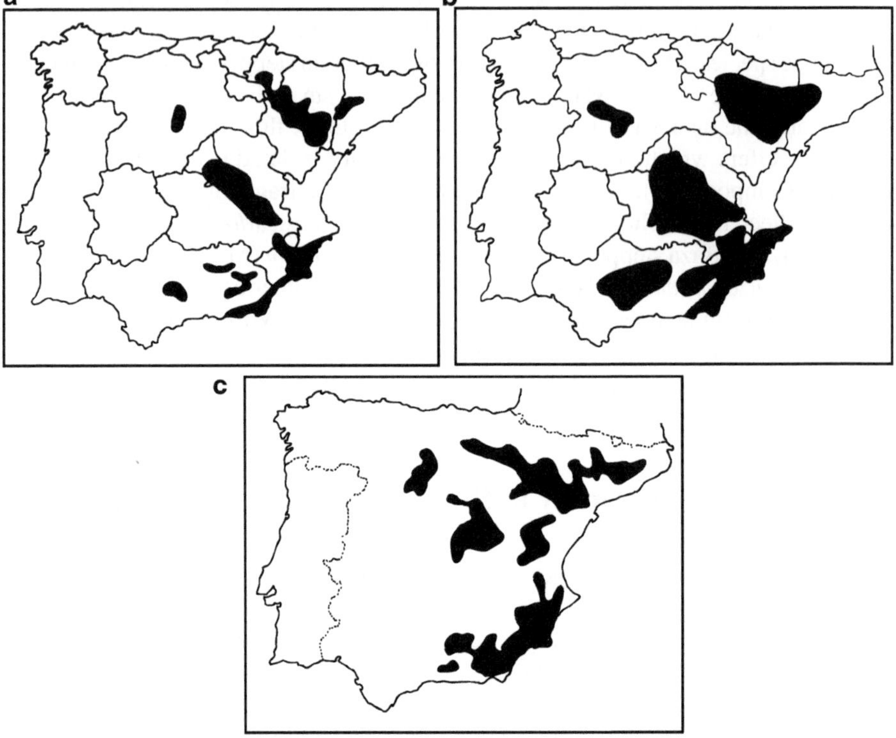

Fig. 9.1 (**a**) Distribution of the Iberian steppes according to Willkomm (1896), and (**b**) Reyes Prósper (1915), (**c**) distribution area of *Ononis tridentata*, a gypsophilous legume, which closely coincides with the distribution of the edaphoxerophilous steppes

include the 'parameras' of the Iberian System, the sparse grasslands of Extremadura and the Sierra Morena, and the region of the Serena or the valley of Alcudia, and also some cereal steppes. In the Iberian peninsula there are large areas where cereal crops are extensively cultivated once in 2 years. This practice results in wide open spaces that are fallow in the year between cultivations ('Iberian dry farming' in the terminology proposed by Dantín Cereceda 1916). These terrains are very favourable for the maintenance of steppe birds.

9.3 Particularities of the Various Iberian Steppes

9.3.1 The Valley of the Ebro

This is one of the most extensive and characteristic steppe areas, although it is very fragmented by cultivation. The main factors determining the steppe character here are continentality, rain shadow, saline substrates. In the flora the Irano-Turanian

Photo 9.1 'Espartales' (*Stipa tenacissima* steppes) and greenhouses in the Campo de Nijar (Almeria) (Photo by H. Sainz Ollero)

element (*Krascheninnikovia ceratoides*, *Microcnemum coralloides*, *Gypsophila perfoliata*) is represented.

Among what appear to be better conserved areas we mention the Bardenas Reales de Navarra; the bad-lands of the Monegros (Alfajarín, Valmadrid, Saso de Osera, being small areas in between crop fields); the plains on the right bank of the Ebro (refuge of the Lomaza of Belchite and the ornithological reserve "El Planerón", protected now as a Special Protection Area (SPA, Birds Directive EEC/79/409)); "Tomillar de Alfés & Mas de Melons" in Lleida, one of the few steppe areas of Catalonia; the sector of the endorheic saline Bujaraloz-Sástago and Chiprana.

Irrigation programmes, promoted by the Government of Aragon, have distorted much of the best steppe areas of the Monegros and in the area of the Bujaraloz-Sástago salt complex. These irrigation projects use the surplus of water in the basin of the Ebro, coming from the Pyrenees.

9.3.2 Semi-arid Southeast: Murciano-Almeriense Sector and Albacete

In these areas the most typical climatic steppes are located on a wide range of substrates. There are interior sub-desert areas (Tabernas); outcrops of gypsum and salts (Sorbas); areas of quartz or slate (Sierra de Cabrera, Filabres, higher part of Alhamilla); marine sediments (Campo de Nijar, nowadays covered by greenhouses, Photo 9.1);

Photo 9.2 Expansion of cereal crop fields in the 'espartales' (*Stipa tenacissima* steppes) of Albacete (Photo by H. Sainz Ollero)

"dalias" which are fields known to the Romans as "campus spartarius" for their extensive formations of esparto (*Stipa tenacissima*) (Photo 9.2); continental sedimentary rocks of lime and clay (Hellín, Albacete); dolomite and lime stone areas (Sierra Gador); and volcanic areas (Cabo de Gata).

Characteristic of the flora is a marked endemism (*Euzomodendron bourgeanum*, *Antirrhinum charidemi*, *Teucrium intricatum*, *T. turredanum*) and notable North African or Saharo-Sindian influences at the level of shrubs or bushes (*Periploca laevigata*, *Maytenus senegalensis*, *Ziziphus lotus*, *Stipa tenacissima*, *Whitania frutescens*, *Launaea arborescens*) and succulents or herbaceous species (*Androcymbium gramineum*, *Fagonia cretica*, *Caralluma europea*, *Mesembryanthemum cristallinum*).

The Murcian steppes extend inland into the basin of the Segura (Cieza, Yecla) reaching the 'espartales' (Stipa communities) and 'yesares' (gypsum vegetation) of the province of Albacete (Hellín) and the salt marsh of Cordovilla. These are forested steppes in which the 'espartales' occur between stands of *Pinus halepensis*, with *Sideritis serrata* as one of the most notable endemic species. This area extends to Almansa, where the steppes are colder, and reaches to the formations of the 'parameras' of the Iberian System with thorny-cushion species, such as *Genista pumila*.

9.3.3 *Depressions of Baza and Guadix in the Province of Granada*

These are steppe areas formed by the orographic isolation imposed by various Betic mountain ranges (Sierra Nevada and Baza to the South, Cazorla, la Sagra and Revolcadores to the North, Maria, Estancias and Filabres to the East). Beside these, the marine and continental sediments with a predominance of marl-chalky material, and endorheic saline areas, where halophytic formations are very well developed, are also important. To be mentioned is the presence of shrubs stands of great biogeographical significance, such as *Vella pseudocytisus* or *Krascheninnikovia ceratoides.*

9.3.4 *La Mancha and the Basin of the Tagus*

Nowadays the steppe areas of La Mancha are very much reduced due to the extension of crop cultivation. They only persist on the gypsum areas of lower agricultural interest, on steep slopes in the basin of the Tagus (Aranjuez) and near some salty lakes (in the la Sagra area). In general the steppes of this zone are considered as secondary, as oak forests appear to be the potential climax vegetation, except in the Campo de Montiel (with 'sabinares' of *Juniperus thurifera*) or on the Quaternary sands of river basins or at Ciudad Real (with 'pinares' of *Pinus pinea*).

9.3.5 *La Serena (Badajoz), the Valley of Alcudia (Ciudad Real) and Other Parts of Extremadura ('llanos' of Trujillo and Cáceres)*

These steppe areas are located on early Paleozoic substrates and are very different from the rest of the Iberian peninsula. They consist of poor pastures of therophytes or hemicryptophytes, in which woody species are rare. It is an area mostly covered by 'dehesas' (savanna-like landscapes) and forest stands of *Quercus* in which the steppes are small enclaves. They occur on poor soils (lithsols), or at higher altitudes, which are relatively cold and affected by drying winds ('paramos').

9.3.6 *The Duero Basin, Tierra de Campos and the Environment of Villafáfila*

These steppes are very limited in area, occurring among extensive croplands, on sedimentary substrates or Tertiary evaporites (marl-chalky and salty) which are unfavorable for forests. Their uniqueness was already noted by Willkomm (1852).

9.3.7 'Paramos' of the Iberian System and the Basin of the Duero: Soria, Palencia ("El Cerrato"), Burgos, ("La Lora, Masa and La Bureba"), Segovia, Guadalajara, Cuenca, Zaragoza (Gallocanta Environment) and Teruel

The steppes of the 'paramos' are among the most original and better represented landscapes of the Iberian peninsula. These 'paramos' are high (900–1,200 m), more or less flat zones on carbonate substrates of the Tertiary (Messinian limestone, formed some 6 million years ago by the silting of the interior lakes of the Iberian depression), or on Mesozoic carbonate substrates (Jurassic dolomites and lime stones) which surround the Iberian System and the eastern part of the Central System. The vegetation consists of a mosaic of thorny bushes ('aulagares', *Genista scorpius*, *G. pumila*, *Erinacea anthyllis*) or dwarfshrub and grassy formations ('tomillares', 'salviares', 'esplegares', with *Thymus vulgaris*, *T. praecox*, *Salvia lavandulifolia*, *Lavandula latifolia*, *Helianthemum spp.*, *Teucrium polium*, *Poa ligulata*, *Festuca hystrix*, *Koeleria vallesiaca*) that appear next to forest stands of *Juniperus thurifera*, and sometimes of *Quercus faginea* or *Q. ilex* subsp. *ballota*. It is a terrain with fossilized landscapes since the last ice age in which the cold Mediterranean steppes and the juniper forests ('sabinares') have undergone few changes. The harshness of the weather conditions (continentality, cold and drying winds), together with a karstic lithology that does not favor the retention of moisture, explains why there is little competition with deciduous trees. These appear in the periphery of the 'paramos', in warmer or less arid areas.

9.4 Changes in Use of the Landscape

Most characteristically the steppes have been used for herding sheep and, in the dryer areas, goats. Centuries ago, the usage was more intensive than it is today. Often the usage for herding has been complemented with dry agriculture. Of course this cultivation has been restricted to parts that, because of their topography, are suited, e.g. valleys. Besides that, hunting for small animals has been popular, because of the abundance of hare and partridge, and other steppe birds. Nowadays, in some regions, hunting is even one of the most profitable usages. Other kinds of minor usage, as for example the picking of 'esparto' (*Stipa*), have been restricted to some local areas.

Concerning the Ebro valley, it has been said that the landscape in Roman times was formed of open boscages of pines and junipers with enclaves of halophytic scrub around salines. From old ages, its usage was confined to herding, agriculture, hunting and gathering of wild plants. This changed profoundly during the Middle Ages, and apparently the boscages had disappeared by then. The role of transhumance in the disappearance of trees in these landscapes has been subject of debate; the usage of agriculture was subordinated to that of herding, but it is not clear if this was also the case in Aragon. Besides that, there were strict rules concerning the use of the grazing lands.

After the Mesta, a powerful medieval sheep holders association, was abolished in 1835/1836 the use for agriculture intensified, which put an end to the remaining trees. The ‘pastos’ and ‘dehesas’ in the Monegros, for example, were transferred into the cultivation of grain on a large scale. Nevertheless, it was obligatory to cultivate the land only once every 2 or more years, and leaving the borders between plots uncultivated.

Since the 1970s the introduction of mechanization and the use of fertilizer and pesticides allowed an intensification of agricultural use. As a consequence the diversity of the landscape has further diminished. Other areas have been put to industrial use, house-building and irrigation.

The development of the ‘paramos’ has been more or less the same, although the point of departure here was different, with more variation in the type of boscages and the abundance of different kinds of tree species. The usage also was more variable and extensive, with a growing importance of herding in accordance with the preponderance of the Mesta. From the still remaining boscages the changes in usage for herding can be analyzed, along with their clearance and their use for firewood.

This change did not begin with the Middle Ages. According to Fidalgo Hijano (1987) already in the twelfth century laws were introduced to protect boscages. But in the eighteenth century the deforestation had already advanced considerably. The expropriation of church property in the period 1837–1855 (known as Mendizabal y Madoz) worked out similarly here as in the Ebro Valley and proliferated the use of dry cultivation (Bauer Manderscheid 1980). In aerial views these changes in the landscape, brought about by the increase of dry cultivation, still can be observed. Yet, the intensification brought about by mechanization, fertilization, etc. did not happen here, due to the low earning capacity of the small and difficult-to-reach plots of land. Only the change in the system of use from once every 2 years into yearly production has influenced the landscape to some extent.

Other influences have come with changes in the way the sheep are herded in some areas. Herds have grown bigger, and sometimes they are confined into restricted plots of land where the sheep can roam freely. But in recent years several steppe areas became undergrazed, and locally grazing was completely abandoned because of the poor income it generated. This lack of grazing degraded many steppe areas. Besides the use for herding sheep and goats, until recently the gathering of grasses (*Stipa* – raw material for paper production) has been substantial. The steppe landscape has also altered because of the plantation of boscages of pines and other trees since the 1960s and again recently, by the construction of wind turbines, Other changes in recent decades include the extension of the cultivation area of crop trees (olive, vine), irrigation, disappearance of traditional fallows, and, particularly in coastal areas, uncontrolled urbanization (Bota et al. 2005).

9.5 Conservation Efforts

With the development of the Natura 2000 network, at least on paper, there has been a remarkable progress in cataloging areas for conservation. Numerous steppe areas have been declared as Special Protection Area (SPA, Birds Directive EEC/79/409) and Special Area of Conservation (SAC, Habitats Directive EEC/92/43). This draws a much more encouraging picture than the one which was presented by Suárez et al. (1992) or the one which was presented at the International Symposium for the Conservation of Steppe Birds and their Habitat, held in Valladolid (Castilla and León, Spain) in 1995. The importance of the steppe ecosystem, its unique landscape and relevance to steppe birds, seems to have got through to society and policy makers. There are currently 30–35 areas of the Natura 2000 network that have been proposed for its steppe characteristics, including important areas, such as the Monegros in Aragon, the Bardenas Reales in Navarre, the desert of Tabernas, Cabo de Gata or Sorbas in Almeria, the Altos of Barahona, the junipers of Cabrejas or the moors of Layna in Soria, the moorland of Maranchón in Guadalajara, the valleys of the Cerrato in Palencia-Burgos, La Serena in Badajoz, etc. In addition, many cereal steppes have been designated as SPAs in Castilla y León, Castilla-la Mancha and Madrid. Within the SACs important gypsum areas (Cuenca alcarria, valley of the Tagus) and salinas (Saladares of Cordovilla and Agramón in Albacete; La Mancha lagoons, etc.) are included. However, claims for a National Steppe Park in the valley of the Ebro, or in the semi-arid area of Almeria-Murcia, continue to meet opposition from local communities. This is especially the case in Aragon, where plans for irrigation in the Monegros, or the promotion of a large Casino, collide with conservation interests.

The notion of the importance of these unique ecosystems, depending on traditional use, is recognizable in the statements of environmental policy makers. However, few concrete conservation actions have been taken so far to reverse the processes of degradation and destruction which affect many areas.

9.5.1 New Proposed Project on Steppe Conservation in Aragon

A new conservation and investigation project on the steppes in Aragon has been developed. The objective of this project is to investigate the effect of abandonment of grazing on the biodiversity of the steppe ecosystem, and to apply the Ecosystem Approach for the conservation of biodiversity. More specifically, it is aimed at determining the role of pastoralism in the maintenance of habitats of endangered steppe birds and plants, and investigating the social economic conditions of extensive livestock breeding.

The steppes of Aragon are of great European importance with many bird and plant species included in the Habitat Directive. These steppes are primary grasslands similar to areas in North Africa and the steppes of Central Asia (Suárez et al. 1992).

Photo 9.3 Black-bellied Sandgrouse (*Pterocles orientalis*) breeding on the dry open plains of the Spanish steppes (Photo by A. Portero Garces)

In Aragon the steppes count for the largest steppe bird communities of Spain with highly endangered species such as the great bustard (*Otis tarda*), the little bustard (*Otis tetrax*), the common stone curlew (*Burhinus oedicnemus*), the lesser kestrel (*Falco naumanni*), the ash harrier (*Circus pygargus*) and the Ricotí lark or DuPont lark (*Chersophilus duponti*). These birds are faced with declining populations at a national and/or international level, and are directly dependent on the steppe ecosystems. Among the plant species there is a high level of endemism (42%).

The project is proposed to be carried out on a site in the 'Muelas del Jiloca', a Special Protection Area (SPA, Birds Directive EEC/79/409) and Special Area of Conservation (SAC, Habitats Directive EEC/92/43) in the state of Aragon, in a semi-arid steppe on calcareous and gypsum soils (Braun-Blanquet and De Bolos 1987). The vegetation consists mainly of Mediterranean shrub steppes, and on sites with gypsum soils the vegetation consists of gypsum steppe: *Gypsophiletalia* (1520), a priority habitat type in the Habitat Directive (92/43/EEC). The area harbours a community of steppe birds, such as the skylark (*Alauda arvensis*), the DuPont lark (*Chersophilus duponti*), the spectacled warbler (*Sylvia conspicillata*), the little bustard (*Tetrax tetrax*), the black-eared wheatear (*Oenanthe hispanica*), the short-toed lark (*Calandrella brachydactyla*), and the calandra lark (*Melanocorypha calandra*). Some of them are highly endangered. Steppe birds in general depend very much on the vegetation structure, and are often limited to sites with rather open vegetation of a very specific height (Photo 9.3).

On these steppes there is a long history of extensive grazing by sheep (Photo 9.4), which maintains a rather open vegetation with many typical plants and birds. During the last decade there has been a drastic reduction in the number of herdsmen,

Photo 9.4 Herdsman with sheep on a stubble field in the steppe area (Photo by M.A. van Staalduinen)

because of the bad socio-economic situation of extensive livestock breeding. Without adequate measures, it is foreseen that this negative trend will continue and extensive livestock breeding will disappear. Nowadays, in the rural areas of Aragon only about 4,000 herdsmen remain.

On the Muelas del Jiloca the number of sheep has diminished to 1,650,000, almost half since 1990. But due to the abandonment of grazing the vegetation has changed dramatically. It is to be expected that within the next decades this will lead to a much denser and higher vegetation and, consequently, a loss of habitat for protected plant and steppe bird species.

The general goal of the conservation project is to promote the conservation of the steppes in Aragon, and to improve the livelihoods of herdsmen who are dependent on these ecosystems and vice versa. The project themes are: (1) studying the effects of abandonment of grazing on the biodiversity and functioning of the steppe ecosystem; (2) studying the effects of fire and cutting on the rejuvenation of the steppe vegetation; (3) studying the consequences of the social-economic situation of the herdsmen for the continuation of pastoralism; (4) development of future perspectives and strategies for an Ecosystem Based Adaptation to the abandonment of pastoralism. These themes will be researched by means of inventories, experiments, measurements and interviews.

An innovative aspect of this project is the implementation of the ecosystem approach developed by the IUCN (Shepherd 2004), aimed at promoting the conservation and sustainable use of the ecosystem in an equitable way. In this project the primary stakeholders, the herdsmen, will be involved in the management of the area, by using the grazing of sheep as a tool for nature conservation. Grazing plans will be developed by using the traditional knowledge of the herdsmen together with ecological knowledge of the steppe vegetation. The results of the study will be applied for the development of restoration methods, a monitoring programme, and contracts for environmental services by herdsman. This will finally result in guidelines for conservation management, and an improvement and diversification of income for the herdsman.

This project is seen as important, because solid action is urgently needed in order to prevent a further degradation of the steppes and a total abandonment of extensive livestock breeding in Aragon, Spain.

References

Bauer Manderscheid E (1980) Los montes de España en la historia. Ministerio de Agricultura, Madrid, p 610

Bota G, Morales MB, Mañosa S, Camprodon J (eds) (2005) Ecology and conservation of steppe-land birds. Lynx Edicions & Centre Tecnològic Forestal de Catalunya, Barcelona

Braun-Blanquet J, de Bolos O (1987) Las comunidades vegetales de la depresión del Ebro y su dinamismo. Deleg. Medio Ambiente, Ayto. de Zaragoza, Spain

Costa M, Morla C, Sainz OH (eds) (1997) Los bosques ibéricos. Planeta, Barcelona

Dantín Cereceda J (1916) Dry-Farming ibérico. Cultivo de las tierras de secano en las comarcas áridas de España. Imprenta Gutemberg, Guadalajara

De Juana E, Santos T, Súarez F, Tellería JL (1988) Status and conservation of steppe birds and their habitats in Spain. In: Goriup PD (ed) Ecology and conservation of grassland birds. ICBP technical publication no 7, Cambridge, pp 113–123

Fidalgo Hijano C (1987) La transformación humana del paisaje en la serranía de Atienza. Ediciones de la Universidad Autónoma de Madrid, Madrid, 116 pp

García Antón M, Maldonado J, Morla C, Sainz Ollero H (2002) Fitogeografía histórica de la península Ibérica. In: Pineda FD, de Miguel JM, Casado MA, Montalvo JM (eds) La Diversidad Biológica de España. Pearson educación S.A, Madrid, pp 45–64

Gómez Campo C (1981) Taxonomic and evolutionary relationships in the genus Vella. Bot J Linn Soc 82(2):165–179

González-Sampériz P (2004) Evolución paleoambiental del sector central de la Cuenca del Ebro durante el Pleistoceno Superior y Holoceno. Inst. Pirenaico de Ecología – CSIC. Dpto de Ciencias de la Antigüedad. Univ. de Zaragoza

Jäger EJ (1971) Die pflanzengeographische Stellung der «Steppen» der Iberian Halbinsel. Flora 160:217–256

Mota JF, Sánchez Gómez P, Merlo ME, Catalán P, Laguna E, de la Cruz M, Navarro FB, Marchal F, Bartolomé C, Martínez Labarga JM, Sainz Ollero H, Francisco Valle F, Serra L, Martínez Hernández F, Garrido JA, Pérez García FJ (2009) Aproximación a la checklist de los gipsófitos ibéricos. Anal Biol 31:71–80

Quézel P (1978) Analysis of the flora of Mediterranean and Saharan Africa. Ann Miss Bot Gard 65:479–534

Reyes Prósper E (1915) Las estepas de España y su vegetación. Publicación a costa de la Casa Real, Madrid

Sainz Ollero H (2003) Introducción al paisaje vegetal español. In: Martí R, del Moral J (eds) Atlas de las Aves reproductoras de España. Dirección General de Conservación de la Naturaleza-Sociedad Española de Ornitologia, Madrid, pp 49–53

Sainz Ollero H, Domínguez Lozano F (2011) El contexto geobotánico. In: Domínguez Lozano F, Guzmán Otano D, Moreno Saiz JC (eds) Biología de la Conservación de Vella pseudocytisus subespecie paui, una planta amenazada en Aragón. Consejo de Protección de la Naturaleza de Aragón, Serie Investigación, Zaragoza

Sainz Ollero H, Sánchez de Dios R, García-Cervigón Morales A (2010) La cartografía sintética de los paisajes vegetales españoles: una asignatura pendiente en geobotánica. Ecología 23:249–272

Shepherd G (2004) The ecosystem approach: five steps to implementation. IUCN, Gland/Cambridge

Suárez F, Sainz Ollero H, Santos T, González Bernáldez F (1992) Las estepas ibéricas. Unidades temáticas ambientales, MOPT, Madrid

Walter H (1977) Zonas de vegetación y clima. 1ª reimpr. 1994, Omega, Madrid

Walter H (1984) Vegetation of the Earth. Springer, Berlin

Willkomm M (1852) Die Strand-und Steppengebeite der Iberischen Halbinsel und deren Vegetation W. Engelmann Leipzig

Willkomm M (1896) Grundzüge der Pflanzenverbreitung auf der Iberischen Halbinsel und deren Vegetation. In: Engler EA, Drude O (eds) Die Vegetation der Erde (I). W. Engelmann, Leipzig (reimp. 1976. A R Gantner Verlag K.G.FL. 9490. Vaduz)

WWF-España (2009) Hernández L, Romero F coord. Bosques españoles. Los bosques que nos quedan y propuestas de WWF para su restauración. Madrid (Dep. Legal: M-40080-2009)

Part II
Degradation

Chapter 10
Pastoral Degradation of Steppe Ecosystems in Central Mongolia

S.N. Bazha, P.D. Gunin, E.V. Danzhalova, Yu.I. Drobyshev, and A.V. Prishcepa

Abstract In a close comparison of grazed and protected stands along a long submeridian transect next to the trans-Mongolian railway (Sukhe-Bator – Ulaanbaataar – Dzamyn-Ud) across several subzones of the Mongolian steppes we studied cover, species composition, and aboveground phytomass. In about all main types of steppe overgrazing led to the spread of bushes. In mountain-meadow and meadow steppes overgrazing tends to lead to dominance of *Artemisia frigida* and *Caragana pygmaea;* in true and dry steppes with *C. microphylla* and *Artemisia frigida;* in desertified (dry) and desert steppes with *Caragana stenophylla* and *C. korshinskii.* Subdominants, depend on subzonal type of steppe and can include *Artemisia adamsii*, *Carex duriuscula*, *Convolvulus ammanii*, *Leymus chinensis*, *Potentilla acaulis* and *Sibbaldianthe adpressa.* When grazing loads increase it can to strongly monodominant communities.

10.1 Introduction

In the twentieth century the steppe ecosystems of Eurasia have been subject to intensive anthropogenous influence mainly because their natural properties proved favorable for human settlement and economic development. The result was a practically utter annihilation of steppe ecosystems in Central and Eastern Europe, and a significant transformation of steppe ecosystems in Western Siberia, Kazakhstan and Transbaikalia. In Mongolia steppe ecosystems have not yet lost their natural potential as the country has a scarce population and preserves the traditional

S.N. Bazha (✉) • P.D. Gunin • E.V. Danzhalova • Yu.I. Drobyshev • A.V. Prishcepa
A.N. Severtsov Institute for Ecology and Evolution,
Russian Academy of Sciences, Moscow, Russia
e-mail: sbazha@inbox.ru; monexp@mail.ru; elenadanzhalova@gmail.com;
altanus@mail.ru; monexp@mail.ru

M.J.A. Werger and M.A. van Staalduinen (eds.), *Eurasian Steppes. Ecological Problems and Livelihoods in a Changing World*, Plant and Vegetation 6,
DOI 10.1007/978-94-007-3886-7_10,

cattle-breeding way of life. However, here 50–60% of the pastures are subject to moderate disturbances, and 20–25% to strong disturbances (Gunin et al. 1998). The strongly increased animal load of the past decades has disrupted the natural balance and, because of the high vulnerability of semiarid and arid ecosystems, this promotes their degradation and desertification. This has resulted, first of all, from the increase in the number of cattle breeders in many regions of Mongolia to the extent that one can speak of a "cattle-breeding boom" that, in the opinion of many authors, presently resulted in "the true revolution" in the agricultural sector of the national economy (Muller and Bat-Ochir 1996; Graivoronsky 1997). This is primarily due to the burgeoning numbers of herders and their livestock. From the 1980s till 2008, the total number of livestock has increased almost twice – from 23.0 to 44.3 million (State of the Environment 2002, 2009). At the same time, the herd structure changed: the number of goats increased 3–4 fold. The weak development of an infrastructure for cattle-breeding facilities has, from an ecological point of view, led to an unreasonable increase in the number of goats in the central part of Mongolia. This is particularly obvious in the majority of suoms (districts) of the Selenga, Central and Eastern-Gobi aimags (regions). During the past two decades, the total number of livestock has increased there 1.5–2 fold. At the same time, the number of goats increased 3–4 fold. All this adversely affected the pastures in Mongolia in general, and in its central part in particular. Given this situation we argue that the results of scientific studies into the basic laws of transformation of steppe ecosystems as a result of pastoral use should be used as guidelines; those results can be used to assess the character of the changes in the plant communities as digressive or regenerative processes, and define the conditions of optimal functioning of the steppe ecosystems in view of their economic and nature protection value.

Our main task here was to study the reaction of the structure and efficiency of the plant communities of the basic types of steppe ecosystems of Central Mongolia on the effect of grazing. We consider structure and efficiency the most sensitive parameters that will allow an estimation of pasture condition and anthropogenic disturbance, and an assessment of the character of the transformation of the steppe ecosystems under pastoral use.

As known, an important feature of the Daurian-Mongolian sector of the Eurasian steppes is their cover with bushes, due to the dominance of various species of bushes, subshrubs and dwarf semi-shrubs belonging to the genera *Caragana*, *Artemisia*, *Spiraea*, *Armeniaca*, *Amygdalus*, *Dasiphora*, and others (Karamysheva 1961; Karamysheva and Khramtsov 1995). Therefore, some researchers distinguish a special category of bushy steppes within the Asian steppes (Yunatov 1950; Kuminova 1960; Kalinina 1954) and even classify this group as a special type of vegetation (Bykov and Stepanova 1953). However, none of these authors did see the bushy steppes as a result of pastoral digression (degradation) under influence of wild and domestic animals. Moreover, based on Yunatov's work (1950, 1954), the well-known expert on the vegetation of Mongolia, it was believed that the nomadic type of economy and the huge area of pastures in Mongolia did not promote "a wide-spread pastoral digression" (Yunatov 1950: 121). As a matter of fact, Gorshkova and Lobanova (1972) and Gorshkova (1973) first proved that intensely degrading processes and strongly disturbed pastures occurred in the Transbaikalian steppes, but she limited their distribution to the steppes of Southern Siberia, not including the steppes of Mongolia. The foremost woody species that expands under grazing and forms secondary communities is *Artemisia frigida* (Miroshnichenko 1964; Chogny 1988). Comparative analysis of the environment and floristic structure of steppe communities of the European and Asian parts of the steppe area has ascertained, that the large floristic variety in endemic and relic bushes in the steppes of Mongolia, as compared to the plain steppes of the European part of Russia, is caused not only by variation in the soil-lithological structure of the substrate (Lavrenko et al. 1991; Lavrenko and Karamysheva 1992) and the history of the floral composition (Kamelin and Gubanov 1993; Kamelin 1994, 2004), but also by their long period of intensive pastoral use (Dinesman and Bold 1992).

Based on these findings we assume, that during the more than 1,000 years of pastoral use of the steppes of Mongolia many bush species, which are more xerophytic in comparison with typical species of the steppes (grassy plants such as true grasses and sedges, and forbs), were widely distributed in and characteristic for petrophytic and psammophytic habitats but gradually established themselves in the typical zonal steppe communities as these steppe communities became degraded as a result of grazing pressure. This hypothesis can possibly be confirmed in two ways: by carrying out long-term geobotanical monitoring of the restoration of the disturbed vegetation cover in plots protected from external anthropogenous influence; and by carrying out parallel studies on the vegetation cover in representative areas of steppes that for a long time have been protected from pastoral and agricultural use (e.g. within military polygons, meteorological stations, zones of restriction of railways, etc.).

10.2 Our Transect Study

In regions of nomad cattle breeding the main problem in assessing the influence of grazing on the condition of the pasture ecosystems is the absence of a control to compare the grazed pastures with in order to define the character of the changes. In Mongolia such reference plots with areas of up to several tens of hectares have been found along the Trans-Mongolian railway: Sukhe-Bator – Ulaanbaataar – Dzamyn-Ud and its branch lines to the cities of Erdenet and Sharyn-Gol (Fig. 10.1). These plots have been free from grazing for some 30–50 years.

In 2006 we did simultaneous studies on the transformation of the vegetation cover of grazed and non-grazed pastures in 10 key plots located along a submeridian transect along the railway and crossing the typical communities in the main subzones of the steppe: mountain-meadow steppe, meadow steppe, true steppe, dry steppe, transitional very dry steppe, and desert steppe (Fig. 10.1, Table 10.1). Observations were carried out within a short period (25 July–25 August 2006) during the maximal development of the dominant species, allowing the comparison of the state of the vegetation cover in the different subzones and comparison of grazed and non-grazed plots. In the derivative communities in the grazed plots we registered the following

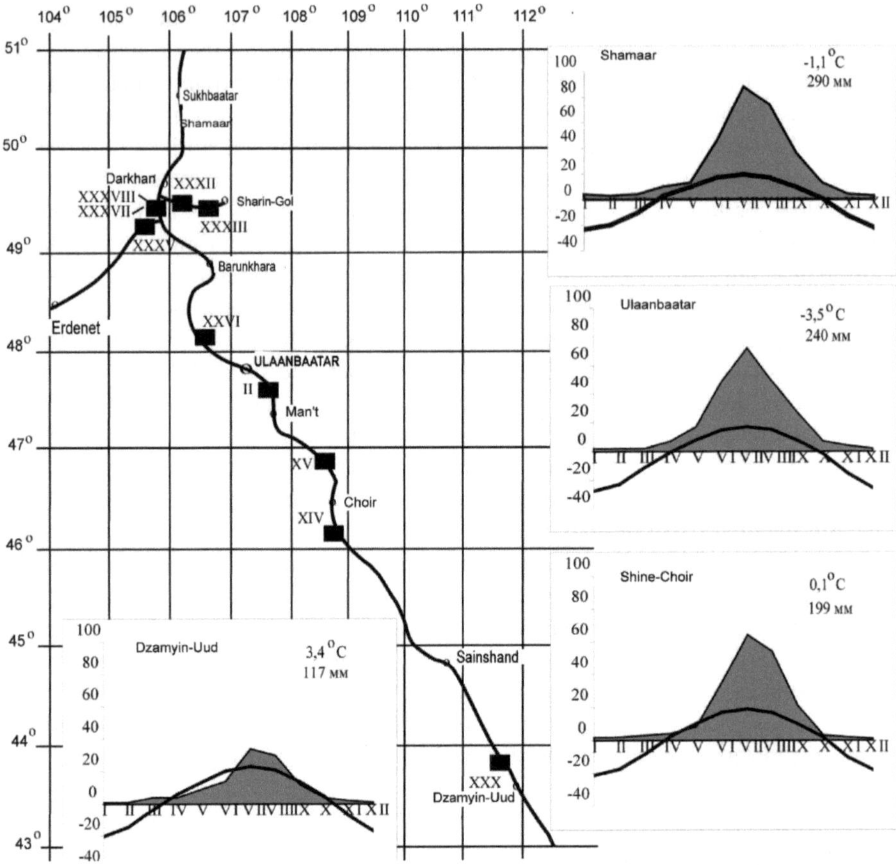

Fig. 10.1 Geographical position of the polygons along the Trans-Mongolian railroad Sukhe-Bator – Ulan-Bator – Dzamyn-Ud and climate-diagrams in the main subzones of the steppe ecosystems (meadow, true, dry, and desert steppe). Polygon labels and communities studied are listed in Table 10.1

qualitative and quantitative changes in structure: change in the phytocoenotic positions of species; nanophytism (reduction in the sizes of plant individuals); amount of herbage, reduction of primary and gross aboveground phytomass; expansion of grass and bush species typical of digressive stages. Grazed and non-grazed (protected) plots were compared and the degree of grazing disturbance (by livestock) was determined by an earlier approved technique (Gunin and Vostokova 1989). Selected plots in a polygon were no more than 10–20 m apart and had identical soil and geomorphological conditions.

Field data were gathered using standard techniques: In each plot detailed geobotanical descriptions in an area of 100 m^2 were made. Calculations of the number, projected aerial cover, and productivity of plants were based on three plots of 1 m^2. The aboveground phytomass was harvested, sorted by species and dried at 105°C.

Table 10.1 Geobotanical characteristics of zonal steppe coenoses along the submeridian transect Sukhe-Bator – Ulan-Bator – Dzamyn-Ud

Steppes	Polygon label	Plant community	Coordinates	Height, m	Grazing regime
Mountain-meadow steppe	XXXIII	[*Caragana pygmaea*]-*Stipa sibirica*, *S. baicalensis* + *Galium verum*	N 49° 23′ 46.0 E 106° 15′ 10.9	885	Non-grazed
		[*Caragana pygmaea*]-*Artemisia frigida* + *Stipa baicalensis*, *S. sibirica* + *Carex duriuscula*	N 49° 23′ 57.9 E 106° 15′ 16.4	882	Grazed
	XXXII	[*Caragana pygmaea*]-*Stipa grandis* + *Artemisia scoparia* + *Galium verum*	N 49° 23′ 24.1 E 106° 14′ 13.7	873	Non-grazed
		[*Caragana pygmaea*]-*Stipa grandis* + *Artemisia scoparia* + *Kochia prostrata* + *Carex korshinskyi*	N 49° 23′ 17.5 E 106° 14′ 05.7	875	Grazed
	II	*Festuca sibirica* + *Stipa krylovii* + *Carex pediformis*	N 47° 37′ 22.3 E 107° 11′ 10.5	1,654	Non-grazed
		Agropyron cristatum + *Stipa krylovii* + *Carex pediformis* + *Koeleria cristata*	N 47° 37′ 22.6 E 107° 11′ 11	1,655	Grazed
Meadow steppe	XXXVIII	*Stipa baicalensis* + *Allium odorum*	N 49° 23′ 35.1 E 105° 55′ 23.3	699	Non-grazed
		Stipa baicalensis + *Cleistogenes squarrosa* + *Carex duriuscula*	N 49° 23′ 35.2 E 105° 55′ 22.3	699	Grazed
	XXXVII	*Allium senescens* + *Cleistogenes squarrosa*	N 49° 23′ 29.0 E 105° 55′ 20.7	703	Non-grazed
		Artemisia frigida, *A. scoparia* + *Cleistogenes squarrosa* + *Carex duriuscula*	N 49° 23′ 22.2 E 105° 55′ 18.3	702	Grazed
True steppe	XXXV	[*Caragana microphylla*]-*Stipa baicalensis*	N 49° 11′ 58.3 E 105° 47′ 14.8	789	Non-grazed
		[*Caragana microphylla*]-*Allium odorum* + *Carex duriuscula* + *Leymus chinensis* + *Stipa baicalensis*	N 49° 11′ 57.5 E 105° 47′ 14.7	789	Grazed
	XXVI	*Stipa baicalensis* + *Convolvulus ammanii* + *Bupleurum scorzonerifolium*	N 48° 04′ 54.7 E 106° 35′ 23.8	1,278	Non-grazed
		Agropyron cristatum + *Stipa krylovii*, *S. baicalensis* + *Leymus chinensis* + *Artemisia adamsii* + *Potentilla acaulis*	N 48° 05′ 07.7 E 106° 35′ 13.8	1,280	Grazed

(continued)

Table 10.1 (continued)

Steppes	Polygon label	Plant community	Coordinates	Height, m	Grazing regime
Dry steppe	XV	[*Caragana microphylla*]-*Stipa krylovii*	N 46° 56′ 41.7 E 107° 43′ 39.9	1,330	Non-grazed
		[*Caragana microphylla*]- *Artemisia frigida* + *Salsola pestifera*	N 46° 56′ 41.5 E 107° 43′ 37.4	1,334	Grazed
Transitional very dry steppe steppery steppe	XIV	[*Caragana stenophylla*]-[*Kochia prostrata*]-*Allium odorum*, *A. bidentatum* + *Cleistogenes squarrosa* + *Stipa gobica* + *S. krylovii*	N 46° 09′ 19.6 E 108° 37′ 14.9	1,221	Non-grazed
		[*Caragana stenophylla*]-*Stipa krylovii* + *Cleistogenes squarrosa* + *S. gobica* + *Allium bidentatum* + *Sibbaldianthe adpressa*	N 46° 09′ 17.7 E 108° 37′ 13.9	1,219	Grazed
Desert steppe	XXX	[*Caragana korshinskii*, *C. stenophylla*]-*Stipa gobica* + *Asparagus gobicus* + *Allium mongolicum*	N 43° 55′ 02.0 E 111° 37′23.2	989	Non-grazed
		[*Caragana korshinskii*, *C. stenophylla*]-*Asparagus gobicus* + *Allium mongolicum* + *Stipa gobica*	N 43° 55′ 09.2 E 111° 37′ 26.4	985	Grazed
		[*Caragana korshinskii*, *C. stenophylla*]-*Asparagus gobicus* + *Allium mongolicum* + *Cleistogenes songorica* + *Stipa gobica*	N 43° 54′ 54.4 E 111° 37′ 07.0	979	Grazed

For bushes we measured height, greatest and smallest diameter of the crown, and calculated the number of branches in subplots of 0.5 and 1.0 ha. Crown projection areas and volumes (based on crown projection and height) of individual bushes were calculated. Total projected cover of the bushes was calculated from crown area and number of individuals. Their aboveground phytomass was calculated in the same way. Weight of the average bush individuals was calculated using the technique of Slemnev (1969). In some plots we only counted the number of bushes and calculated their aboveground phytomass and projected cover from the weight and crown projection area of average individuals of these species as measured in adjacent plots or adjacent polygons (Slemnev et al. 2005). Furthermore, in each plot we collected soil samples from different genetic horizons to determine their natural humidity.

10.3 Dynamics in the Steppes

The submeridian transect is located between 43° and 51° N and crosses Mongolia in its central part. From the orographical point of view, the transect crosses four large geomorphological areas: the Orkhon-Selenga middle-high mountains, the low-mountain massifs of Southwestern Khentei, the central part of the Gobi, and the Eastern-Gobi depression (Geomorphology of MPR 1982). Average absolute heights of the relief decrease from north to south from 2,000 m to 600 m a.s.l. Characteristic elements of relief along the transect are the aeolian sandy deposits on slopes of low mountains, on plains, and in depressions. In most cases they have the character of a turf-covered, ancient relief, accumulated by the wind, and only in windy, sandy corridors (the valley of the Tuul River, Eastern-Gobi depression, etc.) they sometimes serve as the reservoirs of sandy material blown into adjacent landscapes.

Plantgeographically, the territory of the transect lays in the Central-Asian (Daur-Mongolian) subarea of the Eurasian steppe area (Lavrenko 1970). Traditionally the vegetation of the steppe zone in Mongolia is classified into six height-zone and zone-belt types of steppes: "cryophyte-forb-tussock-grass; meadow grass-forb, forb-grass and sedge; forb-tussock-grass; dry tussock-grass and rhizome-grass; desertified tussock-grass and desert semi-shrub-tussock-grass steppes" (Lavrenko et al. 1991: 119–130). Based on data of meteorological stations along the railway (Namkhaijantsan 2002) and an analysis of soil-geomorphological parameters on the map "Ecosystems of Mongolia" (2005) we classified the steppe ecosystems along the transect as: true, dry, desertified (very dry), and desert steppes, which are zonal vegetation types, and mountain-meadow and meadow steppes which occur as separate mosaic units on raised massifs and in intermountain hollows and valleys (Table 10.2).

As Table 10.2 shows, the categories of steppes can differ 2–2.5 times in moisture, and 5–7 times in dryness index. The moisture characteristics of the steppe types (mountain-meadow, meadow, true, dry, desertified and desert) are respectively: humidified, moderately damp, moderately dry, dry, very dry, and extra-dry. Precipitation and evaporation show seasonal and perennial rhythms. Rainfall occurs

Table 10.2 Hydro-thermal regime of steppe ecosystems in Mongolia (According to Beresneva and Rachkovskaya 1978; Nogina 1989; Namkhaijantsan 2002; Gadzhiev 2002)

	Thermal regime					Precipitation	Indices	
The main types of steppe ecosystems	t_I	t_{VII}	t_{year}	A_t	Period with t>5°	X_{year}	R/L_x	Ik
1. **Mountain-meadow steppe** (wet). Cryophyte-forb-grass on chernozem-like and dark-chestnut soils	–21.0	10–12	–3.0	32.5	120–125	>300	1.3–1.5	<80
2. **Meadow steppe** (moderately-wet). Grass-forb on dark-chestnut soils	–22.8	17.5	–1.5	40.3	135–150	250–300	1.7–2.0	80–100
3. **True steppe** (moderately-dry). Forb-tussock-grass on dark-chestnut and chestnut soils	–21.8	20.0	0.2	41.8	160	200–250	2.6–3.0	100–120
4. **Dry steppe**. Tussock-grass on chestnut soils	–20.0	21.5	0.5	41.5	165	150–200	5.0	135–160
5. Very dry steppe. Dwarf semibush-tussock-grass on light-chestnut soils	–18.5	21.7	1.3	40.2	175	120–150	6.0–7.0	>160
6. **Desert steppe**. Dwarf semibush-tussock-grass and tussock-onion on brown soils	–19.0	23.0	3.4	42.5	185	<120	7.5–10.0	>160

Explanation of columns: t_I, annual average air temperature in January; t_{VII}, annual average air temperature in July; t_{year}, annual average air temperature; A_t, amplitude of air temperature in a year; t>5,– period with air temperature>5 (days); X_{year}, annual average sum of precipitation (mm); R/L_x, index of dryness ; Ik, index of winter continentality ($Ik = \Sigma t<0/\Sigma X_h$, where X_h, solid precipitation, $\Sigma t<0$, sum of air temperature for the period with t<0°C)

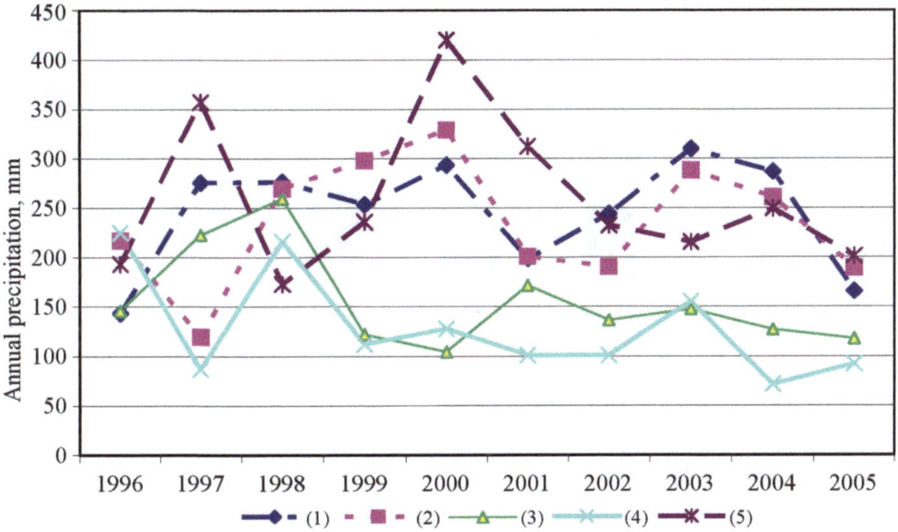

Fig. 10.2 Precipitation dynamics along the Trans-Mongolian railroad. *1*, Barunkhara; *2*, Nalaikh; *3*, Choir; *4*, Dzamyn-Uud; *5*, Shamar

in summer (June–August). Sevast'yanov and Tserensodnom (1994) showed perennial rhythms in northern and southern Mongolia. The data for the last 10 years (1996–2005) (Fig. 10.2) show this too for the northern (Selenga middle-high mountains, southwest spurs of Khentei) and the southern half (Gobi, Eastern-Gobi depression) of the transect.

Aboveground phytomass is closely related to seasonal moisture availability (amount of precipitation and natural humidity of the soils). Therefore, and because of their seasonal and perennial variability, we compare phytomass values of plots under pastoral use and non-grazed plots using the results obtained from 2001 to 2003 as a control (Miklyaeva et al. 2002; Gunin et al. 2003).

10.4 Mountain-Meadow Steppes

Mountain-meadow steppes have been investigated on three polygons. The first polygon (XXXIII) is located on a flat-hilly area south of the railway Darhan – Sharyn-Gol. Absolute heights are 882–885 m (Fig. 10.1, Table 10.1). The non-grazed fragment carried a [*Caragana pygmaea*]-*Stipa sibirica* + *S.baicalensis* + *Galium verum* plant community on dark-chestnut soil. There we counted 36 species; the projected cover was 41.5%, the total phytomass 207.7 g/m^2. Grasses, mostly *Stipa sibirica* and *S. baicalensis*, made up more than 85% of the phytomass. The perennial forbs *Galium verum* and *Veronica incana* were common. *Caragana pygmaea* made up 8.3% of the phytomass (Fig. 10.3, Table 10.3). Onions (*Allium* spp.) and sedges (Cyperaceae) played an insignificant role. Annual plants were absent.

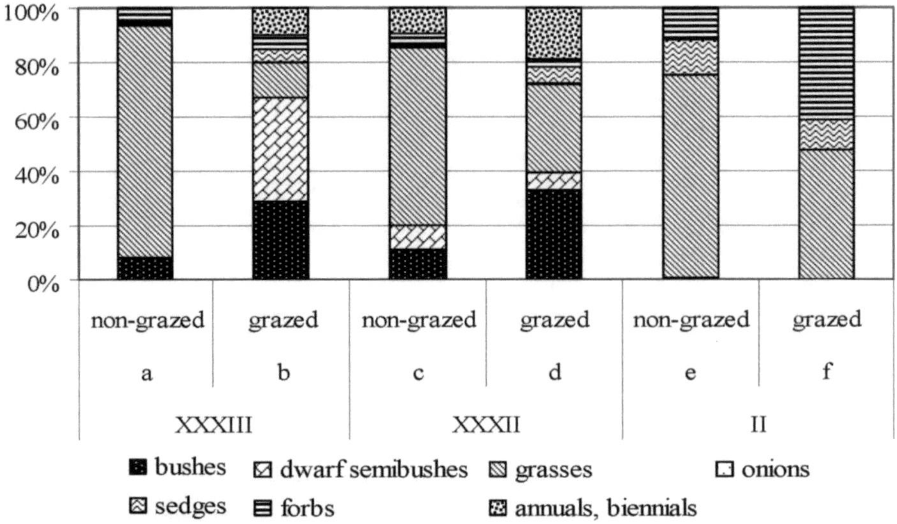

Fig. 10.3 Participation of life forms in the aboveground phytomass in the communities of mountain-meadow steppes (in % of the total aboveground phytomass on a research plot). (**a**) [*Caragana pygmaea*]-*Stipa sibirica*, *S. baicalensis* + *Galium verum*; (**b**) [*Caragana pygmaea*]-*Artemisia frigida* + *Stipa baicalensis*, *S. sibirica* + *Carex korshinskyi*; (**c**) *Festuca sibirica* + *Stipa krylovii* + *Carex pediformis*; (**d**) *Agropyron cristatum* + *Stipa krylovii* + *Carex pediformis* + *Koeleria cristata*; (**e**) [*Caragana pygmaea*]-*Stipa grandis* + *Artemisia scoparia* + *Galium verum*; (**f**) [*Caragana pygmaea*]-*Stipa grandis* + *Artemisia scoparia* + *Kochia prostrata* + *Carex korshinskyi*

In the grazed part of the polygon, the forb-*Stipa* community was transformed into a [*Caragana pygmaea*]-*Artemisia frigida* + *Stipa baicalensis* +*S. sibirica*+ *Carex duriuscula* community. We counted 29 species; cover and total aboveground phytomass were reduced. A dwarf semi-shrub *Artemisia frigida* had become dominant and made up more than 30% of the phytomass and cover (Fig. 10.3, Table 10.3). Here *Caragana* had almost three times more phytomass as compared to the non-grazed plot. The grasses were represented only by *Stipa baicalensis* and *S. sibirica;* their phytomass was reduced 7-fold in comparison with the ungrazed plot. *Carex duriuscula* had increased in phytomass and cover, and there were annual and biennial species (*Artemisia scoparia* and *Dontostemon integrifolia*) present.

The second polygon (XXXII) is located at a height of 873–875 m also to the south of the railway Darhan – Sharyn-Gol (Fig. 10.1, Table 10.1). The non-grazed area carried a [*Caragana pygmaea*]-*Stipa grandis* + *Artemisia scoparia* + *Galium verum* community on thick dark-chestnut soil of loess-like loams. The aboveground phytomass in this community was highest (271.1 g/m^2), the number of species was 35, and the cover reached 35.9%. The dominant *Stipa grandis* contributed 64.2% to the total phytomass. Other grasses present were *Stipa sibirica* and *Cleistogenes squarrosa.* Perennial forbs made up 4.9% of the total phytomass; of these the most widespread was *Galium verum.* The co-dominant and annual sagebrush *Artemisia scoparia* made up about 9% of the phytomass; other monocarpic herbs were scarce.

Table 10.3 Aboveground phytomass of dominant and subdominant species of steppe zonal plant communities along the submeridian transect Sukhe-Bator – Ulan-Bator – Dzamyn-Ud

Polygons	Mountain-meadow						Meadow				True				Dry		Very dry		Desert		
Species	XXXIII-2	XXXIII-1	XXXII-2	XXXII-1	II-2	II-1	XXXVIII-2	XXXVIII-1	XXXVII-2	XXXVII-1	XXXV-2	XXXV-1	XXVI-2	XXVI-1	XV-2	XV-1	XIV-2	XIV-1	XXX-2	XXX-1	XXX-3
1	2	3	4	5	6	7	8	9	10	11	12	13	14	15	16	17	18	19	20	21	22
Bushes:																					
Caragana korshinskii	–	–	–	–	–	–	–	–	–	–	–	–	–	–	–	–	–	–	3.3	3.5	5.4
																			14.4	30.0	51.0
C. microphylla	–	–	–	–	–	–	–	–	–	–	0.4	3.1	–	–	40.8	96.6	–	–	–	–	–
											0.2	2.7			24.8	69.0					
C. pygmaea	17.0	50.6	30.7	76.8	–	–	–	–	–	–	–	–	–	–	–	–	–	–	–	–	–
	8.3	29.1	11.3	32.9																	
C. stenophylla	–	–	–	–	–	–	–	–	–	–	–	–	–	–	–	–	0.4	1.2	1.6	1.2	1.1
																	2.6	11.6	7.0	10.2	10.4
Other bushes	–	–	–	–	–	–	–	–	–	–	–	–	–	–	–	0.6	–	–	–	–	–
(*C. leucophloea*)																0.4					
Dwarf semi-bushes:																					
Artemisia adamsii	–	–	–	–	–	–	–	1.0	–	–	–	–	20.4	25.2	–	–	–	–	–	–	–
								1.2					9.6	24.1							
A. frigida	0.1	65.7	–	–	0.6	–	0.7	2.7	1.6	90.6	–	–	–	0.1	0.9	24.4	–	0.6	–	–	–
	0.05	37.8			0.4		0.4	3.2	0.9	79.6				0.1	0.5	17.4		5.8			
Kochia prostrata	0.2	0.3	23.7	14.9	–	–	–	–	–	–	–	–	–	–	–	–	1.9	0.5	–	–	–
	0.1	0.2	8.7	6.4													12.1	4.8			
Other semi-bushes:	–	–	–	–	–	–	–	–	–	–	–	–	0.2	–	–	–	–	–	–	–	0.1
													0.1								0.9

(continued)

Table 10.3 (continued)

Polygons	Mountain-meadow						Meadow				True				Dry		Very dry		Desert		
Species	XXXIII-2	XXXIII-1	XXXII-2	XXXII-1	II-2	II-1	XXXVIII-2	XXXVIII-1	XXXVII-2	XXXVII-1	XXXV-2	XXXV-1	XXVI-2	XXVI-1	XV-2	XV-1	XIV-2	XIV-1	XXX-2	XXX-1	XXX-3
1	2	3	4	5	6	7	8	9	10	11	12	13	14	15	16	17	18	19	20	21	22
Perennial herbs: grasses																					
Agropyron cristatum	1.5	–	–	–	16.2	19.9	–	–	–	2.3	–	0.3	9.8	19.5	–	–	–	–	–	–	–
	0.7				10.8	20.6				2.0		0.3	4.6	18.6							
Cleistogenes songorica	–	–	–	–	–	–	–	–	–	–	–	–	–	–	–	–	–	–	0.5	0.6	0.5
																			2.1	5.1	4.7
C. squarrosa	–	–	0.1	0.4	–	–	0.2	26.3	43.2	6.8	–	–	–	0.2	0.8	0.2	1.5	0.8	–	–	–
			0.02	0.2			0.1	30.9	24.5	6.0				0.2	0.5	0.1	9.7	7.7			
Festuca sibirica	–	–	–	–	62.4	1.6	–	–	–	–	–	–	–	–	–	–	–	–	–	–	–
					41.5	1.6															
Koeleria cristata	–	–	–	–	4.5	9.3	0.6	–	0.3	–	–	–	–	–	–	–	–	–	–	–	–
					3.0	9.7	0.4		0.2												
Leymus chinensis	5.9	–	–	–	1.2	1.4	–	–	0.1	–	5.3	39.0	1.3	9.4	3.8	1.5	–	–	–	–	–
	2.8				0.8	1.5			0.1		2.0	33.6	0.6	9.0	2.3	1.1					
Stipa baicalensis	44.1	13.9	–	–	–	–	180.7	36.5	–	1.5	234.4	13.0	137.9	14.5	–	–	–	–	–	–	–
	21.2	8.0					96.0	42.9		1.3	89.7	11.2	64.9	13.8							
S. gobica	–	–	–	–	–	–	–	–	–	–	–	–	–	–	–	–	1.3	0.4	13.7	1.0	0.5
																	8.4	4.0	60.0	8.6	4.7
S. grandis	–	–	174.0	75.0	–	–	–	–	–	–	–	–	–	–	–	–	–	–	–	–	–
			64.2	32.1																	
S. krylovii	–	–	–	–	24.9	11.9	–	–	–	–	–	–	–	13.9	106.5	2.7	0.7	0.9	–	–	–
					16.6	12.4								13.3	65	1.9	4.6	8.7			
S. sibirica	124.5	8.2	2.6	1.4	–	–	–	–	–	–	9.9	–	–	–	–	–	–	–	–	–	–
	60.0	4.7	1.0	0.6							3.8										

Onions																					
Allium bidentatum	–	–	–	–	–	–	–	–	–	–	–	–	0.4	–	4.5	4.2	3.5	1.9	–	–	–
													0.2		2.8	3.0	22.4	18.4			
A. mongolicum	–	–	–	–	–	–	–	–	–	–	–	–	–	–	–	–	–	–	1.0	1.2	0.6
																			4.5	10.3	5.7
A. odorum	1.0	–	+	0.2	–	–	5.2	0.4	–	–	–	12.3	1.8	–	–	+	1.6	0.3	–	–	–
	0.5			0.1			2.8	0.5				10.6	0.8				10.3	3.0			
A. senescens	–	–	–	–	–	–	–	–	130.4	+	–	–	–	–	–	–	–	–	–	–	–
									74.0												
Sedges																					
Carex duriuscula	0.9	8.5	–	–	–	–	–	10.2	–	3.7	0.2	48.3	2.4	0.1	–	0.6	0.1	–	–	–	–
	0.4	4.9						12.0		3.3	0.1	41.6	1.1	0.1		0.4	0.7				
C. korshinskyi	–	–	2.0	14.3	–	–	–	–	–	–	–	–	–	–	–	–	–	–	–	–	–
			0.7	6.1																	
C. pediformis	–	–	–	–	19.7	10.8	–	–	–	–	–	–	–	–	–	–	–	–	–	–	–
					13.1	11.3															
Forbs																					
Asparagus gobicus	–	–	–	–	–	–	–	–	–	–	–	–	–	–	–	–	–	–	1.8	2.7	1.0
																			7.8	23.1	9.4
Bupleurum scorzonerifolium	–	–	–	–	1.2	7.3	–	–	–	–	–	–	14.7	–	–	0.5	–	–	–	–	–
					0.8	7.5							7.0			0.4					
Convolvulus ammanii	–	–	–	–	–	–	–	–	–	–	–	–	21.2	0.5	–	–	0.5	0.6	0.1	–	–
													10.0	0.5			3.3	5.8	0.4		
Galium verum	5.2	1.9	8.4	–	4.3	7.9	–	–	–	–	–	–	–	–	–	–	–	–	–	–	–
	2.5	1.1	3.1		2.8	8.1															
Potentilla acaulis	–	–	–	–	–	–	–	–	–	0.9	–	–	–	17.0	–	–	–	–	–	–	–
										0.8				16.2							

(continued)

Table 10.3 (continued)

Polygons	Mountain-meadow						Meadow				True				Dry		Very dry		Desert		
Species	XXXIII-2	XXXIII-1	XXXII-2	XXXII-1	II-2	II-1	XXXVIII-2	XXXVIII-1	XXXVII-2	XXXVII-1	XXXV-2	XXXV-1	XXVI-2	XXVI-1	XV-2	XV-1	XIV-2	XIV-1	XXX-2	XXX-1	XXX-3
1	2	3	4	5	6	7	8	9	10	11	12	13	14	15	16	17	18	19	20	21	22
Sibbaldianthe adpressa	–	–	–	–	–	–	–	–	–	–	–	–	–	0.9	–	–	0.1	1.4	–	–	–
														0.8			0.7	13.7			
Other perennial species (grasses, onions, sedges, forbs)	7.3	7.7	4.7	6.7	15.7	27.0	0.8	3.1	–	–	11.2	–	2.0	3.43	6.4	3.6	1.0	1.4	0.1	0.2	0.6
	3.45	4.4	1.75	2.8	10.2	27.3	0.3	3.6			4.2		1.0	3.2	3.8	2.6	6.7	13.5	0.4	1.6	5.6
Annual and biennial herbs:																					
Artemisia scoparia	–	7.8	24.7	43.6	–	–	–	4.2	–	7.2	–	–	–	–	–	–	–	0.1	–	–	–
		4.5	9.2	18.6				4.9		6.3								1.0			
Other annual and biennial plants	–	9.3	0.2	0.4	–	–	–	0.7	0.6	0.8	–	–	0.3	–	0.5	5.2	2.9	0.2	0.8	1.3	0.8
		5.3	0.03	0.2				0.8	0.3	0.7			0.1		0.3	3.7	18.5	2.0	3.4	11.1	7.6
Total aboveground phytomass (g/m²)	207.7	173.9	271.1	233.7	150.7	97.1	188.2	85.1	176.2	113.8	261.4	116.0	212.4	104.73	164.2	140.1	15.5	10.3	22.9	11.7	10.6

Note: The phytomass is expressed in the numerator in absolute figures (g/m^2); in the denominator in relative figures (in % of the total phytomass of a plot)

Photo 10.1 Polygon II represents a mountain-meadow steppe with northern exposition in the southwestern extremity of Khentei. The non-grazed part has a *Festuca sibirica* + *Stipa krylovii* + *Carex pediformis* and *Agropyron cristatum* community. Under pastoral use (at the upper side of the photo) the dominant *Agropyron cristatum* increased its share in the phytomass and cover. The proportion of *Festuca sibirica* and *Stipa krylovii* became less (Photo by authors)

Kochia prostrata was important and *Caragana pygmaea* formed a bushy layer with more than 11% of the total cover and phytomass. *Carex korshinskyi* and the onions *Allium odorum* and *A. senilis* were rare (Fig. 10.3, Table 10.3).

Under grazing, the community was transformed into a [*Caragana pygmaea*]-*Stipa grandis* + *Artemisia scoparia* + *Kochia prostrata* + *Carex korshinskyi* community on dark-chestnut, carbonate-rich soil on loess-like loams. There were 26 species in the community, and the total phytomass was 233.7 g/m^2. Cover was a little bit higher than in the non-grazed plot due to an increase of *Caragana pygmaea*, which was dominant, and its share in the phytomass had increased 3-fold. Clumps of *Caragana* under grazing were broken up; the majority of individuals contained 1–2 shoots. Large *Stipa* spp. had the same phytomass and cover as in the non-grazed plot, while the annual *Artemisia scoparia* had increased 1.7-fold in phytomass. *Kochia prostrata* had a reduced cover, but formed a significant phytomass. *Carex korshinskyi* had increased, while perennial forbs had diminished in the community. The main species in this category were *Iris tigridia* and *Veronica incana*, but they made up less than 1% of the stand's phytomass. The pasture was used for winter grazing and during our research it was in a good condition.

The third polygon (II) represents an abrupt stony slope with northern exposition in the southwestern extremity of Khentei with a maximal height of 1,655 m (Fig. 10.1, Table 10.1, Photo 10.1). The vegetation here was a *Festuca sibirica* + *Stipa krylovii* + *Carex pediformis* community on soils with rubble. Here we found the highest species number (47); the cover of the community was about 40%, and the aboveground phytomass 150.7 g/m^2. In the community structure grasses dominated forming 75% of the phytomass and more than 67% of the cover. All grasses had a good vital condition; their reproduction frequency ranged from 80% up to 100%. The dominant *Festuca sibirica* made up more than 40% of the cover and phytomass. *Stipa krylovii* and *Agropyron cristatum* were important and *Carex pediformis* was subdominant with more than 13% of the phytomass. Other perennial herbs contributed 11% of the phytomass; the most important being *Thalictrum petaloideum*, *Galium verum*, *Carum carvi*, and *Artemisia dracunculus*. Other species contributed less than 1% to the phytomass (Fig. 10.3, Table 10.3).

Under grazing, in immediate proximity of the ungrazed plot, an *Agropyron cristatum* + *Stipa krylovii* + *Carex pediformis* + *Koeleria cristata* community occurred. It had a cover of 32.7%, and the total phytomass was 1.5 times less than in the ungrazed plot. Grasses made up more than 46% of the phytomass and cover. The dominant *Agropyron cristatum* had increased its share in the phytomass and cover. The share of *Festuca sibirica* and *Stipa krylovii* had become less, as had *Koeleria cristata*. There occurred 19 perennial forb species and their share in the phytomass had increased in comparison with the ungrazed plot. The stand also contained poorly edible and inedible species (*Bupleurum scorzonerifolium*, *Saussurea salicifolia*).

10.5 Meadow Steppes

Meadow steppe was studied in two polygons. Polygon XXXVIII is located on a high, inundated terrace of the Khara-Gol River at a height of 699 m (Fig. 10.1, Table 10.1). The ungrazed plot supported a *Stipa baicalensis* + *Allium odorum* community on inundated, turfen, loamy soil. The community contained 16 species, but with a high cover: 47.7%, and a total phytomass of 188.2 g/m^2. Grasses were predominantly tussock species: *Cleistogenes squarrosa*, *Koeleria cristata*, and *Stipa baicalensis*. *Stipa baicalensis* formed 96% of the phytomass; other grasses less than 1% (Fig. 10.4, Table 10.3). *Allium odorum* was subdominant. Among dwarf semi-shrubs *Artemisia frigida* and *A. adamsii* were conspicuous. Perennial forbs (*Astragalus sp.*, *Thermopsis dahurica*) were not important.

The *Stipa baicalensis* + *Cleistogenes squarrosa* + *Carex duriuscula* community was found under grazing. The species number was low (12 species), and cover and total phytomass were 2.2–2.5 times less than in the ungrazed plot. The tussock grasses *Stipa baicalensis* and *Cleistogenes squarrosa* were dominants, contributing 73.8% to the total phytomass (Fig. 10.4, Table 10.3). *Carex duriuscula* played a significant role and the dwarf semi-shrub *Artemisia frigida* was more common than in the ungrazed plot. The perennial *Potentilla bifurca* occurred, as well as the annual, resp. biennial, *Artemisia scoparia* and *Dontostemon integrifolius*.

Polygon XXXVII is located 8–10 km from Darhan on a high terrace at 703 m (Fig. 10.1, Table 10.1). In the ungrazed plot an *Allium senescens* + *Cleistogenes squarrosa* community occurred on a chestnut sandy soil. The community contained

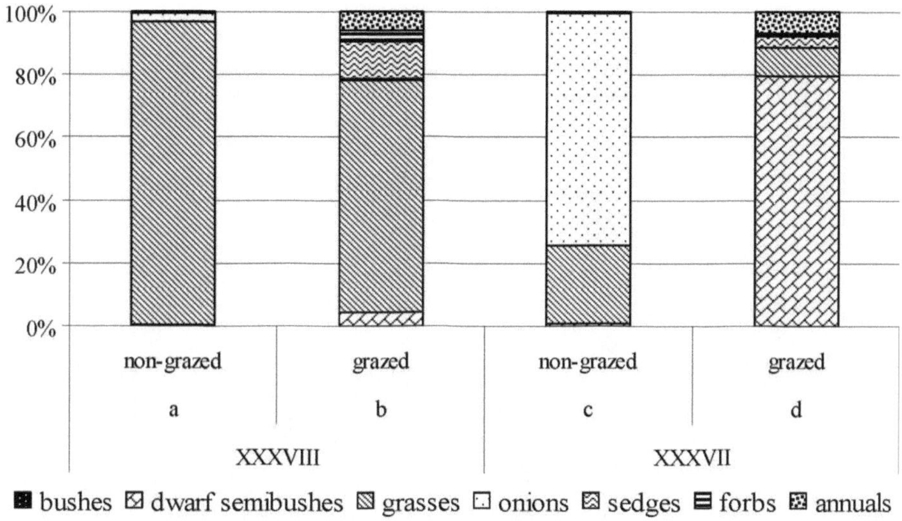

Fig. 10.4 Participation of life forms of plants in aboveground phytomass in the communities of meadow steppes (in % of the total aboveground phytomass on a research plot). (**a**) *Stipa baicalensis* + *Allium odorum*; (**b**) *Stipa baicalensis* + *Cleistogenes squarrosa* + *Carex duriuscula*; (**c**) *Allium senescens* + *Cleistogenes squarrosa*; (**d**) *Artemisia frigida*, *A. scoparia* + *Cleistogenes squarrosa* + *Carex duriuscula*

31 species, had a cover of 31.5%, and a phytomass of 176.2 g/m². The dominant *Allium senescens* made up 74% of the phytomass and 63.5% of the cover. The subdominant *Cleistogenes squarrosa* accounted for 24.5% of the total phytomass (Fig. 10.4, Table 10.3). Participation of other species was less than 1%.

Under grazing, in the immediate proximity of the fence, the *Allium senescens* + *Cleistogenes squarrosa* community was transformed into an *Artemisia frigida* + *A. scoparia* + *Cleistogenes squarrosa* + *Carex duriuscula* community on a typical chestnut light-loamy soil. Species number was reduced to 20 and cover was high (more than 55%). Phytomass was 1.5 times less than in the non-grazed plot and amounted to 113.8 g/m². The dominants were *Artemisia* species. The dwarf semishrub *Artemisia frigida* made up about 80% of all phytomass and 62.8% of the cover, while the annual *Artemisia scoparia* contributed 6.3% to the phytomass. Participation of onions and grasses was sharply reduced. The subdominant grass *Cleistogenes squarrosa* formed 6% of the phytomass; *Agropyron cristatum* and *Stipa baicalensis* were scarce. *Carex duriuscula* formed a significant part of the cover. Of the perennial plants *Potentilla acaulis* was common.

10.6 True Steppes

In the true steppe two polygons were studied. Polygon XXXV is located on the highest terrace of the Khara-Gol River south of the railway Darhan – Erdenet at 789 m (Fig. 10.1, Table 10.1, Photo 10.2). The soil is dark-chestnut. The non-grazed

Photo 10.2 True steppe polygon XXXV on the terrace of the Khara-Gol River. The non-grazed fragment (*right side*) is covered by *Stipa baicalensis* with *Caragana microphylla*. Under pastoral use (*left side*) *Carex duriuscula* and *Leymus chinensis* are dominants (Photo by authors)

fragment of this steppe carried a [*Caragana microphylla*]-*Stipa baicalensis* community consisting of 15 species. Its cover was 46.8%, and its phytomass 261.4 g/m^2. About 90% of the cover and phytomass was formed by *Stipa baicalensis*. *Leymus chinensis* and *Stipa sibirica* were also present. Participation of *Carex duriuscula* was insignificant (less than 1%). The forb *Cymbaria daurica* occurred and the shrub *Caragana microphylla* formed only 0.4% of the phytomass (Fig. 10.5, Table 10.3).

Under pastoral use, outside the fence, a [*Caragana microphylla*]-*Allium odorum* + *Carex duriuscula* + *Leymus chinensis* + *Stipa baicalensis* community had developed on dark-chestnut light-loamy soil. It contained 19 species and its total phytomass was 2.3 times less than in the non-grazed plot. Its cover had increased, basically on account of *Carex duriuscula* and *Leymus chinensis*, which formed 65% and 18.8% of the total cover, respectively (Fig. 10.5, Table 10.3). They dominated the phytomass. The role of dense-tussock grasses was reduced: the phytomass of *Stipa baicalensis* 18-fold and *S. sibirica* was totally absent in the grazed community. Phytomass of *Caragana microphylla* had increased to 2.7%.

Polygon XXVI in the true steppe was 5 km northeast of the Arashant railway station at the bottom part of a slope at 1,280 m (Fig. 10.1, Table 10.1). The soil is chestnut with rubble and carbonates. The non-grazed plot carried a *Stipa baicalensis* + *Convolvulus ammanii* + *Bupleurum scorzonerifolium* community with 31 species, a cover is 36.4% and a total phytomass of 212.4 g/m^2. *Stipa baicalensis* made up more than 60% of the phytomass (Fig. 10.5, Table 10.3). The tussocks of

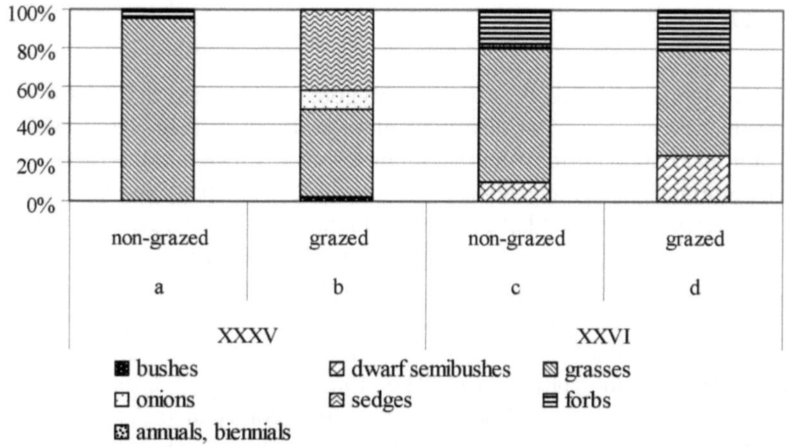

Fig. 10.5 Participation of life forms of plants in aboveground phytomass in the communities of true steppes (in % of the total aboveground phytomass on a research plot). (**a**) [*Caragana microphylla*]-*Stipa baicalensis*; (**b**) [*Caragana microphylla*]-*Carex duriuscula* + *Leymus chinensis* + *Stipa baicalensis* + *Allium odorum*; (**c**) *Stipa baicalensis* + *Allium anisopodium* + *Bupleurum scorzonerifolium*; (**d**) *Agropyron cristatum* + *Stipa krylovii*, *S. baicalensis* + *Leymus chinensis* + *Artemisia adamsii* + *Potentilla acaulis*

Stipa baicalensis were well represented and all shoots were reproductive and reached a height up to 100 cm. Perennial forbs formed 18% of the phytomass. *Bupleurum scorzonerifolium* and *Convolvulus ammanii* were well developed. *Artemisia adamsii* contributed 9.6% to the phytomass, while the other dwarf semi-shrubs *Alissum lenense* and *Artemisia glauca* occurred here and there. All species showed a good vital condition and flowered or set fruit.

Under grazing the vegetation was an *Agropyron cristatum* + *Stipa krylovii*, *S. baicalensis* + *Leymus chinensis* + *Artemisia adamsii* + *Potentilla acaulis* community on light-brown, thin soil on very rubbly, loamy proluvium. Its cover was only 28.3% with 104.7 g/m^2 of phytomass. There were 28 species. *Artemisia adamsii* dominated; *A. glauca* and *A. frigida* were quite rare. Grasses made up more than half of the phytomass. *Stipa baicalensis* was 9.5 times less and the number of tussocks had strongly decreased but, nevertheless, this grass played an appreciable role in the community structure (13.8% of the phytomass) (Fig. 10.5, Table 10.3). While *Stipa krylovii* was as abundant as in the non-grazed plot, *Cleistogenes squarrosa* and *Koeleria macrantha* contributed less than 1%, and *Agropyron cristatum* had increased in comparison with the non-grazed plot, as well as *Leymus chinensis*. *Convolvulus ammanii*, *Cymbaria dahurica*, *Heteropappus hispidus*, *Iris tigridia*, *Potentilla acaulis*, *P. bifurca*, *P. multifida*, *Sibbaldianthe adpressa*, and *Taraxacum dealbatum* occurred, with *Potentilla acaulis* being abundant (26.8% of the cover and 16.2% of the phytomass). Annual and biennial species were rare.

10.7 Dry Steppes

In the dry steppe we studied only one polygon: XV, located at 138 km south of Ulaanbaataar on a nearly flat to rolling plain at 1,334 m (Fig. 10.1, Table 10.1). Under non-grazed conditions the community was a [*Caragana microphylla*]-*Stipa krylovii* community of 27 species with a cover of 70% and phytomass of 164.2 g/m^2. *Stipa krylovii* was dominant both in cover and mass. It was in a good vital condition with all shoots reproducing and up to 80 cm tall. *Caragana microphylla* formed more than 24% of the phytomass and 26% of the cover; the average diameter and height of bushes were 112 and 29 cm, respectively. Practically all individuals showed rich fructification. The onions *Allium bidentatum* and *A. tenuissimum* formed 3.1% of the phytomass. The forbs, including *Cymbaria daurica*, *Haplophyllum davuricum*, and *Scorzonera capito*, contributed 3.4% to the phytomass and the dwarf semi-shrub *Artemisia frigida* less than 1% (Fig. 10.6, Table 10.3). Annual and biennial plants were insignificant.

Under grazing, close to the non-grazed plot near the fence, the vegetation was a [*Caragana microphylla*]-*Artemisia frigida* + *Salsola pestifera* community with 21 species, a sparse cover and 1.2 times less aboveground phytomass. *Caragana microphylla* had become a dominant, contributing 69% to the total phytomass (mainly consisting of perennial branches) and more than 28% of the cover (Fig. 10.6, Table 10.3). Its clumps had broken up and as a whole they were smaller in size. *Artemisia frigida* had become a subdominant. Grasses were very poorly developed and their contribution in the community was insignificant. The mass of *Stipa krylovii* was 39 times less in comparison to the non-grazed plot, its individuals were

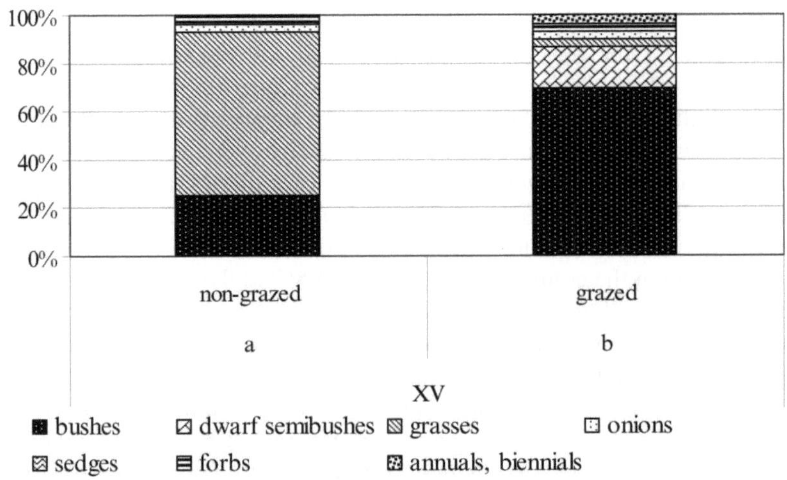

Fig. 10.6 Participation of life forms of plants in aboveground phytomass in the communities of dry steppes (in % of the total aboveground phytomass on a research plot). (**a**) [*Caragana microphylla*]-*Stipa krylovii*; (**b**) [*Caragana microphylla*]-*Artemisia frigida* + *Salsola pestifera*

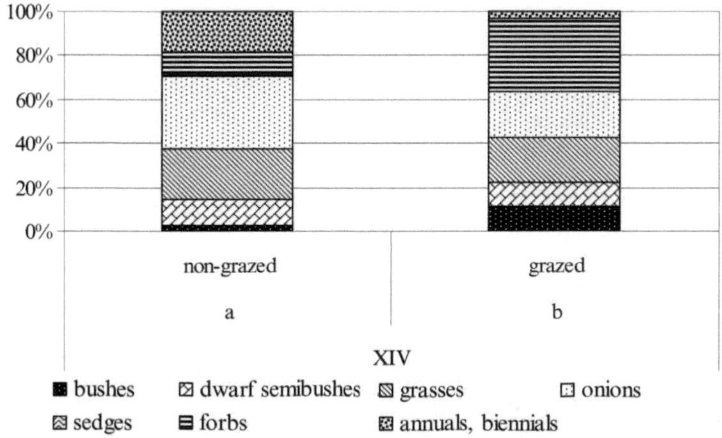

Fig. 10.7 Participation of life forms of plants in aboveground phytomass in the communities of transitional very dry steppes (in % of the total aboveground phytomass on a research plot). (**a**) [*Caragana stenophylla*]-*Allium odorum*, *A. bidentatum* + *Cleistogenes squarrosa* + *Stipa gobica* + *S. krylovii*-*Kochia prostrata*; (**b**) [*Caragana stenophylla*]-*Stipa krylovii* + *Cleistogenes squarrosa* + *S. gobica* + *Allium bidentatum* + *Sibbaldianthe adpressa*

undersized and most of their shoots were not reproducing. Cover and phytomass of *Allium bidentatum* had remained the same. The group of forbs were enriched by *Astragalus sp.*, *Bupleurum scorzonerifolium*, and *Potentilla bifurca* and annual and biennial species had increased as well as sedges; *Salsola pestifera* dominated. In this plot we clearly observed the encroachment of bushes.

10.8 Transitional Very Dry Steppes

In the transition zone between dry and desert steppes polygon XIV was studied, located near to the top of a flat hill at 1,220 m a.s.l., and 28 km south of Choir city (Fig. 10.1, Table 10.1). The non-grazed plot supported a [*Caragana stenophylla*]-*Allium odorum*, *A. bidentatum* + *Cleistogenes squarrosa* + *Stipa gobica* + *S. krylovii* + *Kochia prostrata* community with 31 species, a cover of 13.2%, and the very low phytomass of only 15.7 g/m^2. The community is formed by species characteristic of the dry steppe (*Stipa krylovii*, *Allium bidentatum* and *A. odorum*) and some dominants of the desert steppe (*Stipa gobica*, *Allium mongolicum* and *A. polyrrhizum*). Tussock grasses made up 22.7% of the total phytomass. The share of onions was 33.4%, and the good vital condition of *Allium odorum* and *A. bidentatum* was conspicuous. These species dominated in terms of cover (29 and 15.2% respectively). Perennial forbs contributed 10% of the phytomass and annual and biennial species 18.5%. *Caragana stenophylla*, with 2.6% of the phytomass, formed the bushy layer. The dwarf semi-shrub *Kochia prostrata* was subdominant and made up 12.1% of the phytomass (Fig. 10.7, Table 10.3).

Photo 10.3 Desert steppe polygon XXX, located on a flat sandy plain, 30 km north of the Mongol-Chinese border. Its vegetation on the non-grazed plot is a [*Caragana korshinskii, C. stenophylla*]-*Stipa gobica* + *Asparagus gobicus* + *Allium mongolicum* community (Photo by authors)

Under grazing the community was transformed into a [*Caragana stenophylla*]-*Stipa krylovii* + *Cleistogenes squarrosa* + *S. gobica* + *Allium bidentatum* + *Sibbaldianthe adpressa* community with 29 species. The grassy cover was sparse (9.3%) and the total phytomass was 1.5 times less than in the non-grazed plot (10.3 g/m^2). Grasses formed 20.4% of that phytomass, and most of it was *Stipa krylovii*. *Cleistogenes squarrosa* also was common. As in the ungrazed plot *Allium bidentatum* was also dominant in this community, but its cover and phytomass was less. *Caragana stenophylla* made up 11.6%; semi-shrubs 10.6%. Forbs contribute up to 34% to the phytomass. *Sibbaldianthe adpressa*, which is an indicator of vegetation degradation, was in good vital condition. *Artemisia frigida* was a common dwarf semi-shrub.

10.9 Desert Steppes

In the desert steppe we studied polygon XXX, located on a flat sandy plain, 30 km north of the Mongol-Chinese border, at 979–990 m a.s.l. (Fig. 10.1, Table 10.1, Photo 10.3). Its vegetation was a [*Caragana korshinskii, C. stenophylla*]-*Stipa gobica* + *Asparagus gobicus* + *Allium mongolicum* community on brown desert-steppe soil with 11 species, a cover of 16.3% and a phytomass of 22.4 g/m^2. As a result of the droughty conditions in the Gobi-part of Mongolia during the first half of the summer of 2006, the retarded development of the vegetation was obvious, and

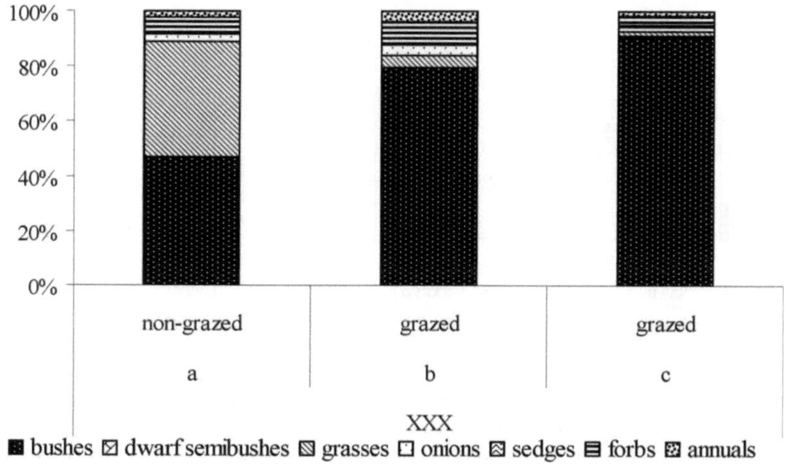

Fig. 10.8 Participation of life forms of plants in aboveground phytomass in the communities of desert steppes (in % of the total aboveground phytomass on a research plot). (**a**) [*Caragana korshinskii, C. pygmaea*]-*Stipa gobica* + *Asparagus gobicus* + *Allium mongolicum*; (**b**) [*Caragana korshinskii, C. pygmaea*]-*Asparagus gobicus* + *Allium mongolicum* + *Stipa gobica*; (**c**) [*Caragana korshinskii, C. pygmaea*]-*Asparagus gobicus* + *Allium mongolicum* + *Cleistogenes songorica*+ *Stipa gobica*

many species were in poor vital condition. *Caragana stenophylla* and *C. korshinskii* were well developed and made up 21.4% of the phytomass. In a plot of 50 × 50 m we counted 32 specimens of *C. korshinskii* having an average diameter of 76 cm and a height of 52 cm; in that area we also counted 232 specimens of *Caragana stenophylla* of 53 cm high and with a ratio of live shoots to dry ones of 7:10. 78.2% of the total phytomass consisted of herbaceous plants. The dominant in cover and phytomass was *Stipa gobica;* it contributed more than 60% to the total phytomass (Fig. 10.8, Table 10.3). Of the forbs *Asparagus gobicus* was the most important species with 8%. In this community, only *Allium mongolicum* was in a rather good vital condition. *Cleistogenes songorica* and the annual *Tribulus terrestris* had plenty of shoots.

Under grazing, the vegetation had strongly changed. Here we investigated two plots (at the eastern and western sides of the railway). On the eastern side the vegetation was a [*Caragana korshinskii, C. stenophylla*]-*Asparagus gobicus* + *Allium mongolicum* + *Stipa gobica* community on a poor, brown, sandy desert-steppe soil. There were 16 species. The herbage was scarce, its projective cover being 14.4%, and its phytomass 11.7 g/m^2. The bushy layer formed 40.2% of the phytomass. *Caragana* species were in good vital condition. The number of *Caragana korshinskii* shrubs in the plot area had increased 4-fold, but their size had decreased (their average diameter 1.5-fold, their height 1.7-fold). The number of *C. stenophylla* shrubs had remained the same as in the ungrazed plot, but their height was 1.2 times smaller, and the total number of shoots had increased. Perennial forbs made up most of the phytomass, particularly *Asparagus gobicus* with 23.1%. The annual *Tribulus terrestris*, an indicator of grazing, was common. *Allium mongolicum* was in good

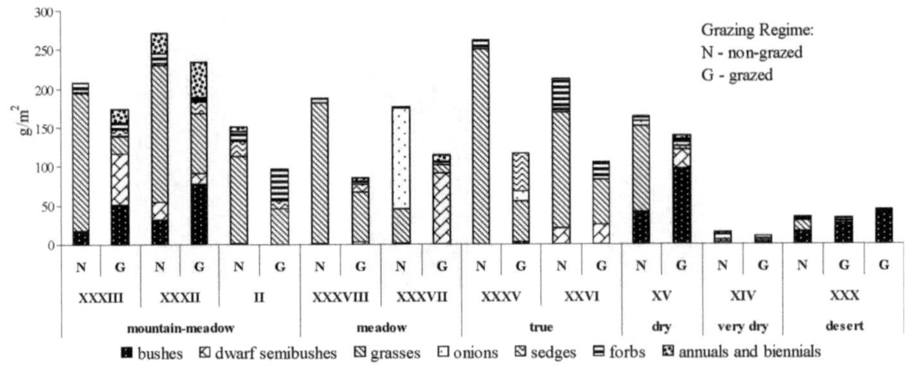

Fig. 10.9 Aboveground phytomass of the zonal communities along the Trans-Mongolian railroad Sukhe-Bator – Ulan-Bator – Dzamyn-Ud

vital condition and made up 10.3% of the phytomass. *Stipa gobica* 8.6% but it was everywhere grazed off and its tussocks were trodden. *Cleistogenes songorica* and *Haplophyllum davuricum* were common.

On the western side of the railway, on a strongly sandy, brown desert-steppe soil, the pasture was a [*Caragana korshinskii, C. stenophylla*]-*Asparagus gobicus* + *Allium mongolicum* + *Cleistogenes songorica* + *Stipa gobica* community with 14 species, a low cover of 11.8%, and a phytomass of 10.6 g/m^2. *Caragana* species formed more than 60% of the phytomass. The number of *Caragana korshinskii* shrubs was 8 times higher than in the ungrazed plot, but their average diameter was 1.6 times smaller and their height almost 2 times. The number of *C. stenophylla* shrubs had remained the same, but their total number of shoots had increased. From the herbaceous plants, *Asparagus gobicus* dominated and made up 9.4% of the phytomass and more than 8% of the cover. *Cleistogenes songorica* formed 18.6% of the projective cover, but was bitten off everywhere, and it contributed only 4.7% to the phytomass (Fig. 10.8, Table 10.3). *Stipa gobica* had kept the same mass, but its tussocks were trampled and severely fragmented. Among the annuals *Tribulus terrestris*, *Setaria viridis*, and *Salsola collina* were found.

10.10 Conclusion

Our researches in ten polygons along the Trans-Mongolian railway, covering the main types of steppe, have revealed the structure and aboveground phytomass of the plant communities and their dominants and subdominants. We found a reduction in phytomass in ecologically analogous communities as a result of (over)grazing even though the stage of grazing disturbance was assessed as of medium intensity in all these steppe communities. Thus, in the grazed vegetation along the railway, the aboveground phytomass was reduced 1.2–2.0 fold as compared to the non-grazed equivalent plots (Fig. 10.9).

In the steppe communities along our transect, when protected from grazing, the dominants, both in terms of cover and phytomass, are *Stipa grandis*, *Allium senescens*, *S. baicalensis*, *S. krylovii*, *Allium odorum*, *S. gobica*, respectively. Grazing causes two important processes to happen. On the one hand, herbaceous plants are replaced by woody bushes (*Caragana microphylla*, *C. pygmaea*, *C. stenophylla*, *C. korshynskii*) and dwarf semi-shrubs (*Artemisia frigida* and *A. adamsii*) and thus pasture lands are encroached by bushes. On the other hand, perennial and annual species that thrive well in degrading pasture land invade. Our observations showed that in grazed plots *Artemisia dracunculus*, *A. palustris*, *Carex duriuscula*, *Convolvulus ammanii*, *Potentilla acaulis*, *Sibbaldianthe adpressa* and *Veronica incana* become important in different sections along the transect. These species are known as indicators of pastoral digression (degradation) in Mongolia (Yunatov 1950; Chogny 1988; Gunin and Miklyaeva 2006). Thus, our results correspond to those of Graivoronsky (1997) for Transbaikalia and of Chogny (1988) and Miroshnichenko (1964) for Mongolia.

In **mountain-meadow steppe** at medium intensity of disturbance dominants are replaced by subdominants: *Festuca sibirica* was replaced by *Agropyron cristatum*, and in the *Stipa grandis* community *Caragana pygmaea* increased and *Stipa grandis* became a subdominant. In the community dominated by *Stipa sibirica* bushes and dwarf semi-shrubs (*Caragana pygmaea*, *C. microphylla*, and *Artemisia frigida*) increased to dominance and *Stipa sibirica* was strongly declining.

In **meadow steppe** grazing triggered the xerophytization processes and the increase of *Artemisia frigida*. Onion vegetation could be replaced completely by *Artemisia frigida*, but *Stipa baicalensis* communities proved more stable against pastoral influence and changes were not cardinal. *Stipa baicalensis* remained a dominant species, but its phytomass became sharply reduced.

In **true steppe** grazing induced an increase of *Leymus chinensis* and *Carex duriuscula*, while the role of *Stipa baicalensis* was reduced, and in the forb-sagebrush-grass community *Artemisia adamsii* and *Potentilla acaulis* increased, probably because of strong grazing pressure, but also the occurrence of many rodents in the community may stimulate this development.

In **dry steppe,** under long-term grazing, there was a total change in subdominants: semi-shrubs and dwarf semi-shrubs strongly increased and herbaceous species decreased. As the top soil became drier under grazing, euxerophytes, such as *Caragana microphylla* and *Artemisia frigida*, increased

The strong expansion of *Caragana* under grazing was clearest in **desert steppe,** where *Caragana korshinskii* and *C. stenophylla* even could become dominants, particularly on sandy soils.

10.11 Assessing Pastural Conditions of Mongolian Ecosystems

As has been shown (Bazha et al. 2008; Danzhalova 2009) the use of direct field measurements (species composition, cover, aboveground phytomass, growth form of dominant species) for the assessment of the degree of degradation of the

Table 10.4 Coefficients of bush encroachment, invasiveness, and palatability in the main types of steppe communities in Central Mongolia

		Coefficient of bush encroachment		Coefficient of invasiveness		Coefficient of palatability	
Type of steppe	Polygon number	Non-grazed	Grazed	Non-grazed	Grazed	Non-grazed	Grazed
Mountain-meadow	XXXIII	0.1	2.0	0.2	6.7	33.0	6.2
	XXXII	0.2	0.6	0.5	2.1	8.2	3.6
	II	–	–	0.2	1.3	15.0	3.9
Meadow	XXXVIII	–	–	0.01	1.3	234.2	8.4
	XXXVII	0.01	3.9	0.4	74.9	292.6	11.7
True	XXXV	–	–	0.07	3.4	–	–
	XXVI	0.1	0.3	0.4	2.7	3.8	1.2
Dry	XV	0.3	6.5	0.5	17.9	25.5	14.9
Very dry	XIV	0.2	0.3	0.7	1.3	2.5	2.2
Desert	XXX	0.9	3.8–9.6	1.1	5.9–19.8	11.1	7.2–19.8

vegetation due to grazing and anthropogenic disturbance leads to ambiguous results. Morphometric and reproductive parameters of the dominant species change as the role of the species in a community changes under the influence of grazing, especially in the final stages of degradation and are therefore only of limited use. For a reliable assessment it proved necessary to apply additional criteria. We suggest to use ratios of field-measured variables to quantify bush encroachment, invasiveness, and palatability.

Bush encroachment. In order to quantitatively assess bush encroachment in degrading vegetation under prolonged grazing pressure, we suggest to use the weight ratio of species with woody shoots (bushes, semi-shrubs, and dwarf semi-shrubs) to that of herbaceous species.

The values of that ratio in the ungrazed steppe communities discussed here are always lower than 1.0 and range from 0.01 in meadow steppes to 0.9 in desert steppes. In the grazed equivalent communities the values are always higher and vary from 0.3 to 9.6 (Table 10.4).

On this basis is it possible to evaluate bush encroachment. Considering a ratio of 2.0 or higher as typical of a very high degree of disturbance, we classify the following communities into that category: [*Caragana pygmaea*]-*Artemisia frigida* + *Stipa baicalensis*, *S. sibirica* + *Carex duriuscula* community in the mountain-meadow steppes (XXXIII); *Artemisia frigida*, *A. scoparia* + *Cleistogenes squarrosa* + *Carex duriuscula* community in the meadow steppes (XXXVII); [*Caragana microphylla*]-*Artemisia frigida* + *Salsola pestifera* community in the dry steppes (XV); [*Caragana korshinskii*, *C. stenophylla*]-*Asparagus gobicus* + *Allium mongolicum* + *Stipa gobica* and [*Caragana korshinskii*, *C. stenophylla*]-*Asparagus gobicus* + *Allium mongolicum* + *Cleistogenes songorica* + *Stipa gobica* communities in the desert steppe (XXX) (Table 10.4).

Invasiveness. The role of typical steppe dominants and species that typically indicate vegetation degradation varies in communities under grazing pressure

(Miklyaeva and Fakhire 2004; Miklyaeva et al. 2004). It is most reliable to evaluate invasiveness by the weight ratio of invasive and native species.

All our grazing-protected communities (except the desert steppe) score values of invasiveness of less than 1.0; in fact, the value ranges between 0.01 in the meadow steppes and 1.1 in the desert steppes. In the equivalent grazed communities values are higher and vary from 1.3 to 75.

Using the ratio of invasiveness, and considering a value of 5.0 or higher as indicative of highly invaded communities, we classify the following communities into that category: [*Caragana pygmaea*]-*Artemisia frigida* + *Stipa baicalensis*, *S. sibirica* + *Carex duriuscula* community in the mountain-meadow steppes (XXXIII); *Artemisia frigida*, *A. scoparia* + *Cleistogenes squarrosa* + *Carex duriuscula* community in the meadow steppes (XXXVII); [*Caragana microphylla*]-*Artemisia frigida* + *Salsola pestifera* community in the dry steppes (XV); [*Caragana korshinskii*, *C. stenophylla*]-*Asparagus gobicus* + *Allium mongolicum* + *Stipa gobica* and [*Caragana korshinskii*, *C. stenophylla*]-*Asparagus gobicus* + *Allium mongolicum* + *Cleistogenes songorica* + *Stipa gobica* communities in the desert steppes (XXX) (Table 10.4).

Palatability. An analysis of the fodder quality of plant communities is necessary for the economic evaluation of steppe ecosystems. The factor of palatability estimates the proportion of edible and inedible species in the total phytomass. It is defined as the weight ratio between well edible and a satisfactorily edible species on the one hand and poorly edible and inedible species on the other. For the evaluation of palatability of the steppe species we applied the recommendations and data published by well-known experts (Tsatsenkin and Yunatov 1951; Yunatov 1954; Gunin and Vostokova 1989; Kurkin 2005).

The values of phytomass palatability in our steppe communities (Table 10.4) vary strongly, from 1.2 up to 292.0 irrespective of community type and grazing regime. Apart from the fact that in all grazed communities the values are lower (except for the desert steppe), it seems not possible to clearly categorize our communities in terms of disturbance based on the palatability ratio. Most likely, better data on the fodder value of these species are needed, first of all on the contents of digested protein, alkaloids, and etherial oils (Ermakova and Mikheev 1969; Nechaeva 1970; Mirkin et al. 1988). On the other hand, palatability of plants is an ambiguous variable which depends on plenty of factors: the kind of grazing animal and their number, the season of the year, climatic conditions, and the geographical locations of the pastures (Nechaeva 1987; Kurkin 2005). But animal husbandry specialists have shown that such data are not sufficient (Shagdarsuren 2005).

Based on the above results, in the investigated plant communities five stages of pastural degradation of the steppe have been detected. To these five stages we should add another one, the sixth stage of domination of annual plants and little participation of species of the previous stages. These six stages make up three levels of disturbance as shown in Table 10.5.

We have shown that it is possible to quantitatively assess the stages of pastural degradation in steppe communities and that coefficients of bush encroachment and invasiveness are the most effective for diagnosis. The coefficient of bush

Table 10.5 Criteria of disturbance of steppe plant communities in Central Mongolia

		Stage of disturbance					
		Partial disturbance	Pastural degradation			Transformation	
Process	Parameter	I – very weak	II – weak	III – moderate	IV – strong	V – very strong	VI – trampled
Change in dominants	Composition of dominants and subdominants	Native edificators	Joint dominance of native edificators and secondary species	Species typical as subdominants of very weakly disturbed communities become dominants	Degradation-indicating species dominate and native edificators subdominate	Degradation-indicating species dominate and native edificators present in small numbers	Annual species
Bush encroachment	Coefficient of bush encroachment	0–0.3		0.3–1.0		>2.0	–
Invasiveness	Coefficient of invasiveness	0–0.5	0.5–1.0	1.0–2.0	2.0–5.0	>5.0	–

encroachment is most useful to distinguish the higher degrees of degrading disturbance, while the coefficient of invasiveness proves its functionality in all stages of pastural disturbance.

We conclude to observe that, as a result of increased grazing pressure in Central Mongolia during the past 20 years, plant communities with strongly to very strongly disturbed vegetative covers have begun to prevail. The main cause for this degradation of the pastures is overgrazing. Currently, the existing animal loads on the pastures far exceeds their natural forage capacity. To remedy this situation it is necessary to introduce measures which reduce the loads. The first measure should be to organize monitoring of pasturelands and on this basis rotate grazing. In addition, a complex of social and economic activities should be realized:

1. Regulation of grazing loads by implementing stricter control of pastures.
2. Optimization of herd composition: return to the balanced five-species herd, along with the reduction of the number of goats.
3. Introduction of progressive taxation for herders.
4. Repair and restoration of the network of wells and water points.

Finally, in terms of developing actions to counter the expansion of areas with low-productive, inedible or poisonous species, and their invasion into the plant communities of the steppe zone, it is necessary to set up a scientific, experimental research station for pasture improvement.

References

Bazha SN, Gunin PD, Danzhalova EV, Kazantseva TI (2008) Diagnostical parameters of pastural digression of steppe plant communities of the Mongolian biogeographical province of Palearctica. Volga Ecol J 4:251–263, [in Russian]

Beresneva IA, Rachkovskaya EI (1978) On the factors of zoning in the southern part of the MPR. Probl Desert Devel J A:19–29, [in Russian]

Bykov BA, Stepanova EF (1953) Bush steppes as a type of vegetation. News Rus Geogr Soc 85:6 [in Russian]

Chogny O (1988) The laws of pasqual digression and pasqual demutation of pastures. Transactions of the Joint Soviet-Mongolian Complex Biological Expedition, Moscow, pp 45–81, [in Russian]

Danzhalova EV (2009) Pasqual digression of plant communities of steppe ecosystems of the Central Mongolia. Ph.D. thesis, Moscow, 27 pp, [in Russian]

Dinesman LG, Bold G (1992) The history of grazing and pasqual digression development in the steppes of Mongolia. Historical ecology of wild and domestic ungulates: The history of pastoral ecosystems. Science Publ, Moscow, pp 172–216, [in Russian]

Ecosystems of Mongolia, Atlas (2005) Ecore Scient. Press, Moscow, 48 pp

Ermakova IA, Mikheev GD (1969) On nutrient quality of the principal pastoral forage of Turkmen SSR. News Acad Sci TSSR Ser Biol 6:119–128, [in Russian]

Gadzhiev IM (2002) Soil cover. Steppes of Central Asia. SO RAN Publ, Novosibirsk, pp 8–44, [in Russian]

Geomorphology of Mongolian Peoples Republic (1982) Scient. Press, Moscow, 259 pp, [in Russian]

Gorshkova AA (1973) Pasturelands of Transbaikalia. East Sib. Book Publ, Irkutsk, 157 pp, [in Russian]

Gorshkova AA, Lobanova IN (1972) Change of ecology and structure of steppe communities of Transbaikalia under influence of pastoral regime. Reports of the Institute of Geography of Siberia and Far East, Irkutsk, pp 38–43, [in Russian]

Graivoronsky VV (1997) Modern arts of Mongolia. Social problems of the transition period 1980–1995. East Lit Publ, Moscow, 184 pp, [in Russian]

Gunin PD, Miklyaeva IM (2006) Contemporary processes of degradation and desertification of ecosystems of the East-Asian sector of steppes and forest-steppes. Contemporary global changes of the environment. MSU Publ, Moscow, vol 2, pp 389–412, [in Russian]

Gunin PD, Miklyaeva IM, Bazha SN, Slemnev NN, Cherdonova VA (2003) Features of plant communities degradation and desertification, in forest steppe and steppe ecosystems of Southern Transbaikalia. Arid Ecosystems 9:7–21, [in Russian]

Gunin PD, Vostokova EA (eds) (1989) Methodical recommendations for evaluation and mapping of the present state of ecosystems. GUGK MPR, Ulaanbaataar, 108 pp, [in Russian]

Gunin PD, Vostokova EA, Matyushkin EN (1998) Protection of ecosystems of Inner Asia. Science Publ, Moscow, 220 pp, [in Russian]

Kalinina AV (1954) Stationary researches of pastures of Mongolian People Republic. Science Publ, Moscow, 128 pp, [in Russian]

Kamelin RV (1994) Geography and phytocoenology of *Armeniaca sibirica* (L.) Lam Plant Resourc 30:3–25, [in Russian]

Kamelin RV (2004) Geography and phytocoenology of Prunus (Amygdalus) pedunculata (Rosaceae). Bot J 80:400–425, [in Russian]

Kamelin RV, Gubanov IA (1993) Flora of Mongolia and its analysis: heterogeneity of composition, geographical elements, spectrum of biomorphes. J Gen Biol 54:59–71, [in Russian]

Karamysheva ZV (1961) Vegetation of North-Eastern part of Central-Kazakhstan low-mountains. In: Proceedings of Botanical Institute, Series 3. Geobotany, Issue 13, Leningrad, pp 464–486, [in Russian]

Karamysheva ZV, Khramtsov VN (1995) The steppes of Mongolia. Braun Blanq 17:1–70

Kuminova AV (1960) Plant cover of Altai. News of Siberian Branch of Academy of Sciences USSR, 450 pp, [in Russian]

Kurkin KA (2005) Method for estimation of pastures' productivity in the Oka River floodplain on the base of eco-genetical classification. Bot J 5:768–778, [in Russian]

Lavrenko EM (1970) The provincial division of the Central Asian subregion steppe regions of Eurasia. Bot J 12:609–625, [in Russian]

Lavrenko EM, Karamysheva ZV (1992) Steppes of the former Soviet Union and Mongolia. In: Coupland RT (ed) Ecosystems of the world, vol 8. Elsevier, Amsterdam, pp 3–59

Lavrenko EM, Karamysheva ZV, Nikulina RI (1991) Steppes of Eurasia. Science Publ, Leningrad, 145 pp, [in Russian]

Miklyaeva IM, Fakhire A (2004) Distribution of successional explerents over dry steppe pastures in the Central Mongolia. Materials of Russian Geographic Society. Biogeography 12:36–42, [in Russian]

Miklyaeva IM, Gunin PD, Bazha SN (2002) New data on century dynamics of pastoral ecosystems of Mongolia. Materials of Russian Geographic Society. Biogeography 10:41–53, [in Russian]

Miklyaeva IM, Gunin PD, Slemnev NN, Bazha SN, Fakhire A (2004) Disturbance of vegetation of steppe ecosystems. Arid Ecosyst 10:24/35–25/46, [in Russian]

Mirkin BM, Kashapov RSh, Alimbekova LM (1988) Geobotanical and economical character of hay fields and pastures of the «Tevshrulekh» goskhoz. Phytocoenotical bases of improvement of natural forage lands of MPR. Moscow, pp 7–44, [in Russian]

Miroshnichenko YuM (1964) On distribution of Artemisia frigida Wild. in MPR. Bot J 50: 420–425, [in Russian]

Muller F-V, Bat-Ochir B (1996) On the necessity of regulations for pastoral land use in Mongolia. Appl Geogr Dev 48:29–51

Namkhaijantsan G (2002) Technological documents of Mongolian railway climate. General Department of Mongolian Railway, Ulaanbaatar, 271 pp

Nechaeva NT (ed) (1970) Vegetation of the Central Karakums and its productivity. Yilym Publ, Ashkhabad, 171 pp, [in Russian]
Nechaeva NT (1987) Reserves to increase forage and products of cattle breading in a desert. Bull Agric Sci 3:75–81, [in Russian]
Nogina NA (1989) Originality of soils and processes of soil formation in the Central-Asian facia (taiga, steppe, desert). Soil Sci 9:5–14, [in Russian]
Sevast'yanov DV, Tserensodnom J (1994) Climate and lakes. Limnology and paleolimnology of Mongolia. Science Publ, Saint-Petersburg, pp 27–37, [in Russian]
Shagdarsuren O (2005) Biology of Mongolian pastures in recent years. MUIS Publ, Ulaanbaatar, 304 pp, [in Mongolian]
Slemnev NN (1969) Phytomass increment and photosynthesis of needles in pine forest stands of different density and forest types. Ph.D. thesis, Leningrad, 18 pp, [in Russian]
Slemnev NN, Kamelin RV, Gunin PD, Bazha SN (2005) Bush societies and their role in successional processes in the forest-steppe zone of Mongolia. Bot J 90:481–508, [in Russian]
State of the Environment: Mongolia (2002) UNEP RRC.AP, 79 pp
State of the Environment: Mongolia (2009) UNEP RRC.AP, 83 pp
Tsatsenkin IA, Yunatov AA (1951) Natural forage resources of MPR. Science, Moscow, 350 pp, [in Russian]
Yunatov AA (1950) The main features of plant cover of Mongolian People Republic. Science, Moscow, 223 pp, [in Russian]
Yunatov AA (1954) Fodder plants of pastures and hay fields of MPR. Science, Moscow-Leningrad, 351 pp, [in Russian]

Chapter 11
Plant Functional Types Across Dune Fixation Stages in the Chinese Steppe Zone and Their Applicability for Restoration of the Desertified Land

Ming Dong, Jianjiang Qiao, Xuehua Ye, Guofang Liu, and Yu Chu

Abstract Desertification is among the most serious environmental and socio-economic problems in the world. In northern China, aeolian desertification has resulted in dunes in the Chinese steppe zone. It is not only an issue of ecological and environmental problems, but also of social economic problems, which lead to soil erosion, degradation of vegetation, destruction of ecosystems, loss of biodiversity, decline of steppe productivity and descent of ecosystem services. We found that compositions and diversities of plant species and functional types (PFT) differed between the dune fixation stages in the Chinese steppe zone. The different PFTs may represent different strategies adaptive to the local habitats, and thus, may play different roles in communities. Therefore, our results may be applied for the restoration of fixed habitats from shifting sands.

11.1 Desertification Problem in the Chinese Steppe Zone

11.1.1 Definition of Desertification

The United Nations Convention to Combat Desertification (UNCCD) defined desertification as land degradation in arid, semi-arid, and dry sub-humid areas resulting from various factors, including climatic variations and human activities. It encompasses both biophysical and social factors (Thomas 1997; Reynolds 2001; Reynolds et al. 2007). Desertification has been high on the agenda of global

M. Dong (✉) • J. Qiao • X. Ye • G. Liu • Y. Chu
State Key Laboratory of Vegetation and Environmental Change, Institute of Botany, Chinese Academy of Sciences, Beijing 100093, China
e-mail: dongming@ibcas.ac.cn; jianjiangqiao@126.com; yexuehua@ibcas.ac.cn; liuguofang2004@126.com; chuyu@ibcas.ac.cn

M.J.A. Werger and M.A. van Staalduinen (eds.), *Eurasian Steppes. Ecological Problems and Livelihoods in a Changing World*, Plant and Vegetation 6,
DOI 10.1007/978-94-007-3886-7_11, © Springer Science+Business Media B.V. 2012

environmental issues since the 1970s, though the concept was first used in 1940s (Thomas 1997). And nowadays, it is one of the most serious environmental and socio-economic problems at global scale, especially in arid, semi-arid and dry sub-humid areas of Africa, Central Asia, Australia and northern China (Liu and Wang 2007). It is reported that more than 250 million people are directly influenced by desertification worldwide. In addition, about one billion people in 110 countries are at risk of being affected by desertification (Ayoub 1998; Guo et al. 2010).

11.1.2 Desertification in Northern China

China is one of the countries mostly affected by desertification (Guo et al. 2010). In China, desertification has been considered in terms of water erosion, salinization and aeolian desertification (Wang and Zhu 2003). Aeolian desertification, specifically, is defined as land degradation in arid, semi-arid, and parts of dry sub-humid areas, characterized by former non-desert areas being replaced by desert-like landscapes with sand drift activities, caused by excessive human activities under susceptible ecological conditions (Zhu and Liu 1984). If not otherwise specified, here "desertification" refers to aeolian desertification.

In northern China, aeolian desertification is the most common type of desertification, and it occurs widely (Zhao et al. 2005a, b). During the past 50 years, desertified land increased rapidly, with a rate of 1,560 km^2 $year^{-1}$ in the 1960s–1970s (Zhu et al. 1981), 2,100 km^2 $year^{-1}$ in the 1980s (Zhu and Wang 1990) and 3,600 km^2 $year^{-1}$ in the 1990s (Wang et al. 2003), amounting to 38.57×10^4 km^2 and accounting for 15% of the northern Chinese territory (Wang et al. 2004). The direct economic loss caused by desertification is approximately 54.1 billion RMB Yuan $year^{-1}$ (Zhang et al. 1996).

Steppe in northern China (Photo 11.1), as a land and vegetation type, mainly is used for the development of animal husbandry, including meadow steppe, typical steppe and desert steppe. In China this occurs in Inner Mongolia, Ningxia, Shaanxi, Hebei, Shanxi, Gansu, Jilin, Heilongjiang, Qinghai and Tibet and it extends over about 3.1×10^6 km^2 (Lü et al. 2005). In this region, the environment is ecologically fragile. Moreover, steppe degradation and desertification is a common phenomenon because of human-mediated activities, such as overgrazing, land reclamation and mining (e.g. Zhu et al. 1981; Zhang 1994; Wang and Zhu 2001; Li et al. 2002; Zhao et al. 2005a, b), as well as of climatic factors (Hai et al. 2003; Sun and Li 2002). According to statistics, degraded steppe in China accounted for 50.2% of the available steppe area nationwide in 1995. Degraded steppe in the northern pasturing area of China amounted to around 10% in the 1970s, 30% in the 1980s, and 50% in the 1990s. Today, steppe in China has been degrading rapidly at a speed of 20,000 km^2 $year^{-1}$ (Lü et al. 2005). It is reported that, in 2003, the area of desertified (severely degraded) steppe in China (mainly occurring in northern China) was 450,806 km^2, accounting for 32.9% of the whole country's steppe, among which 7,1919 km^2 in Gansu, 117,460 km^2 in Qinghai and 258,040 km^2 in Inner Mongolia (Lü et al. 2005).

Photo 11.1 Steppe grassland near Ordos, Inner Mongolia, China (Photo by M. Dong)

11.1.3 Sandlands in the Chinese Steppe Zone

Speaking of the desertification problem in the Chinese steppe zone, the inland dune ecosystems in China have to be mentioned, as they represent the most desertified regions in the Chinese steppe zone. The term "sandland" is used specifically for these inland dune areas in the steppe zone. Among these inland sandland ecosystems, the most representative and area-largest sandland ecosystems are: Mu Us Sandland (Photo 11.2), Otindag Sandland, Horqin Sandland and Hulunbeir Sandland, all of which occur in the semi-arid temperate steppe region of northern China historically covered by highly productive grasslands (Fig. 11.1) (Chinese Academy of Sciences 1985; Li 2006). Presently, however, they are covered by aeolian mobile dunes (with a vegetation cover <15%), semi-fixed dunes (with a vegetation cover of 15–40%) and fixed dunes (with a vegetation cover >40%) (Wang et al. 2009). The four sandlands cover in north–south direction about 10° of latitude, and stretch east–west across a longitude of 16°. Mean annual precipitation ranges from 250 to 450 mm, mean annual temperature from −2.5°C to 7.7°C. The soil type is mainly an aeolian sandy soil. The vegetation of the sandlands is non-zonal. It contains or is even dominated by many shrub species, for instance, those of genera *Artemisia*, *Caragana*, *Hedysarum* and *Salix* (Photo 11.3).

Photo 11.2 Desertified lands in Mu Us Sandland, Inner Mongolia, China (Photo by M. Dong)

11.1.4 Possible Reasons for Desertification in Northern China

The desertification in the Chinese steppe zone is not only an issue of ecological and environmental problems, but also of social economic problems, which lead to soil erosion, degradation of vegetation, destruction of ecosystems, loss of biodiversity, decline of steppe productivity and descent of ecosystem services. Furthermore, it threatens the living conditions of residents and people in nearby areas, especially those in the Beijing-Tianjin Region, and hampers the sustainable development of the regional economy and society (Liu 2004; Chu 2005; Chu et al. 2006; Cheng et al. 2007). The large amount of dust transported into eastern China and the North Pacific is caused by the rapid desertification in the sandland region (e.g., the dust particles transported over long-distances from Otindag Sandland accounted for 20.9% of the dust mass in Beijing during dusty days) (Chinese Academy of Sciences 1985; Zhuang et al. 2001; Cheng et al. 2005).

Many factors may have contributed to the degradation of the steppe in northern China, including changes in land use, overgrazing, and climatic fluctuations. In

Sandlands	Mu Us	Otindag	Horqin	Hulunbeir
Longitude (E)	107°20′-111°30′	112°22′-117°57′	118°35′-123°30′	117°10′-121°12′
Latitude (N)	37°30′-39°20′	41°56′-44°24′	42°41′-45°15′	47°20′-49°59′
Area (km2)	40,000	21,400	42,300	13,100
Precipitation (mm·a-1)	250-450	320-450	343-450	280-400
Temperature (°C·a-1)	7.7	1.6	5.8-6.4	-2.5-0
Soil type	aeolian sandy soil	aeolian sandy soil	aeolian sandy soil	aeolian sandy soil

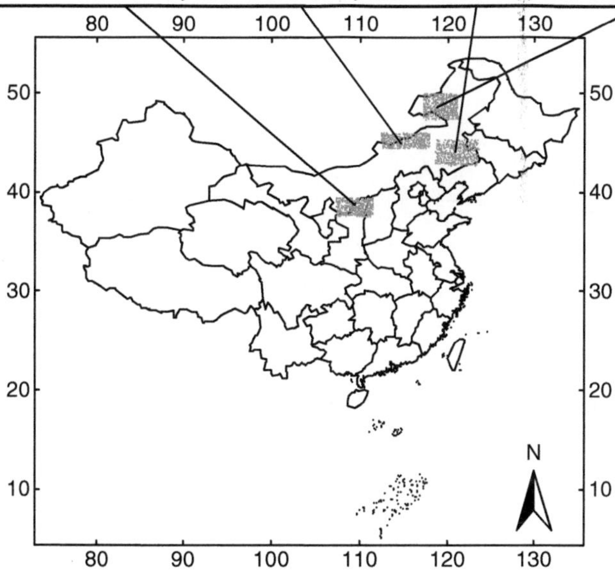

Fig. 11.1 The environmental features of Mu Us Sandland, Otindag Sandland, Horqin Sandland and Hulunbeir Sandland (From Qiao et al. 2011)

Otindag Sandland, both the densities of grazing animals and human beings have exceeded the threshold of the carrying capacity of the grassland (Jiang 2002); in the Xilin River Basin, anthropogenic factors (especially overgrazing) are primarily responsible for the steppe degradation (Tong et al. 2004).

The sandland in the Chinese steppe zone is a particular land type and also a special habitat for plants (Photos 11.3 and 11.4). Changes in the habitat are accompanied by vegetation succession (Chinese Academy of Sciences 1985). Therefore, research on vegetation changes on the different habitats (mobile dunes, semi-fixed dunes and fixed dunes) of the sandland is essential for understanding their causes and dynamics and helpful to the rational use of their resources (Zhao et al. 2007).

Photo 11.3 Degraded steppe land near Ordos, Inner Mongolia, China, with different degrees of dominance of *Artemisia ordosica* and, in the foreground, *Hedysarum laeve* (Photo by M.J.A. Werger)

Photo 11.4 *Psammochloa villosa*, a pioneer clonal grass species in Mu Us Sandland, Inner Mongolia, China (Photo by M. Dong)

Table 11.1 Plant functional types (PFTs) in northern Chinese sandlands, based on life span, photosynthetic pathway, reproductive mode and growth form

Plant functional type (PFT)	Life span	Photosynthetic pathway	Reproductive mode	Growth form
Annual-C_3-nonclonal-forb	Annual	C_3	Non-clonal	Forb
Annual-C_4-nonclonal-forb	Annual	C_4	Non-clonal	Forb
Annual-C_4-nonclonal-grass	Annual	C_4	Non-clonal	Grass
Perennial-C_3-clonal-forb	Perennial	C_3	Clonal	Forb
Perennial-C_3-clonal-grass	Perennial	C_3	Clonal	Grass
Perennial-C_3-clonal-shrub	Perennial	C_3	Clonal	Shrub
Perennial-C_3-nonclonal-forb	Perennial	C_3	Non-clonal	Forb
Perennial-C_3-nonclonal-shrub	Perennial	C_3	Non-clonal	Shrub
Perennial-C_4-clonal-grass	Perennial	C_4	Clonal	Grass

11.2 Plant Functional Types Across the Dune Fixation Stages in the Chinese Steppe Zone

A plant functional type (PFT) is a group of species which share similar traits (morphological and physiological attributes), function in similar ways and show similar responses to environmental conditions. It provides a logical link between the physiological characteristics and life-history strategies at the individual level and the ecological processes that operate at the ecosystem level (Walker 1992; Chapin 1993; Noble and Gitay 1996; Paruelo and Lauenroth 1996; Diaz and Cabido 1997; Gitay and Noble 1997). In the past decades, a lot of studies on vegetation succession and restoration of sandlands have been done, usually with the focus on the variation in species composition during these processes (e.g. Chang and Wu 1997; Li et al. 1999, 2001; Shen 1999; Guo 2000; Guo et al. 2000; Zhao et al. 2003, 2007; Liu et al. 2004; Zhang et al. 2004) but rarely from the angle of PFTs (but see Chu et al. 2006). Understanding the functional aspects of PFTs, as based on life span, photosynthetic pathway, reproductive mode and growth form (Table 11.1) and the PFT diversity across dune fixation stages of sandland helps to clarify the dynamics of the vegetation and to assess the potential of the constituent species for the restoration of the habitat and vegetation of the desertified sandland.

Our field investigation for the four sandlands, which was carried out in August of 2007, revealed that the semi-fixed and fixed dunes shared the same PFT richness and diversity (Shannon index) and both were significantly higher as compared to the mobile dunes (Fig. 11.2).

A correspondence analysis (CA), one of many ordination techniques, indicated that different dune fixation stages filtered distinct PFTs. The mobile dunes of the four sandlands all shared the same dominant PFT annual-C_4-nonclonal-forb, in terms of richness and Importance Value (IV) (Figs. 11.3 and 11.4). In addition, annual-C_4-nonclonal–grass was also a dominant PFT in the mobile dunes at Otindag and Hulunbeir Sandlands while perennial-C_3-clonal-shrub was a subdominant PFT in the mobile dunes at Horqin Sandland (Figs. 11.3b, c, d and 11.4b, c, d).

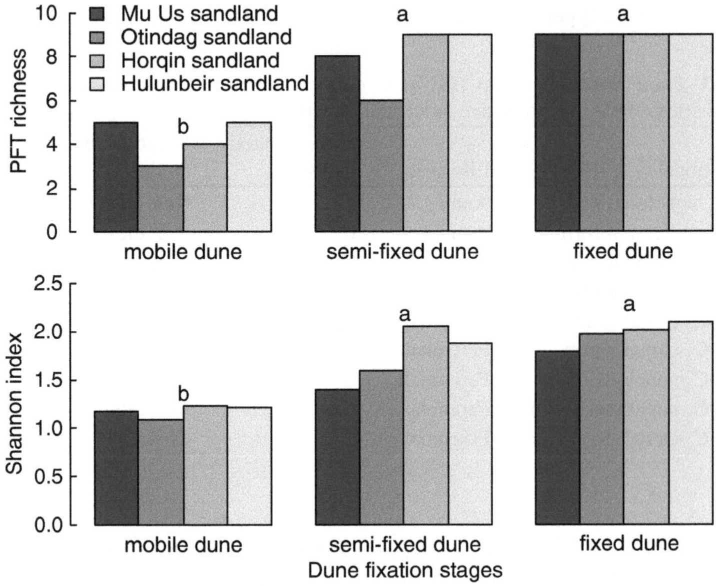

Fig. 11.2 PFT richness (percentage of the total number of species number in PFT) and Shannon index value across dune fixation stages in Mu Us Sandland, Otindag sandland, Horqin sandland and Hulunbeir sandland. Bar chunks sharing the same letter are not significantly different among mobile dune, semi-fixed dune and fixed dune at $P < 0.05$

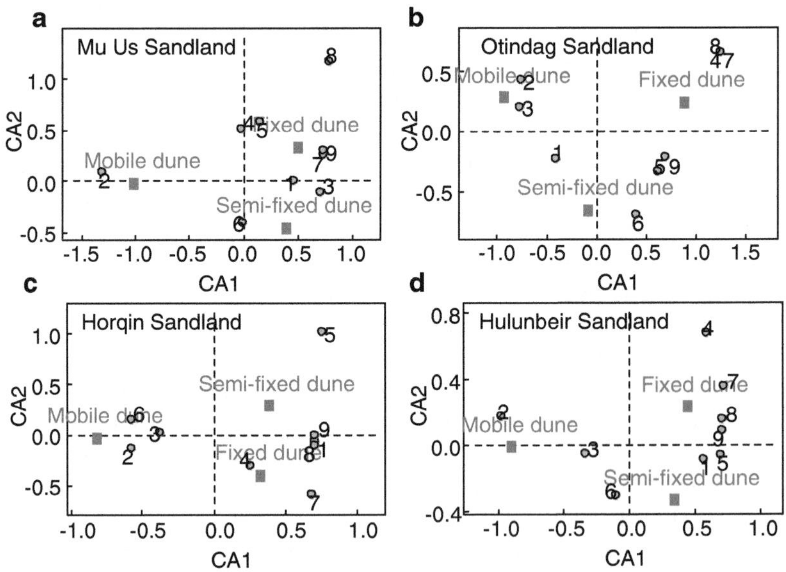

Fig. 11.3 Correspondence analysis (*CA*) biplot for the percentage of species number per PFT (%) on mobile dune, semi-fixed dune and fixed dune in (**a**) Mu Us Sandland, (**b**) Otindag sandland, (**c**) Horqin sandland, and (**d**) Hulunbeir sandland. *Note: 1* Annual-C_3-nonclonal-forb, *2* annual-C_4-nonclonal-forb, *3* annual-C_4-nonclonal-grass, *4* perennial-C_3-clonal-forb, *5* perennial-C_3-clonal-grass, *6* perennial-C_3-clonal-shrub, *7* perennial-C_3-nonclonal-forb, *8* perennial-C_3-nonclonal-shrub, *9* perennial-C_4-clonal-grass

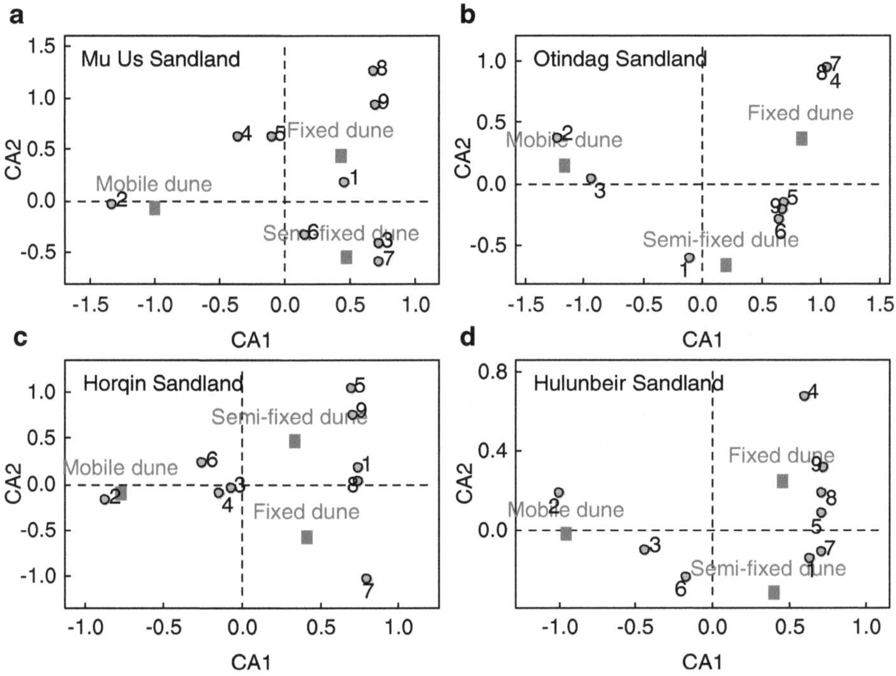

Fig. 11.4 Correspondence analysis (*CA*) biplot for importance values of PFTs on mobile dune, semi-fixed dune and fixed dune in (**a**) Mu Us Sandland, (**b**) Otindag sandland, (**c**) Horqin sandland, and (**d**) Hulunbeir sandland. *Note: 1* Annual-C_3-nonclonal-forb, *2* annual-C_4-nonclonal-forb, *3* annual-C_4-nonclonal-grass, *4* perennial-C_3-clonal-forb, *5* perennial-C_3-clonal-grass, *6* perennial-C_3-clonal-shrub, *7* perennial-C_3-nonclonal-forb, *8* perennial-C_3-nonclonal-shrub, *9* perennial-C_4-clonal-grass

The semi-fixed and fixed dunes of the sandlands showed locality-specific characteristics in richness and IV of dominant PFTs, probably because they possess an intense spatial heterogeneity of habitat compared to the homogeneous environment of the mobile dunes. In semi-fixed dune stages in Mu Us Sandland, perennial-C_3-nonclonal-shrub and annual-C_4-nonclonal-grass were the dominant PFTs (Fig. 11.3a), while in the Otindag, Horqin and Hulunbeir Sandlands the PFT annual-C_3-nonclonal-forb was dominant (Fig. 11.3b, c, d). In terms of IV the PFT perennial-C_3-clonal-shrub was important in the Mu Us, Otindag and Hulunbeir Sandlands, and in Horqin Sandland it was the PFT perennial-C_3-clonal-forb.

In fixed dune stages all four sandlands had the same prominent PFT in terms of richness and IV, namely perennial-C_3-nonclonal-shrub (Figs. 11.3 and 11.4). There were general trends showing that perennial-C_3-clonal-shrubs were major components of the vegetation in semi-fixed dunes and perennial-C_3-nonclonal-shrubs in fixed dunes, while annual-C_4-nonclonal-forbs dominated at mobile dunes, both in terms of IV and percentage of the total number species per PFT. Additional

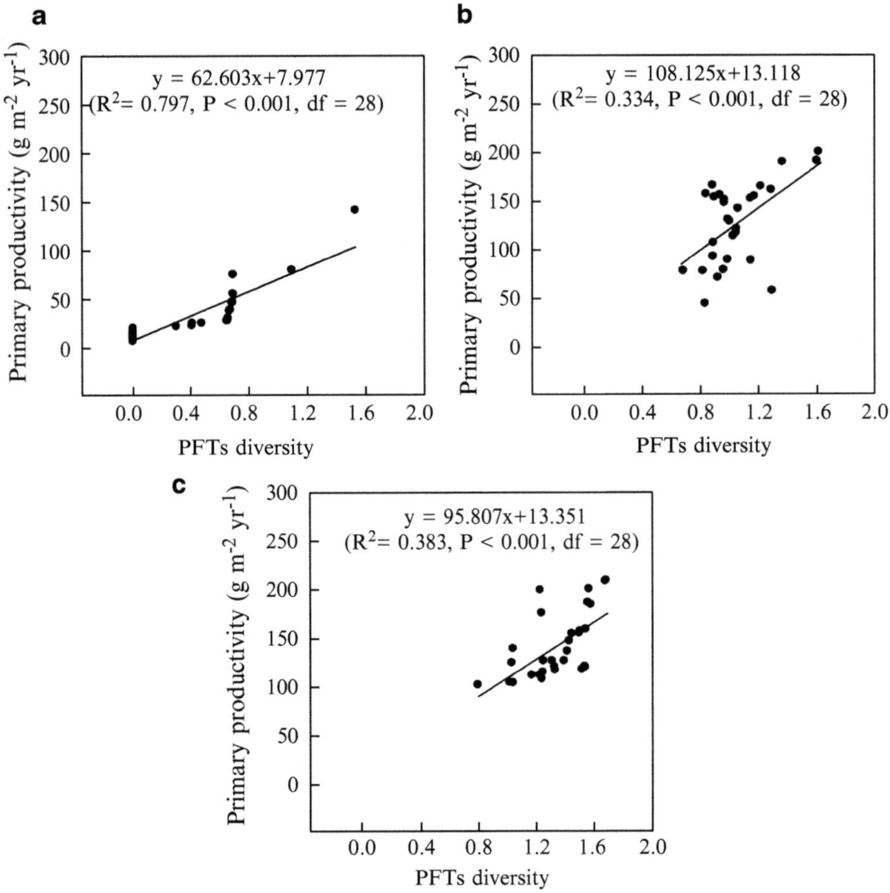

Fig. 11.5 Relationships between PFT diversity and primary productivity of vegetation across dune fixation stages in Mu Us Sandland. (**a**) mobile dune, (**b**) semi-fixed dune, and (**c**) fixed dune

field investigations at Mu Us Sandland, which were performed in August of 2007, showed that the primary productivity of the vegetation was positively correlated with PFT diversity in all three dune fixation stages (Fig. 11.5).

11.3 Plant Functional Types Applicable to Restoration of Shifting Sands

The findings in Mu Us Sandland, Otindag Sandland, Horqin Sandland and Hulunbeir Sandland suggest that not only species but also PFT composition and diversity of vegetation differ between the dune fixation stages in the Chinese steppe zone. Different PFTs may represent different strategies adaptive to the local habitats, and thus, may play different roles in communities (Song and Dong 2002; Chu et al. 2006). Therefore, our results may be applied for the restoration of fixed habitats from shifting sands.

Perennial C_3 clonal shrubs (the PFT of perennial-C_3-clonal-shrub) seemed to have a more stable performance in terms of important values and the percentage of species number per PFT, and were relatively more important, and therefore probably better adapted for growth in mobile dune, semi-fixed dune and fixed dune habitats. That may be a consequence of their growth habits. That is, in mobile dunes, the instability of the sandy substrate together with the dry and windy climate makes it difficult for plants to establish and survive (Liu et al. 2005). Perennial C_3 clonal shrubs, such as *Artemisia halodendron*, *A. intramongolica*, *A. ordosica*, *A. sphaerocephala*, *Hedysarum leave*, *Periploca sepium* and *Salix psammophila*, have many properties, which make them adaptive to the harsh environmental conditions in mobile dunes (e.g. Fahn 1964, 1990; Fahn and Cutler 1992; Danin 1983, 1996; Huang and Gutterman 1998, 1999, 2000; Yuan et al. 2008). For instance, all of them have dense brushy growth forms that are helpful in reducing wind speed and fixing-sand effectively; they have strong survival characteristics under conditions of sand burial and wind erosion, e.g. the ability to sprout adventitious roots from the stems buried by sand (Li et al. 2010a, b).

At the stage of semi-fixed dunes, perennial C_3 clonal shrubs are the dominants (Fig. 11.3). These dominants can induce small-scale heterogeneity in soil nutrients, forming small fertile islands, where litter and nutrients accumulate, and the structure of topsoil under their canopies improves (Li et al. 2008). This, in turn, is more and more beneficial for the growth of perennial herbs, mosses and algae. Gradually, the shifting sands become fixed, and the composition and functioning of the vegetation tends to that of the climatic climax (steppe) (Guo 2000; Piao et al. 2006). Thus, perennial C_3 clonal shrubs can play a vital role in the early restoration stages of sandland vegetation in the Chinese steppe zone. Proper manual planting of these species in mobile and semi-fixed dunes can speed up the restoration of shifting sands to steppe on the sands of the northern Chinese steppe zone.

References

Ayoub AT (1998) Indicators of dryland degradation. In: Squires V, Sidahmed AE (eds) Drylands, sustainable use of rangelands into the twenty first century. International Fund for Agricultural Development, Rome, pp 11–24

Chang XL, Wu JG (1997) Species diversity during desertification on Kerqin Sandy Land. Chin J Appl Ecol 8:151–156

Chapin FS (1993) Functional role of growth forms in ecosystem and global process. In: Ehleringer JR, Field CB (eds) Scaling physiological processes. Academic, San Diego, p 388

Cheng TT, Lü DR, Zhang WX, Xu YF, Li XR (2005) The contribution of Hunshandake Sandland during dusty days in Beijing April 2000. J Yunn Univ 27:245–251

Cheng X, An S, Chen J, Li B, Liu Y, Liu S (2007) Spatial relationships among species, above-ground biomass, N, and P in degraded grasslands in Ordos Plateau, northwestern China. J Arid Environ 68:652–667

Chinese Academy of Sciences (1985) Vegetation of Inner Mongolia. Scientific Press, Beijing

Chu Y (2005) Plant functional type diversity in the Otindag Sandland. Ph.D. thesis, Institute of Botany, Chinese Academy of Sciences, China

Chu Y, He WM, Liu HD, Liu J, Zhu XW, Dong M (2006) Phytomass and plant functional diversity in early restoration of the degraded, semi-arid grasslands in northern China. J Arid Environ 67:678–687

Danin A (1983) Desert vegetation of Israel and Sinai. Cana Publishing House, Jerusalem

Danin A (1996) Plants of desert dunes. Springer, Berlin

Diaz S, Cabido M (1997) Plant functional types and ecosystem function in relation to global change: a multiscale approach. J Veg Sci 8:463–474

Fahn A (1964) Some anatomical adaptations of desert plants. Phytomorpbology 14:99–102

Fahn A (1990) Plant anatomy. Pergamon Press, Oxford

Fahn A, Cutler DF (1992) Xerophytes. Gebruder Borntraeger, Berlin

Gitay H, Noble IR (1997) What are functional types and how should we seek them? In: Smith TM, Shugart HH, Woodward FI (eds) Plant functional types. Cambridge University Press, Cambridge, pp 3–19

Guo K (2000) Cyclic succession of *Artemisia Ordosica* Krasch community in the Mu Us sandy grassland. Acta Phytoecol Sin 24:243–247

Guo K, Dong XJ, Liu ZM (2000) Characteristics of soil moisture content on sand dunes in Mu Us sandy grassland: why *Artemisia Ordosica* declines on old fixed sand dunes. Acta Phytoecol Sin 24:275–279

Guo J, Wang T, Xue X, Ma SX, Peng F (2010) Monitoring aeolian desertification process in Hulunbir grassland during 1975–2006, northern China. Environ Monit Assess 166:563–571

Hai CX, Fu JS, Wang XM (2003) Influence of climate change and anthropogenic activity on soil erosion and desertification in Fengning county, Hebei Province. J Arid Land Resources Environ 17:69–76

Huang ZY, Gutterman Y (1998) *Artemisia monosperma* achene germination in sand: effects of sand depth, sand/water content, cyanobacterial sand crust and temperature. J Arid Environ 38:27–43

Huang ZY, Gutterman Y (1999) Water absorption by mucilaginous achenes of *Artemisia monosperma*: floating and germination as affected by salt concentrations. Israel J Plant Sci 47:27–34

Huang ZY, Gutterman Y (2000) Comparison of germination strategies of *Artemisia ordosica* with its two congeners from deserts of China and Israel. Acta Bot Sin 42:71–80

Jiang GM (2002) Strategy of regeneration of the degraded ecosystem in Hunshandak Sandland. Chin Sci Technol Forum 15(3):13–15 (in Chinese with English abstract)

Li G (2006) Biomass, net primary productivity, and ecological restoration of sparse forest grassland ecosystem of Hunshandake Sandyland. Ph.D. thesis, Institute of Botany, Chinese Academy of Sciences, China

Li XR, Zhao YX, Yang ZZ, Liu HP (1999) Study on evolution of air-seeding vegetation and habitat in Maowusu sandland. Acta Phytoecol Sin 23:116–124

Li QF, Hu CY, Wang MJ (2001) Analysis on the causes of eco-environmental deterioration in Hunshandake Sandland region and Countermeasures. J Arid Land Resources Environ 15:9–16

Li QF, Li FS, Wu L (2002) A primary analysis on climatic change and grassland degradation in Inner Mongolia. Agric Res Arid Areas 20:99–100

Li PX, Wang N, He WM, Krüsi BO, Gao SQ, Zhang SM, Yu FH, Dong M (2008) Fertile islands under *Artemisia ordosica* in inland dunes of northern China: effects of habitats and plant developmental stages. J Arid Environ 72:953–963

Li SL, Werger MJA, Zuidema PA, Yu FH, Dong M (2010a) Seedlings of the semi-arid *Artemisia ordosica* are resistant to moderate wind denudation and sand burial in Mu Us Sandland, China. Trees 24:515–521

Li SL, Zuidema PA, Yu FH, Werger MJA, Dong M (2010b) Effects of denudation and burial on growth and reproduction of *Artemisia ordosica* in Mu Us Sandland. Ecol Res 25:655–661

Liu HJ (2004) Ecological adaptation of vegetation and characteristics of plant resources in Otindag Sandland. M.Phil. thesis, Institute of Botany, Chinese Academy of Sciences, China

Liu S, Wang T (2007) Aeolian desertification from the mid-1970s to 2005 in Otindag Sandy Land, Northern China. Environ Geol 51:1057–1064

Liu MZ, Jiang GM, Yu SL, Li YG, Gao LM, Niu SL, Jiang CD, Peng Y (2004) Dynamics of plant community traits during an 18-year natural restoration in the degraded sandy grassland of Hunshandake Sandland. Acta Ecol Sin 24:1731–1737

Liu ZM, Jiang DM, Yan QL, Li XH, Li RP, Luo YM, Wang HM (2005) Study on dispersal biology of common species of Flora of the Horqin Steppe. Acta Pratacult Sin 14:23–33

Lü ZJ, Lu XS, Xin XP (2005) Present situation and trend of grassland desertification of northern China. Acta Agrest Sin 13:24–27

Noble IR, Gitay H (1996) A functional classification for predicting the dynamics of landscapes. J Veg Sci 7:329–336

Paruelo JM, Lauenroth WK (1996) Relative abundance of plant functional types in grasslands and shrublands of North America. Ecol Appl 6:1212–1224

Piao SJ, Wang ZJ, Yan XL, Zhang BC, Yin H (2006) Analysis of niche fitness of *Artemisia halondendron* population on Horqin Sandyland. J Plant Ecol 30:593–600

Qiao JJ, Zhao WW, Xie XF, Liu GF, Ye XH, Chu Y, He H, Dong M (2011) Variation in plant diversity and dominance across dune fixation stages in the Chinese steppe zone. J Plant Ecol-UK. doi: 10.1093/jpe/rtr032

Reynolds JF (2001) Desertification. In: Levin S (ed) Encyclopedia of biodiversity. Academic, San Diego, pp 61–78

Reynolds JF, Smith DMS, Lambin EF, Turner BL II, Mortimore M, Batterbury SPJ, Downing TE, Fernández HDRJ, Herrick JE, Huber-Sannwald E, Jiang H, Leemans R, Lynam T, Maestre FT, Ayarza M, Walker B (2007) Global desertification: building a science for dryland development. Science 316:847–851

Shen WS (1999) Successional stage and rate of the aerial seeding vegetation in the Maowusu Sandyland. Sci Silv Sin 35:103–108

Song MH, Dong M (2002) Clonal plants and plant species diversity in wetland ecosystems in China. J Veg Sci 13:237–244

Sun W, Li BS (2002) The relation between coupling among the principal components of desertification factors and desertification in rear hills of Bashang since 1950. Geogr Res 21:392–397

Thomas DSG (1997) Science and the desertification debate. J Arid Environ 37:599–608

Tong C, Wu J, Yong S, Yang J, Yong W (2004) A landscape-scale assessment of steppe degradation in the Xilin River Basin, Inner Mongolia, China. J Arid Environ 59:133–149

Walker B (1992) Biodiversity and ecological redundancy. Conserv Biol 6:18–23

Wang T, Zhu ZD (2001) Some problems of desertification in northern China. Quaternary Sci 21:56–65

Wang T, Zhu ZD (2003) Study on aeolian desertification in China – 1. Definition of aeolian desertification and its connotation. J Des Res 23:209–214

Wang T, Wu W, Xue X (2003) Time-space evolution of desertification land in northern China. J Des Res 23:230–235

Wang T, Wu W, Xue X (2004) Spatial-temporal changes of aeolian desertified land during last 5 decades in northern China. Acta Geogr Sin 59:203–212

Wang XM, Yang Y, Dong ZB, Zhang CX (2009) Responses of dune activity and desertification in China to global warming in the twenty-first century. Global Planetary Change 67:167–185

Yuan PF, Ding GD, Zhao K (2008) The impacts of sand burial on the growth characteristic of *Salix psammophila* in mobile sand dune. Res Soil Water Conserv 5:53–55

Zhang XS (1994) Principles and optimal models for development of Maowusu sandy grassland. Acta Phytoecol Sin 18:1–16

Zhang Y, Ning D, Smil V (1996) An estimate of economic loss for desertification in China. China Pop Resources Environ 6:45–49

Zhang JY, Zhao HL, Cui JH, Li YL, Yang JD (2004) Development process of the vegetation and the change of species composition in the Horqin Sandland. Arid Zone Res 21:72–75

Zhao HL, Zhao XY, Zhang TH, Zhou RL (2003) Study on damaged process of the vegetation under grazing stress in sandy grassland. Acta Ecol Sin 23:1505–1511

Zhao WZ, Xiao HL, Liu ZM, Li J (2005a) Soil degradation and restoration as affected by land use change in the semiarid Bashang area, northern China. Catena 59:173–186

Zhao HL, Zhao XL, Zhou RL, Zhang TH, Drakeb S (2005b) Desertification processes due to heavy grazing in sandy rangeland, Inner Mongolia. J Arid Environ 62:309–319

Zhao CY, Wang T, Dong ZB, Zhao XY, Xue X (2007) Plant diversity and its relationship with habitat in Horqin Sandland. Pratacult Sci 24:10–18

Zhu ZD, Liu S (1984) The concept of aeolian desertification and the differentiation of its development. J Des Res 4:2–8

Zhu ZD, Wang T (1990) An analysis of the trend of land desertification in northern China during the last decade based on examples from some typical areas. Acta Geogr Sin 45:430–440

Zhu ZD, Liu S, Xiao LS (1981) The characteristics of the environment vulnerable to desertification and the ways of its control in steppe zone. J Des Res 1:2–12

Zhuang GS, Guo JH, Yuan H, Zhao CY (2001) Composition, source and grain size fraction of the dust storms in China in 2000 and its influence to the global environments. Chin Sci Bull 46:191–197

Chapter 12
Population Dynamics of a Key Steppe Species in a Changing World: The Critically Endangered Saiga Antelope

Valery M. Neronov, Anna A. Lushchekina, Tatiana Yu. Karimova, and Nadezhda Yu. Arylova

Abstract At the end of 1950s the population of the saiga antelope (*Saiga tatarica*) in the North-West Pre-Caspian region counted over 800,000 individuals. Estimates of 2008 suggested that about 18,000 saigas were left, in a reduced range area, and after the severe 2009–2010 winter the population has declined to only 8,000 animals. To understand the present conditions a detailed study was conducted using our own field observations (2003–2008), all available literature data over more than 50 years, and long-term data of several meteorological stations located in the area of saiga habitation. Results show that the main reasons for the catastrophic decline in saiga numbers were decreases in the number of adult males and in general fertility of the females. Juvenile mortality in recent years stayed within the limits of long-term fluctuations. The weight of newborn saigas does not differ (and presently is even higher) from that at the end of the 1950s. Comparison of our observations with data on fluctuations in average annual temperature and rainfall showed that there was no direct effect of climatic factors on the state of this population. Accordingly, in the North-West Pre-Caspian region additional measures should be urgently taken to secure the survival of this unique ungulate.

V.M. Neronov (✉) • A.A. Lushchekina • T. Yu. Karimova
A.N. Severtsov Institute of Ecology and Evolution,
Russian Academy of Sciences, Moscow, Russia
e-mail: rusmabcom@gmail.com

N.Yu. Arylova
Center for Wild Animals of the Republic of Kalmykia, Elista, Russia
e-mail: arylova@gmail.com

M.J.A. Werger and M.A. van Staalduinen (eds.), *Eurasian Steppes. Ecological Problems and Livelihoods in a Changing World*, Plant and Vegetation 6,
DOI 10.1007/978-94-007-3886-7_12, © Springer Science+Business Media B.V. 2012

12.1 Introduction

In 1992, at the World Summit in Rio de Janeiro, the majority of the countries of the World have signed the Convention on biological diversity (CBD) (Convention on… 1995). Despite a certain progress in realizing the requirements of the CBD since then, rates of biodiversity reduction and infringement of functions of various ecosystems continue to cause serious concerns to the world community. As the international project "Millennium ecosystem assessment" (2001–2005) has shown it is urgently necessary to take additional measures without which the tasks of sustainable development cannot be implemented. One such "tool" can be the application of an ecosystem approach for conservation of biodiversity, as was decided upon at the Second meeting of the Conference of parties of the CBD (Jakarta 1995). As intended the ecosystem approach aims to guarantee the achievement of three purposes facing the CBD, namely: preservation, sustainable use and fair and equal distribution of all benefits received from the genetic resources. Implementing the ecosystem approach, it is necessary to take into the account the spatial and temporary changes in the ecosystem and the certain inertia of some ecological processes. Furthermore, this approach envisages the maintenance of interaction of all sectors of the economy and society and it was proposed to decentralize systems of management in order to provide the operatively necessary balance of most of the various interests. The IUCN (Shepherd 2004) has pointed out that such an approach requires that attention is given to five stages: (1) Identification of the interested parties and specification of the sizes of ecosystems; (2) the characteristics of structure, functions and management of ecosystem; (3) identifying the contrasting economic interests that have effects on the ecosystem and its populations; (4) verifying the possible influences of ecosystems on adjacent territories and introduction of adaptive management; (5) a clear choice of the long-term purposes and flexible approaches to their achievement. We try to consider these five stages in our assessment of the dynamics of the Saiga antelope, a key steppe species, and suggest measures to solve the urgent problems on conservation of this antelope in the North-West Pre-Caspian region (Neronov et al. 2005; Bol'shakov et al. 2009).

The Saiga antelope (*Saiga tatarica*) is a critically endangered migratory ungulate of the steppes and semi-deserts of Eurasia. Since the collapse of the Soviet Union in 1991, numbers of saiga populations have declined by more than 90%. The international awareness of the critical situation concerning the saiga led to the listing of this species in Appendix II of CITES in 1995 and as a critically endangered species in the IUCN Red List in 2002. In 2004 the critical situation of the saiga in its whole range was reflected in resolutions adopted by CITES COP-13 and IUCN WCC held in Thailand.

Two subspecies of the saiga are currently recognized: *Saiga tatarica mongolica,*[1] which inhabits the small steppe area in the west of Mongolia, and *Saiga tatarica*

[1] The Appendix-II listings of saiga antelope by the CMS and CITES have been updated to include two species, *Saiga tatarica* and *Saiga borealis* in accordance with Wilson and Reeder (2005). However, this proposal has to be evaluated in the light of further genetic and morphological studies of modern and fossil specimens of saigas.

tatarica, which occupies the vast plains of Central Asia and the Pre-Caspian region. Massive seasonal migrations have been described particularly for *Saiga tatarica tatarica* in Central Asia.

During its evolution the saiga became very well adapted to the harsh and unpredictable conditions of an extreme environment. Despite its rather sheep-like body shape, the saiga antelope is one of the fastest terrestrial vertebrates, capable of reaching speeds of up to 80 km per hour. Individuals have a short life span and adults achieve high reproductive rates, adaptations that allow for rapid demographic recovery following particularly severe climatic episodes. The males are crowned with a pair of light yellow and waxy horns they use as effective weapons in their fights with other males. By the 1920s, over-harvesting had almost completely eliminated the saiga from most of its range. Intensive hunting and land-use development in steppe in the late nineteenth and early twentieth centuries reduced global saiga populations to only a few hundreds. Before the 1990s, the Soviet Union's closed borders supported the antelope population by cutting off international trade routes. Saiga hunting was banned from 1919 until the 1950s, allowing numbers to recover to nearly two million, and sending the antelope from near extinction to the most numerous ungulate in the Soviet Union (Lushchekina and Struchkov 2001). The amazing recovery was in part due to the animal's high fecundity – saiga females begin breeding in the first year of life and give birth to their first lamb during the second year. Older females are capable of bringing two and even three lambs at once.

12.2 Four Populations and Their Historical Dynamics

In addition to a population in the North-West Pre-Caspian region, three further populations of *S. t. tatarica* are known in Central Asia: the Ural, the Ustyurt and the Betpakdala populations. It is known that some herds from Ustyurt in Kazakhstan can migrate to adjacent territories in Uzbekistan and Turkmenistan. Saiga herds of the North-West Pre-Caspian population can migrate from Kalmykia to Dagestan and other adjacent territories in Russia. During the period between 1980 and 1994, total saiga numbers fluctuated at around 670,000–1,251,000 individuals. During the same period, single population estimates were as follows: North-West Pre-Caspian population, with 142,000–430,000; Ural population, with 40,000–298,000; Ustyurt population, with 140,000–265,000; and Betpakdala population, with 250,000–510,000 individuals. All four *S. t. tatarica* populations experienced severe population declines after 1998. The annual rate of population decline during 1998–1999 was roughly 35%, and reached a dramatic 56% drop during 1999–2000. The 2002 census revealed the following numbers for the North-West Pre-Caspian, Ustyurt, Ural and Betpakdala populations: 19,500, 19,100, 6,900 and 4,000, respectively (Milner-Gulland et al. 2001, 2003).

Saiga aggregations vary in size throughout the year (Photo 12.1). Larger herds are recorded during the reproductive season, but they also have been observed during other parts of the year. During winter, large herds are better able to remove

Photo 12.1 Thirsty saigas at watering place on a hot summer day (Photo by Tatiana Karimova)

the superficial layer of snow and reach the needed forage resources. During summer, large herds may offer individual animals temporary relief from the massive attacks by blood-sucking insects. Most importantly, the herding behavior of saigas offers better protection and early warning against predators, particularly wolves that are common in many parts of the Eurasian steppes.

12.3 Decline of the Saiga in the North-West Pre-Caspian Region

At the beginning of the twenty-first century, the saiga's range within the North-West Pre-Caspian region shrank very much. Not so long ago many herds, some with more than a 1,000 animals, of this unique species freely moved across the grassland habitats in the territories of the Republic of Kalmykia and Astrakhan oblast. Sometimes they even went to Dagestan Republic, Rostov and Volgograd oblasts. In the middle of 1980s, during the cold winters with lots of snow, animals occurred within 10–15 km from the city of Astrakhan. Today it is a very rare occasion to see saigas, even in very remote parts of the steppe. Therefore, information about its present migrations is almost lacking. In the late 1950s the North-West Pre-Caspian population of the saiga antelope counted over 800,000 individuals. Population estimates from 2008 suggested that approximately 18,000 saiga remained in the region (Neronov and Lushchekina 2009). And on top of that, recent media reports indicate that the severe 2009–2010 winter has catastrophically reduced the region's herds to only 8,000–9,000 animals.

The reasons for such a decline are numerous. Some authors (Bukreeva 2002; Sidorov and Bukreeva 2007; Bliznyuk 2009) relate it to long-term cyclic fluctuations in reaction upon gradual changes in the habitat and range of the species. Danilkin (2005) explains it as a result of uncontrolled and disastrous exploitation of the species and the absence of appropriate protection. Based on the results of studies into the saiga's feeding it was assumed, that the decrease in numbers results from shortage of preferred forage because of changes in vegetation cover in the main places of their distribution area (Abaturov 2007; Larionov 2008). These hypotheses demand additional study and verification. Anyhow, we believe that the development of agriculture and particularly the construction of irrigation channels had a serious impact and drastically diminished saiga numbers and changed its nomadic behavior.

12.3.1 *The Effects of Irrigation Works*

The development of irrigation systems and other large infrastructural works usually have a very great impact on the habitats and landuse of an area, and therefore on its wildlife. For example, the lake-rich Sarpa Depression had been good for the saiga, especially so in the spring, when it would grow lush with ephemeral plants and where female saiga preferred to deliver their young, became largely isolated in the early 1960s due to the construction of Sarpa Irrigation Facilities. In 1981–1982 the Sarpinskaya and Chernozemelskaya irrigation systems were put in operation. Their total length is about 500 km and the irrigated area measures about 62,000 ha. As a result the saiga range has been decreased to 20,000–23,000 km^2 and the number of animals at that time was estimated at about 160,000–200,000 (as against 600,000–700,000 in 1977–1978).

The spasm of economic reconstruction in the region was growing more and more frenetic. In the late 1970s, the Sarpa Depression saw the launch of a new irrigation project, the Kalmyk–Astrakhan Facilities intended to transform this area into a rice estate. While this mammoth establishment was rapidly devouring what was left of the saiga's favorite spring pastures, the remaining steppe tracts in the western parts of Kalmykia and adjacent provinces were turned into a uniform mass of plowed fields, interrupted only by the newly-built canals and roads. As a result, its usual summer retreats in such areas as the Ergeni Heights and the Kuma—Manych Interfluve were practically eliminated (Lushchekina and Struchkov 2001). The construction of the Volga-Chograi canal started in the second half of 1980s (80 km of its bed was excavated to a depth of 20 m). This also brought changes in the patterns of migration routes and in saiga's numbers. The impact of this huge canal, if it had been completed, could have further decrease the range of the saiga and their numbers. Fortunately, the construction of the canal was stopped and only 80 km of very steep slopes of part of the canal remind us of this groundless project (Lushchekina and Struchkov 2001).

The development of irrigated agriculture in these regions, coupled with the building of artificial waterways and reservoirs to "improve" the quality of natural pastures in the drier central and southern parts of Kalmykia, affected its saiga population in a number of ways. Apart from depriving it of important habitats, irrigation facilities reduced its usual summer mobility by creating additional sources of drinking water. And since the animals came to stay there in large numbers, the nearby pastures were increasingly overgrazed. Furthermore, an expanding network of water distribution channels (which eventually reached a total of over 1,300 km) vastly impaired their seasonal migrations. Built without paying heed to the animal's existence, the irrigation trenches are known to have caused heavy casualties among the saiga, most of which occurred among the newborns accompanying their mothers on the way from the birth sites. In May of 1977, for example, over 14,000 saiga (most of them lambs of 3–10 days old) were found dead along a 5 km-long portion of an irrigation trench in central Kalmykia, having failed to make it through the water that was being pumped into its bed while they tried to cross it. As the irrigation network was growing longer and denser, so was the network of transportation routes. In 1960, the total length of paved roads in Kalmykia was just about 100 km; by 1986, it was 1,604 km. Like canals, the roads increasingly hampered the animal's migrations and became a powerful eliminative factor.

12.3.2 Effects of Changes in Livestock Keeping

On the eve of the Great October Revolution, sheep comprised about two-thirds of the entire livestock population in the North-West Pre-Caspian region, followed by cattle (~20%) and horses (~13%). During the 1920s, the sheep section grew up to 74%, while those of cattle and horses dropped, respectively, to 16% and 7.5%. The collectivization drive of the 1930s furthered this tendency a little bit, by pushing sheep up, and horses down, a few percent more. But it was during the 1960s that Kalmykia was subjected to the most dramatic changes in the structure of its livestock population ever since the advent of the Soviet power: in just one decade, the percentage of sheep reached 85 or more, while the fraction of horses sunk below the level of 1%. There was an unprecedented explosion of sheep in the area. In the early 1960s, their total number was somewhere about two million head, over two times more than it had been one decade before. This soon led to forage deficits, which provoked haphazard attempts to "improve" the natural pastures by turning them into fields to cultivate more fodder. During the 1960s, Kalmykia saw over 150,000 ha of its pastureland plowed under in pursuit of this goal. Of course, it resulted in the fragmentation of the saiga range, competition for forage with livestock and in isolation of some sub-populations.

In the early 1970s, under state controlled plan-based economic policies, sheep numbers more than doubled, and stayed at over 800,000 heads for about 20 years. During these years, the bulk of the sheep industry was represented by large agricultural enterprises, which overexploited the grasslands. Grazing intensified up to a point where some pastures were grazed year-round. The intensive grazing caused widespread wind erosion and vegetation degradation (Zonn 1995). Due to overgrazing, southern Russia was called "Europe's first anthropogenic desert" in the mid-1990s (Saiko and Zonn 1997). It is believed that the degradation of pastures has been caused by the large number of Merino sheep which have some morphological and behavioral features unsuitable for grazing on poor dryland sand pastures all year around. Many traditional breeds of livestock used in the territory of the Republic of Kalmykia before the Second World War have disappeared during the mass evacuation of Kalmyks to Siberia in 1943. Currently there are attempts, supported by the Government of the Republic of Kalmykia, to restore traditional animal husbandry and some original mutton-wool fat tail sheep previously used by Kalmyks have been imported from Mongolia and China for this purpose.

However, following the breakdown of the USSR in 1991 livestock populations dropped by almost an order of magnitude and remained low until around 2000. Large collective and state-owned farms were no longer subsidized after 1991, resulting in broad-scale de-collectivization and abandonment. And drastic socioeconomic changes often lead to changes in disturbance agents. In arid grassland this can be the substitution of grazing by fire. And this is precisely what happened in the North-West Pre-Caspian region in later years.

12.3.3 *Effects of Fires*

Fires shape vegetation structure and composition, and may result in ecosystem degradation, hydrologic changes, soil disturbance, and shrub encroachment. The restoration, conservation and management of arid grasslands thus require solid information on their fire regime and its changes over time.

The livestock decline in Kalmykia after 1991 allowed the vegetation to recover and led to an increase of grassy fuels and ultimately in an increase in grassland fires. A study using remote sensing provides estimates of the trend in burned area for the arid grasslands of the North-West Pre-Caspian region. Its main finding (Dubinin et al. 2010) was a dramatic increase in burned areas starting in the late 1990s. The area burned each year jumped from almost no fires in the 1980s to large-scale burning, covering up to 20% of study area in a single year after the mid-1990s. Since 1998, on average 1,381 km^2 burned per year (9% of the study area). At this rate, the entire area will burn every 11 years. Fires occurred almost exclusively during the driest season of the year. Unfortunately, no data on lightning occurrences and dry thunderstorms exist for the area, but our field experience suggests that they are rather rare. Thus we assume that most of the fires are human-caused, resulting from carelessness (e.g., widespread use of old machinery, smoking, etc.), transportation, mainly to and from local herding enterprises, hunting activities, including techniques of illegal poaching of the saiga antelope.

12.3.4 Poaching and Hunting

Since the early 1980s, saiga populations have suffered from illegal poaching and trade. The demographic effect of periodic summer droughts, occasional severe winters, the spread of some acquired diseases, and pressure of predators, have been magnified by heavy hunting activity with dramatic results for the majority of herds. Hunting and particularly poaching were connected with a demand for saiga horns for the Chinese traditional medicine market. Such trade through "open" frontiers was a source of hard currency for a number of people involved in this illegal business (Sokolov et al. 1991). More recently, in Russia, against a background of drastic declines in livestock, saigas have been targeted for meat consumption and this added to the overall pressure on their populations. In the 1990s, during the period of increased poaching, the calculated number of annual losses in the saiga population amounted to approximately 20,000–25,000 individuals (Bukreeva 2002). Because of selective shooting of saiga males their numbers were reduced from 16.7–30% in favorable years till to 6.4–8.2% in 1992–1993, and from 12.3% to 0.9% in 1995–2000. As a result the sex-age structure of the saiga population was essentially altered and its reproduction became reduced.

12.3.5 Consequences

Under all these anthropogenic impacts the North-West Pre-Caspian saiga population has changed to a semi-settled way of life and in recent years practically all year-round they live within the limits of strictly protected areas – the Biosphere Reserve "Chernye Zemli" (Republic of Kalmykia) and the Sanctuary "Stepnoi" (Astrakhanskaya oblast). Our observations and interviews with local people and staff of different conservation organizations showed radical changes in the migrations of saigas in their territory of the North-West Pre-Caspian region and a new pattern of concentration at more or less safe, strictly protected areas. The herds make small movements which have a circular character and are difficult to call real migrations. Very often poaching and wild fires push such nomadic roamings from place to place. In August 2002 it was the first time during the recent 10 years that several herds (in total more than 5,000 animals) migrated over a long distance (about 150 km) from the southeast to the north. Return migration of these herds happened in October 2002 but unfortunately, due to poaching, they lost very many adult males. This has altered sex-age structure of the saiga population and affected its reproduction.

12.4 Rutting Dynamics of the Saiga

For an estimation of the present condition of the saiga population, that is important to give strength to a strategy of conservation and sustainable use of the resources of this valuable game species, we need solid and detailed scientific data, including those on breeding processes during different periods. Using literature data (Bannikov et al. 1961; Tsapliuk 1966; Bliznyuk 1982; Zhirnov 1982; Sidorov and Bukreeva 2007) and our own observations (2003–2008) we analyse long-term changes in the periods and places of mass rutting of saigas. In this analysis we extensively used various kinds of maps (vegetation, soil cover, density of roads and others) and we paid special attention to the landscape structure of the North-West Pre-Caspian region (Fig. 12.1).

At the end of the 1950s, the ruts of saigas in the North-West Pre-Caspian region were observed practically on all territories of the Chernye Zemli (Fig. 12.2, Photos 12.2 and 12.3). In the 1970s, due to the anthropogenous transformation of the Sarpa lowland and the consequent increase in unrest there, saiga ruts had shifted to separated, not connected sites in the eastern, southern and southwest parts of the Chernye Zemli (Bukreeva 2002; Bliznyuk 2009). In the 1990s, during the rut, saigas began to concentrate in one, rather large aggregation, in a southeastern part of the Chernye Zemli, and during the last decade ruts of saigas were observed in the same territory – where the saiga also lambed – i.e. within the limits of the Biosphere reserve "Chernye Zemli" and adjacent to it the Sanctuary "Stepnoi". Our analysis of the monthly averages of the meteorological stations of Yusta and Komsomolsky (Fig. 12.3), located in the distribution area of the saigas, showed that the amount of precipitation in December in the Chernozemelsky district in the 1990s (on average 11.8 ± 7.2 mm) has decreased compared to the 1950s (on average 24.6 ± 9.5, Mann–Whitney $U = 14.0$, $P = 0.01$). During the same period the precipitation seems to have increased in the Yustinsky district (on average 16.6 ± 11.1 mm in the 1950s against 21.2 ± 12.9 mm in the 1990s) though here the result is statistically not significant (Mann–Whitney $U = 36.5$, $P = 0.31$). These changes, in our opinion, can have led to displacement of places of the saiga's rut from the north (Yustinsky) to more southern areas of the Chernye Zemli where the proportion of snow in the precipitation is rather insignificant. Within 2001–2007 the amount of precipitation in December in the Yustinsky (23.2 ± 16.7 mm) and Chernozemelsky (22.7 ± 6.9 mm) districts was approximately similar. Nevertheless, the rutting places of the saiga, remained concentrated in the southern areas of the Chernye Zemli and this is, most likely, due to the absence of unrest and the availability of suitable habitats.

Unlike the location of the rut, the periods of the rut did not undergo essential changes, and if some changes occurred in particular years it is possible to explain these by weather conditions. In 1957 mass mating was observed from December 16–17 till December 24–26 and lasted 5–7 days. In harems there were from 2–3 up to 25–26 females, and the harems occurred at 100–150 m from each other (Bannikov et al. 1961). In 1959 mass mating took place later – from December 21–22 to December 28–30 (Zhirnov 1982). Bliznyuk (1982) reported that in the period 1972–1977 mass rut came to an end by December 20–25, and this means that it occurred

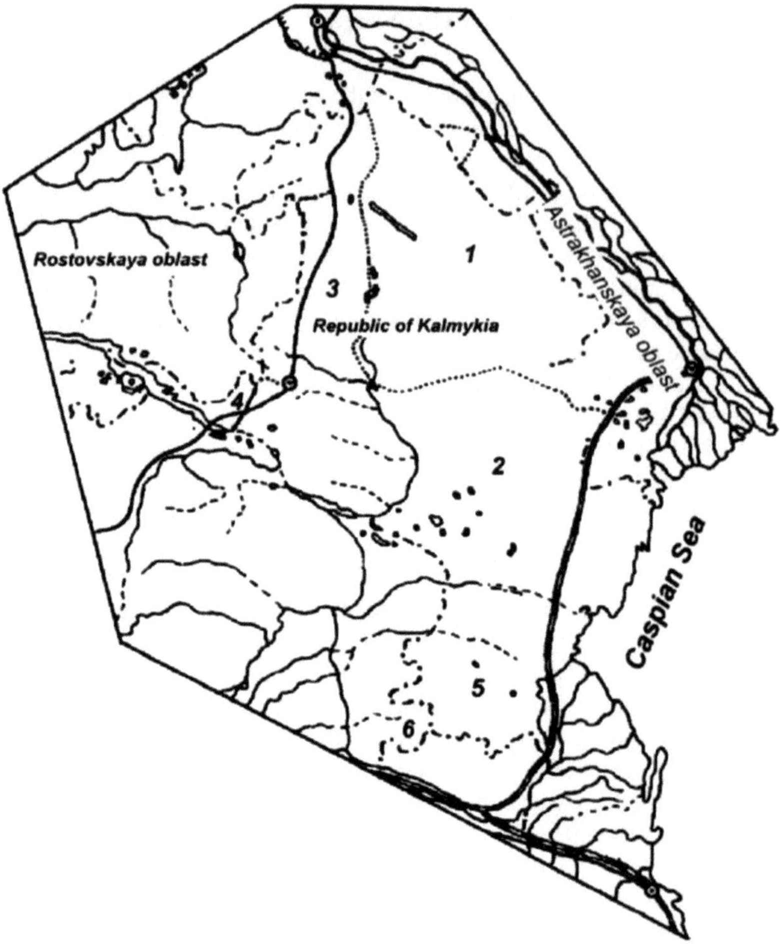

Fig. 12.1 Landscape-geographical districts of the Republic of Kalmykia and adjacent territories (After Zonn 1995). Northern: clay-solonchakous; *2* Southern: sandy-loamy ("Chernye Zemli"); *3* Pre-Ergenyisky: light chestnut – solonetz – loamy; *4* Western: agricultural (chernozem); *5* North-Dagestansky: sandy-loamy; *6* Western: plain-loamy, agriculture and pasturable

at approximately the same time as in the end of the 1950s (the second decade of December). 1959 was an exception when, because of winter harvesting, the rut of saigas shifted to the end of December (Bannikov et al. 1961). In the 1990s the rut was also observed in the second decade of December, except for 1995 and 1996 when mass mating was observed in the beginning of December – from December 5–10 and from December 8–14, respectively. In 1998 mass mating happened later – from December 28 till January 8 (Bukreeva 2002). During the last 8 years the rut occurs at approximately the same time – in the second decade of December (from 12 to 25), though 2006 was again an exception, as the rut was earlier – from December 9–20.

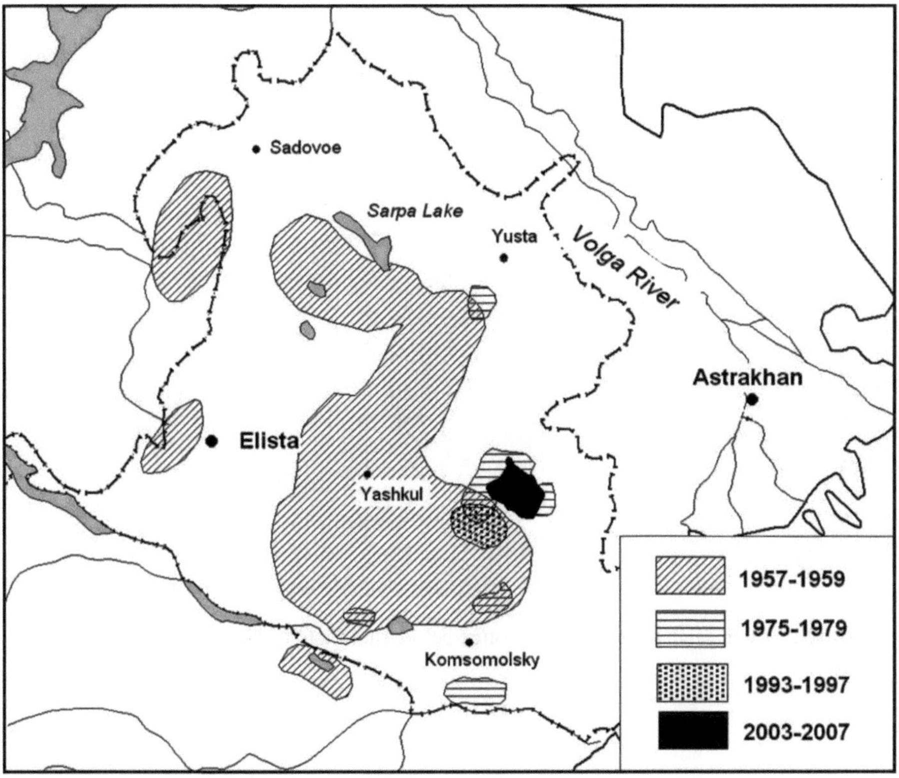

Fig. 12.2 Rut sites of the saiga

12.5 Fertility of Saiga Females

As for the rut, our data show that, since 2003, lambing of the saigas also occurs in one aggregation located in the territory of the Biosphere Reserve "Chernye Zemli" and adjacent to the Sanctuary "Stepnoi", that are quiet and do not make the saigas nervous and that harbour good habitat conditions for them. The birth of the young usually occurs at daily average temperatures around +15.9°C (according to the Utta meteorological station). The date of beginning of mass lambing directly depends on the dates of the rut ($r=0.83$, $p < 0.001$) and is less dependent of the weather conditions during that period.

The general fertility of the females in period 1999–2003 and their participation in breeding have considerably declined. Dryness of adult and young females increased: in 1999 10% and 48.0%, respectively, in 2000 8.8% and 50.0%, in 2001 76.1% and 97.2%, in 2002 22.2% and 66.7%, in 2003 53.8% and 77.8% (Bannikov et al. 1961; Bliznyuk 2009, 1982; Zhirnov 1982; Zhirnov and Maksimuk 1998; Bukreeva 2002). The reduction in the proportion of pregnant females is correlated with the low number of males in the population during the rut. For example, in 2000 the proportion of

Photo 12.2 Adult saiga male at winter pasture before rut (Photo by Nadezhda Arylova)

Photo 12.3 Saiga females at spring pasture before lambing (Photo by Valery Maleev)

males during the rut was only 0.9%, and in 2002 about 0.6% (Sidorov and Bukreeva 2007). The past few years the proportion of adult males in the population increased – up to 12.5% in 2006–2007 but the proportion of pregnant females continues to remain low. Analysis suggested that the low fertility of saigas in 2001 and 2003 may be connected with droughty conditions during the whole of 2000 and in the summer months of 2002, which has led to deterioration of forage resources. Thus, the decrease in fertility of saiga females may result both from insufficient numbers of males in the population and rather adverse weather conditions.

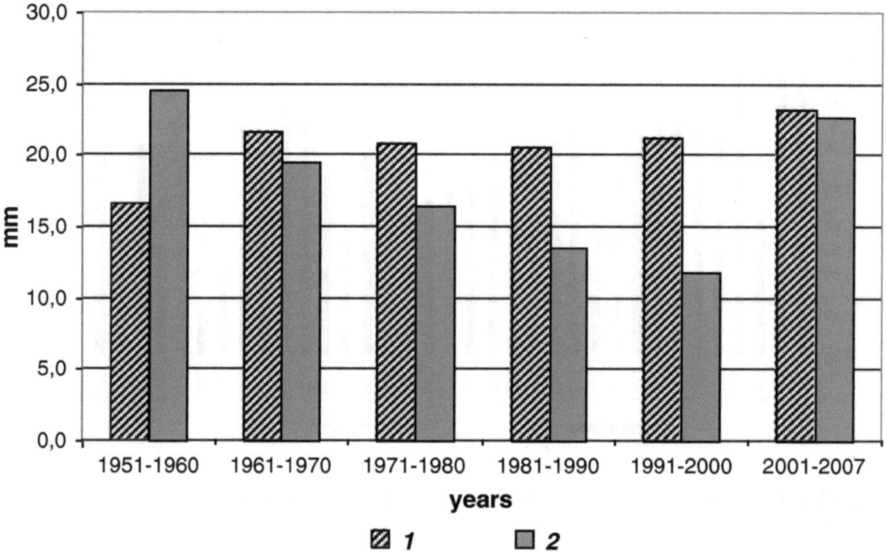

Fig. 12.3 Average monthly precipitation in December 1951–2007 (*1* meteostation Yusta, *2* meteostation Komsomolsky)

For last 10 years the production of young animals (in the age of up to 4 months) per female has almost halved in comparison with the 1950s (0.7 and 1.2, respectively) (Fig. 12.4), and death rates of the young in their first days of life in the period 1999–2008 was on average 11.1%. That is quite comparable to values of the end of the 1950s – 9.5%. The mortality rates of the young in the first days after birth reflect the viability of the newborns (if it did not result from extreme weather conditions during the lambing) (Bliznyuk 2009; Kühl et al. 2009b). Data obtained by us indirectly testify to normal current embryogenesis in the saiga females. Only during some years (1994, 1998–2000 and 2007) a significant percentage of the mortality of newborn saigas (on average 16.4%) have been caused by adverse weather conditions (a cold snap and rains) during the lambing (Bliznyuk 2009). Average weight of newborn saigas during the last years (1998–2008) was 3.49 ± 0.22 kg which is significantly more than during the earlier period (1957–1997: 3.19 ± 0.28 kg) (Mann–Whitney U = 35.5, P = 0.01). The weight of newborn males in the period 1998–2008 (3.6 ± 0.2 kg) is also significantly more (Mann–Whitney U = 6, P = 0.03) than before 1997 (3.26 ± 0.3 kg). Weights of females for these periods are statistically not different (3.39 ± 0.17 kg and 3.11 ± 0.26 kg, respectively) (Mann–Whitney U = 7.5, P = 0.053) (Fig. 12.5). Minimal average weights of newborn males and females (3.02 kg and 2.9 kg, respectively) were observed in 1994 (Bliznyuk 2009; Bukreeva 2002), and maximal ones in 2008 (3.9 ± 0.4 kg (n = 89) and 3.7 ± 0.4 kg (n = 67), respectively).

Since the course of embryonal development of the future offspring is defined by the feeding conditions of pregnant saiga females as from the second month after

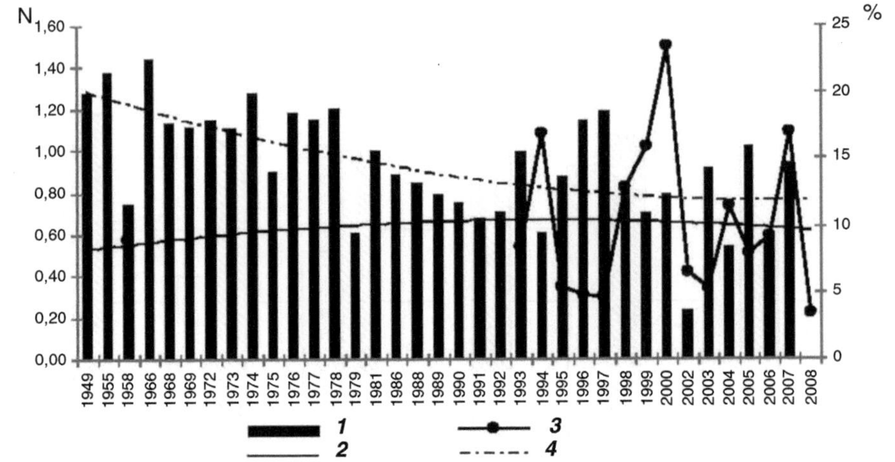

Fig. 12.4 Lamb crop per female in the population (*1*, *2*) and mortality of the newborn animals during their first days of life (*3*, *4*)

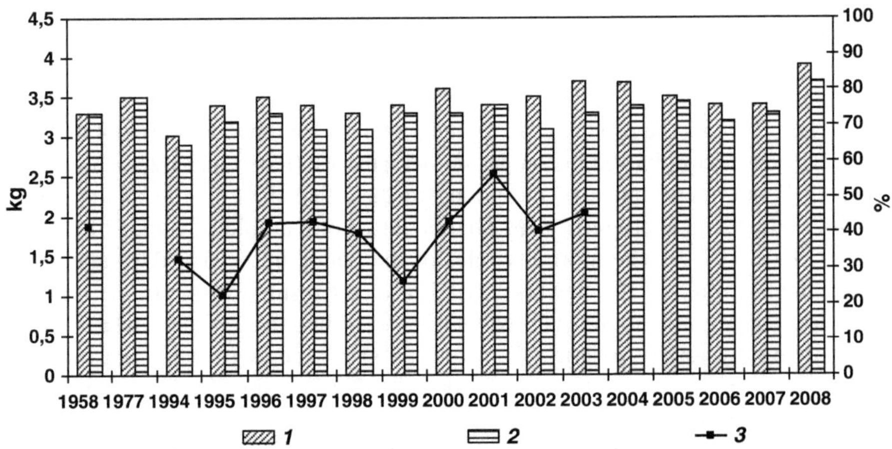

Fig. 12.5 Weight of newborn saiga (*1*: males, *2*: females) and proportion of females with twin embryos (*3*) in the period 1958–2008

fertilization (Abaturov 2007; Bliznyuk 2009), our results indirectly testify that in the zone of saiga habitation the provision of forage from the middle of the last century has not undergone essential changes. Published data on fertility show that the percentage of females with double embryos (without taking dry females into account) at the end of the 1950s and in the period 1993–2003 was approximately identical: 39.1 ± 9.3% (see Fig. 12.5). Thus, the stability in weight of the newborn saigas, most likely, is not related to an increase in the percentage singles (which obviously have more weight than lambs from twins).

12.6 Sex Rates

From 1998 to the present males among newborns are significantly more frequent (on average 52.7%) than in the period 1957–1997 (on average 49.4%) (t_{27}=−2, 4, n_1=18, n_2=11, P=0.024). Earlier it was supposed, that some prevalence of males among newborns is characteristic for periods of high population numbers and that they are less when population numbers are low (Bannikov et al. 1961; Zhirnov 1982). Our observations of the last decade have not confirmed this hypothesis. In the period 1999–2008 males prevailed among newborn saigas, while population numbers still remain at a low level. It is possible in this species numbers are regulated by intrapopulation mechanisms and that a shortage of males in the population leads to a higher frequency of them among the newborn. A number of publications (Bannikov et al. 1961; Zhirnov 1982; Bliznyuk 1982; Zhirnov et al. 1998; Bukreeva 2002) and our own observations have shown, that from the 1970s till now there is a serious disbalance in the sex ratio of adult animals in the saiga population (Fig. 12.6). Up to the 1990s the proportion of adult males in the population remained rather high (about 18–19%), however, the second half of the 1990s is characterized by a significant decrease in males (up to 10% on average) though the proportion of newborn males essentially did not vary. At the moment of the beginning of the rut the share of adult males, as a result of selective hunting at the end of the 1990s, considerably decreased (on average up to 4%) and in 2000 the share of males was only 0.9% (Bukreeva 2002). During the last years females prevail in saiga herds, also in young animals, but the share of adult males in the population has started to increase gradually: up to 12.5% over 2006–2007 .

Thus, our analysis shows that, while saiga numbers decreased within the limits of the North-West Pre-Caspian region, the proportion of adult males decreased catastrophically and there was an increase of dryness and decrease in the general

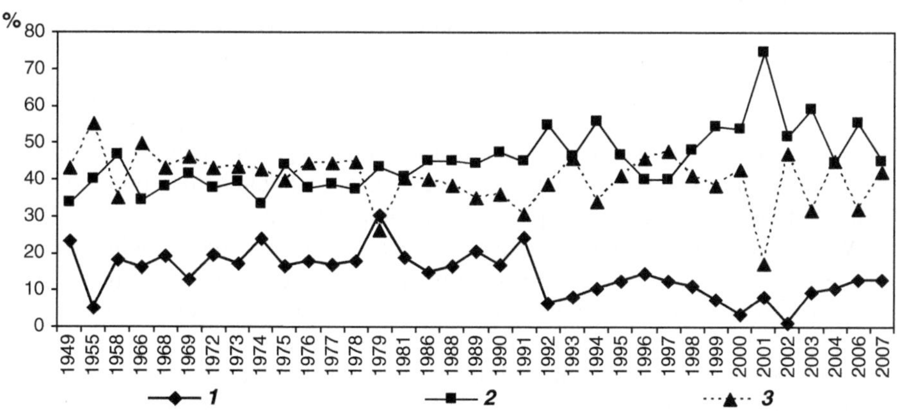

Fig. 12.6 Dynamics of the sex-age structure of the saiga population in the North-West Pre-Caspian region (August, 1949–2007). *1*: proportion of males; *2*: proportion of females; *3*: proportion of youngsters born per year, both sexes

fertility of the females. The additional increase in a number of dry females in separate years seems to be related to extremely dry conditions during the vegetative period. However, even under these conditions, breeding females maintain a high fertility. Nevertheless, the production of young animals over all females in the population halved in comparison with more favorable periods, but it does not result from higher mortality of the young during the first days of their life. Young mortality stayed within the limits of long-term fluctuations and even decreased during the last years. The weight of newborn saigas practically does not differ (and presently is even higher) from that at the end of the 1950s when the numbers of saigas were high.

12.7 Current Constraints of Saiga Conservation

In order to save the saiga's genepool the Government of the Republic of Kalmykia has given the Center for Wild Animals of Kalmykia 800 ha of land in the Yashkul district (near the Ermeli settlement) with good natural pastures, for the construction of enclosures and to start a program of saiga captivity breeding. The Ermeli breeding center now has a good number of saigas of different ages, several enclosures have been fenced, a laboratory has been build and a permanent water and energy supply (including from renewable sources) has been installed. Still a lot more land belonging to this Center is available for grazing and haymaking. Since nearby there are some villages promoting the traditional husbandry and sustainable use of saigas ideally a model farm could be established, which might produce (following to traditional technologies) milk products, wool, meat and also handicrafts as an additional income for rural people within the saiga range. Numerous ecotourists and groups of students/school children visiting the Center would be good customers for the above products.

In cooperation with the Center, and using its enclosures, it will be possible to conduct some experiments on the rational use of pastures capacity by livestock and saigas to prevent future desertification and wild fires as well. Demonstrating a successful business model at the farm should attract rural people from neighboring areas and provide the know-how to apply in their own lives to improve their living standards. Such combined experience of saiga breeding and traditional sheep and cattle husbandry with intensive use of fenced pastures should help to promote saiga conservation and to provide people with additional income.

Within the frame of the international project supported by the Darwin Initiative, a team of specialists interviewed people living in villages within the saiga's range. Results showed that local people have very little knowledge about the role of the saiga in the steppe ecosystems and their conservation status. Seventy-five percent of the respondents stated that unemployment was the primary factor driving saiga exploitation. The dramatic decline in saiga numbers due to poaching appeared to be directly linked to the collapse of rural economies within the Republic. Among local people, 41% admitted to poaching saiga; of this group, 65% said that they hunted saiga for their own meat consumption and 35% hunted to sell the horns.

In connection with the developed situation, in 1998 the Order of the President of Republic Kalmykia "On strengthening protection of the saiga, prohibition of trade and sports hunting up to 2001" has been published. By a Decree of the President of Republic Kalmykia (# 133 on 25.07.2001) "On emergency measures on protection and conservation of a Kalmyk population of saiga" this interdiction has been prolonged up to 2010, and in May, 2009 the new Decision (# 155) on banning commercial and sports hunting on saiga up to 2013 has been adopted. However these legislative measures yet have not brought desirable results: among wild ungulates in Kalmykia illegal hunting on saiga continues to be at the first place (Namrova 2000). Social and economic surveys of settlements in the region of Chernye Zemli has shown (Medzhidov et al. 2005; Kühl et al. 2009a) that the very low standard of living of the population in rural areas in Kalmykia and absence of real sources of earnings force people to engage in poaching, because on the "black" market horns and meat of saiga are in great demand. Besides survey results have shown that the local population is badly informed and does not know that hunting on saiga is forbidden (only 46% of people heard about it) (Medzhidov et al. 2005; Kühl et al. 2009a). In our opinion, these facts only show a limited part of the practice of poaching and the damage thereof, unfortunately, is much larger.

Besides the impact of poaching on the saiga population, it is necessary to mention the considerable rise in the numbers of wolves. This now has become the most important factor in the regulation of saiga numbers in the North-West Pre-Caspian region. Taking into account the very low saiga numbers, if wolves take even only 5–6 saigas per day this causes a significant damage to the population and can lead to the full disappearance of this species from the territory of Russia.

Accordingly, the development and application of new radical and effectual measures for its saving are required. First of all, some strengthening anti-poaching measures and not only by means of law enforcement bodies, but also with help of the local population should be introduced. Keeping in mind the present status of the saiga population in the North-West Pre-Caspian region, it is very necessary to regularly monitor the situation. To this end it is necessary to use new methods to avoid any additional negative impacts on the animals. Quantitative estimation of numbers of animals is a major need for an adequate management of the population's status. The latest aerial counts of saiga numbers in the North-West Pre-Caspian region were conducted 5 years ago. Since then numbers and also the sex-age structure of the population have been determined by estimations of experts during ground observations on the pedestrian and automobile routes with an extrapolation towards uninspected territories. Considering the shyness of the saiga, its gregarious instinct, its fast moving over long distances, and also its non-uniformity of distribution over the territory, it is difficult to guarantee precise observation for a long time and from close distance. Therefore, expert estimations are subjective and contain serious errors. Besides, since the 1970s aerial counts of the saiga are being conducted during the lambing season (Maksimuk et al. 1977) when about 90% of the saiga are concentrated in a limited territory. The low-flying aircraft or helicopter causes huge damage to the population, making the animals nervous in such important period of their life. This cause flights of animals from the lambing grounds, also by pregnant females, and this adversely affects their condition and future fertility (frequently females gave birth 'on the run').

For reliable counts we believe it necessary to implement the use of infrared shooting which is possible even at night. It has already been shown that it is the most reliable and safe method for wild life counts (Chernook 2009). To assess the patterns of saiga migration during different seasons it is possible to tag animals with satellite collars (Dubinin et al. 2009). In order to investigate saiga nutrition it is possible to use cuticular-corpological (Larionov 2008) and phytolithic analyses of their excrements (Karimova et al. 2010). The hormonal analysis of their excrements can be used also for the determination of their fecundity and the proportion of females participating in the reproduction, instead of shooting pregnant females (Arylova et al. 2006). Reproductive success can be estimated by observations along transects crossing the lambing grounds. By using such methods of inventory, developed and used by us in the period 2003–2008 during implementation of the Darwin Initiative project, we obtained valuable data suitable for their further statistical processing.

Monitoring the saiga population in this way does not only provide the necessary scientifical data on the conditions of the population but also to create conditions for growth of its numbers. As soon as number of animals has reached a level, at which a diminishing of their numbers does no longer threaten the population's existence, it will be possible to introduce the sustainable use of these valuable resources. Basing on experience of other countries (Pirot et al. 2000; Tishkov and Petrova 2002), we consider that to this purpose it is the extremely important to start the introduction of the ecosystem approach without further delay in the North-West Pre-Caspian region. This should be combined with a territorial form of wildlife management, first of all in existing protected areas. Presently, various protected areas have been created within the North-West Pre-Caspian region (The compiled list of specially protected natural territories of the Russian Federation 2001). Protecting different habitats within the Chernye Zemli Biosphere Reserve established in 1990 has been an important step in conserving the saiga population. Three sanctuaries with lower levels of protection were also created, restricting land use and access of motorized vehicles. These lands provide refuges to the saiga during the lambing period in May, when more than 70% of the North-West Pre-Caspian population gathers there to give birth. And here the animals are relatively safe from poachers. Yet the existing protected areas are not enough to save the saiga. Not only that they cover too small a portion of the animal's range. But they also are fixed in space, while the herds of animals are not. Considering the character of the saiga it is expedient to connect these protected areas into an ecological network that will provide safety for saiga herds even during their migrations (Lushchekina et al. 2005). It is absolutely necessary to establish ecological corridors between protected areas. In some cases it will be necessary to restore the previous capacity of pastures within the saiga range and make safe crossings of some obstacles (cannels, roads, pipelines, etc.). In order to create such an ecological network based on the requirements of the ecosystem approach and the preservation of the saiga population additional legislative decisions and coordinated actions of the nature protection organizations within the North-West Pre-Caspian region are required. Without these conditions the stated tasks will not be realized. Uncoordinated protection measures, as practised by various organizations and departments in the Republic of Kalmykia until recently, cannot save the saiga. In order to successfully implement the ecosystem approach, it is necessary that management should be decentralized to a integration level that will allow to provide the necessary balance of the various interests. Considering that many ecosystems earlier were used by the saiga underwent essential transformations during the agricultural development, priority should be given to habitat restoration of places used for breeding and seasonal migrations of the saiga. Now the responsibility for the protection and monitoring of the saiga population is transferred to the Ministry of Natural Resources, Protection of the Environment and Development of Energetics of the Republic of Kalmykia. For the preservation and restoration of saiga numbers it is extremely important to achieve a close interaction of this Ministry with other republican departments and with corresponding federal territorial institutions. Another major condition for the successful implementation of all actions is the support of the local population, which should be directly interested in the preservation of the saiga. While expanding the programs of education for sustainable development within the framework of the United Nations Decade (2005–2014), efforts of local administrations and education systems should be focused in such way that a wide participation of the public and youth organizations is guaranteed.

12.8 Outlook

Experience from former years, when legal saiga harvesting brought essential income to the State budget, convince us that now, with the system of a hunting economy in place, the creation of a special Saiga Watch Service is necessary. Powers of this Service should be stipulated by the Law on the Animal World or in special acts, and the Service itself should have the resources necessary for the implementation of a complex of actions. In order to work successfully the Service should maintain close contacts with various ministries and with other organizations responsible for nature conservation. Applying an ecosystem approach for conservation of biodiversity with widening of the participation of the local population we should learn from the experience in other countries, also with respect to different forms of game ranching on grounds which are not used in agriculture. For conservation, restoration and sustainable use of the saiga a project as CAMPFIRE project, begun in Zimbabwe in 1989 is obviously possible for us (Neronov et al. 2005). This project provided transfer of the rights to using bioresources from federal bodies directly to rural communities and covered a territory with more than 250,000 residents. The latest publications on this project state that in the area not only the infrastructure of the rural settlements, planning of land tenure and well-being of the local residents have been improved, but also an increase in the number of game animals was observed. Successful introduction of the ecosystem approach in the North-West Pre-Caspian region could become a concrete contribution to the CBD and directly to the National strategy on conservation of biodiversity (National Strategy… 2001). Owing to the many different values of the saiga, the strongly increased interest in this unique species from the side of the international organizations and local authorities and potentials of applying an ecosystem approach for the conservation of fragile arid ecosystems in the North-West Pre-Caspian region, UNDP/GEF in May 2010 launched a project entitled "Improving the system and mechanisms for management of protected areas in the steppe biome of Russia" in which the Chernye Zemli ecoregion is selected as one of the pilot sites and the saiga is a key species of this project. We hope that this project, in combination with all other efforts mentioned above, will help to restore the numbers of the saiga in the North-West Pre-Caspian region to the higher level observed some years ago and that it will permit to start a sustainable use of its resources.

Acknowledgments During our long-term studies we had help from many colleagues and local people. We are very thankful to them and also mentioned with appreciation the generous support of different funds: the Darwin Initiative "Using saiga antelope conservation to improve rural livelihoods", INTAS "Reproductive ecology of the Critically Endangered saiga antelope", UK Government's DEFRA Small Environmental Projects Scheme, "Rotating cows as a tool for saiga antelope conservation", the Darwin Initiative "Evaluating approaches to public engagement in saiga conservation", the UNESCO MAB/Secretariat, the Fundamental Research Programme of the Department of Biology Sciences, Russian Academy of Sciences, "Russia's Biological Resources: Conditions assessment and Monitoring Fundamentals" (Project II.10).

References

Abaturov BD (2007) Saiga population in Russia and conservation problems. Vestnik Russ Acad Sci 77:785–793 (in Russian)

Arylova NYu, Voznesenskaja AE, Lushchekina AA, Voznesenskaja VV (2006) A new method of non-invasive monitoring of the reproductive status of saiga as alternative nature protection technology./ In: Materials of 10th Puschino conference of young scientists "Biology as a science of 21 century", Puschino, 252, (in Russian)

Bannikov AG, Zhirnov LV, Lebedeva LS, Fandeev AA (1961) The saiga biology. Published by Agriculture literature, Moscow, 336 pp (in Russian)

Bliznyuk AI (1982) Ecology and rational utilization of the saiga population in the North-West of the Pre-Caspian Region. Ph.D. thesis in biology, Moscow, 129 pp, (in Russian)

Bliznyuk AI (2009) Saiga of the Kalmyk population. Dzhangar Publishing House, Elista, 544 pp, (in Russian)

Bol'shakov VN, Lushchekina AA, Neronov VM (2009) Biodiversity conservation: from ecosystem to the ecosystem approach. Russ J Ecol 40:73–80

Bukreeva OM (2002) European saiga (*Saiga tatarica* L.) population and factors determining its status. Ph.D. thesis in biology, Moscow, 22 pp, (in Russian)

Chernook VI (2009) The application of thermal imaging to saiga counts. Saiga News 8:15–16

Convention on Biological Diversity. Text and Annexes (1995) UNEP/CBD. Geneva, 34 pp, (in Russian)

Danilkin AA (2005) Hollow-horned ruminants (Bovidae). KMK Scientific Press, Moscow, 550 pp, (in Russian)

Dubinin MYu, Lushchekina AA, Radeloloff FK (2009) Experience of satellite telemetry of saiga in the North-West Pre-Caspian region. Saiga News 8:7–8

Dubinin M, Potapov P, Lushchekina A, Radeloff VC (2010) Reconstructing long time series of burned areas in arid grasslands of southern Russia by satellite remote sensing. Remote Sens Environ 114:1638–1648

Karimova TYu, Kiseleva NK, Arylova NYu (2010) Study of the saiga antelope's spring diet with phytolith analysis. Herald of Game Management, Moscow 7:29–34 (in Russian)

Kühl A, Balinova N, Bykova E, Arylov Yu, Esipov A, Lushchekina A, Milner-Gulland EJ (2009a) The role of saiga poaching in rural communities: linkages between attitudes, socio-economic circumstances and behaviour. Biol Conserv 142:1442–1449

Kühl A, Mysterud A, Grachev IuA, Bekenov AB, Ubushaev BS, Lushchekina AA, Milner-Gulland EJ (2009b) Monitoring population productivity in the saiga antelope. Anim Conserv 12:355–363

Larionov KO (2008) A nutrition and provision of fodder for saiga (*Saiga tatarica*) depending on features of vegetation on a pasture. Dissertation Cand. Biol. Sci., Moscow, 129 pp, (in Russian)

Lushchekina A, Struchkov A (2001) The saiga antelope in Europe: once again on the brink. Open Country 3:11–24

Lushchekina AA, Neronov VM, Badmaev VS, Khludnev AV (2005) The organization of territorial protection of nature and prospect of preservation of saiga population on the right bank of Volga. Volga Region Ecol Mag, Saratov 1:80–85 (in Russian)

Maksimuk AV, Zuev EYa, Abaturov BD, Zhirnov LV (1977) The report on development of ecologo-economic bases of protection and rational use of the saiga in RSFSR for 1977, Hunting Management Department of Ministry of Agriculture, Moscow, 72 pp, (in Russian)

Medzhidov RA, Kaminov YuB, Obgenova OB (2005) Local population and decrease of numbers and opportunities of preservation of saiga in Kalmykia. Steppe Bull 18:18–21 (in Russian)

Milner-Gulland EJ, Kholodova MV, Bekenov A, Bukreeva OM, Grachev Yu, Lushchekina AA, Amgalan L (2001) Dramatic declines in saiga antelope populations: saiga antelope declines. Oryx 35:340–346

Milner-Gulland EJ, Bukreeva OM, Coulson T, Kholodova MV, Lushchekina AA, Bekenov AB, Grachev YuA (2003) Reproductive collapse in saiga antelope harems. Nature 422:135

Namrova VG (2000) Criminally-legal problems in the struggle against illegal hunting in the Republic of Kalmykia. Bull Kalmyk State Univ, Elista 1:160–164 (in Russian)

National Strategy of Biodiversity Conservation of Russia (2001) Moscow: Rus Acad Sci Min Nat Resources of RF, 63 pp, (in Russian)

Neronov VM, Lushchekina AA (2009) The saiga antelope in dryland ecosystems of Eurasia: urgent actions to ensure its sustainable future. In: Proceedings of the 29th international congress of game biologists, Moscow, 17–22 Aug 2009, Part 2, pp 14–21

Neronov VM, Lushchekina AA, Arylov YuN (2005) Ecosystem approach for conservation and restoration of the saiga population in the lower Volga River basin. In: Proceedings of the international conference "Bioresources and biodiversity of the Volga River basin ecosystems", Saratov, 24–28 Apr 2005, pp 164–166, (in Russian)

Pirot J-Y, Meynell PJ, Elder D (2000) Ecosystem management: lessons from around the world. A guide for development and conservation practitioners. IUCN, Gland/Cambridge, 132 pp

Saiko T, Zonn I (1997) Europe's first desert. In: Glantz M, Zonn I (eds) Scientific, environmental and political issues in the circum-Caspian region. Kluwer Academic Publishers, Dordrecht, pp 141–144

Shepherd G (2004) The ecosystem approach: five steps to implementation. IUCN, Gland, 30 pp

Sidorov SV, Bukreeva OM (2007) The saiga population status in the North-West of the Pre-Caspian Region, its development and utilization. Abstracts on Intern. Conf. "Theriofauna of Russia and neighboring territories", Rus Acad Sci Moscow, 452 pp, (in Russian)

Sokolov V, Zhirnov L, Yeremeyev A, Shuvaev Yu, Visyaschev G (1991) A currency fever and saigas. Journal of Hunting and Hunting Economy, Moscow 4:1–3 (in Russian)

The compiled list of specially protected natural territories of the Russian Federation (2001) All-Russian Center of forest resources, Moscow, 452 pp, (in Russian)

Tishkov AA, Petrova TE (2002) Ecosystem approach to conservation of biodiversity on regional and local levels. Transition to sustainable development: global, regional and local levels. Foreign experience and problems of Russia, KMK Scientific Press, Moscow, pp 267–286, (in Russian)

Tsapliuk OE (1966) Sexual rhythms of saigas. In: Proceedings of the Institute of Zoology, Acad. Sci. Kazakh SSR, vol 26, pp 193–211, Almaty, (in Russian)
Wilson DE, Reeder DM (2005) Mammal species of the world: a taxonomic and geographic reference, 3rd edn. John Hopkins University Press, Baltimore, 2142 pp
Zhirnov LV (1982) Restored to life (ecology, protection, utilization of the saiga), Forest Industry Press, Moscow, 224 pp, (in Russian)
Zhirnov LV, Bekenov AB, Grachev YA (1998) Abiotic or weather-climatic factors. In: Sokolov VE, Zhirnov LV (eds) The saiga: phylogeny, taxonomy, ecology, protection and utilization. Publ Rus Acad Sci, Moscow, pp 233–241, (in Russian)
Zhirnov LV, Maksimuk AV (1998) The saiga population size in the 20th century in the North West Pre-Caspian Region. In: Sokolov VE, Zhirnov LV (eds) The saiga: phylogeny, taxonomy, ecology, protection and utilization. Moscow, pp 219–225, (in Russian)
Zonn SV (1995) Desertification of nature resources in Kalmykia during last 70 years and protection measures. In: Biota and environment of Kalmykia, Korkis Scientific Press Elista, Moscow, pp 12–52, (in Russian)

Chapter 13
The Przewalski's Horse and Its Reintroduction in the Steppe of Hustai National Park, Mongolia

Piet Wit, Namkhai Bandi, Inge Bouman, and Marja A.van Staalduinen

Abstract The Przewalski's horse – Takh in Mongolian – is the last surviving genuine wild horse, and a close relative of the domestic horse. The Takh once roamed the steppes of Central Asia and Europe, but since the 1960s has gone extinct from the wild. The Reintroduction Project for the Takh was set up to bring back the species to its Mongolian homeland. A breeding programme was started to build a new population and secure its genetic basis. In Germany and the Netherlands several semi-reserves were established, where the Takh were able to adapt to live in large open territories. The best adapted second-generation Takh were chosen to be released back into the wild. The first transport of 16 Takh took place in 1992. Ultimately, 84 Takh have been re-introduced to Hustai National Park, and they adjusted well. Each year new foals added to the population and in 2009 the population had grown to 260 horses. Wolves are the main predators in Hustai National park.

13.1 Introduction

When the last wild tarpan, a species of the wild horse, died in the wild in Ukraine in 1879, scientists in Europe declared that this was the end of the existence of the wild horse on earth. It was a sensation, however, when in 1881, the Russian scientist

---

P. Wit • I. Bouman
Foundation for the Preservation and Protection of the Przewalski Horse, Klaaswaal, The Netherlands
e-mail: wit@syzygy.nl; frph@antenna.nl

N. Bandi
Hustai National Park, Ulaanbaatar, Mongolia
e-mail: takhi@hustai.mn

M.A. van Staalduinen (✉)
Ecology & Biodiversity, Utrecht University, Utrecht, The Netherlands
e-mail: M.A.vanStaalduinen@uu.nl

M.J.A. Werger and M.A. van Staalduinen (eds.), *Eurasian Steppes. Ecological Problems and Livelihoods in a Changing World*, Plant and Vegetation 6,
DOI 10.1007/978-94-007-3886-7_13,

L.S. Poliakov announced that there were still wild horses to be found in Central Asia (Wit and Bouman 2006). Poliakov had analyzed the hide and skull, which the Russian colonel Nikolai Michailovich Przewalski in 1878 had brought from his expedition in the border area of Kazakhstan, Kyrgizia, China and Mongolia, in Inner Asia. After a detailed study of the specimen and a comparison of the skull and hide with 22 species of *Equidae*, Poliakov came to the conclusion that it was indeed an unknown wild horse. In honour of its discoverer the species was given its official name, *Equus przewalskii* Poliakov 1881.

13.2 Historical Distribution

In prehistoric times there were many different species of wild horses. They had a wide distribution covering the entire Eurasian steppe belt from Europe to eastern Asia (Bouman and Bouman 1994; Wit and Bouman 2006). The earliest graphic accounts of Przewalski-like wild horses stem from about 20,000 years ago, from cave paintings in Italy, western France and northern Spain. This is an indication that Przewalski-like horses might have been rather abundant at that time.

After the last ice age, at around 10,000 BC, the steppes across large parts of Europe and Asia became afforested. Some of the wild horses adapted to these changes, while other horses retreated further into the steppe. In the Mesolithic (10,000–8,000 BC) and Neolithic (8,000–3,000 BC) horses became rare. They had been wiped out in most of southern Europe, surviving in small populations in central Europe and in larger numbers on the eastern European plains. During the third millennium BC, several of these wild horses were domesticated in the steppes east of the Dnjepr.

After the domestication of wild horses and the rise of many equestrian nomadic peoples on the plains of Eurasia, the numbers of wild horses decreased drastically and only a number of marginal populations survived. In eastern Europe the tarpan remained, while Przewalski's horses survived east of the Urals in the steppes of Asia. Persistent hunting by man ultimately caused Przewalski's horses only to survive in the most remote areas of the present border zone of Mongolia and China.

The very first written account of Przewalski's horses originates from Tibet. The monk Bodowa recorded their existence around 900 AD. By medieval times the Takh or Takhi – as the Mongolians call this wild horse – must have been a rarity in Mongolia and in the Chinese provinces of Xinjiang and Gansu. In 1500 an official decree was issued that prohibited hunting of the Takh.

Being very hard to catch, the rare and agile Takh has been an object of reverence to the equestrian Mongolians for a long time. Mongolian Khans gave dignified visitors noble riding horses or sometimes a genuine Takh caught in the wild.

13.2.1 Last Refuge

In the beginning of the twentieth century the Przewalski's horses had retreated from the rich steppe areas to the semi-deserts in the south, their last refuge. Unlike the wild asses, they do not really thrive in semi-deserts and deserts, the steppe being their original habitat. In 1926 hunting Takh in its last remaining area was officially prohibited. But this law was not sufficiently enforced and could not save the species in the wild.

The years following the Second World War saw a drastic decline in numbers of Przewalski's horses in the wild. In the 1960s and 1970s many Mongolian and Russian expeditions were sent out to locate any remaining wild Takh, but to no avail. The very last observation of Takh in the wild was in 1969 by the Mongolian scientist Dovchin at the Gun Tamga well in the Dzungarian Gobi in southwest Mongolia.

The Dzungarian Gobi, a vast stretch of desert and desert-steppe enclosed by mountains, is not the optimum environment for wild horses, as it is situated at the fringe of the horse's natural range (Wit and Bouman 2006). Here, in the period October–March the nomadic herdsmen descended with their livestock from the Altai mountains in the north to the Gobi desert. During the winter their animals grazed on the arid steppes on the lower mountain slopes, where the domestic animals not only competed for food with Przewalski's horses, wild asses and Mongolian gazelles, but also for the few watering places at distances of up to 30 km. Until 1950 it was forbidden to herd livestock in the area south of the Honin-Us source, that lies some 40 km north of the border with China. In the border zone of Mongolia and China competition was mainly with the livestock owned by the border armies as the local nomads were not allowed to enter this militarised zone.

The survival situation for the Takh changed dramatically for the worse when in 1945 and 1947 permission was granted to Kazakhs from China to relocate in southwestern Mongolia. Having no livestock of their own the Kazakhs hunted to satisfy their needs, and unlike the Mongolians, they relish horsemeat. Thus, hunting pressure increased. While out in the wide open steppe with plenty of streams the shy Takh could stay out of reach of hunters, they were unable to avoid ambush at the scarce sources of water that they had to frequent. Moreover, the extremely cold winters of 1945, 1948 and 1956 also took their toll. Millions of cattle died and wildlife in general suffered greatly. For Przewalski's horses, already a rarity, the situation was disastrous. Its numbers shrunk to near extinction, and modern weaponry killed off the last specimens in the wild.

13.3 Reintroduction of Takh

After the discovery of the Takh by colonel Przewalski, between 1898 and 1903 several attempts were made to capture Przewalski's horses and bring them to Europe (Wit and Bouman 2006). However, it turned out to be impossible to catch adult horses: they were too shy and too fast. Only foals could be caught, with many stallions and mares being killed when they tried to defend their young. The transport to Europe was long and strenuous and many foals perished. In the end only a mere 54 horses reached their destination alive. Some of the foals were sold to the owners of large estates such as Baron F.E. von Falz-Fein (who in 1874 had established a large natural reserve at Askania Nova, the Ukraine, on which the Przewalski's horses roamed as in the wild), the Duke of Bedford (Woburn, United Kingdom), and Frans Ernst Blaauw (Gooilust, The Netherlands). Others were sent to zoos in Europe and North America. In the period 1930–1950 a few more Przewalski's horses were captured, but most soon died.

When it became evident that no Przewalski's horses had survived in the wild, the preservation of the entire species became fully dependent on the population in captivity. Back in the early 1900s when the wild horses were first brought into captivity, it became clear that the foals that had been captured had great difficulty adapting to zoo life. Moreover, only a few of the mares produced foals and thus the numbers grew very slowly. The captive population suffered severe blows during the Second World War: just 31 horses survived and it seemed that the species was going to extinction. But then a few horse lovers decided to take action.

In the 1950s Dr Erna Mohr (Hamburg) initiated the International Studbook for Przewalski Horses, that was later continued by Prague Zoo. In 1974, Jan Bouman and Annette Groeneveld used the studbook as their guiding principle when they began to construct pedigrees of all living Przewalski's horses, back to the first foals caught in the wild (Bouman 1998; Wit and Bouman 2006). Now, for the first time, it became clear that out of the original 54 wild foals that had reached Europe alive, the genetic characteristics of only 12 animals were present in the entire world population. Various breeding lines had died out and inbreeding had increased to an alarming degree.

In 1976 the Third International Symposium for the Preservation of the Przewalski horse was organised by Munich Zoo. Jan and Inge Bouman ventilated their grave concern regarding the preservation and the continued existence of the species in captivity and called for international cooperation by all Przewalski's horse breeders. The Munich symposium recommended all interested parties to consult the 'Bouman studbook inventory system' for breeding Przewalski's horses in zoos.

In 1977 the Foundation for the Preservation and Protection of the Przewalski's Horse (FPPPH) was founded. The main focus of interest of the new foundation was the reintroduction of captive-bred Przewalski's horses into the wild. In 1979 the reintroduction plan for the Przewalski's horse started as an official project. In that year the North American zoos that kept Przewalski's horses decided to collaborate in a cooperative breeding programme and in 1986 the European zoos followed suit. The first phase of the reintroduction project consisted of the establishment of semi-reserves in 1980 by the Foundation Reserves for the Przewalski's Horse (FRPH) (Fig. 13.1). This new foundation was a joint initiative of the WWF-The Netherlands and the FPPPH. The objective was to breed a reservoir of Przewalski's horses with a diverse genetic background obtained from West-European, North American and Russian zoos. The purpose of the programme was to reduce inbreeding, to improve the fitness of Przewalski's horses and to increase their chances of surviving in the wild. Their release in nature reserves in The Netherlands and

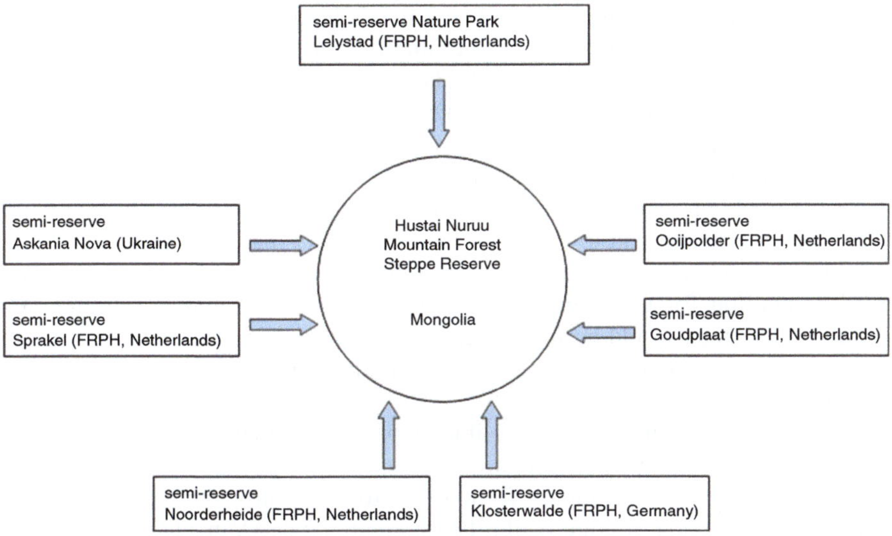

Fig. 13.1 The locations of the semi-reserves that provided the Przewalski's horses that were sent to Hustai Nuruu

Germany would give them the chance to run wild. The breeding in the semi-reserves proved to be very successful. Within 10 years the Foundation had more than 60 Przewalski's horses in the different semi-reserves.

Remarkable progress was made in the 1980s and 1990s, also in zoos. The renewed attention for cooperative breeding programmes in zoos, with the emphasis on strategic breeding, resulted in a population 'explosion' of the Przewalski's horse in captivity. In 1990 the number of horses had increased to 961. Some zoos were even confronted with limiting facilities to accommodate their population of Przewalski's horses. The time was ripe to reintroduce the Przewalski's horse into the wild.

The search for suitable areas for reintroducing Przewalski's horse would focus on areas in Soviet Inner Asia and Mongolia (Bouman 1998; Wit and Bouman 2006). Three essential criteria were used for site selection: (1) fulfilment of the habitat requirements of the Takh, such as sufficient forage, water and shelter; (2) existing predator populations should not impose a too high predation pressure; (3) no possibility of contact with feral or domestic horses, with which Przewalski's horses readily hybridise. In cooperation with Prof. Vladimir Sokolov a total of seven potentially suitable areas were identified: in Ukraine, eastern Kazakhstan and two in Siberia, and in Mongolia the Dzungarian Gobi, western Zhavkhan Province, and Hustai Nuruu reserve.

After visiting several sites in Soviet Inner Asia that proved to have overgrazed grasslands, the area at Zaisan in the steppe of Tarbagatai in eastern Kazakhstan appeared to meet many of the set requirements. However, there happened to be little local interest in reintroducing Przewalski's horses at this site, and the idea was rapidly abandoned. In Mongolia the situation was different. Here the Tahk was a national symbol, praised in many ballads and folk tales for its swiftness and force, and the Mongolians were very keen on the return of the wild horse. In Mongolia first the Dzungarian Gobi, the desertic area surrounded by mountains and which had been the last refuge of the Przewalki's horses before their extinction from the wild, was assessed. However, although the area was still sparsely populated, the conclusion had to be drawn that the factors that had contributed to the wild horses' extinction were still present. In that area with little natural resources the competition for grass and water with wild asses, Mongolian gazelles and domestic livestock would be very tough. The Zavkhan Province site also was discarded as it was hard to access and bring the Takh. Thus, Hustai Nuruu emerged as the most suited site.

Hustai Nuruu is an easily accessible steppe area situated at a 100 km south west of Ulaan Baatar, the capital of Mongolia. The reserve area occupies 50,600 ha at elevations ranging from 1,100 to 1,840 m above sea level and is located at the southern edge of the forest-steppe zone (Van Staalduinen 2005; Wallis de Vries et al. 1996). The area comprises mountains, plains, dunes and a river valley, and several vegetation types of the forest-steppe and typical steppe zones can be found. The largest part of Hustai Nuruu is covered with steppe vegetation with several grass species, among others *Stipa krylovii*, and a variety of herbs and legume dwarf-shrubs. Only 5% of the total area of Hustai National Park is covered by forest, *Betula platyphylla* forest, which grows on northern slopes of the mountains. Tiny streams flow from these mountains and fuse lower down in larger valleys.

It was concluded that in Hustai Nuruu the Przewalski's horse would stand a better chance to survive than in the dry Dzungarian steppe with practically no water (Wit and Bouman 2006). The good accessibility was an advantage for building up an adequate reserve staff, an effective control system and good co-operation with Mongolian scientific institutes, and later also proved favourable to attract tourists, and thus income for the reserve. MACNE (the national organization of nature conservation in Mongolia) was positive about the proposal and the government of the province offered its fullest support.

In June 1992 the first transport of Takh from The Netherlands to Mongolia took place, consisting of 8 horses from the FRPH semi-reserves in The Netherlands (Bouman 1998) and 8 from the semi-reserve in Askania Nova (Ukraine). In 1994, 1996, 1998 and 2000 another 68 Przewalski horses arrived in Hustain Nuruu (Fig. 13.1). All transports were organized and financed by the Foundation FRPH (Photos 13.1 and 13.2). After arrival the Prezwalski horses were released into three large enclosures for an initial adaptation period of 1–2 years. After that the horses were released and lived as free roaming groups (Photos 13.3 and 13.4). From 1992 to 2000 there have been five transports with a total of 84 Przewalski's horses to Hustai Nuruu. In 1998 Hustai Nuruu received the status of National Park.

Photo 13.1 Transport of the crates with Przewalski's horses to Hustai Nuruu (Photo by I. Bouman)

Photo 13.2 Release of a Przewalski's horse in Hustai Nuruu (Photo by I. Bouman)

Photo 13.3 Prezwalski's horses in autumn in Hustai Nuruu National Park (Photo by I. Bouman)

Photo 13.4 The Prezwalski's horses seem well adapted to the severe, cold winter (Photo by I. Bouman)

13.4 Population Dynamics

The dynamics of a population are driven by two major sets of factors: those that contribute to its increase ('gain components') and those that cause it to decline ('loss components'). In the case of the Przewalski's horses at Hustai National Park a completely new population had to be built up. Gain components comprised the addition of imported horses from the semi-reserves in Europe and the additions to the population by births (Boyd and Bandi 2002) (Photo 13.5). Conversely, the loss components consist of reduction due to mortality. From 1993 until the end of 2004 237 foals were born (Wit and Bouman 2006). In the same period the mortality of the imported Przewalski's horses was 57% (48 dead horses out of a total of 84) and the mortality of the wild-born horses (foals, juveniles and adults) 49% (116 deaths out of a total of 237).

In 2009 the population had grown to 260 horses. The growth percentage of the Takh population during 2001–2009 (the period without new import) was 9.1%. Compared to the annual growth rate of 20% for feral horses in the USA (Berger 1986) and to the average growth rate of 32.5% for the Camargue horses (Duncan 1992) the growth of the population of free living horses in Hustai National Park is slow. However, Duncan (1992) considered the growth rate of the Camargue horses as very high and argues that this is due to the high annual survival rates of the horses

Photo 13.5 A foal accompanied by mares (Photo by I. Bouman)

in this predator-free environment. Duncan (1992) also stated that ungulates such as wildebeest and buffalo in the Serengeti have much lower growth rates with the populations increasing by 10% per year after the rinderpest had been eradicated (Sinclair and Norton-Griffiths 1979).

The high mortality rate of foals in Hustai National Park, which averaged 40.2% from 1993 to 2007, is a matter of concern since foals are the only gain component nowadays. This mortality rate is much higher than for foals from North American feral horse populations, which average 8–10% a year. Factors influencing foal mortality in Mongolia are:

- *General condition of the foals.* The winters in Mongolia are long and severe. When the foals are weakened they are more susceptible to infection and become an easy prey for wolves.
- *Wolf pressure.* Predation by wolves on foals is increasing.
- *Individual variation.* When captive-bred animals are released into the wild, it cannot be expected that the off-spring of each animal is equally successful in adapting to the wild.
- *Foaling season.* Foals born in early spring, in late autumn or during winter have a lower chance of survival.

Predation by wolfs on foals is increasing (Hovens and Tungalaktua 2005). In the period from 1993 to 2004 of the total of 237 foals 59 (25%) were killed by wolves or died from injuries caused by wolves. In the first 6 years only 14 foals (16%) were killed by wolves. In 2002 12 foals (32%) were killed, in 2003 14 (48%) and in 2004 10 (27%). This high percentage of kills by wolves is a reason for concern now that the Takh population is still relatively small (Photo 13.6).

The Wolf (*Canis lupus*) and the Lynx (*Lynx lynx*) are the only large predators in Hustai National Park (Wit and Bouman 2006). Lynxes are rare but wolf density is relatively high. The number of wolves which den in the park and the direct environment was estimated at some 25 animals (2004). While the presence of wolves has been a matter of concern since the start of the reintroduction project, because wolf pressure seems to be a limiting factor for building up the Takh population, in the future, when the number of Takh has reached the carrying capacity of the Hustai National Park (estimated at some 500 Takh), the presence of wolves may be an asset in keeping the Takh population healthy.

Several ecological factors may play a role in the occurrence of years of high wolf predation on foals, such as the large number of foal killed by wolves in 2004. This might be explained by the drought of the year before. Because of a very dry summer which resulted in low forage availability, many tarbagans (marmots, *Marmota sibirica*) were in a bad condition before hibernating. During their winter sleep many died in their burrows, because of insufficient fat reserve. The subsequent shortage of marmot prey probably strengthened the wolves' pressure on foals.

In Hustai National Park the hunting of wolves is prohibited. In the surrounding buffer zone, herdsmen and hunters from Ulaan Baatar like to hunt them. As long as wolves carry a risk for the slowly growing Takh population the park staff decided that considerable attention will be given to preventive actions in order to control undesirable wolf impact.

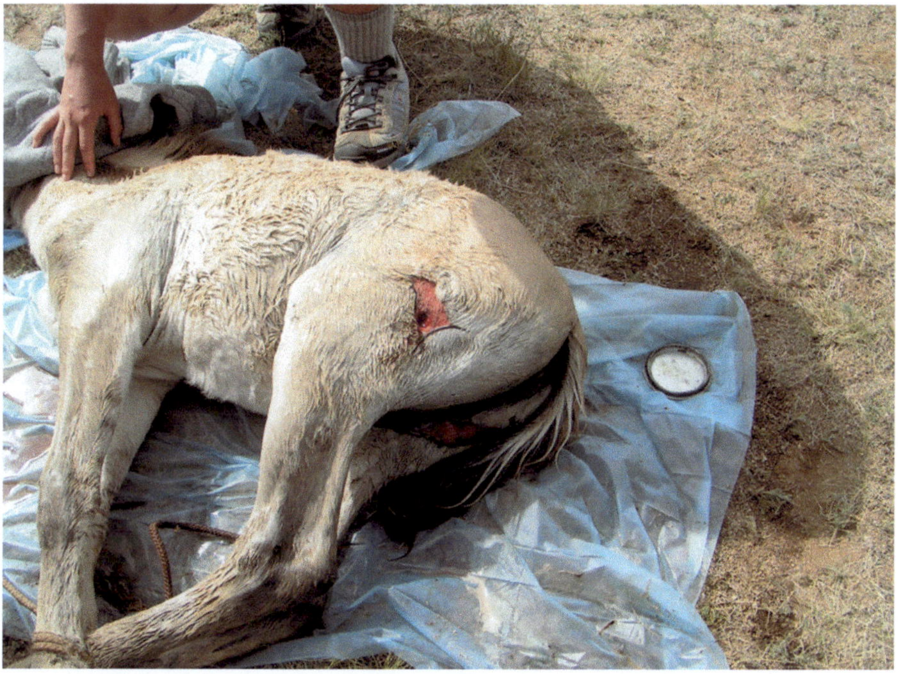

Photo 13.6 A foal that died of injuries caused by a wolf (Photo by I. Bouman)

The reintroduction project of the Takh in Hustai National Park shows to be a great success in nature conservation in the steppe. An important steppe animal, on the brink of its extinction, is brought back to the wild, and an ecological key component of the local steppe ecosystem is restored.

References

Berger J (1986) Wild horses of the Great Basin: social competition and population size. Wildlife and behavior and ecology. University of Chicago Press, Chicago, pp 283–316

Bouman I (1998) The reintroduction of Przewalski horses in the Hustain Nuruu Mountain forest-steppe reserve in Mongolia. An integrated conservation development project. Netherlands Commission for International Nature Protection. Mededelingen 32:1–50

Bouman I, Bouman JG (1994) The history of Przewalski's horse. In: Boyd L, Houpt KA, Boyd L, Houpt KA (eds) Przewalski's horse, the history and biology of an endangered species, SUNY Series in endangered species. State University of New York Press, Albany, pp 5–38

Boyd L, Bandi N (2002) Reintroduction of Takhi, Equus ferus przewalskii, to Hustai National Park Mongolia: time budget and synchrony of activity pre- and post-release. Appl Anim Behav Sci 78:87–102

Duncan PB (1992) Horses and grasses: the nutritional ecology of equids and their impact on the Camargue. Springer, Berlin, p 287

Hovens JPM, Tungalaktua KH (2005) Seasonal fluctuations of the wolf diet in the Hustai National Park (Mongolia). Mam Biol 70:210–217

Sinclair ARE, Norton-Griffiths M (eds) (1979) Serengeti: dynamics of an ecosystem. University of Chicago Press, Chicago

Van Staalduinen MA (2005) The impact of herbivores in a Mongolian forest-steppe. Ph.D. thesis, Utrecht University, Utrecht

Wallis de Vries MF, Manibazar N, Dügerlham S (1996) The vegetation of the forest-steppe region of Hustain Nuruu, Mongolia. Vegetation 122:111–127

Wit P, Bouman I (2006) The tale of the Przewalski horse. KNVV Publishing, Utrecht

Part III
Climate Change

Chapter 14
Transformation of Steppe Communities of Yakutia Due to Climatic Change and Anthropogenic Impact

E.I. Troeva and M.M. Cherosov

Abstract In the Late Pleistocene vast steppe landscapes covered the territory of Yakutia. Global climatic changes in the Quaternary have transformed the vegetation cover, and now only isolated "islands" of steppes remain, being unique objects of Yakutian nature and very attractive for scientific investigation. They are confined to prominent parts of terraces above floodplains, south-facing slopes of river valleys, and xerophytic belts of depressions in the taiga formed as a result of thermokarst processes (alases). Used as a natural fodder land, Yakutian steppes experience significant anthropogenic stresses. They suffer from overgrazing and recreation load and gradually transform into ruderal communities. This can be aggravated by modern changes in climate, since warming and raising humidification are fatal for xerophytic vegetation formed under the harsh conditions of an ultra-continental climate. For conservation of the unique extrazonal steppe landscapes special recovery methods for disturbed vegetation are required as well as the establishment of steppe reserves in the regions with the highest biodiversity of steppe communities.

14.1 Introduction

Steppes are unique objects of Yakutian nature and very attractive for scientific investigation. In Yakutia they represent patches of extrazonal vegetation, i.e. vegetation that is common in another climatic zone but occurs outside that zone in landscapes with special environmental features. The isolated "islands" of steppes in Yakutia (Fig. 14.1) are confined to prominent parts of terraces above floodplains, south-facing

E.I. Troeva (✉) • M.M. Cherosov
Institute for Biological Problems of Cryolithozone,
Russian Academy of Sciences, Siberian Branch, Yakutsk, Russia
e-mail: etroeva@mail.ru; cherosov@mail.ru

M.J.A. Werger and M.A. van Staalduinen (eds.), *Eurasian Steppes. Ecological Problems and Livelihoods in a Changing World*, Plant and Vegetation 6,
DOI 10.1007/978-94-007-3886-7_14,

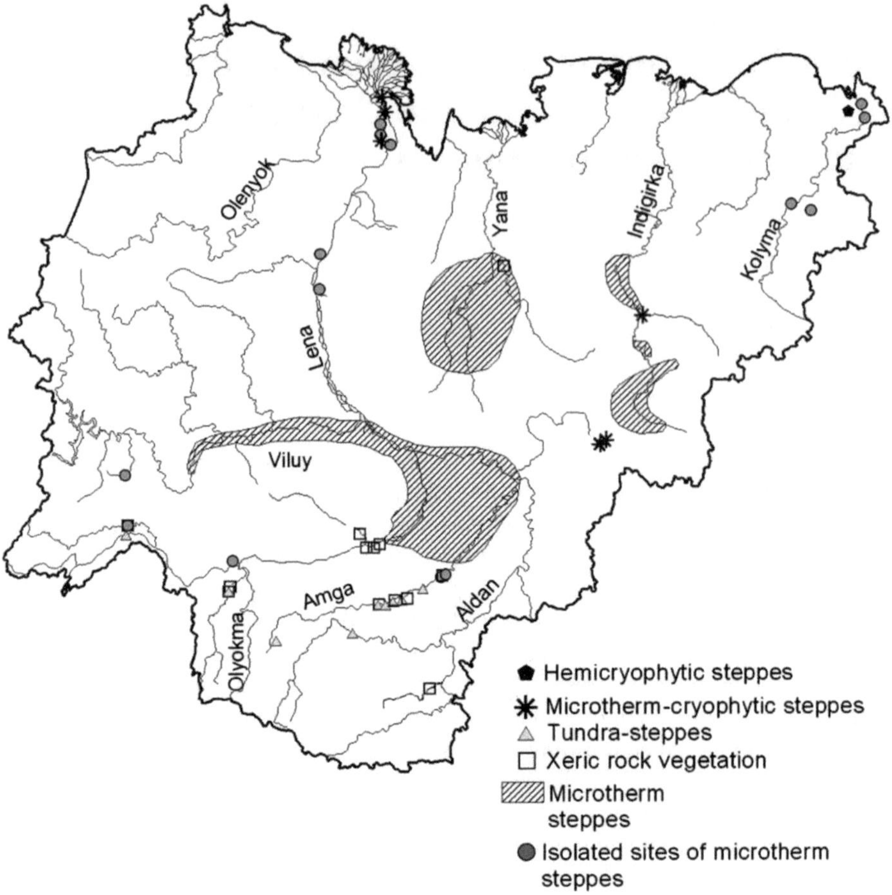

Fig. 14.1 Distribution of steppe vegetation in Yakutia

slopes of river valleys, and xerophytic belts of depressions in the taiga formed as a result of thermokarst processes (alases).

In Yakutia steppes occur in regions characterized by an arid climate and loamy soils underlain by calcareous bedrocks: the Central-Yakutian Lowland and mountain-valley landforms of the Yana and Indigirka River basins. A small forest-steppe "island", situated opposite the Olyokma River mouth (South-West Yakutia), is considered to be a connecting link between Yakutian and Prebaikalian steppe landscapes. Steppe vegetation is also recorded beyond the Polar Circle on properly insolated parts of relief in the Lena, Olenyok, and Middle and Lower Kolyma River valleys (Skryabin and Karavaev 1991). The so-called tundra-steppes (Yurtsev 1981; Koropachinsky 1996) should be mentioned as well. They are confined to calcareous screes on the slopes of river valleys (Fig. 14.1).

The soils under steppes in the central regions are of the frozen chernozem type. Usually they represent complicated soil complexes covered with mozaics of steppe and meadow steppe vegetation. The steppe soils of the Yana-Kolyma region (North-East Yakutia), on the contrary, are more homogenous, mostly belonging to a peculiar type of 'mountain shallow chernozems' (Skryabin and Karavaev 1991).

Floristically, the Yakut steppes have many features in common with Prebaikalian and Transbaikalian-Mongolian steppes, having borrowed many plant species from there. Partially, species from xeric vegetation of the north-western regions of North America also contributed to the steppe flora of Yakutia. This especially refers to the Indigirka and Kolyma River basins.

14.2 Steppe Types of Yakutia

Classification of steppe vegetation is based on floristic composition, life forms, and response to hydrothermal conditions. Accordingly, the Yakut steppe may be divided into true and meadow types; microtherm, hemicryophytic and microtherm-cryophytic types; bunchgrass and rhizomatous types (Andreyev et al. 1987; Yurtsev 1981). According to an ecological-floristic classification, the Yakut steppes belong to the class *CLEISTOGENETEA SQUARROSAE* Mirkin et al. 1986 ex Korolyuk 2002.

Most xeric and poor in species are the bunchgrass steppe communities (order *Stipetalia krylovii* Kononov et al. 1985) with predomination of *Psathyrostachys caespitosa, Agropyron cristatum.* The forb component is represented by *Alyssum obovatum, A. lenense, Eritrichium sericeum, Chamaerodos erecta.* Such communities occupy insignificant areas on dry, properly insolated, eroded slopes of river valleys or alas depressions (Photo 14.1).

Stipa spp. prevail in coenoses growing on sodded soils. The calcareous slopes in the middle reaches of the Lena River (300–200 km south of Yakutsk) bear rocky variants of steppe vegetation of the order *Stipetalia krylovii*, i.e. communities of *Stipa krylovii* and *Psathyrostachys caespitosa* with participation of petrophytic species (*Artemisia santolinifolia, Youngia tenuifolia, Patrinia rupestris, Orostachys malacophylla, Thalictrum foetidum*, etc.).

More common are the forb-bunchgrass steppes growing on humus-rich meadow-chernozem soils with predomination of *Festuca lenensis* (*Festucetalia lenensis* Mirkin in Gogoleva et al. 1987) in aggregation with other steppe grasses: *Koeleria cristata, Poa botryoides, Stipa spp.*, sometimes *Cleistogenes squarrosa.* The abundant forb component consists of xeric and mesoxeric species: *Veronica incana, Artemisia commutata, Silene repens, Onobrychis arenaria, Potentilla nivea, Carex duriuscula, Potentilla bifurca* and *Thymus spp.*

Similar steppe formations are observed in the North-East of the republic, in the Yana, Indigirka and, partially, Kolyma River basins (Photo 14.2). Starting from the upper reaches of all their numerous tributaries, this steppe vegetation is traced along the Yana and Indigirka valleys reaching 69 NL and 68 NL, respectively. The climatic and orographic peculiarities of this region are reflected in the botanical composition of those steppes. First of all, they contain a cryophilous component represented by *Calamagrostis purpurascens, Dracocephalum palmatum, Poa glauca*, etc.

Photo 14.1 *Psathyrostachys caespitosa* community on an eroded slope of an alas (Photo by E. Troeva)

Cold-resistant mountain steppe species (*Helictotrichon krylovii*, *Poa botryoides*, *Festuca lenensis*, *Artemisia bargusinensis*) form communities as high as 1,200 m above sea level.

The order *Festucetalia lenensis* also comprises meadow steppes confined to the landscapes with more favourable moisture conditions: ridges of river terraces, forest margins. These communities combine the features of both forb-*Festuca* steppes and forb meadows. The grass component still consists mainly of the same xerophilous grasses, but the forb component is dominated by mesic and xeromesic species: *Pulsatilla flavescens*, *Carex pediformis*, *Bromopsis pumpelliana*, *B. korotkiji*, *Agrostis trinii*, *Aster alpinus*, *Lychnis sibirica*, *Dianthus versicolor*, *Euphorbia discolor*, *Veronica longifolia*, *Sanguisorba officinalis*, *Geranium pratense*, etc. (Photo 14.3). In the North-East rich forb meadow steppes are found in river valleys (*Bromopsis pumpelliana*, *Pulsatilla patens s.l.*, *Delphinium grandiflorum*, *Heteropappus biennis*, *Castillea rubra*, *Poa botryoides*, etc.). The Indigirka River terraces bear tracts of meadow steppes composed of the endemic species *Helictotrichon krylovii* (Karavaev 1958; Skryabin 1968). The forests are fringed with meadow steppe communities on slope summits (*Helictotrichon krylovii*, *Calamagrostis purpurascens*, *Arnica iljinii*, *Pulsatilla patens s.l.*, etc.). Apparently, such meadow variants of steppes with a cryophylous component occur even farther North, in the Subarctic tundra region (middle reaches of the Anabar River,

Photo 14.2 *Festuca lenensis* steppe on the south-facing slope of the Adycha River valley (North-East Yakutia) (Photo by E. Troeva)

North-West Yakutia, 71° NL). In the Upper Kolyma and Middle Indigirka River basins bunch sedge steppes are found with *Carex pediformis* with participation of *Poa glauca*, *Eremogone tschuktschorum*, *Dianthus repens*.

Fragments of hemicryophytic steppes (according to Yurtsev's (1981) interpretation) were discovered in the lower reaches of the Kolyma River. They are confined to slopes of hydrolaccolites and lake depressions in the lowland boggy tundra. The communities are composed of *Carex spaniocarpa* with participation of *Trisetum spicatum*, *Potentilla hookeriana*, *Armeria scabra*, *Polemonium boreale*, *Arnica iljinii*, *Koeleria asiatica*, *Astragalus alpinus*, etc. These coenoses are characterized by a pronounced moss-lichen layer of *Polytrichum piliferum*, *Cetraria cucullata*, *C. islandica*, *Cladonia cocciferra*, *Alectoria ochroleuca*, *Dactylina arctica*. Data on such steppe communities are very scarce, but according to their floristic composition they are meadow steppes, and their moist conditions impede the growth of bunchgrass species there. Anthropogenic activity and invasion of tundra species have resulted in degradation of these unique landscapes of cold steppes (Yurtsev 1981; Andreyev and Perfilyeva 1975; Andreyev and Galaktionova 1981; Koropachinsky 1996).

Rhizomatous *Carex* steppes represent strongly degraded landscapes with predomination of *Carex duriuscula*. They occupy vast territories in the most populated areas of Central Yakutia and the Middle Indigirka River basin. Cherosov et al. (2005) proposed to consider the *Carex duriuscula* communities with a pronounced

Photo 14.3 Rich forb meadow steppe in the Amga River valley (Central Yakutia) (Photo by E. Troeva)

anthropogenic component (*Artemisia jacutica*, *Lappula squarrosa*, *Lepidium densiflorum*, *Leptopirum fumarioides*) as vegetation of the synanthropic class *ARTEMISIETEA VULGARIS* Lohmeyer et al. in R. Tx. 1950 (alliance *Artemisio-Caricion duriusculae* Czerosov 2005, order *Onopordetalia acanthii* Br.-Bl. et R. Tx. ex Klika et Hadac 1944). The diagnostic species of the alliance *Artemisio-Caricion duriusculae* (Gogoleva et al. 1987; Cherosov et al. 2005) are synanthropic species, as well as a number of steppe species that withstand overgrazing and thus occur both in typical and pasture variants of steppe communities and synanthropic coenoses: *Artemisia commutata*, *Carex duriuscula*, *Koeleria cristata*, *Leymus chinensis*, *Poa transbaicalica*, *Potentilla bifurca*, *Potentilla conferta*, *Veronica incana*, *Heteropappus biennis*, with *Carex duriuscula* and *Potentilla bifurca* often becoming dominant. Disturbed steppe vegetation also features rather common species of the class *ARTEMISIETEA VULGARIS* (*Carduus crispus*, *Leonurus quinquelobatus*, *Melilotus albus*, *Melilotus officinalis*, *Tanacetum vulgare*), and the order *Onopordetalia acanthii* (*Astragalus danicus*, *Carduus nutans*, *Descurainia sophia*, *Lappula squarrosa*, *Rumex thyrsiflorus*) (Cherosov et al. 2005).

Communities with equal participation of *Carex duriuscula* and natural xerophilous species (*Artemisia commutata*, *Koeleria cristata*, *Veronica incana*, etc.) should, however, be included in the alliance *Festucion lenensis* of the class *CLEISTOGENETEA SQUARROSAE*.

Photo 14.4 Petrophytic steppe communities of *Elytriga jacutorum* on calcareous slopes of the Tokko River valley (South-West Yakutia) (Photo by E. Troeva)

Petrophytic steppes on calcareous bedrocks are an intrinsic component of the mountainous dry landscapes of Yakutia. First of all, they are the communities of *Elytrigia jacutorum* in the river basins of the Aldan Tableland (the Tokko, Amga, Buotama, Maya and other rivers) (Photo 14.4). *Elytrigia jacutorum* forms various associations with steppe xerophytes and mesoxerophytes (*Thesium refractum*, *Artemisia commutata*, *Silene repens*, *Euphorbia discolor*, *Carex pediformis*, *Galium verum*, etc.) and with participation of petrophytic forbs (*Youngia tenuifolia*, *Artemisia santolinifolia*, *Thalictrum foetidum*, *Orostachys spinosa* and the shrub *Juniperus sibiricus*) (Skryabin 1976; Koropachinsky 1996; Zakharova 2005; Zakharova et al. 2010).

14.3 The Effects of Changes in Climate

The precise distribution of steppe landscapes over Yakutia is of special interest in the light of global climate change. It is known that the spatial dynamics of their distribution on a geochronological scale coincides with changes in the Quaternary warming and cooling periods. Many publications have been devoted to the reconstruction of landscapes of the past and their dynamics due to periodical climate change in the territory of modern Yakutia (Tolmachev 1954; Skryabin and Karavaev 1991; Giterman 1985; Kozhevnikov and Ukraintseva 1996; Tomskaya 2000; Ukraintseva 2002; etc.). Interpretation of such paleodata helps to predict the possible spatial-temporal dynamics of the steppe vegetation. Studies on the reconstruction of Pleistocene vegetation are based on analyses of spores and pollen spectra buried in the soils, as well as on plant macro-remains found in stomachs of specimens of the mammoth fauna.

One view states that the most cold-resistant mountain steppe plants (*Helictotrichon krylovii*, *Poa botryoides*, *Festuca lenensis*, etc.), reaching to 1,200 m above sea level on south-facing slopes, evolved during the Neogene or Early Anthropogene in alpine regions of northern Eurasia (Skryabin and Karavaev 1991; Sakalo 1963). Belova (1970) stated that petrophilous steppes became an integral part of the vegetation of mountain systems of Prebaikalia during the Zyrian cryochrone (70–10 Ka BP), whereas the thermophilic xerophytes (*Stipa krylovii*, *Agropyron cristatum*, *Koeleria cristata*, *Psathyrostachys caespitosa*) migrated to the North from zonal southern steppes.

According to Skryabin and Karavaev (1991), penetration of zonal steppe species to Central Yakutia and their further advance to the North-East took place during one of the interstadials characterized by a warm and dry climate, while the following glaciation eliminated pre-glacial and interglacial vegetation over vast territories. This explains the isolated remains of steppe landscapes in separate sheltered habitats, hundreds of kilometres away from their modern zonal distribution area.

However, having reviewed published data, Protopopov and Protopopova (2010) prove that the peak of steppe distribution in Yakutia falls in the period of the Sartan glaciation (30–10 Ka BP), when the climate was much colder and more arid. Then, summer temperatures where higher than at present, and air humidity was much less. Forest tracts decreased abruptly and changed into meadow steppe landscapes bordering the tundra in the North. In the Yana-Kolyma Lowland the only forest forming species was *Larix cajanderi*. In Central Yakutia there were islands of *Pinus-Larix* forests with participation of *Betula*.

A number of the steppe species in Central Yakutia originate from the Transbaikalian and Mongolian steppes. The steppe vegetation east of the Verkhoyansk Range is genetically related to the North American xerophytic landscapes via the Beringian Bridge. The species growing there (*Koeleria cristata*, *Artemisia frigida*, *Carex duriuscula*) are main edificators in the prairies of North America (Yurtsev 1974, 1981; Skryabin and Karavaev 1991). Meadow steppe associations, in combination with *Salix* shrubberies, represented rich fodder lands for large herbivorous representatives of the mammoth fauna (mammoths, bisons, woolly rhinoceros) that are now extinct. During that period steppes became a zonal vegetation type in Yakutia. Bordering on the tundra in the North, they formed a unique type, the Arctic steppe. That combination was possible under the special climatic conditions of that period. Hot summers and extremely cold winters most likely induced rigorous thermokarst processes resulting in changes in microrelief. Dry air and desiccating soil on prominent parts of the relief favoured the growth of xerophytes, while intensive melting of frozen grounds saturated soils with moisture in light depressions

Table 14.1 Average winter and January temperatures in North Yakutia during the Late Pleistocene and Holocene as compared to modern temperatures, °C

	Late Pleistocene, 40–10 Ka BP	Holocene, 10–5 Ka BP	Modern period[a]
Winter	−29 to −33	−25 to −27	−28.0 to −30.3
January	−45 to −48	−38 to −40	−29.3 to −31.9

According to Vasilchuk and Kotlyakov (2000), NASA Langley Research Center: http://eosweb.larc.nasa.gov
[a]Tertile temperature values at a height of 2 m

providing suitable conditions for the growth of tundra-bog species. Probably, the modern associations with *Carex spaniocarpa* in the Lower Kolyma River basin represent relic remains of such landscapes. During the warming and climatic optimum of the Holocene, the vegetation in the territory of Yakuita obtained features that strongly ressemble modern ones.

The results of numerous studies on paleoclimate (Monin and Shishkov 1979; Budyko 1980; Tomirdiaro 1980; Giterman 1985; Tomirdiaro and Chyornenky 1987; Konischev 1995; Stuiver et al. 1995; Solomina 1999; Vasilchuk and Kotlyakov 2000; Alfimov et al. 2003; Kienast et al. 2005; Andreyev et al. 2007; etc.) showed that there is a difference of 6–12°C between temperatures in the Holocene and minimum temperatures in the Pleistocene. Data for North Yakutia (zones of tundra and near-tundra sparse forests) (Table 14.1) show that the winter temperatures of the Late Pleistocene were 4–8°C lower as compared to those in the Holocene.

As to the summer temperatures, Alfimov et al. (2003) proved that Late Pleistocene July temperatures in northern Yakutia were 2–4°C higher than those of today. This yielded increased evaporation rates and climate aridization. Thus, the climate of the Late Pleistocene was sharply continental and very arid. These very conditions became optimal for the development of microtherm steppes on a scale that made them a zonal vegetation type. And the subsequent reduction in area occupied by steppe landscapes down to the isolated islands during several tens of millennia can be explained by climate change. Similar processes of cold steppe reduction were described for Prebaikalia (the south of East Siberia) (Belova 1985).

As shown in Table 14.1, the temperatures of modern Yakutia approach those of the Holocene. And the warming process is still in progress. The global warming observed in the twentieth century is apparent in all regions of Russia. According to the Climate Change Bulletin the warming in all Russia, from the end of nineteenth to the end twentieth century, manifested itself as an overall rise in the mean annual air temperature by approximately 10°C during this period. Data from meteorological and geocryological stations of the former USSR show a trend of climate warming in most regions of the North during the last three decades (Israel et al. 1999). Discussing climate warming, Maximov et al. (2010) cited data of investigations based on the model of paleoclimatic reconstruction based on CO_2 concentration measurements (Budyko et al. 1991; Gavrilova 1993). The results strongly suggest that the most

Table 14.2 Climatic characteristics of Transbaikalia and Yakutia

Selenga-Onon Province	Central Yakutia	Verkhoyansk Region
January		
−24°C to −28°C	−38°C to −44°C	−38°C to −48°C
July		
16°C to 18°C	16°C to 18°C	12°C to 14°C
Number of days with snow cover		
Up to 160	Up to 240	Up to 240–280
Annual precipitations, mm		
Up to 400	Up to 300	Up to 400
Precipitation-evaporation ratio		
−200	−100	100

http://www.ecosystema.ru/08nature/world/geoussr/2-2-2.htm

drastic warming will take place in the northern parts of the continents, i.e. at high latitudes, particularly in wintertime. Thus, if the global warming amounts to 2°C, then the average January temperature at 40–80° NL will rise by 2–12°C; in case it amounts to 4°C the January rise is expected to be 6–20°C. The warming in summertime will be inconspicuous (2–3 times as less, i.e. a rise of 4–6°C). Since the winter at high latitudes is longer than the summer, this will affect the annual temperatures.

Does this mean that the areas occupied by steppe landscapes will continue to reduce in area? We think that the reason of such a reduction during the Late Pleistocene-Holocene was not temperature rise as such, but the decrease in aridity. Although the Holocene featured some increase in humidity, the estimated average annual total of precipitation during the Sartan cryochrone reached 190–200 mm (Belova 1985), which is even less than the modern values of the most arid regions of Yakutia.

The data from Table 14.2 may serve as an indirect confirmation. They represent climatic indices of the regions where the extrazonal Yakut steppes occur, as well as the zonal steppes in the south of East Siberia (Selenga-Onon Province), which are very similar in floristic composition.

According to physiographic regionalization, the steppes of Transbaikalia (or Dauria) belong to the Central Asian steppe subzone and are situated in the Selenga-Onon province of the forest-steppe zone of the Transbaikalian mountain country (Fig. 14.2) (see also Kirilyuk et al. this volume). The winters in the south of East Siberia are milder as compared to those in Central and North-East Yakutia. However, the climate aridity as expressed by the precipitation-evaporation ratio, is two or more times higher. Hence, provided climate warming is accompanied with an increase in humidity and a decline in continentality, the steppe areas of Yakutia may face the risk of extinction. Besides, intensive thawing of frozen grounds will change the moisture conditions of the soils, making them more favourable. Under such conditions, the true steppes will gradually transform into their meadow variants. The xerophytic vegetation will remain in microrefugia at the most prominent parts of the relief with

Fig. 14.2 The Steppe belt in Russia. The small box indicates the Transbaikalian steppes of the Selenga-Onon Province. The homogeneous *grey colour* defines the borders of the continental part of Yakutia

proper drainage and insolation. After final retreat of the cryolithozone and desiccation of the soils the steppe landscapes may restore, but only provided that the hydrothermal coefficient will remain at the same level as the present one. However, such vegetation will be similar to that of southern regions, since cold-resistant xerophytes will never find proper conditions for growth. To confirm this hypothesis additional complex investigations are required, including microclimatic studies of steppe landscapes. On the other hand, even provisional geobotanical and floristic analyses of steppe vegetation may partly give an answer to this matter.

14.4 Climatic Effects on Geobotanical and Floristic Patterns

The climate in Yakutia changes considerably from south to north, and this is reflected in coenotic and floristic patterns in the steppes. To illustrate this, the transformation in communities in which *Festuca lenensis* is important was analysed along a south-west – north-east climatic gradient. In Table 14.3 the basic climatic parameters of the regions in which these communities were studied are given, acquired from satellite data for the period of the 1980s to the 2000s. As shown in the table, climatic parameters become harsher in south-west – north-east direction. The most arid and cold conditions are observed at the site in North-East Yakutia, whereas the site in South-West Yakutia features a mild climate. *Festuca lenensis* steppes are best developed in Central Yakutia, being confined to ridges of terraces above floodplains, dry belts of alas depressions, and insolated valley slopes, where anthropogenic impact

Table 14.3 Long-term climatic averages of the geobotanical study sites

Region (GPS coordinates where geobotanical data were collected)	Average maximum temperature of July, °C	Average minimum temperature of January, °C	Monthly averaged insolation incident on a horizontal surface (kWh/m²/day)	Monthly averaged precipitation (mm/day)
North-East Yakutia (68 N, 135 E)	21.05	–39.69	5.12	272.27
Central Yakutia (62 N, 129 E)	25.04	–37.52	5.61	329.38
South-West Yakutia (60 N, 120 E)	23.77	–31.89	5.37	390.53

NASA Langley Research Center, http://eosweb.larc.nasa.gov

is observed. As the abundance of this species diminishes, other xerophytic grasses become dominant. In the north-east and south-west, the steppe communities with *Festuca lenensis* occur only on warm slopes with a southern aspect. Besides, in South-West Yakutia *Festuca lenensis* never plays a dominant role.

Change in floristic composition due to climatic differences can be demonstrated by a cluster analysis of the coenofloras of steppes with *Festuca lenensis* along the gradient (south-western, central and north-eastern regions) (Fig. 14.3). The analysis involved nearly 80 vegetation plots of steppes with *Festuca lenensis*, situated on properly insolated slopes of valleys or alas depressions and experiencing a minimum of anthropogenic impact (valley communities were not included due to strong grazing effects). The cluster analysis yielded five specific coenofloras which did not differ in geographical elements (according to Sekretareva 2004): all the clusters feature predomination of boreal-steppe and boreal latitudinal elements, as well as Eurasian and Asian longitudinal elements in the flora. The diagram shows the distinct character of the mesoxerophytic coenoses of South-West Yakutia (cluster 1). They are rather rich: 68 characteristic species, including species growing in Southern and Central Yakutia (*Goniolimon speciosum*, *Allium ramosum*, *Heteropappus biennis*, etc.). However, in Central Yakutia they occur in other types of plant communities. *Festuca lenensis* does not dominate in the steppes of South-West Yakutia; it is a subordinate species in *Stipa*, *Psathyrostachys* or *Agropyron* communities. In contrast with similar coenoses in Central Yakutia, these communities are less xeric, probably because there is more precipitation in the south-west. The Central Yakutian steppes with *Festuca lenensis* consist of two different coenofloras. The xeromesophytic communities (cluster 2) are confined to forest edges where moisture conditions are more favourable and more stable. Accordingly, they are rich meadow steppes (103 species) and characterized by a pronounced mesophytic element (*Poa pratensis*, *Hordeum brevisubulatum*, *Potentilla stipularis*). Since Central Yakutia is a densely populated region, anthropogenic effects are apparent even in relatively intact landscapes. Hence, the vegetation cover there often contains

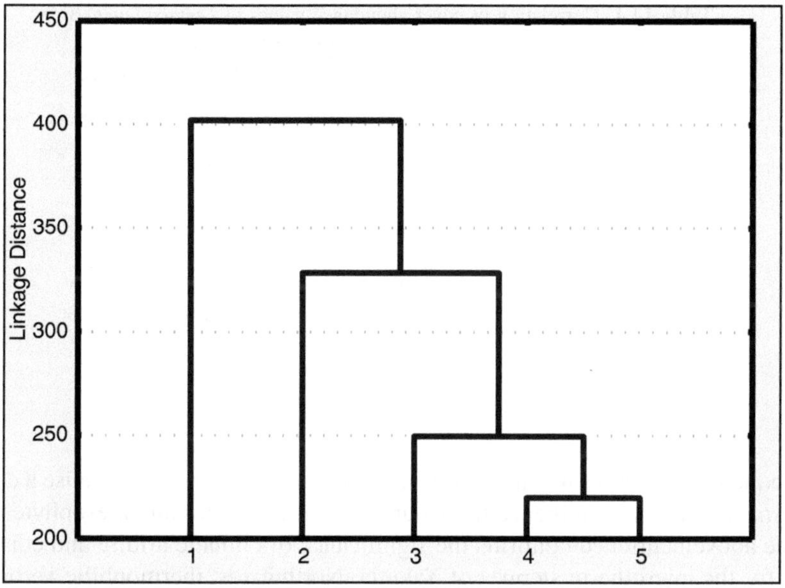

Fig. 14.3 Results of Ward's analysis (Euclidean distances) for the coenofloras of steppes with *Festuca lenensis* in Yakutia. *1* – mesoxerophytic steppes of South-West Yakutia, *2* – xeromesophytic steppes of Central Yakutia, *3* – mesoxerophytic steppes of Central Yakutia, *4* – mesoxerophytic steppes of North-East Yakutia, *5* – xerophytic steppes of North-East Yakutia

ruderal species (*Saussurea amara*, *Artemisia mongolica*, *A. jacutica*). The role of *Festuca lenensis* is least pronounced in these steppes. The mesoxerophytic steppes with *Festuca lenensis* in Central Yakutia (cluster 3) are found on open and properly drained habitats. This explains the impoverishment of its coenoflora (58 species) at the expense of mesophytes. Under drier conditions, the role of *Festuca lenensis* increases and, sometimes, it becomes dominant. However, this may also result from anthropogenic effects. Cluster 4 (53 species) represents the mesoxerophytic variants of steppes with *Festuca lenensis* in North-East Yakutia. They differ from central and south-western steppes by the presence of such species as *Artemisia lagopus*, *Dracocephalum palmatum*, *Calamagrostis purpurascens*, which indicate the cryophytic character of these communities. On the other hand, the petrophytic component (*Thalictrum foetidum*, *Orostachys spinosa*, etc.) makes them related to the steppes of South-West Yakutia. Xeric variants of *Festuca lenensis* steppes (cluster 5, 10 species) are confined to the steepest and most eroded slopes. Under such conditions only the core species of the Yakutian *Festuca lenensis* steppes occur: *Veronica incana*, *Artemisia commutata*, *Carex duriuscula*, *Euphorbia discolor*.

Comparison of cover-abundance values and moisture conditions (Table 14.4) proves that *Festuca lenensis* becomes a dominant species under the xeric and cold conditions of the north-east. This can be explained by the disability of other xerophytic bunch grasses to compete with *Festuca* under such unfavourable conditions.

Table 14.4 Correlation of cover-abundance values of *Festuca lenensis* and moisture conditions in steppe coenoses in Yakutia (Ya)

	Ecological indices values (moisture)[a]	Cover-abundance values[b]
NE Ya, xeric	49.4	4
NE Ya, mesoxeric	52.7	3–4
C Ya, mesoxeric	52.7	1–3
C Ya, xeromesic	55.4	+–2
SW Ya, mesoxeric	51.8	1–2

[a] Based on the ecological indices for Siberia by Korolyuk et al. (2005)
[b] Based on the Mirkin's (Mirkin et al. 2001) scale (from '+' = up to 1% through '5' = 50–100%)

Consequently, increases in temperature and moisture conditions will cause a decline of *Festuca lenensis*' dominance in favour of xeromesophytes and mesophytes.

The abovementioned confirms the significance of climatic aridity and continentality for the microtherm steppes of Yakutia. Northwards, thermophilic xerophytic grasses (*Stipa spp.*, *Psathyrostachys caespitosa*, *Elytrigia jacutorum*) gradually decrease giving way to such cold-resistant species as *Festuca lenensis*. However, the main core of xerotherm and xeropetrophytic species (*Pulsatilla patens s.l.*, *Veronica incana*, *Carex duriuscula*, *Euphorbia discolor*, *Ephedra monosperma*, *Allium spp.*, *Thalictrum foetidum*, *Orostachys spinosa*, *Alyssum obovatum*, *Agropyron cristatum*, etc.) remains present throughout the whole latitudinal gradient, provided the climate is rather dry there. In the North they find favourable habitats in the most insolated and dry parts of the relief. Decrease of aridity is accompanied by enrichment of steppe communities by xeromesophytic and mesophytic species which, with further decline of continentality and rise of precipitation, force out the xerophytes transforming these landscapes from steppes into meadows.

14.5 Effects of Elevation

Climatic (particularly temperature) parameters strongly correlate with altitudes above sea level. Hence, the analysis of floristic change in steppe coenoses along an altitudinal gradient can demonstrate possible transformations of steppe coenoflora due to global warming or cooling. Yakutia is a territory with a rather complex orographic structure and ranges in elevation from lowland tundra (0 m above sea level) to mountain systems in the south and north-east of the republic (the highest point is Pobeda Mnt., 3,147 m a.s.l. in the Chersky Range). The steppe formations occur both in lowlands and on mountain slopes. However, it should be taken into consideration that climatic characteristics of regions at a height of, for instance, 600 m a.s.l. for latitudes 58 NL and 68 NL are not comparable due to strong effects of latitude on temperature. For example, the uppermost sites where the cryoxerophyte *Festuca lenensis* is found are at about 1,200 m a.s.l. at 63 NL, 480 m at 65 NL, and 255 m at 68 NL. Therefore, we decided to focus on a certain region and not to involve in the analysis the whole vast territory of the republic (over 3 million km^2). The middle reaches of the Indigirka River (65–66 NL, 143 EL) were selected as we had sufficient geobotanical data along an altitudinal gradient from 190 to 1,020 m a.s.l. for this region. Considering that the average vertical temperature gradient is 6.5 K km^{-1} (Khromov and Petrosyants 1994), the difference in temperature along the gradient is about 5.4° K (or 5.4°C). The steppe vegetation here occupies both river valleys and slopes of southern aspects. The valley is home for meadow steppes predominated by *Pulsatilla multifida* and *Helictotrichon krylovii* with participation of xeromesophytic forbs (*Castillea rubra*, *Cerastium arvense*, *Linum perenne*, *Oxytropis spp.*, etc.); degraded landscapes are covered with *Carex duriuscula* communities. The slopes bear more xeric variants of steppe coenoses with *Koeleria cristata*, *Orostachys spinosa*, *Selaginella rupestris*, etc. At the border of steppe and forest, where forest communities stabilize moisture conditions, *Helictotrichon krylovii* meadow steppes appear again.

About 50 plots were used in the analysis. Simple correlation of floristic data and altitudinal values yield a pattern of plant response to change in temperature. For the most characteristic species of the local steppe flora (occurring with a constancy of over 50%) the maximum and minimum altitudinal levels of occurrence were recorded (Fig. 14.4). There is no strict confinement of species to a certain altitude. However, the diagram shows four clear patterns in the altitude-temperature relations of the coenoflora of the Indigirka basin steppes.

Pattern I is shown only by *Pulsatilla multifida*. It covers the widest range in altitude (and therefore temperatures), and runs from valley landscapes (190 m a.s.l.) to subalpine shrubberies of *Pinus pumila* (960 m a.s.l.). Pattern II comprises the species that occur from the valleys to the upper limits of the steppe communities, i.e. about 800 m a.s.l. Pattern III comprises the species that occur from 400 to 800 m a.s.l. It should be noted that the uppermost river terraces in this region are at 400–450 m a.s.l., marking the transition from valley floor to slopes. For instance, the endemic species of North-East Siberia *Helictotrichon krylovii* appears from about 400 m, covering wide terraces and climbing up to the upper limits of the steppe communities. These are the species of warm and light slopes. Pattern IV comprises the arctic-alpine species *Dracocephalum palmatum* occurring from 470 m a.s.l. up into the belt of subalpine shrubberies (1,096 m a.s.l.).

In sum, at altitudes below 400 m rich meadow steppes occur growing under favourable temperature and moisture conditions, while over 400 m the air temperature is so low that the steppe communities occupy the most insolated and dry slopes. Low temperatures at this altitude are confirmed by the appearance of the arctic-alpine species *Dracocephalum palmatum* and *Calamagrostis purpurascens*. These are the so-called microtherm-cryophytic steppes. Beyond 800 m a.s.l., where it is still colder, the forest-steppe landscape is gradually replaced by subalpine vegetation in which only isolated patches of steppe vegetation occasionally occur. We have as yet no precise data on limiting temperature values, but based on the results

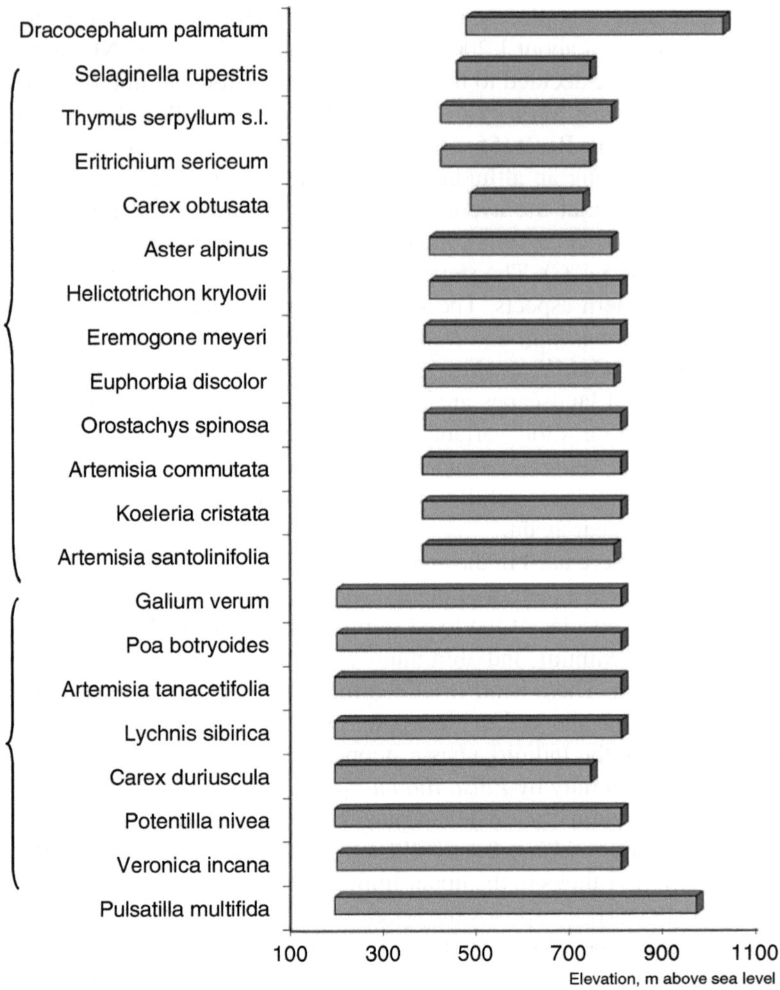

Fig. 14.4 Altitudinal ranges for some representative steppe species of the Indigirka River basin (North-East Yakutia)

given above it seems that a difference of about 2°C is enough for the replacement of prevailing vegetation types.

14.6 Anthropogenic Impacts

Another factor affecting the spatial dynamics and qualitative composition of steppe communities is anthropogenic impact. Since the eleventh to twelfth centuries, the times when the stock-raising tribes settled in Yakutia (Basharin 1956), steppe landscapes have served as natural fodder lands, mostly as pastures. The extreme climatic conditions (cold winters lasting 9 months, short droughty summers, cold soils, and

Photo 14.5 An alas vegetation in Central Yakutia. Xerophytic communities are confined to the upper dry belt (slopes) of the depression (in the foreground of the photo) (Photo by M.J.A. Werger)

perennially frozen grounds) forced the nomadic migrants to change their lifestyle, switching to settled cattle-breeding. During their centuries-old history the Yakuts have worked out the system of scattered stock keeping in small alases and in river valleys. The size of the livestock pool was determined by the capability to maintain them in droughty years. And today, like in ancient times, herds of horses and cattle graze on grasslands of the valleys of the Lena, Tatta and Amga Rivers, in taiga alases (Photo 14.5), as well as in steppe landscapes of North-East Yakutia (the Yana and Indigirka River basins) (Gavrilyeva et al. 2010; Andreyev 1974). Unfortunately, nowadays the fodder lands are misused. Due to absence of management they often experience year-round grazing loads. This leads to degradation of the vegetation of pastures aggravating both their quantitative and qualitative characteristics. For instance, studies of the early twentieth century (Dolenko 1913, Abolin 1929) show the predominance of *Festuca-Stipa* communities on the above-floodplain terraces of the Middle Lena River (Yakutsk vicinity), while presently only fragments of these communities occur there, most of the area being covered with degraded anthropogenic vegetation with *Carex duriuscula* as the dominant species.

The problem of pasture degradation in the valley landscapes of Yakutia was first raised in the 1960s–1970s (Ivanova 1967, 1981; Ivanova and Perfilyeva 1972). Grazing effect on alas steppe vegetation was studied in the 1990s (Mironova 1992; Gavrilyeva 1998; Ivanov et al. 2004). Ivanova (1981) assumed that primary steppe formations, like in South Siberia, represented complex communities containing four grass species: *Stipa krylovii*, *Festuca lenensis*, *Koeleria cristata*, and sometimes *Cleistogenes squarrosa* being observed in the Lena River valley and its slopes. Grazing would lead to transformation of multidominant steppe communities into less complex, monodominant communities. Ivanova supported this theory by observations on steppe formation on high floodplains. But original monodominant *Festuca lenensis* steppes are found in North-East Yakutia (the Adycha River basin). In our opinion this is not a result of grazing but it is explained by the rather harsh environmental conditions in that region.

When grazing starts, *Stipa krylovii* is eliminated first, being most vulnerable to trampling and most preferred by the animals. *Festica lenensis*, being more tolerant to unfavourable anthropogenic influence, often becomes the only dominant species in the steppe communities. Further load leads to replacement of *Festica lenensis* by *Carex duriuscula.*

For alas vegetation Mironova (1992) and Gavrilyeva (1998) defined three stages in the degression series indicative of light, moderate and strong grazing loads, respectively. Accordingly, under light grazing the xerophytic vegetation of the upper belts of alases typically support a *Koeleria cristata-Carex duriuscula* formation, the other species still being vital and reproductive. The next stage signifies replacement of the abovementioned communities by *Elytrigia repens-Taraxacum ceratoforum-Carex duriuscula* featuring invasion of syńanthropic species, more tolerant to grazing (*Artemisia spp.*, *Saussurea amara*, *Plantago depressa*, etc.). Strong grazing results in remarkable changes in floristic composition with predomination of synanthropic species (*Artemisia jacutica*, *Descurainia sophia*, *Lappula squarrosa*, *Polygonum aviculare*). Grass species are eliminated, often bare ground and solonetzic spots appear.

Ivanova (1981) offered another degression series for pastures of valley landscapes. She discerned five stages that are worth of detailed consideration (Fig. 14.5).

I. *Lack of grazing*. The soil is always covered with abundant litter of dead plant parts. The grass stand is represented by 3–4 grass species of which the tall bunch grasses *Stipa krylovii*, *Poa botryioides* or *Elytrigia jacutorum* often dominate. The well-developed litter layer favours moisture accumulation, which in turn determines the presence of mesic species.

II. *Moderate grazing*. Trampling results in the gradual transition of the litter layer into the humus soil horizon. This ensures soil richness. The stage is also characterized by the destruction of the original four-grass floristic structure of the steppes. Communities become monodominant (*Stipa krylovii*, *Agropyron cristatum*, *Koeleria cristata*, *Festuca lenensis*, etc.). Pasture species appear, though in low abundance.

III. *Strong grazing*. Tall grasses that are characteristic for stage I disappear. Pasture species are regular and abundant. Trampling results in soil packing that upsets its capillary conductance and, consequently, yields desiccation and salt accumulation. The steppe has a more xerophytic appearance with participation of halophytes. *Festuca lenensis* still plays the dominant role in this stage. Along with the *Festuca lenensis* formations, *Koeleria cristata* steppes also indicate strong grazing. The first develop on light soils with proper drainage, while the

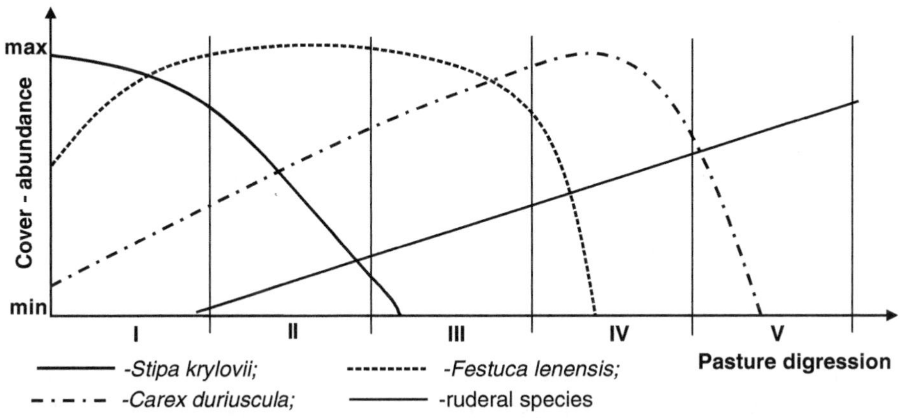

Fig. 14.5 Scheme of dominant species replacement in the degression series of steppe communities (After Ivanova 1981)

latter occur on heavier substrate. The *Koeleria* formations are typically poorer in species and often contain ruderal plants (*Lepidium densiflorum*, *Lappula squarrosa*) (Zakharova et al. 2010).

IV. *Overgrazing*. *Festuca lenensis* and *Koeleria cristata* give way to *Carex duriuscula*. Modern valley landscapes in the middle reaches of the Lena River predominantly bear *Carex duriuscula* anthropogenic steppes, while *Festuca lenensis* communities are of secondary importance.

V. *Grass elimination*. *Carex* steppes are replaced by communities composed of ruderal species.

In our view dominance of *Koeleria* and *Carex duriuscula* indicates a rather strong grazing load, not light grazing. While the classification of Mironova (1992) and Gavrilyeva (1998) explicitly addresses alas vegetation, Ivanova's (1981) classification refers to degradation of steppe vegetation in valley landscapes, but applies to xeric alas vegetation as well, since communities with *Stipa krylovii*, *Psathyrostachys caespitosa* and *Fesuca lenensis* (stages I and II according to Ivanova) are common on the slopes of alas depressions (Gogoleva et al. 1987). Thus, Mironova-Gavrilyeva's stages 1–3 correspond to Ivanova's stages III to V, respectively. Therefore we consider Ivanova's classification more comprehensive and we will use it here.

Grazing results in a change of floristic composition and physical characteristic of soils. Let's discuss both factors separately.

Quantitative changes in vegetation due to pasture degradation were studied for alas vegetation (Gavrilyeva 1998; Ivanov et al. 2004). Alases with the following usage regimes were considered: no grazing – strict grazing control – lack of grazing control – overgrazing. Over this grazing gradient species composition impoverished and species richness dropped from 16–28 to 7–14 species in steppe communities that were year-round used for grazing. Productivity decreased more than two-fold, basically because of the elimination of natural forb species (*Dianthus versicolor*, *Thalictrum simplex*, *Linum perenne*, *Campanula glomerata*, etc.). Xerophytes

(*Carex duriuscula, Koeleria cristata*) and ruderals (*Polygonum aviculare, Lepidium densiflorum, Lappula squarrosa, Potentilla bifurca*, etc.) increased. The sharp change in floristic composition is also expressed in the low values of Jaccard's (1901) coefficient (Table 14.5). Significant pasture loads led to decreasing cover values and shortening of the vertical profile with most phytomass concentrating nearer to the soil surface (Figs. 14.6 and 14.7). As the diagrams show, the quantitative characteristics of communities with strict grazing control exceed those for intact pastures. This proves that total absence of grazing is not the best for a proper soil-vegetation cover. Despite the rich floristic composition, such ungrazed vegetation is ecologically not stable but experiences shrub invasion, and litter accumulation negatively affects the grass stand.

Pasture and recreational degradations feature similar ecological consequences. However, each type of anthropogenic impact has its own peculiarities. Recreation usually does not lead to loss of phytomass and grazing leads to more pronounced changes in floristic composition (Seledets 1976).

Grazing and trampling modify the structure and physical characteristics of the soil. Packed soils feature poor capillary conductance resulting in desiccation and salt accumulation in the root-inhabiting layer. Table 14.6 shows that soils below pastures with strict grazing control have the best water permeability, even better than intact communities. The same pattern is shown in soil density along the grazing gradient (Ivanov et al. 2004). Similar transformation of steppe communities is observed in valley landscapes as well. Over half a century ago Alyokhin (1950) favoured moderate grazing, since cattle destroys last-year's plant remains by consumption or trampling, thus favouring better growth of plants the following spring. Hence, controlled browsing and regular surface improvement of pastures provide optimal conditions for retaining the ecological balance in steppe communities and their natural self-regulation (Ivanov et al. 2004).

While pasture rotation and regulation of pasture load is important, reconstruction of degraded pastures needs top soil improvement by mechanical treatment together with application of mineral fertilizers (Matveyev 1981). But nevertheless, also then ruderal species, especially of *Artemisia*, increase. Thus, additional sowing of desirable grasses is required. Since the 2000s the agrosteppes method (Dzybov and Denschikova 2003) designed for the steppe zone of Russia, seems an attractive alternative. The method implies recovery of disturbed stands by sowing of mixtures of natural grass seeds collected from intact communities with a rich species diversity and high cover values. An advantage of this method is that a minimal technogenic treatment is applied. The results of a 5-year experiment (Nezdiyminoga 2009, 2010) under the conditions of Central Yakutia (hot and arid summer, high soil salinity) demonstrated the recovery of the vegetation in degraded landscapes, with increased cover and predomination of natural species from intact communities, provided that similarity in ecological conditions and species composition of the degraded and intact communities (cf. Korolyuk et al. 2005) is observed (Table 14.7). The agrosteppe method has proved its effectiveness for impoverished xeric vegetation situated in the urban area under conditions of insufficient moisture and soil salinity on dry terraces of the Middle Lena River valley. The recovery succession appeared to take 4–6 years under Yakutian conditions compared to 2–3 years in southern Russia.

Table 14.5 Change in basic indices of the vegetation of xeric belts of alases in the course of pasture degradation

		Productivity		Agrobotanical composition, in 100 kg.ha–[1]			Vertical structure			
Degression stage[a]	Cover, %	In 100 kg.ha–[1]	%[b]	Grasses	Sedges	Forbs	Total height, in cm	A[c]	Number of species per 10 m^2	Jaccard, %
II	45	8.2	100	1.6	1.4	5.2	50	33% – 10–20	21	II–III: 19.3
III	45	3.8	46.3	1.6	1.7	0.5			17	III–IV: 52.3
IV	50	2.8	34.1	0.9	1.2	0.7	15	68% – 0–5	15	II–IV: 9.4

According to Gavrilyeva (1998)
[a]According to Ivanova's (1981) classification
[b]Decrease in productivity as a percentage of the productivity in stage II
[c]Phytomass (%) in an aboveground vegetation layer (cm)

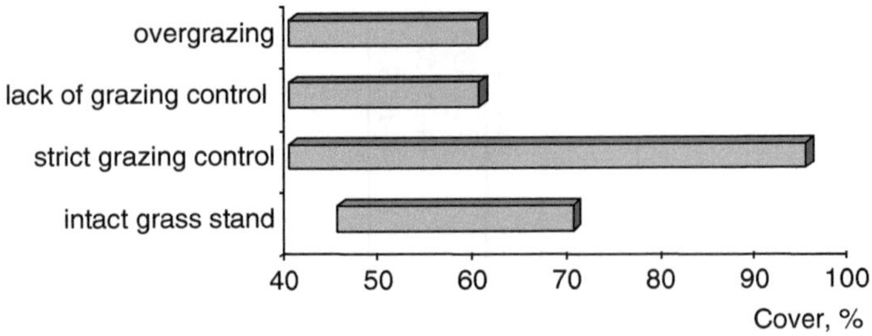

Fig. 14.6 Cover values of xerophytic communities in alases under various grazing loads

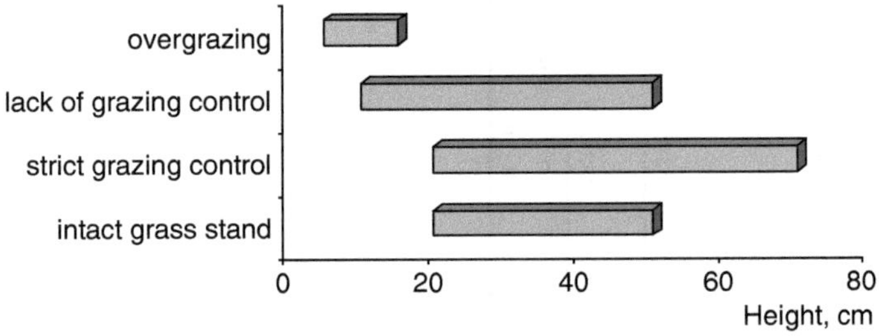

Fig. 14.7 Height of grass stand of xerophytic communities in alases under various grazing loads

Table 14.6 Water permeability of frozen alas soils under pastures with various loads of grazing, mm/h

	Pastures with various regimes of usage			
Hour of observation	Year-round grazing	Lack of grazing control	Strict grazing control	Intact meadow steppe
1st	77	46	320	147
2nd	64	26	244	215
3rd	45	33	133	88.5

According to Ivanov et al. (2004)

Table 14.7 Cover value dynamics of steppe communities reconstructed by the agrosteppe method

	Years				
Parameters	2004	2005	2006	2007	2008
Number of species	19	9	9	9	9
Cover, %					
Total	60	60	70	75	80
Steppe and meadow grasses	25	40	55	75	80
Weed species	35	20	15	+	+

From Nezdiyminoga (2009)

14.7 Outlook

After having become a dominating landscape and then having declined as a result of climatic fluctuations during the Quaternary, the remaining steppe vegetation of Yakutia experiences significant anthropogenic stress. Thanks to the fact that some remaining steppe landscapes are located in remote and uninhabited areas, the biodiversity of the steppe vegetation has a chance to be preserved. However, the most common types of steppe (especially the meadow and true bunchgrass variations) suffer from overgrazing and recreation load and gradually transform into ruderal communities.

We believe that aridity of the climate is one of most important and essential conditions for the existence of steppes. Hence, changing climatic factors aggravated by anthropogenic impact may result in serious and sometimes irrevocable and unpredictable consequences, from the elimination of steppe species from communities to a strong reduction of the area of steppe landscapes due to forest expansion. Modern geobotanical studies in the north-east of Yakutia (the Indigirka River basin) prove that forest communities give way to steppe vegetation in the course of post-fire succession affected by anthropogenic and zoogenic impact (Troeva et al. 2010). However, despite this, inevitable temperature increases and rise of air humidity will yield environmental conditions that are unfavourable for the growth of xerophytic plants. And this will lead to a reduction in biodiversity both at the floristic and the community levels.

For conservation of the unique extrazonal steppe landscapes, remnants of the Late Pleistocene, establishment of steppe reserves is required in the regions with the highest biodiversity of steppe communities, i.e. the Indigirka and Yana River basins. Such protected areas should be of small size and under controlled and restricted human activity.

References

Abolin RI (1929) Geobotanicheskoye i pochvennoye opisaniye Leno-Viluyskoy ravniny. In: Trudy komissii po izucheniyu Yakutskoy ASSR 10. Leningrad. – Geobotanical and soil description of the Lena-Viluy Plain (in Russian)

Alfimov AV, Berman DI, Sher AV (2003) Tundrostepnye gruppirovki nasekomykh i rekonstruktsia klimata pozdnego pleistotsena nizoviy Kolymy. Zoologichesky zhurnal 82:281–300. – Tundra-steppe groupings of insects of Late Pleistocene in the Lower Kolyma (in Russian)

Alyokhin VV (1950) Geographiya rasteny (osnovy fytogeografii, ekologii i geobotaniki). Uchpedgiz, Moscow. – Plant geography (principles of phytogeography, ecology and geobotany) (in Russian)

Andreyev VN (ed) 1974 Tebenyovochnye pastbischa Severo-vostoka Yakutii. Yakutskoye knizhnoye izd-vo, Yakutsk. – Winter pastures of North-eastern Yakutia (in Russian)

Andreyev VN, Galaktionova TF (1981) Relictovye tundrostepnye gruppirovki v Nizhnekolymskoy tundre. In: Andreyev VN (ed) Rastitelnost Yakutii i eyo okhrana. Izd-vo YaF SO AN SSSR, Yakutsk. – The relic tundra-steppe groups in the Lower Kolyma tundra (in Russian)

Andreyev VN, Perfilyeva VI (1975) Fragmenty stepnoy rastitelnosti na Severo-Vostoke Yakutii. In: Byulleten NTI: Biologicheskiye problemy Severa. Izd-vo YaF SO AN SSSR, Yakutsk:14–16. – The fragments of steppe vegetation in North-East Yakutia (in Russian)

Andreyev VN, Galaktionova TF, Perfilyeva VI, Scherbakov IP (1987) Osnovnye osobennosti rastitelnogo pokrova Yakutskoy ASSR. Izd-vo YaF SO AN SSSR, Yakutsk. – Basic features of vegetation cover of the Yakutian ASSR (in Russian)
Andreyev AA, Schirrmeister L, Grosse G et al (2007) Rastitelnost i klimat v raione morya Laptevykh v pozdnem pleistotsene i golotsene po pyltsevym dannym. In: Fundamentalnye problemy kvartera: itogi izucheniya i osnovnye napravleniya dalneyshikh issledovany. Proceedings of 5th all-Russian meeting on the study of Quaternary. GEOS, Moscow: 12–13. – Vegetation and climate in the region of Laptev Sea during Late Pleistocene and Holocene according to pollen data (in Russian)
Basharin GP (1956) Istoriya agrarnykh otnosheniy v Yakutii. Moscow. – The history of agrarian relations in Yakutia (in Russian)
Belova VA (1970) Nekotorye osobennosti formirovaniya rastitelnosti Baikalskoy kotloviny v pleistotsene. In: Zapadno-Zabaikalsky filial Geographicheskogo obschestva SSSR. Collection of articles. Irkutsk. Issue 51:56–61. – Some peculiarities of vegetation development in Baikal depression in Pleistocene (in Russian)
Belova VA (1985) Rastitelnost i klimat pozdengo kainozoya yuga Vostochnoy Sibiri. Nauka, Novosibirsk. – Vegetation and climate of Late Cainozoic in the south of East Siberia (in Russian)
Budyko MI (1980) Klimat v proshlom i buduschem. Gidrometeoizdat, Leningrad. – Climate in the past and the future (in Russian)
Budyko MI, Israel YuA, McCracken MS, Hecht AD (eds) (1991) Predstoyascheye izmeneiye klimata. Sovmestny sovetsko-amerikansky otchyot o climate i ego izmeneniyakh. Gidrometeoizdat, Leningrad. – Forthcoming climate changes. Joint Soviet-American report on climate and its changes (in Russian)
Cherosov MM, Sleptsova NP, Mironova SI et al (2005) Sintaksonomiya sinantropnoy rastitelnosti Yakutii. YaF Izd-va SO RAN, Yakutsk. – Syntaxonomy of the synanthropic vegetation of Yakutia (in Russian)
Dolenko GI (1913) Dolina reki Leny bliz Yakutska. In: Predvaritelny otchyot po organizatsii i issledovaniyu rabot po issledovaniyu pochv Aziatskoy Rossii v 1912 godu. Saint-Petersburg. – The Lena River valley in Yakutsk vicinity (in Russian)
Dzybov DS, Denschikova TYu (2003) Osnovy biologicheskoy rekultivatsii narushennykh zemel. GUP "Stavropolskaya kraevaya tipografia", Stavropol. – Principles of biological revegetation of disturbed lands (in Russian)
Gavrilova MK (1993) Climate and permafrost. Permafrost Periglac Process 4:99–111
Gavrilyeva LD (1998) Pastbishnaya digressiya i ratsionalnoye ispolzovaniye rastitelnosti alasov Leno-Amginskogo mezhdurechya. Abstract of Thesis for a Candidate degree. Yakutsk. – Pasture degradation and rational use of alas vegetation in the Lena-Amga Interfluve (in Russian)
Gavrilyeva LD, Sofronov RR, Arzhakova AP, Barashkova NV, Ivanov IA (2010) Hayfields and pastures. In: Troeva EI et al (eds) The Far North: plant biodiversity and ecology of Yakutia. Series plant and vegetation, vol 3. Springer Science+Business Media, Dordrecht, pp 275–281
Giterman RE 1985 Istoriya rastitelnosti Severo-Vostoka SSSR v pliotsene i pleistotsene. Nauka, Moscow. – History of vegetation of the USSR during Pliocene and Pleistocene (in Russian)
Gogoleva PA, Kononov KE, Mirkin BM, Mironova SI (1987) Sintaksonomiya i simfitosotsiologiya rastitelnosti alasov Tsentralnoy Yakutii. Izd-vo Irkutskogo GU, Irkutsk. – Syntaxonomy and symphytosociology of alas vegetation of Central Yakutia (in Russian)
Israel YA, Pavlov AV, Anokhin YA (1999) Analiz tekuschego i ozhidaemogo izmeneniya klimata i kriolotozony v Rossiyskoy Arktike. Meteorologiya i gydrologiya 3:18–27. – Analysis of current and expected change of climate and cryolithozone in the Arctic of Russia (in Russian)
Ivanova VP (1967) O vliyanii vypasa na stepnuyu rastitelnost v doline r. Leny. In: Lyubite i okhranyaite prirodu Yakutii (Proceedings of 4th Republican meeting on nature conservation in Yakutia). Yakutskoye knizhnoye izdatelstvo, Yakutsk:86–93. – Grazing effect on steppe vegetation in the Lena River valley (in Russian)
Ivanova VP (1981) Tipchakovye stepi – odin iz etapov pastbischnoi digressii rastitelnosti v doline srednei Leny. In: Andreyev VN (ed) Rastitelnost Yakutii i eyo okhrana. YaF SO AN SSSR,

Yakutsk:37–56. – Festuca lenensis steppes as one of stages of pasture digression in the Middle Lena Valley (in Russian)

Ivanova VP, Perfilyeva VI (1972) Sokhranit kovylnye stepi Yakutii In: Priroda Yakutii i eyo okhrana (Proceedings of 6th Republican meeting on nature conservation in Yakutia). Yakutskoye knizhnoye izdatelstvo, Yakutsk:116–121. – Conserving Stipa steppes of Yakutia (in Russian)

Ivanov AA, Mironova SI, Savvinov DD (2004) Alasnye luga Leno-Viluyskogo mezhdurechya Tsentralnoy Yakutii pri razlichnykh rezhimakh ispolzovaniya. Nauka, Novosibirsk. – Alas meadows of the Lena-Viluy interfluve of Central Yakutia under various regimes of use (in Russian)

Jaccard P (1901) Etude comparative de la distribuition florale dans une portion des Alpes et des Jura Bull Soc Vaudoise Sci Nat 37:547–579 (in French)

Karavaev MN (1958) Fragmenty reliktovykh stepey s *Helictotrichon krylovii* (N. Pavl.) Henrard v Yakutii. Botanichesky zhurnal 43:481–489. – Fragments of relic steppes with *Helictotrichon krylovii* (N. Pavl.) Henrard in Yakutia (in Russian)

Khromov SP, Petrosyants MA (1994) Meteorologiya i klimatologiya. Izd-vo MGU, Moscow. – Meteorology and climatology (in Russian)

Kienast F, Schirrmeister L, Siegert C, Tarasov P (2005) Paleobotanical evidence for warm summer during the last cold stage. Quaternary Res 63:283–300

Konischev VN (1995) Evolyutsia kriosfery v razlichnye istoricheskie etapy Zemly. Vestnik MGU Series Geographicheskaya 1:8–15. – Cryosphere evolution during different historical stages of the Earth (in Russian)

Korolyuk AYu, Cherosov MM, Troeva EI et al (2005) Ekologicheskaya otsenka flory i rastitelnosti Tsentralnoi Yakutii. Izd-vo YaNTs SO RAN, Yakutsk. – Ecological evaluation of flora and vegetation of Central Yakutia (in Russian)

Koropachinsky IYu (ed) (1996) Zelyonaya kniga Sibiri: redkiye i nuzhdayuschiesya v okrane rastitelnye soobschestva. Nauka, Novosibirsk. – Green Book of Siberia: Rare and protection-requiring plant communities (in Russian)

Kozhevnikov YuP, Ukraintseva VV (1996) Tundrostepi pleistotsena: argumenty “za” i “protiv”. Izvestiya RAN (geographiya) 3:96–110. – The tundra-steppes of the Pleistocene: arguments pro and contra (in Russian)

Matveyev IA (1981) Effektivnost intensifikatsii selskogo khozyaistva v Yakutii. Kolos, Moscow. – Effectivity of intensification of agriculture in Yakutia (in Russian)

Maximov TCh, Kononov AV, Petrov KA, Ivanov BI (2010) Structural and functional peculiarities of the plants of Yakutia. In: Troeva EI et al (eds) The Far North: plant biodiversity and ecology of Yakutia. Series plant and vegetation, vol 3. Springer Science+Business Media, Dordrecht, pp 317–355

Mirkin BM, Naumova LG, Solomesch AI (2001) Sovremennaya nauka o rastitelnosti. Logos, Moscow. – Modern science of vegetation (in Russian)

Mironova SI (1992) Pastbischnaya digressia alasnykh lugov. In: Zakharov ID et al (eds) Intensifikatsia lugovogo kormoproizvodstva Yakutii (Collection of articles). Izd-vo SO RASKHN, Novosibirsk:73–80. – Pasture digression of alas grasslands (in Russian)

Monin AS, Shishkov YuA (1979) Istoriya klimata. Gidrometeoizdat, Leningrad. – History of climate (in Russian)

Nezdiyminoga OV (2009) Botanicheskiye i populyatsionnye parametry dinamiki vosstanovleniya rastitelnosti narushennykh uchastkov doliny sredney Leny metodom agrostepey. Vestnik Buryatskoy gosudarstvennoy selkhozakademii 3(16):94–98. – Botanical and population parameters of restoration dynamics of vegetation of disturbed sites of the Middle Lena River valley by use of the agrosteppe method (in Russian)

Nezdiyminoga OV (2010) Effektivnost metoda agrostepey dlya vosstanovleniya rastitelnosti narushennykh kormovykh ugodiy Tsentralnoy Yakutii. Agrarnaya nauka 5:4–5. – Agrosteppe method efficiency for restoration of natural fodder lands in Central Yakutia (in Russian)

Protopopov AV, Protopopova VV (2010) History of vegetation development. In: Troeva EI et al (eds) The Far North: plant biodiversity and ecology of Yakutia. Series plant and vegetation, vol 3. Springer Science+Business Media, Dordrecht, pp 151–156

Sakalo DI (1963) Ekologicheskaya priroda stepnoy rastitelnosti Evrazii i eyo proiskhozhdenie. In: Materialy po istorii flory i rastitelnosti SSSR, vol 4. Izd-vo AN SSSR, Moscow, Leningrad, pp 407–425. – Ecological nature of steppe vegetation of Eurasia and its origin (in Russian)

Sekretareva NA (2004) Sosudistye rasteniya Rossiyskoy Arktiki i sopredelnykh territory. Tovarischestvo nauchnykh izdaniy, Moscow. – Vascular plants of Russian Arctic and adjacent territories (in Russian)

Seledets VP (1976) Vliyaniye antropogennykh faktorov na rastitelnost okrestnostey Yakutska. Byulleten Glavnogo botanicheskogo sada 99:70–75. – Effect of anthropogenic factors on vegetation in the vicinities of Yakutsk (in Russian)

Skryabin SZ (1968) Stepnaya rastitelnost v srednem techenii Indigirki. Abstract of Thesis for a Candidate degree. Yakutsk. – The steppe vegetation in the Middle Indigirka River (in Russian)

Skryabin SZ (1976) Fragmenty gornoy petrofitnoy stepi v nizovye reki Tokko (Yuzhnaya Yakutia). Botanichesky zhurnal 61:1748–1751. – The fragments of a mountainous petrophytic steppe in the lower reaches of the Tokko River (South Yakutia) (in Russian)

Skryabin SZ, Karavaev MN (1991) Zelyony pokrov Yakutii. Knizhnoye izd-vo, Yakutsk. – The green cover of Yakutia (in Russian)

Solomina ON (1999) Gornoye oledeneniye severnoy Evrazii v Golotsene. Nauchny Mir, Moscow. – Alpine glaciation in northern Eurasia (in Russian)

Stuiver M, Grootes PM, Braziunas TF (1995) The GISP2 18O climate record of the past 16,500 years and the role of the sun, ocean and volcanoes. Quaternary Res 44:341–354

Tolmachev AI (1954) K istorii vozniknoveniya i razvitiya temnokhvoynoy taigi. Izd-vo AN SSSR, Moscow-Leningrad. – On the history of the origin and development of the dark coniferous taiga (in Russian)

Tomirdiaro SV (1980) Lyossovo-ledovaya formatsiya Vostochnoy Sibiri v pozdnem pleistotsene i golotsene. Nauka, Moscow. – Loess-ice formation of East Siberia during Late Pleistocene and Holocene (in Russian)

Tomirdiaro SV, Chyornenky BI (1987) Kriogenno-eolovye otlozheniya Vostochnoy Sibiri Subarktiki. Nauka, Moscow. – Cryogenic-eolian sediments of Subarctic East Siberia (in Russian)

Tomskaya AI (2000) Kormovaya baza mamonta v pozdnem pleistotsene Yakutii. Yakutsk. – Fodder base of mammoths in the Late Pleistocene in Yakutia (in Russian)

Troeva EI, Cherosov MM, Sergeyev VA (2010) Kharakteristika tipov kormovykh ugodiy doliny Indigirki (na uchastke Khonuu-Inyali). In: Remigailo PA (ed) Geobotanicheskiye i resursovedcheskiye issledovaniya v Arktike (Collection of articles). YaNTs SO RAN, Yakutsk, pp 174–180. – Description of fodder land types of the Indigirka River valley (section Khonuu-Inyali) (in Russian)

Ukraintseva VV (2002) Rastitelnost i klimat Sibiri epokhi mamonta. Izd-vo Polikom, Krasnoyarsk. – Vegetation and climate of Siberia during the mammoth epoch (in Russian)

Vasilchuk YuK, Kotlyakov VM (2000) Osnovy izotopnoy geokriologii. Izd-vo MGU, Moscow. – Principles of isotopic geocryology (in Russian)

Yurtsev BA (1974) Stepnye soobschestva chukotskoy tundry i pleistotsenovaya "tundrostep". Botanichesky zhurnal 59:484–501. – The steppe communities of the Chukchi tundra and the Pleistocene "tundra-steppe" (in Russian)

Yurtsev BA (1981) Reliktovye stepnye kompleksy Severo-Vostochnoy Azii: problemy rekonstruktsii kriokseroticheskikh landshaftov Beringii. Nauka, Novosibirsk. – The relic steppe complexes of the North-East Asia: problems of reconstruction of the cryoxeric landscapes of Beringia (in Russian)

Zakharova VI (2005) K flore petrophitnykh stepey okhranyaemykh territory. In: Cherosov MM, Gogoleva PA, Simonova SV (ed) Flora i rastitelnost kriolitozony 2: Rastitelnost kriolitozony. Izd-vo YaNTs SO RAN, Yakutsk. – On the flora of the petrophytic steppes of protected territories (in Russian)

Zakharova VI, Cherosov MM, Troeva EI, Gogoleva PA (2010) Steppes. In: Troeva EI et al (eds) The Far North: plant biodiversity and ecology of Yakutia. Series plant and vegetation, vol 3. Springer Science+Business Media, Dordrecht, pp 193–198

Chapter 15
Influence of Climate Change on Vegetation and Wildlife in the Daurian Eco-region

Vadim E. Kirilyuk, Viktor A. Obyazov, Tatyana E. Tkachuk, and Olga K. Kirilyuk

Abstract One of the most remarkable natural peculiarities of Dauria is an alteration of dry and wet stages within about 30-year climate cycles, which affect habitats and biota very strongly. During dry stage biomass and bioproductivity decrease, some habitat types diminish or disappear. Most of species both aquatic and terrestrial survive drought using different adaptations. The vegetation is adapted to climate cycles and resiliently reacts with fluctuations and cyclical successions. The distribution areas of many ground vertebrates pulsate in concord with the climate cycles. But the continuing warming gradually destroys the complete reversibility of these processes and leads to aridization: the dry stage of the cycle has mostly negative consequences including a decreasing sustainability of the ecosystem complexes and shifts in the range limits and migration routes of animals. Many vertebrate species barely survive.

V.E. Kirilyuk • O.K. Kirilyuk (✉)
Daursky State Nature Biosphere Reserve, Chita, Russia

Institute of Nature Resources, Ecology and Cryology, Siberian Branch of the Russian Academy of Sciences, Chita, Russia
e-mail: vkiriliuk@bk.ru; kiriliuko@bk.ru

V.A. Obyazov
Institute of Nature Resources, Ecology and Cryology, Siberian Branch of the Russian Academy of Sciences, Chita, Russia

T.E. Tkachuk (✉)
Daursky State Nature Biosphere Reserve, Chita, Russia

Zabaikalsky State Humanitarian Pedagogical University, Chita 672020, Russia
e-mail: tetkachuk@yandex.ru

M.J.A. Werger and M.A. van Staalduinen (eds.), *Eurasian Steppes. Ecological Problems and Livelihoods in a Changing World*, Plant and Vegetation 6, DOI 10.1007/978-94-007-3886-7_15,

15.1 Introduction

In the last decades considerable climate changes of a global character are taking place. However, at the regional level they differ much from the general global level, and have peculiar features per region. Even greater differences between regions are observed in the ecosystems's reactions to climatic variability, which to a great extent depend on the natural and economic conditions of the regions.

One of the regions that is ecologically strongly dependent on climate changes is the Daurian steppe. Dauria lies in the northern part of Central Asia. Most of the Daurian steppe area is situated in North-East China and East Mongolia; the Russian part is confined to Zabaikalsky Krai and Buryat Republic. In terms of the WWF Terrestrial Ecoregions of the World the Daurian Forest-Steppe covers the Nenjiang River grassland, the Daurian forest-steppe, the Mongolian-Manchurian steppe, and the Selenge-Orkhon forest-steppe eco-regions (Olson and Dinerstein 1998). These grassland areas are united by geographic location, annual and multi-year rhythms in ecological factors, and structure and composition of communities (Tkachuk et al. 2008).

The region lies at 600–800 m above sea level, comprises mainly plains and rolling relief, and has an ultra-continental climate. Low winter temperatures (in the Russian part the average January temperature is –25°C) result in deep freezing of the soils and the formation of permafrost pockets. This favors the cryophytic character of the Daurian steppes. The spring is cold, windy and dry, while most of the rainfall coincides with the highest annual temperatures during the second half of the summer. This leads to a highly intensive cycling of nutrients in the short summer period and, as a result, to the formation of primarily poor, shallow soils. Widespread occurrence of stony soils explains the important role of long-rooted herbs, not only in meadow steppe, but also in typical steppe communities (Gadgiev et al. 2002).

The Central Asian Steppe Sub-Region of the Eurasian Steppe Region (Lavrenko 1970) has a flora that is notably different from that of the steppes to the west, with genera such as *Cymbaria*, *Saposhnikovia*, *Filifolium*, *Panzeria*, *Schizonepeta*, *Stellera*, *Lespedeza*, etc. The *Stipa* spp. dominating the Daurian steppes belong to *Capillatae*, different from those of western steppes. Communities dominated by *Filifolium sibiricum (Asteraceae)*, *Leymus chinensis*, *Stipa baicalensis (Poaceae)* are not found in other grassland areas (Karamysheva 1993). The most typical steppe communities are dominated by *Stipa krylovii*, *S. baicalensis*, *Leymus chinensis*, *Artemisia frigida*, etc. Shallow salt lakes with halophytic vegetation around them are also characteristic of the region. Grass-dominated grassland communities are intermingled with other vegetation types (wetlands, saline vegetation, forest groves, bush, etc.) and should be described and preserved only in a broader landscape context.

The animal world of Dauria is diverse. There are species here whose main distribution area is much farther to the east, west or south (Japanese or Red-crowned crane (*Grus japonensis*), Manchurian tsokor (*Myospalax psilurus epsilanus*), Mongolian gerbil (*Meriones unguiculatus*), Common crane (*Grus grus*), etc.). Among the endemics of Dauria are the Mongolian gazelle (*Procapra gutturosa*) (up to 90% of the world population of the species lives here), Daurian hedgehog

(*Mesechinus dauuricus*), Daurian souslik (*Spermophilus dauricus*), Mongolian lark (*Melanocorypha mongolica*), Upland buzzard (*Buteo hemilasius*) and others. A peculiar feature of the animal world is its variety of bird species determined by the narrowing of the global migration routes of birds in the Dalainor-Torey depression (Goroshko 2009). The number of transitory migrants in the region's ornithofauna is not less than 45%. Here more than 40 bird species are registered, which are listed both in the Red List of IUCN and the national Red Data Books of Russia, Mongolia, and China.

On the whole an important peculiarity of the region is that the amount of endemic natural communities in Dauria, which have been formed with participation of different floras and faunas under conditions of permanent climate changes, is much higher than the amount of endemic species.

Daurian steppes, which were included by WWF into the list of 200 globally important eco-regions, possess a high level of biodiversity for a steppe zone. Under the conditions of global warming and the cyclical changes in moisture availability characteristic for the region, appreciable changes in species composition, abundance and spatial distribution of wildlife take place. Here we discuss the direct and indirect impacts of the main consequences of climate changes on communities of wildlife and vegetation, and give examples.

15.2 Peculiarities of the Region's Climate Change

The climate of the eco-region is discussed taking the Onon-Argun inter-river area, which has been monitored instrumentally for a long time, as an example. The climate of the adjacent parts of the eco-region located in Mongolia and China alter accordingly, but there is a tendency of a considerable decrease in the annual amount of precipitation from the east (foothills of the Great Hingan) to the west. Average annual air temperatures in the Onon-Argun area change throughout the territory from low positive degrees to the degrees below −3°C. Long-term changes in air temperature at different meteorological stations vary correspondingly. The longest series of observations at the meteo-station Nerchinsky Zavod gives the possibility to estimate century changes in air temperature (Fig. 15.1).

In the chart of average annual temperature (Fig. 15.1) three periods can be distinguished. From the middle of the nineteenth century till the beginning of the twentieth century there was an increase in temperature. Then up to the middle of the twentieth century the trend was about zero. In the early 1950s a new period of warming began. The rise in temperature in that period is characteristic of the entire area between the Argun and the Onon rivers (Fig. 15.2). Moreover, at the transition from the 1980s to the 1990s a leap in temperature occurred and the last two decades stand out with especially high temperatures. In that period during 7 years (1990, 1993, 1995, 1998, 2002, 2007, 2008) the average annual temperature exceeded the highest values of the previous year. The highest temperatures were registered in 2007 and 2008, but over the entire two decades on average the annual temperature

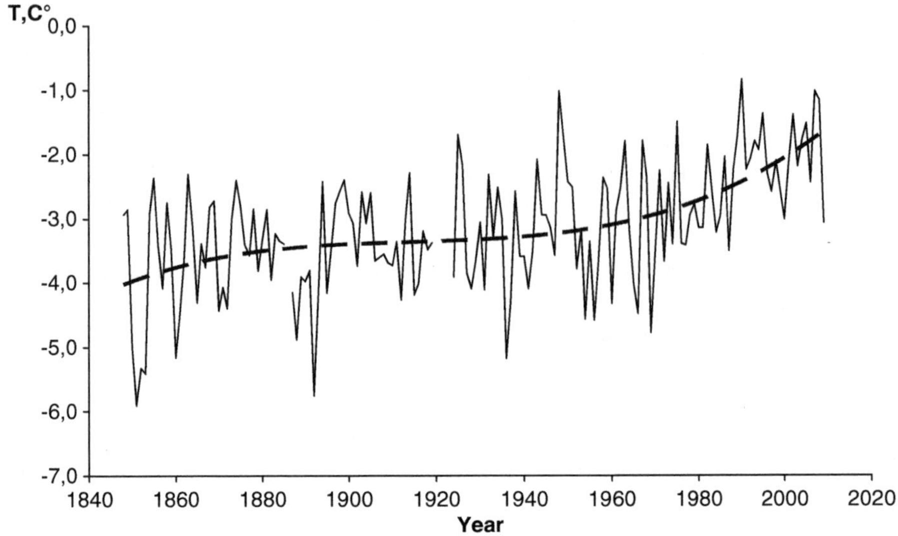

Fig. 15.1 Long-term changes in average annual air temperature according to the data of the Nerchinsky Zavod meteo-station during the period of 1848–2009. Original series and polynomial trend

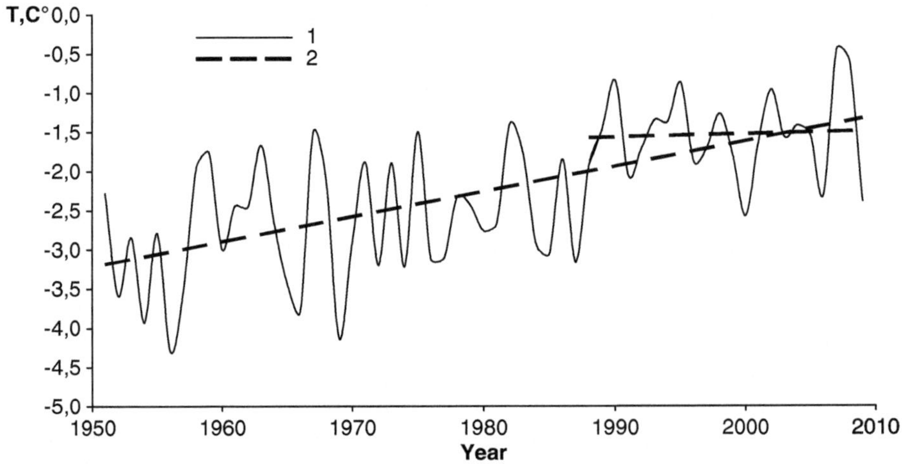

Fig. 15.2 Long-term changes in average annual air temperature for the Onon-Argun inter-river area in the period 1951–2009. *1* original series, *2* linear trend

stayed equally high. This pattern in temperature changes is seen more or less equally in the data of all observation points located in different parts of the Onon-Argun inter-river area.

In the period 1951–2009 the average annual temperature in the study area increased by 1.9°C, and in different parts of the study area the linear trend showed

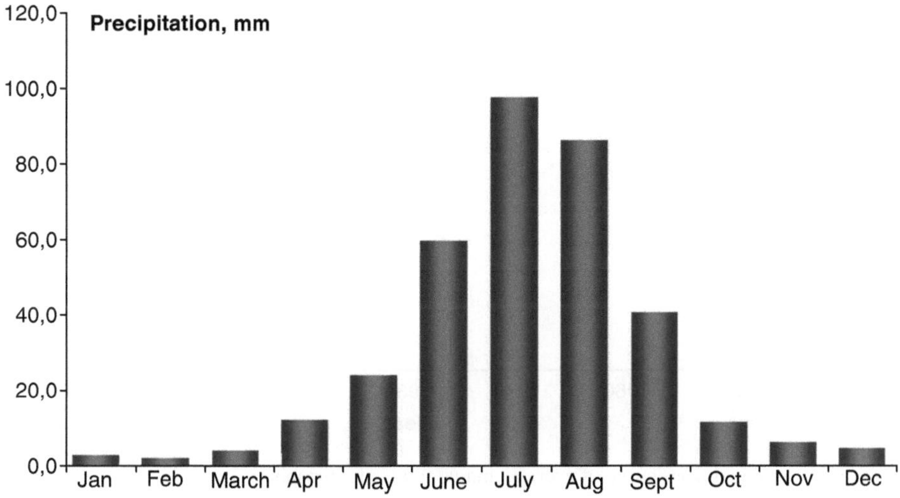

Fig. 15.3 Annual distribution of the precipitation in the Onon-Argun inter-river area

increases from 1.5°C up to 2.2°C during 59 years (Fig. 15.2). This trend is statistically significant ($p<0.05$). It led to an increase of the period with positive temperatures in the northern part of the Daurian steppe from 165–167 to 173–179 days.

The strongest increase has been registered in February, and only slightly lower in March and April. These 3 months take half of the total increase in average annual temperatures. The lowest figures are for the period of October–December. Thus the strongest increase is observed in spring, the least in autumn. For the cold period of the year (October–April) the rise in air temperature amounts to 2.4°C over 59 years; for the warm period of the year (May–September) to 1.3°C ($p<0.05$).

The amount of precipitation in the Onon-Argun inter-river area is uneven. Its annual quantity alters throughout the study area from less than 300 mm in the environs of the Torey lakes to more than 400 mm in the north-east of the area.

The distribution of precipitation in the Onon-Argun inter-river area over the year is extremely uneven (Fig. 15.3). The largest amount falls in summer and is on average about 68% of the annual total, and the largest quantity comes in July (around 27% of annual total). During the 3 winter months only 3% of the precipitation falls, with in January and February each less than 1% of the annual total (or around 3 mm per month).

The amount of precipitation in the Onon-Argun inter-river area decreased since the middle of last century by 45 mm on average, which is about 13% of the annual total. But the trend is statistically not significant at the 5% level.

Long-term changes in precipitation are cyclic, and in South-eastern Transbaicalia cycles are most vividly apparent within the timespan of a century (Obyazov 1994). From 1955 to 1963 there was a period with above average precipitation, followed, up to 1982, by a below average period. In 1983 a wet period started again, lasting to 1998. 1999 was the beginning of a new dry period, and it is as yet hard to say when

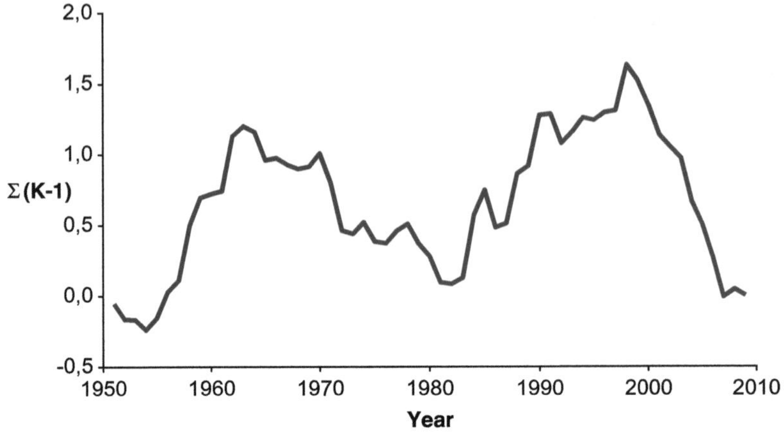

Fig. 15.4 Residual mass curve of average annual totals of atmospheric precipitation in the Onon-Argun inter-river area

that ends. In any case during 2008–2010 no clear changes in river flow and lake levels were noticeable.

Thus, up to two full cycles in precipitation occurred in the Onon-Argun inter-river area over the past 60 years (Fig. 15.4). Apart from these decennia-long cycles also cycles of another wave length occur in the area, including a 4–5 year cycle (Obyazov 1994). Cycles identified are based on classical statistical methods for analyzing climate change, in particular by the construction of residual mass curves (Vladimirov 1990).

As regards the annual distribution of the precipitation, in the entire Zabaikalsky krai precipitation in the cold period increased since 1955, causing a rise in snow cover of 2–7 cm, or 20–50% of the average snow height. Another clearly recognizable trend is that in the dry phases of the 30-year humidification cycle the average snow height is higher than in wet phases. At the station in Nizhny Tsasuchey, in two analysed dry periods (1964–1982 and 1999–2010) in the winter months (November–March), there was 31.7% more precipitation (20.17 mm versus 15.31 mm) than in the corresponding wet periods (1955–1963 and 1983–1998).

The correlation between quantity of precipitations and air temperature is very weak.

Long-term changes in river flows in the Onon-Argun inter-river area are highly correlated. The average correlation coefficient for the study area is 0.8, the highest ones, exceeding 0.9, being characteristic of the rivers of the Onon basin, and also between the Onon and the Shilka rivers, and the lowest (less than 0.6) between the rivers of the Argun and the Onon basins. This synphasic and cyclic pattern is clearly shown in Fig 15.5. In the period 1956–2008 two full cycles can be distinguished: from 1956 to 1964 a full-flow phase of 9 years, followed by a low-flow phase of 18 years (from 1965 to 1982), and again a full-flow phase of 15 years (1983–1998). In 1999 a low-flow phase began and is still continuing.

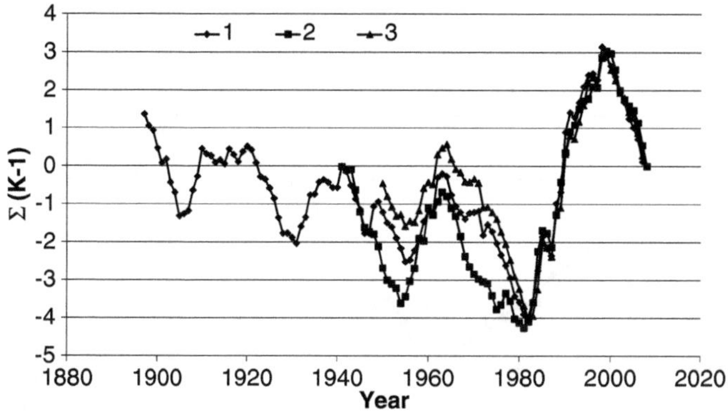

Fig. 15.5 Long-term changes in flow of the Shilka near Sretensk town (*1*), the Argun near Olochi village (*2*) and the Onon near Chiron village (*3*), apparent in residual mass curves

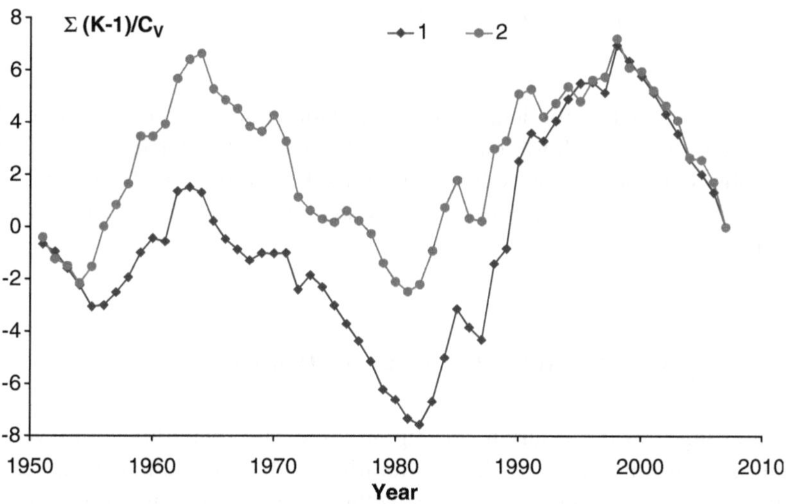

Fig. 15.6 Long-term flow changes of the Shilka river (*1*) and precipitation totals (*2*) generalized for the Onon-Argun inter-river area (residual mass curves)

Inter–year changes in annual flow of most rivers in the Onon-Argun inter-river area show a negative trend. The annual mode of flow from 1956 to 2008, generalized for the area, decreased by 0.53 $l/c_{*}km^2$, or by 21%. The inter–year flow changes in the warm period of the year (May–September) are strongly determined by the amount of precipitation in this period.

Inter–year changes in precipitation and flow occur synphasically (Fig. 15.6) and the correlation coefficient between the observation series of flows of all the rivers in the Shilka basin and the average total of precipitation exceeds 0.7.

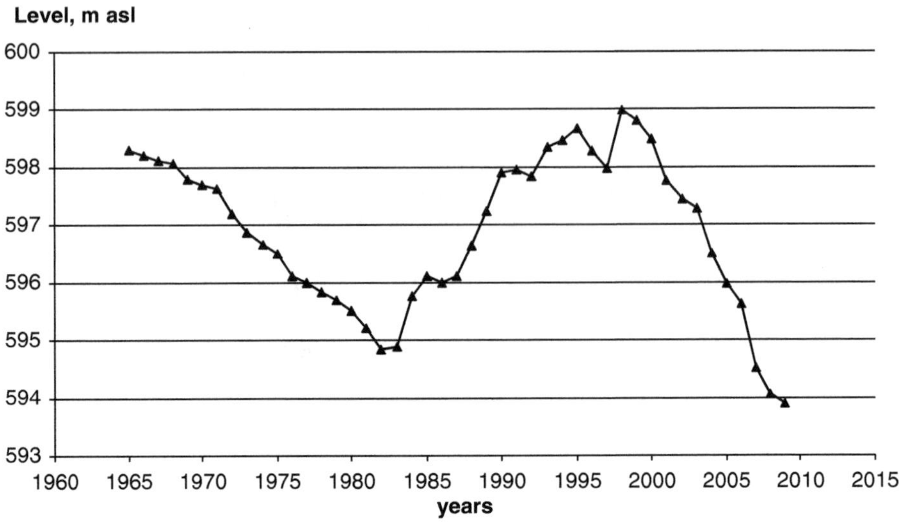

Fig. 15.7 Long-term changes in the level of Lake Barun-Torey

Lake levels also directly depend on precipitation and correlate with river flows. For instance, Fig. 15.7 shows clear phases in measured water levels in the Torey lakes. It should be noted that by June 2009 Lake Barun-Torey had completely dried up, while just before its floor contained only shallow large puddles, which appeared after rains.

15.3 Influence of Climate Cycles on Habitats

Climate changes in the Daurian eco-region, especially humidification cycles and continuing warming, cause habitat alteration and even habitat disappearance. The most intensive changes occur in wetlands, which are important intrazonal biotopes of the steppe zone. In dry phases of the cycles all small rivers and most of the springs, and up to 90–98% of lakes dry up. Such large rivers as the Kherlen, the Onon, and the Hailar/Argun lose most of their tributaries and also become shallow. By 10 June 2009 one of the largest water bodies of the region – Lake Barun-Torey – had completely dried up (except for small puddles at the spots of spring outlets) (Fig. 15.8). At that time the levels of Lakes Dalainor, Buir-Noor, and Khukh-Noor also fell considerably.

In Lake Barun-Torey the saline that formed after drying up gradually is getting overgrown and turning into a meadow and steppe (this will be described further on). The lake depression, devoid of water (551 km^2 in 1999), instead of being an aquatic ecosystem inhabited by aquatic organisms and organisms that live near water, including thousands of tons of fish, (Photo 15.1) became part of the terrestrial ecosystem. Presently, it harbors animals that are absolutely not associated with water, for

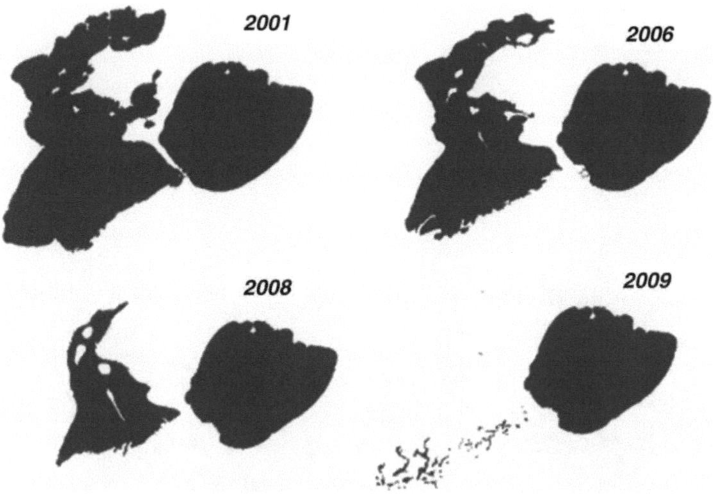

Fig. 15.8 Changes in area of Lake Barun-Torey (*left*) and Lake Zun-Torey (*right*) from 2001 to 2009

example, Mongolian gerbils and gazelles. Disappearance of water sources in steppes makes steppe habitats unsuitable for animal species that need waterings. We address this further on. Following disappearance of the lakes, or their substantial shallowing and salinization, near-water macrophytes also degrade, in particular *Phragmites australis*, and shrubs wither. The same occurs in the drying rivers. Thus, tall shore and floodplain vegetation turns into meadow and steppe habitats. Islands and spits of the lakes that were nesting areas for loads of waterfowl and near-water birds, including colonial ones (only on the Torey lakes in wet periods there were up to 17 islands inhabited by masses of birds), cease to be such and lose their significance.

In the beginning of dry periods tall herbaceous vegetation, including floodplain vegetation, disappears faster due to the increasing intensity of steppe fires. Thus, in 1996–1997 50–70% of the steppes and floodplains burned. Large tussock grasses and bushes re-grow slowly after fires. Increasing evaporation on burnt areas in dry periods contributes to further drying of floodplains, and the level of groundwater falls as well. Therefore, restoration of the lake's level and of the river's flow with the start of a wet phase slows down.

Increased number of fires in dry phases of the cycles limits spreading of forest habitats. Fires to a great extent annihilate insular pine forests (*Pinus sylvestris*) in the steppe zone, and move the forest boundary to the north (in part of the region under consideration) or upwards (on mountain ridges). For instance, in the Tsasuchey pine forest, in the course of the current dry phase the number of fires has grown sixfold in comparison with the previous wet phase. Changes in temperature and moisture regime caused repeated increase of windy days in the pine forest and lengthened fire-danger periods. As a result, emergence of high fire danger occurred not only in spring but also in September–October and in 1999–2010 more than 75% of the pine forest burned out. The number of forest fires in the region as a whole

Photo 15.1 Dead fish on the bank of Torey lakes in the dry period of climatic cycle (Photo O. Goroshko)

Photo 15.2 Consequences of drought and overgrazing in the Daurian steppe (Dornod aimag, Mongolia) (Photo V. Kirilyuk)

known data on withering of forest tracts and belts in the Russian and the adjacent Mongolian territory.

The impacts of dry phases in the 30-year humidification cycle on steppe habitats proper reveal in general a decline in their sheltering features (lower height and density of vegetation, almost complete disappearance of long-stemmed grass plants) and foraging qualities (lower biomass of vegetation). And, especially in the southern part of the study area, aridization leads to total xerophytization of the plant cover with, in some communities, a decrease in species. Also, salt-tolerant species spread to dried-up water basins and flows. Some aspects of these processes are considered in more details in the next section.

Similar to the pulsation of the boundary between the steppe and forest-steppe zones, shifts of the steppe – desert boundary take place. Changes in steppe habitats towards the end of the dry phase can cause desertification. However, if they result from relatively short cycles these processes are usually reversible. Most steppe species have adaptations to survive unfavorable periods and subsequently quickly restore their abundance. Yet, in some periods, vast stretches of steppe become almost uninhabitable for most steppe animals, as they approach desert conditions (Photo 15.2).

It is extremely difficult to assess whether these changes are due to natural factors or not, as considerable parts of the steppes of the Daurian eco-region are under severe pressure of pasture cattle-breeding. Numbers of free-grazing livestock, though subject to regulation by natural factors, are also artificially maintained high by humans, and this harms the ecosystems.

With the beginning of a wet phase rivers and lakes fill with water, and this process goes much faster than drying. Then ground vegetation, growing densely in river beds and on lake floors, is flooded and rots. The processes of destruction of large amounts of organic matter with simultaneous re-filling and warming up of the water in the shallow basins give a sharp rise in the reproduction of plants and animals of all trophic levels. In the steppe zone many intrazonal aquatic and near-water habitats with rich foraging bases re-appear. The foraging capacity of the steppe biotopes sharply increases, many water sources appear, and shelter conditions improve. In the most humid part of the steppe zone the area with bush and tree vegetation starts increasing.

The most important limiting factor for many non-migrating and non-hibernating animals is the (height of the) snow cover. Areas with a continuous snow cover of 20 cm or more thick, and areas with a firm and ice-encrusted snow cover of 10–12 cm thick, are uninhabitable for most steppe species.

On the whole, the habitats of the Daurian eco-region are subject to cyclic changes with strong amplitudes. Accordingly, the amount of biomass of living organisms differs many-fold between dry and wet phases. Thus, food supply and other conditions for life and reproduction differ. During the dry phase, which is the most critical for many vertebrate species, a few habitat refugia remain, e.g. large rivers and river floodplains, lakes that don't dry up, southern slopes of hills with outcrops of rocks and ravines in the north of the steppe zone and northern slopes in the south of it. In the unbroken landscapes of the steppe such relatively favorable habitats depend on variation in local conditions.

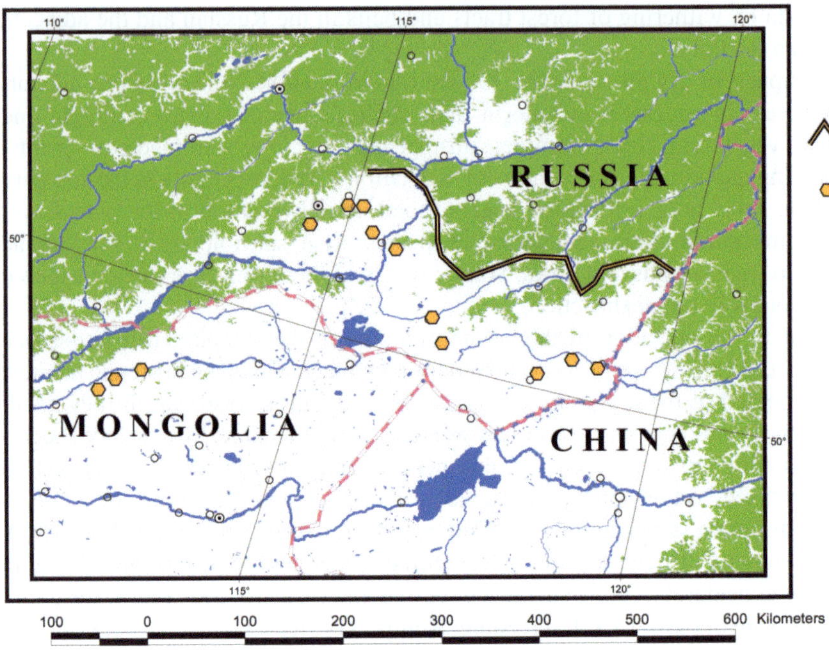

Fig. 15.9 Withering of forest belts and tracts of forest on southern slopes in the forest-steppe zone; *1* boundary of the expansion of withering of small-leaved species (*Betula pendula*, *Populus suaveolens*, *P. tremula*) in tree stands of the forest-steppe according to Annenhonov (2008); *2* points of complete or partial (more than 50%) withering of tree stands in forest belts or tracts of forest on southern slopes according to data of our research in 2008–2010

reached a maximum by the middle of the dry period. In 2007, in Zabaikalsky krai only, according to official statistics, about 2,500 forest fires were registered, which destroyed more than 350,000 ha of forest.

Drought also directly influenced the forests. Isotherms correlated with the distribution of permafrost (−2.5°C is the southern boundary of the area of insular permafrost, −7.5°C the boundary of the continuous permafrost area) were actively moving to the north. In the past 40–50 years the shift amounted to several hundreds of kilometers in some places. The southern boundary of the permafrost area moved further north than 52° northern latitude, having come close to the limit of permafrost retreat in the known geological periods of the interglacial (Shesternev et al. 2008). Retreat of permafrost is accompanied by subsidence of the groundwater level in the current dry period, which together with insufficient humidity of the air causes further drying of soils and a gradual increase of xerophytic species in plant communities. Because of moisture deficits, not only in the areas of former permafrost, partial or complete withering of *Pinus*, *Larix*, and even more, of *Betula* and some other small-leaved trees occurred in a strip of up to 100–150 km wide along the southern boundary of the forest zone. As a result steppes are shifting north more rapidly. A clear indicator of this is the general withering of artificial forest belts in South-East Transbaikalia (Annenhonov 2008). Figure 15.9 summarizes the

The high dynamics in habitats in the Daurian eco-region are accompanied by strong leaps in bioproductivity, and these support the high biodiversity of the Daurian steppes, including the abundance of many mammal and bird species.

15.4 Influence of Climate Cycles on Vegetation

The indigenous vegetation consists mainly of steppes, forests and tree groves, and vegetation of river valleys and lake depressions. Their spatial pattern depends on climate and soil conditions.

In the northern part of the Daurian region, along the foothills of the Hentei, Great Hingan and other mountain ranges, so-called "exposition forest-steppe" occurs where forests cover northern slopes, while grasslands and xerophytic shrubs occupy southern ones. *Ulmus pumila* also forms groves on southern slopes. Forests consist of *Larix gmelinii*, *Pinus sylvestris*, *Betula pendula* and *Populus tremula*. *P. sylvestris* can form large forest islands on sands along river valleys. Savanna-like landscapes with sparse trees of *U. pumila* or *P. sylvestris* occur in some places (Photo 15.3).

Photo 15.3 Small-tussock grass – herbaceous steppe with *Pinus sylvestris* ("Pine savanna") in eastern part of Mongolia (Photo T. Tkachuk)

Photo 15.4 *Filifolium sibiricum-Koeleria cristata* steppe at Russian part of Argun river basin. Inset: *Filifolium sibiricum* (Photo T. Tkachuk)

Bush-steppe communities with *Ulmus macrocarpa*, *Spiraea aquilegifolia*, *S. media*, *S. pubescens*, *Armeniaca sibirica* and also *Artemisia gmelinii* communities are derivatives of the broad-leaved forests of past geological periods and typical of contemporary Daurian landscapes. Their grass layer contains both xero- and mesophytic species, which alternately prevail in years with different humidity. Such a dual composition is typical of plant communities in Dauria (Dulepova 1993).

Steppe is the main zonal vegetation type in Dauria. It is subdivided according to degrees of humidity into three subtypes: meadow steppes, typical steppes and dry steppes.

Meadow steppes are found mostly in the north and north-east of Dauria. For example *Stipa baicalensis* steppes occur in the east along the Argun river and its tributaries. Steppes dominated by the rhizome graminoid *Leymus chinensis* are more abundant in southern valleys of Transbaikalia and North-East Mongolia (Grubov 1959), especially in lake depressions. Another characteristic formation is the *Filifolium* steppe (Photo 15.4). They occupy mountain slopes and planes mostly in forest-steppe landscapes. *Filifolium sibiricum* (*Asteraceae*) forms both meadow and typical steppe communities and is perfectly adapted to the ultra-continental climate. Often herbaceous and grass-herbaceous meadow steppes occur.

Typical steppes occupy vast areas. Small-tussock steppes with *Festuca litvinovii*, *F. valesiaca*, *Poa botryoides*, *P. attenuata*, *Cleistogenes squarrosa* are wide-spread

Photo 15.5 *Stipa krylovii* steppe at Daursky Biosphere Reserve (Russia) (Photo T. Tkachuk)

both in forest-steppe and steppe landscapes. Large-tussock steppes dominated by *Stipa krylovii* (Photo 15.5) occupy plains in the south of Transbaikalia and North-Eastern Mongolia.

Dry steppes form a continuous belt only in Central-Southern Mongolia. Here prevail mainly gramineous dry steppe species: *Stipa krylovii*, *Agropyron cristatum*, *Cleistogenes squarrosa*, *Koeleria macrantha*, *Artemisia frigida*, as well as species of the desert steppe: *Stipa klementzii*, *S. glareosa*, *Cleistogenes songorica*, etc. (Karamysheva 1993).

Petrophytic variants of all steppe subtypes are wide-spread and occur mostly on tops of steppe hills (Photo 15.6). Such species as *Chamaerhodos trifida*, *Arctogeron gramineum*, *Orostachys spinosa*, *Stellaria scherleriae*, *Eremogone capillaris*, etc. are typical for them.

Complex relief oversets latitudinal zonal and subzonal borders and makes them difficult to recognize. So, enclaves of forest-steppe are spread far southwards, and also patches of dry steppes are found on southern slopes in the forest-steppe zone in Russian Dauria.

In floodplain wetlands (Photo 15.7) meadows with grasses (of the genus *Calamagrostis*) and tussock sedges (*Carex schmidtii* and others) prevail as well as groves of *Salix* spp. *Phragmites australis* communities and several species of wild fruit trees (*Crataegus* sp., *Padus avium*, *Malus baccata*, etc.) are also common.

Photo 15.6 Petrophytic steppe with *Scutellaria baicalensis* in the Daursky Reserve (Photo T. Tkachuk)

Photo 15.7 Floodplain of Onon river with complex of meadows with thickets of *Salix* spp. and other bushes (Photo I. Glushkov)

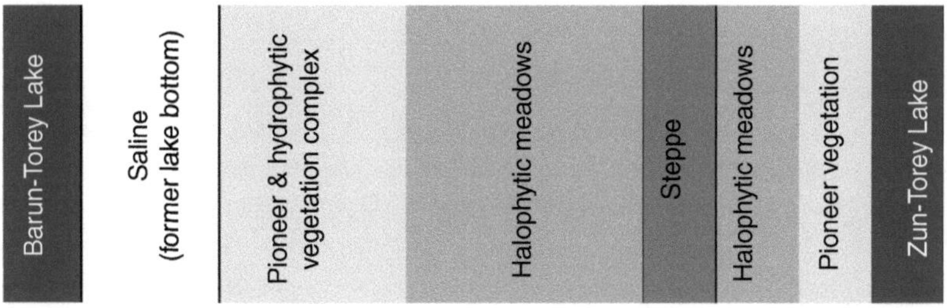

Fig. 15.10 Scheme of vegetation distribution along the fixed transect between the Zun-Torey and Barun-Torey lakes. Relative width of the fields in the diagram corresponds to width of the vegetation zones in the transect

The bottoms of small depressions and shores of salt and brackish lakes are occupied by salines. Patches of small salines also spread in steppe areas. Their vegetation is dominated by annual chenopods (*Suaeda corniculata*, *Kochia densiflora*, *Atriplex sibirica*, *A. patens*, etc.), perennials such as *Artemisia anetifolia* and *A. laciniata*, and the dwarf shrub *Kalidium foliatum*. In big depressions they are surrounded by meadows with *Puccinellia* spp., *Hordeum brevisubulatum*, *Leymus chinensis* and *Iris lactea* and also *L. chinensis* steppes. Communities of the large tussock grass *Achnatherum splendens* are common in depressions and may also fringe lower slopes of steppe hills, where ground water is shallow. *Achnatherum* steppes usually include some halophytic species (*Limonium aureum*, *Saussurea amara*, *Iris lactea*).

As the pattern of vegetation types mentioned closely depends on small differences in relief and some related habitat conditions, fluctuations in climate considerably affect the temporal and spatial diversity, changing the distribution of species and composition of communities over time in a cyclic manner.

We studied vegetation dynamics along a fixed transect between the Torey lakes in the *Daursky* Biosphere Reserve during 2002–2010 (Tkachuk and Zhukova 2010). The transect is about 4 km long, and includes steppe, halophytic meadows, pioneer halophytic vegetation and dense stands of hydrophytes (*Phragmites australis* and *Bolboschoenus planiculmis*).

The sequential distribution of the vegetation along the transect is shown in Fig. 15.10. The lowest, recently dried-up parts of the transect carry halophytic pioneer vegetation with patches of hydrophytes (Photo 15.8). The 1st lake terrace and lake shores, which became dry 2–4 years ago, support halophytic meadows, and the highest part of the transect, on the 2nd and 3rd terraces, steppe vegetation. The vegetation belts are much wider at the Barun-Torey depression than at that of Zun-Torey.

Vegetation dynamics along this transect during 9 years are illustrated in Fig. 15.11. The main trends are the areal increase of all vegetation types together and the disappearance of the shore line vegetation. The strong increase of pioneer vegetation and hydrophytes in 2004 is caused by the rapid retreat of the shore-line and connected elongation of the transect. The hydrophytes later decrease as the ground water level falls. Halophytic meadows increase from year to year, especially

Photo 15.8 Patches of *Bolboschoenus planiculmis* and sparse *Suaeda corniculata* on the dry bottom of Barun-Torey lake (Photo T. Tkachuk)

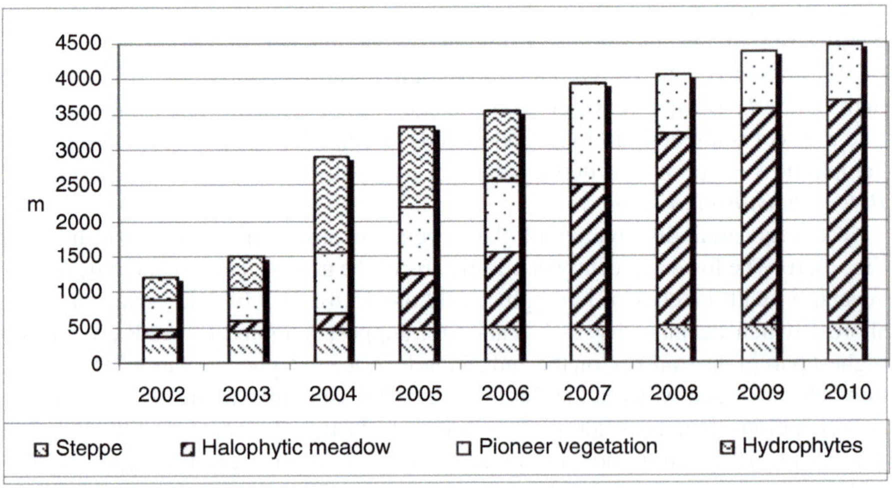

Fig. 15.11 Change of different vegetation types in the monitored transect in the period 2002–2010

Photo 15.9 *Puccinellia tenuiflora* meadow on the bottom of the dried-up Barun-Torey lake (Photo T. Tkachuk)

since 2005. This sudden increase comes with a lag of 1 year behind that of the pioneer vegetation. The increase in steppe vegetation is insignificant.

The dynamic successional vegetation pattern between the Torey lakes (as well as at other drying lakes) is similar to that of other salty steppe lakes in Central Asia (Vostokova 1983; Zhao et al. 2010), though some details are typical for the Daurian steppe region.

Newly dried-up sites are first colonized mainly by a few annual chenopod species and, nearly at the same time, by *Tripolium vulgare*, *Argusia rosmarinifolia*, *Knorringia sibirica*, *Puccinellia tenuiflora* and *P.macranthera.* In the second year *Suaeda corniculata* fully covers the sites, except where some large, local patches of *Phragmites australis* and *Bolboschoenus planiculmis* already existed. Dense patches of *Ph. australis* outside the lake zone remain for some 3–5 years or more and those of *B. planiculmis* for 1–2 years, though these species survive in communities indefinitely in low abundance or as tubers (*B. planiculmis*).

The 1st and 2nd lake terraces, as well as shores dried-up for 3 or more years, carry halophytic meadows which rapidly expand from year to year: in our transect from 90 m in 2002 to 3,140 m in 2010. In the next few years they turn into meadows, mostly *Puccinellia tenuiflora* meadows (Photo 15.9), and locally also *Carex reptabunda* meadows. Some of these meadows persist already for 8 years. The *Puccinellia* meadows are succeeded by *Hordeum brevisubulatum* meadows. The zone of *Hordeum*

Photo 15.10 *Artemisia frigida – Stipa krylovii* steppe on the terraces of the Torey lakes (Photo T. Tkachuk)

meadows shifts only meters per year at the expense of the *Puccinellia* meadows, and in turn is replaced by *Leymus chinensis* meadow and than by *L. chinensis* steppe. This *Leymus* belt shifts slowly down the terrace slope: at the bottom displacing meadows while at the top being displaced by typical steppe.

The typical steppe area on the 2nd and 3rd lake terraces (Photo 15.10), dominated by *Stipa krylovii*, *Artemisia frigida* and *Allium polyrhizum*, increased hardly (from 460 m in 2002 to 550 m in 2010). Long-term climatic cycles that are thought to have caused the drying-up do not immediately affect these communities and fluctuations in species abundance seem reactions on short-term fluctuations in precipitation, and perhaps also to grazing intensity. Climate-dependent fluctuation dynamics are characteristic of steppe vegetation, and Dulepova (1993) described these for the northern part of Dauria (Eastern Transbaikalia). They are brought about by various mechanisms, e.g. fluctuations in the amounts of seeding and germination, survival of seedlings, different numbers of annual shoots in tussocks and on other perennial plants that change the competition between species differing in drought tolerance but coexisting in the same communities. Some annual chenopods become very abundant in some rainy years but remain in the community only in the seed bank during a sequence of dry years. Overgrazing also can affect abundances, especially in dry years. And other species, with perennating organs such as bulbs and rhizomes (*Lilium pumilum*, *Gagea pauciflora*, *Bolboschoenus planiculmis*), can stay dormant during unfavorable years and re-appear aboveground in better years.

The vegetation of the Daurian steppe and lake complexes seems adapted to cyclical climate changes and is resilient as regards species composition and biological features, as apparent from vegetation fluctuations and cyclical succession. The strongest effects of climate cycles on the vegetation are observed at lake depressions where a cyclical succession with a change in vegetation types takes place. The steppe vegetation is more resistant to climate changes and shows mostly short-term fluctuations.

15.5 Influence of Climate Cycles on Wildlife

Climate changes influence vertebrate animals both through transformation of habitats and directly. A convincing example of indirect impact is the drying up and shallowing of water basins and courses, which are the natural habitat for aquatic animals and the main habitat for animals that live near water. In such cases aquatic organisms, including fish, die completely (Photo 15.1) or survive in near-dormant regenerative stages, including larvae and roe at remaining survival stations. Also the Golden Carp (*Carassius auratus gibelio*) can preserve its vitality for some time after the water basin has dried up by burying itself in the silt. In the Uldza river basin, where in fact no fish is left due to drying, carps remain so far only in Lake Khukh-Noor – the deepest reservoir of the basin.

Changes in depth and mineralization of the water of the lakes and rivers also lead to the loss of suitable nesting sites, including islands, and changes in the availability of foraging items. This actually caused considerable changes in the composition of waterfowl and near-water birds nesting at the Torey lakes (Fig. 15.12).

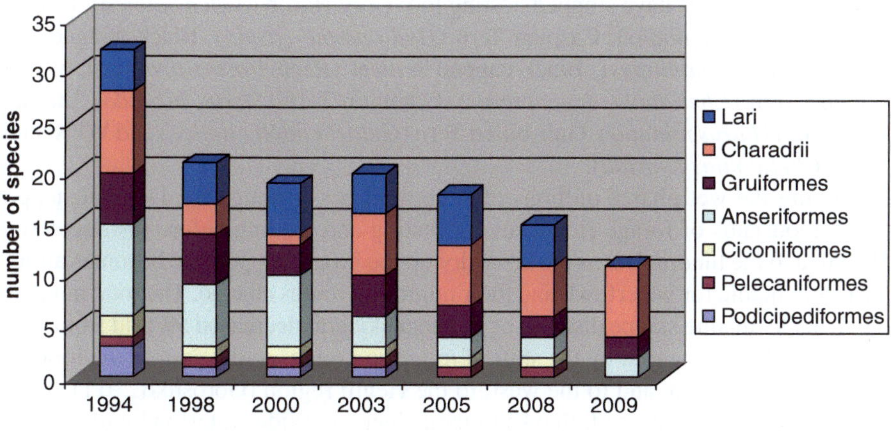

Fig. 15.12 Change in species composition of the nesting fauna for the main groups of waterfowl and near-water birds at the Torey lakes in the period 1994–2009

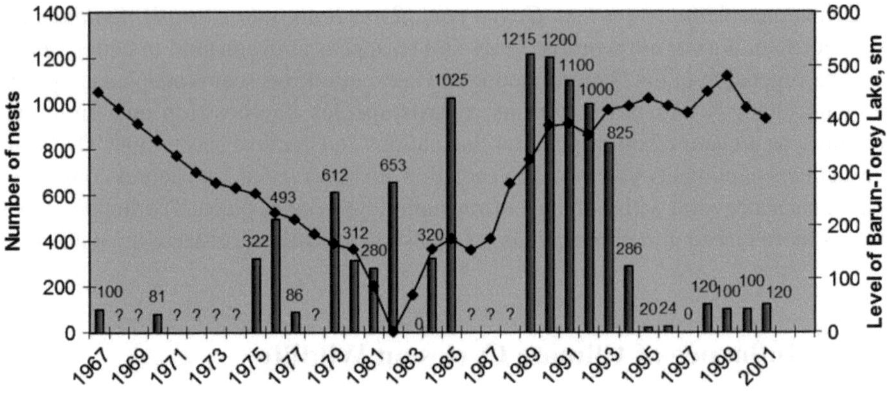

Fig. 15.13 Number of nests of *Larus relictus* at the Torey Lakes in the period 1967–2001 (Tkachenko and Obyazov 2003)

With transformation of the lake ecosystem numbers of individuals change at even higher rates than species composition, including the numbers of the very rare Relict Gull (*Larus relictus*), for which low-water periods in the filling stage are more favorable (Fig. 15.13).

Great Cormorants (*Phalacrocorax carbo)*, whose numbers on the Torey lakes reached a maximum of more than 2,200 in 2001 (Tkachenko and Obyazov 2003), had stopped nesting altogether by 2010. This was due first to dying of fish (2006–2007) and then the disappearance of the last island. After that cormorants spread without forming large colonies over the vast area of forest-steppe and southern taiga, up to the Baikal where they hadn't been for some decades. With the disappearance of islands and the impoverishment of forage availability at the Torey lakes other colonial birds have stopped nesting too: Gray Heron (*Ardea cinerea*), Herring Gull (*Larus cachinnans*), Caspian Tern (*Hydropogne caspia*), Black-winged Stilt (*Himantopus himantopus*), Black-capped Avocet (*Recuvirostra avosetta*), White-winged Tern (*Chelidonias leucopterus*), Common Tern (*Sterna hirundo*), the rare Little Gull (*Larus minutus*), Gull-billed Tern (*Gelochelidon nilotica*) and Whiskered Tern (*Chelidonias hybrida*).

During the wet phases millions of waterfowl pass through the Daurian steppes using thousands of forage-rich lakes for resting and feeding before rushing across the taiga to the tundra. But as the lakes dry up the broad steppe zone becomes almost insurmountable for waterfowl, and their migration routes change. The total numbers of all ducks in nine steppe districts of Zabaikalsky krai decreased 59-fold from 1999 to 2009, which is caused by the shift of migration routes to the east, to the foothills of the Great Hingan, and to the west, to the Hentii region (Goroshko 2011). These ridges stretch far south and, thanks to their rivers, provide waterfowl with suitable conditions for a stop-over during their migration. It makes the flight across dry steppe and desert shorter.

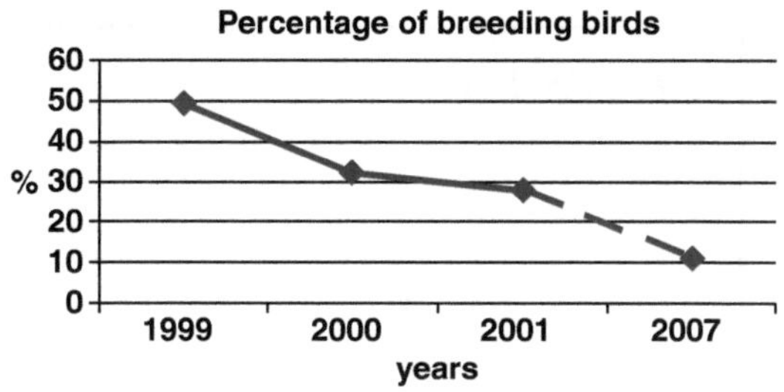

Fig. 15.14 Change in percentage of breeding White-naped Cranes (*Grus vipio*) in the region in 1999–2007 (Goroshko 2007)

With the rivers and lakes drying up almost all White-naped Cranes (*Grus vipio*) have moved from the steppe to the forest-steppe for nesting. Total numbers of nesting White-naped Cranes during the last dry cycle fell sharply (Fig. 15.14), which led to a general reduction of the western population of the species.

The only steppe crane species in the region, the Demoiselle (*Anthropoides virgo*), although not inhabiting wetlands, also greatly depends on water sources for raising its nestlings. Demoiselles nest along river floodplains, lakes shores, at springs or wells. Sharp reductions of water bodies during drought lead to a decrease in total numbers and in the numbers of breeding birds. While in the previous wet phase the population density in the Torey depression and the basin of the Onon river in its middle reaches was 1.1 individuals per km^2 (Goroshko 2002), and in the Kherlen basin and to the south of it 0.3–0.8 individuals per km^2, by the end of the current dry phase, in 2007–2010, these figures were only 0.1–0.3 and 0.01–0.05 individuals per km^2, respectively. Thus its population density has decreased from 4-fold to 250-fold. At the same time the number of Demoiselles has increased remarkably in the forest-steppe zone.

The Great Bustard (*Otis tarda dybowskii*), an inhabitant of meadow steppes and lake depressions covered with tall plants, moves to the forest-steppe zone in dry periods, and through wide-open river valleys and across agricultural lands also to the southern part of the forest zone. That also caused considerable numbers of bustards to move from Mongolia to Russia in the first decade of the 2000s.

Populations of the most typically steppe birds also experience dips in dry periods. Especially hard conditions can occur in huge parts of the southern dry steppes, where the grass cover degrades due to drought and high densities of livestock. In June 2008, in the south of Sukhebaatar aimak in Mongolia, for example, where almost no grass cover was left, the total population density of all counted birds was 7.9 individuals per km^2, most of them small species, and the proportion of Asian Short-toe Lark (*Eremophila alpestris*) was 44%. Among the larger birds along 273 km of census routes the population density of the Demoiselle was only 0.01 individuals per km^2, that of the Saker Falcon, Upland Buzzard and Raven also very low at about 0.004 individuals per km^2, and that of the Steppe Eagle (*Aquila rapax*) even lower at 0.002 individuals per km^2. At the same time in the south-west of Eastern aimak, where vegetation was relatively abundant as a result of scattered rains, the density of the total bird population exceeded 1,000 individuals per km^2.

During dry periods the distribution areas and patterns of many mammal species changes considerably. The home range of the Raccoon Dog (*Nyctereutes procyonoides*)

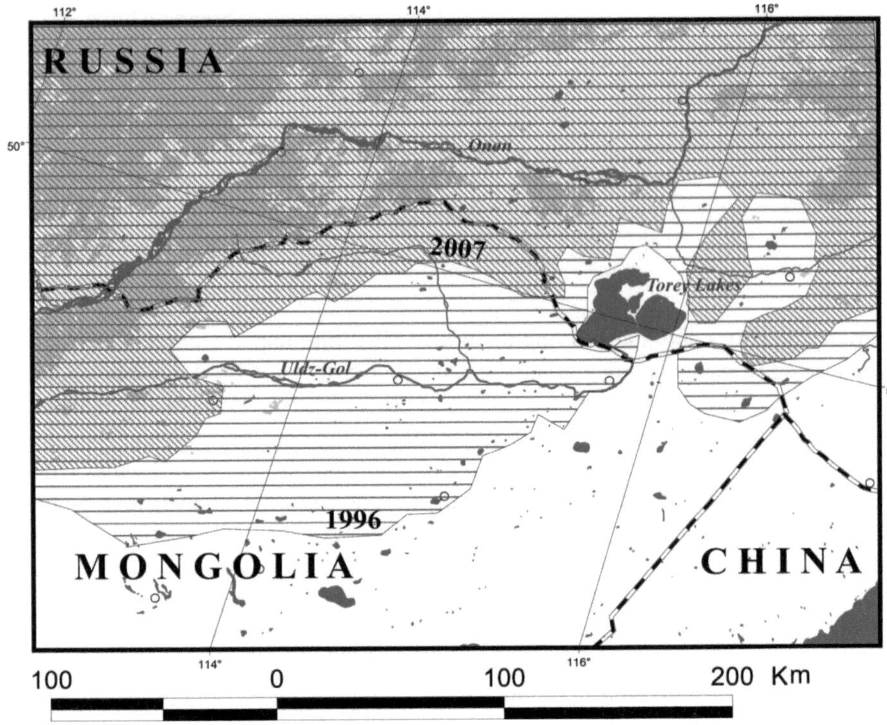

Fig. 15.15 Southern limit of the range of the Siberian Roe Deer (*Capreolus pygargus*) in 1996 and in 2007

very strongly so. This species expanded to the eastern part of the Daurian steppes in the middle of the twentieth century having moved from the east (Peshkov 1967). By 1999, the Raccoon Dog, though preferring rivers and lakes, was common almost everywhere and occupied meadow steppe, forest and shrub habitats in the Russian and adjacent parts of the eco-region. At the Torey lakes its numbers amounted to some thousands. By 2008 there were no dogs left there. On the whole their numbers and distribution in this part of Dauria reduced very drastically and the Raccoon Dog now only inhabits the floodplains of large rivers, such as the Onon and the Argun. Moreover, its population density in survival stations has not increased.

In the south of the region the distributions of the Fox (*Vulpes vulpes*), Eurasian Badger (*Meles meles*) and Wolf (*Canis lupus*) have become diffuse, as these species need watering places. Most of the southern occurrences of the Long-tailed Souslik (*Spermophilus undulates*) on the right bank of the Onon have disappeared. The Siberian Roe deer (*Capreolus pygargus*) penetrated deep into the steppe zone in wet periods, and depending on the availability of waterings it occurred not only in insular forests, groves and floodplains, e.g. along the Uldza river, but also in hilly herbaceous steppes with *Salix*-scrub in depressions. By 2007–2008 the limits of its range had moved up to 100 km to the north (Fig. 15.15). Similarly, following the disappearance of habitat fragments with trees and shrubs, as well as shallow waters

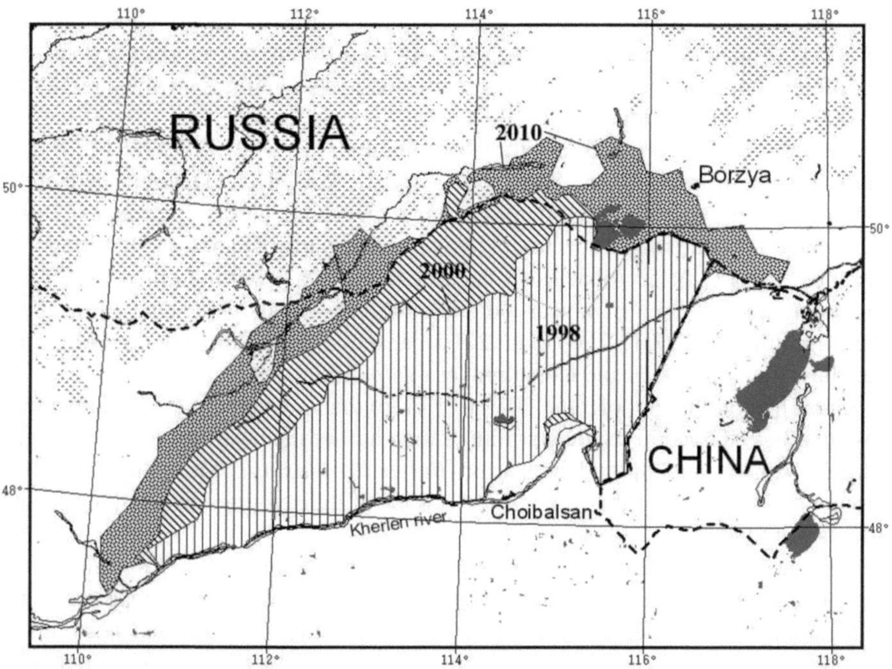

Fig. 15.16 Change in the distribution of the North Kherlen population of the Mongolian Gazelle (*Procapra gutturosa*) from 1998 to 2010

with *Phragmites* and *Salix*, the southern limits of Wild Boar (*Sus scrofa*), Red Deer (*Cervus elaphus*), Eurasian Lynx (*Lynx lynx*) and Mountain Hare (*Lepus timidus*) have noticeably shifted northward.

At the same time, some dry steppe species, particularly Siberian Jerboa (*Alactaga sibirica*) and Mongolian Gerbil (*Meriones unduiculatus*), spread northward during the drought years, though not rapidly: the Gerbil extended its range to 10–15 km to the north of the Torey lakes, and the Jerboa locally several tens of kilometers farther to the forest-steppe through steppe ledges.

Special attention deserves the Mongolian gazelle (*Procapra gutturosa*), the most numerous and typical ungulate species in the region. In the period 1994–2010 we observed and censused the population, mapped calving areas, seasonal distribution and migration routes, and analysed the space-time dynamics of the species. In the studied part of the Daurian steppe eco-region, mainly within East Mongolia, there are two large populations of gazelle, the North Kherlen and Matad ones, comprising from 700,000 to 1,600,000 individuals in spring (Olson et al. 2005; Kirilyuk 2011). Dispersion in connection with drought began in both populations but the Matad one is restricted in its movements by artificial barriers on the Mongolian-Chinese frontier, and thus alterations in the space structure of this population were not natural. The North Kherlen population started to move northward, also in Russia, after 1998, the last year of the wet phase (Kiriliuk 2007). By 2010 the northern boundary of its range had shifted by 70–120 km (Fig. 15.16).

The distance of autumn migrations, the yearly range used by the population, and the locations of their calving areas changed, but to a different degree per year, and due to different environmental factors. These movements sometimes involved more than 50,000–100,000 individuals.

Climatic factors bring direct threats to animals as well. In 1998, because of unusually high rainfall in June and August, Roe deer, Daurian hedgehogs (*Mesechinus dauuricus*), Tolai hares (*Lepus capensis*), Daurian ptarmigans (*Perdix dauurica*) and some rodent species lost considerable parts of their litter. In Mongolian gazelles epizooty of necrobacillosis because of hooves' swelling killed 85–90% of the young and 20–30% of 1-year-old and adult individuals. Abundant snowfalls at the beginning of 2001, 2003 and 2010 caused mass death among Mongolian gazelles, Raccoon Dogs and Corsak foxes (*Vulpes corsac*).

Great heat at the end of June 2010 (for more than a week the day's temperature rose to 43–49°C) led to the death of more than half of the nestlings of steppe eagles (*Aquila rapax*) and upland buzzards (*Buteo hemilasius*). Mongolian gazelles, having mass calving at that time, lost 67–85% of their young, despite the species' adaptation to heat.

Fires accompanying droughty periods are among the most powerful catalysts for the movements of animals. Instantly destroying forage and shelter, steppe fires force wild ungulates to move away. In spring, fires directly kill eggs and nestlings of ground-nestling birds, which prevail in the steppe, as well as young mammals, especially also hedgehogs and other not-so-fast animals. Autumn fires destroy plant forage for 7–8 months to come and have a long-term negative effect. The first large steppe fire caused by an initial manifestation of the coming of the dry phase, in 1996, in East aimak, Mongolia, covered 70% of the territory. Immediately after the fire thousands of Roe deer, Wild boars and Manchurian deer moved to Russia. The following year fires occurred again, and after 1998, when the period of dry years definitely had set in, the movement of these ungulates to Russia continued.

The fact that in the dry phase the snow cover, on average, is thicker than in the wet, has important limiting consequences. Winters with much snow, therefore, occur in the years with a low grass cover, which leads to the death of many birds and animals, and also the numbers of wintering bird species decrease considerably. Deep snow cover, which is not characteristic for the steppe, is a disaster for beasts and birds, because under the influence of winds snow in the steppe very quickly turns to a hard icy crust that blocks access to fodder, and for ungulates it also makes motion more difficult. Deep snow causes high mortality among hares due to greater vulnerability to predators. The distribution limits of Pallas' cat (*Otocolobus manul*) coincide with the boundary of the steppe zone, but in the environs of the Argun, where the snow frequently is deeper, the distribution limit of Pallas' cat is far from the steppe boundary, and is determined by the depth of the snow cover; Pallas' cats cannot survive where the average multi-year maximum snow depth is more than 16–17 cm.

Increased duration of the warm period, on the other hand, improves the living conditions for some species and can facilitate the expansion of their area. For example, the Daurian hedgehog needs a minimum warm period of 160–165 days for successful reproduction, and that is why, with warming and deforestation, it moves northward where its existence is limited by the duration of the warm period or the forest border.

15.6 Conclusions

Due to clear cyclic changes in humidity habitats in the Daurian eco-region change. During dry phases habitats with tall plants, which provide much foraging capacity and good protection, reduce in area, and most of the wetlands vanish completely, while in wet phases they appear again and provide a sharp rise in bioproductivity.

The vegetation of Dauria is adapted to cyclical climate changes and resiliently reacts with fluctuations and cyclical succession.

In dry phases the steppe border in the south gives way to the desert, and in the north it shifts the forest limit; in wet phases the reverse processes take place. Most of the species, both aquatic and terrestrial, survive drought using different adaptation strategies. The most important are: surviving in few refuge habitats; persisting in the dormant phase of the life cycle; survival of non-reproductive adult individuals. The distribution areas of many ground vertebrates pulsate in concord with the cyclic changes in humidity. But the continuing warming gradually destroys the complete reversibility of these processes and leads to aridization.

On the whole, in Dauria the dry phase of the humidification cycle of about 30 years, which occurs against the background of global warming, causes remarkably strong changes in nature, which have mostly negative consequences: the level of biological diversity falls, as well as the sustainability and productivity of natural ecosystem complexes, the biomass of living organisms decreases, the borders of the ranges and the migration routes of mammals and birds shift. Many vertebrate species find themselves on the brink of survival.

Acknowledgement This work was carried out with the support of the project of the Russian Foundation for Fundamental Research "Instruments for estimating the quality of economic growth of the natural-resource region in conditions of climate changes", № 10-06-00060a.

References

Annenkhonov OA (2008) About state of forest components of forest-steppe of Transbaikalia in connection with dynamics of climate. Changes of climate of Central Asia: social-economical and ecological consequences, Chita, pp 149–153

Dulepova BI (1993) Steppes of mountain forest: Steppe of Dauria and their dynamics. Chita State Pedagogical Institute Publishers, Chita, 396 p

Gadgiev IM, Korolyuk AY, Tytlyanova AA et al (2002) Steppes of inner Asia. SB RAS Publisher, Novosibirsk, 299 p

Goroshko OA (2002) State and protection of cranes' and bustards' populations in South-East Transbaikalia and the adjacent areas of Mongolia. Abstract of Dissertation for candidate of biology degree, All-Russian Research Institute for Nature of Ministry of Natural Resources Moscow, 19 p

Goroshko OA (2007) Present status of White-naped Crane in Russia. International Workshop for the North East Asian Crane Site Network. Panjin City, China, unpublished report

Goroshko OA (2009) Migrations of birds. In: Goroshko OA (ed) Concise encyclopaedia of Transbaikalia: natural heritage. Novosibirsk, Nauka Publisher, pp 337–339

Goroshko OA (2011) Analysis of the state and multi-year dynamics of waterfowl in Zabaikalsky krai and recommendations for organizing their rational use. Report of the scientific research, State Biosphere Reserve Daursky, 19 p

Grubov VI (1959) Experience of botanical-geographycal division of inner Asia. Leningrad, Academy of Sciences of USSR, 78 p

Karamysheva ZV (1993) Botanical geography of steppes of Euroasia. Steppes of Euroasia: conservation and restoration problems. RAS Institute of Geography, S. Petersburg/Moscow, pp 6–29

Kiriliuk VE (2007) The first results and perspectives of the Mongolian gazelle (*Procapra gutturosa*) restoration in Russia. Express-Izdatelstvo, Chita, 36 p

Kirilyuk OK, Kirilyuk VE (2011) To assess the effectiveness of territorial protection of the Mongolian gazelle (*Procapra gutturosa*). A diversity of soils, and biota of North and Central Asia, Ulan-Ude, pp 74–76

Lavrenko EM (1970) Province division of the central-Asian sub-region of the Steppe Region of Euroasia. Botanizheski J 55:1734–1747

Obyazov VA (1994) Connection of fluctuations in water fill of the Transbaikalia steppe zone lakes with multi-year hydro-meteorological changes on the example of the Torey lakes. News Russ Geogr Soc 126:48–54

Olson DM, Dinerstein E (1998) The global 200: a representation approach to conserving the Earth's most biologically valuable eco-regions. Conserv Biol 12:502–514

Olson KA, Fulter TK, Schaller GB, Odonkhuu D, Murray MG (2005) The Mongolian gazelle. Oryx 39:164–169

Peshkov BI (1967) Spreading of the raccoon dog in the Chita region. Protection and reproduction of natural resources, Chita, pp 78–79

Shesternev DM, Enikeev FI, Obyazov VA, Chuprova AA (2008) The Cryolithozone of Transbaikalia in conditions of global climate changes: the problems and research priorities. Changes of climate of Central Asia: social-economical and ecological consequences, Chita, pp 46–53

Tkachenko EE, Obyazov VA (2003) Change of the Torey lakes level and nesting of colonial near-water birds. Ground vertebrates of Dauria. Collection of scientific papers of the Daursky State nature biosphere reserve, Chita, pp 44–59

Tkachuk TE, Zhukova OV (2010) Results of vegetation monitoring on the stationary transect at Daursky nature reserve. Nature protection collaboration of Zabaikalsky krai (Russia), Autonomous Province Inner Mongolia (China) and Eastern Aimak (Mongolia) in transboundary eco-regions, Chita, pp 299–302

Tkachuk T, Kirilyuk O, Simonov E (2008) Daurian Steppe. Compendium of regional templates on the status of temperate grasslands conservation and protection. Prepared for the world temperate grasslands conservation initiative workshop: "Life in a working landscape: Towards a conservation strategy for the world's temperate grasslands. Hohhot, China". Canada, 2008, pp 46–61

Vladimirov AM (1990) Hydrological calculations. Gidrometeoizdat, Leningrad, 366 p

Vostokova EA (1983) Ecological rows of vegetation of closed depressions in the Mongol Peoples Republic. Ecological-coenotic and geographical peculiarities of vegetation. Nauka Publisher, Moscow, pp 40–49

Zhao F, Liu H, Yin Y, Hu G, Wu X (2010) Vegetation succession prevents dry lake beds from becoming dust sources in the semi-arid steppe region of China. Earth Surf Processes Landf 36(7):864–871. doi:10.1002/esp.2114

Chapter 16
Changes in the Southern Siberian Forest-Steppes

Irina B. Vorobyeva

Abstract The characteristics of the forest-steppe of Middle Siberia, and its distinctive features in comparison with the forest-steppe within the plains of the European territory of Russia and Western Siberia are discussed and the main features of the vegetation of islands of forest-steppe within the northern and southern subzones are given. We made a spatial and temporal analysis of the present state of the soils of forest-steppe geosystems within the Nazarovskaya Depression in the context of climate change and found regularities in its variability, and determined changes in the differentiation of the organic matter. The dynamics in soil moisture deposition vary according to long-term precipitation cycles. Our experimental data prove that the humus state of the soils changes both in space and time. An increase of the mean annual air temperature over the past two decades under a sufficient amount of precipitation caused the greatest changes in the humus state of dark-grey wood soil and weakly-leached chernozem on slopes with forest-steppe. During the more than 20 years of observation the contents of total carbon, humic acids and insoluble residue have increased up to threefold. These signs give evidence of a tendency in the present development of the southern Siberian island forest-steppe towards the formation of islands of natural steppe, which can be considered a landscape response to global and regional changes in the natural environment.

16.1 Introduction

Southern Siberia has a continental climate, a complicated topography, and a varied pattern of geological and geomorphological structures. As a result the forest-steppes in the area serve as a dynamic model of the interactions between the two major

I.B. Vorobyeva (✉)
V.B. Sochava Institute of Geography, Siberian Branch, Russian Academy of Sciences,
Irkutsk, Russia
e-mail: irene@irigs.irk.ru

M.J.A. Werger and M.A. van Staalduinen (eds.), *Eurasian Steppes. Ecological Problems and Livelihoods in a Changing World*, Plant and Vegetation 6,
DOI 10.1007/978-94-007-3886-7_16,

types of natural environment, namely forest (taiga in the context of Middle Siberia) and steppe. The main characteristic feature of the zone of forest-steppe is that here two types of geosystems (forest and steppe) can co-exist. The forest-steppe in southern Siberia is heterogeneous and encompasses plains as well as mountain expanses.

Within the plains of the European territory of Russia and Western Siberia forest-steppe can occupy large patches on environmentally analogous sites and form ecologically similar biogeocenoses, while in Middle Siberia forest-steppe is different from place to place. While in the West the forest-steppe forms a distinct, continuous zone between the forest zone and the steppe zone, the forest-steppe of Middle Siberia does not form a continuous zone but occurs in the form of isolated islands amid the taiga. Within such islands steppe vegetation and, consequently, steppe soils occupy terraces of valleys of present-day water courses as well as of ancient dry valleys, whereas forest vegetation predominates on watersheds.

Forest-steppes in Middle Siberia occur in the form of islands on 'plakor' areas. A 'plakor' is a flat or gently sloping near-watershed area, characterized by a deep groundwater table and by the absence of significant washout or accumulation, and as a consequence the vegetation and soil correspond closely to the landscape forms. Here the forest-steppes do not consist of an alternation of patches of forest and open steppe, but constitute a combination of park forests on watersheds with steppes occupying steep slopes and dry valleys (Khotinsky 1961). Quite in contrast with the very gradual zonal-geographical transitions from forest-steppe landscapes to either zonal forest or zonal steppe landscapes, the natural boundaries between the components of the forest-steppe happen to be very sharp. This is characteristic for forest-steppe landscapes under a temperate continental climate. With an increase of continentality and aridity the sharp boundaries between separate patches of forests and steppes disappear and become gradual. So far the ecological and landscape characteristics of the southern-Siberian forest-steppes have been poorly investigated.

Within the Middle Siberian Upland three areas form the forest-steppe zone: Krasnoyarsk-Kanskaya, Irkutskaya and Angara-Lenskaya (Richter 1960). The Achinskaya forest-steppe of Middle Siberia is transitional between the forest-steppe of Middle and Western Siberia. The Krasnoyarskaya and Kanskaya forest-steppes occupy the footslopes (piedmont) and intermontane hollows and are separated by the low forested uplifts of the Kemchugskoye Upland and the Southern Yeniseisky Ridge. Middle Siberian forest-steppe depressions are marked by a complex geology, the presence of carbonate rocks, and severe erosional dissections. Apart from the present-day river valleys, a dense network of dry valleys, small ravines and lake beds has developed here. In the zone of the Middle Siberian forest-steppes the proportion of tilled lands approaches 50% of the total area. Natural vegetation is preserved only on steep slopes unsuitable for plowing and in isolated patches on the watersheds and terraces.

The southern Siberian forest-steppes have a special place in the system of landscape zones. They occur in the form of isolated islands among continuous forest areas, where the northern taiga unites with the mountain taiga of southern Siberia. Subtaiga represents a transitional belt of different width from the forest-steppe to the forest zone and belongs to the subzone of the so-called 'grass small-leaved and coniferous forests'. In this landscape zone the greatest changes took place under the influence of human economic activity. In the south the forest-steppe and subtaiga are limited in their occurrence by the mountain systems of the Eastern Sayan and the Kuznetsk Alatau. This area is located close to the center of Asia, at large distances from oceans and seas, in the region of attenuation of Atlantic air streams and of significant influence of Arctic air. It occupies the contact zone of plain and mountain expanses.

16.2 Vegetation Patterns and Steppe Types

Forest-steppes, which occur in the form of islands, are characterized by an intricate pattern of plant communities that are representative of other zones and subzones. The zonal vegetation types (forests with a grass layer and meadow steppes) are confined to watersheds and gentle slopes covered with a layer of Quarternary deposits. Cenoses that differ in environmental and ecological characteristics alternate at short intervals and create a complicated, motley pattern of vegetation cover, which has no parallel elsewhere in the entire area of the forest-steppe zone (Gerasimov et al. 1964).

According to their vegetation cover forest-steppes in the form of islands are divided into northern and southern subzones. In the former one considerable areas of watersheds are occupied by open 'wood-meadow grass forests' of *Betula pendula* Roth., *B. pubescens* Ehrh, *Pinus sylvestris* L., and *Larix sibirica* Ledeb. Steppe cenoses are confined to the dry valleys and southern slopes of dissected watersheds. Belts of dark-coniferous 'spruce (*Picea obovata* Ledeb.)-fir (*Abies sibirica* Ledeb.) or birch (*Betula pendula* Roth. and *B. pubescens* Ehrh) swamp forests' stretch along narrow cold river valleys. As regards soils, grey wood soils and chernozems (leached and ordinary) predominate.

In the southern subzone forests on watersheds disappear. *Pinus sylvestris* L. and *Larix sibirica* Ledeb. give way to *Betula pendula* Roth. and *B. pubescens* Ehrh which grow under conditions of increased moistening on north-facing slopes, in river valleys, in swales and other lowlands. Forest cover amounts to only 5–12%. True steppes, as well as stony and desert steppes, cover the steep southern and western slopes of watersheds. The vegetation cover of the valleys is also diverse: 'birch (*Betula pendula* Roth. and *B. pubescens* Ehrh) woods' and 'aspen (*Populus tremula* L.) meadows' are well-developed in flood plains, while 'steppe, saliferous, and saline meadows' are found on low terraces. In the southern Krasnoyarskaya forest-steppe there are small patches of true steppes, the so-called 'steppe cores' (Cherepnin 1956). Meadow steppes differ greatly in species composition, variation in seasonal aspect, and the developmental rhythm of their vegetation from other steppe types. They are situated on 'plakors' or in topographic depressions under conditions of continuous moistening, which is shown by signs of gleization at a depth associated with the layer of long-term seasonal freezing of the soils (Yerokhina and Lyubimova 1960). The flowering maximum coincides with the temperature maximum, and there is no summer dormancy.

Different conditions prevail on steep convex south-facing slopes under 'small-sod-grass (*Festuca cinerea, Koeleria cristata, Artemisia frigida, Agropyrum cristatum, Poa botryoides, Carex pediformis* C.A.Mey, *Carex duriuscula* C.A.Mey) and large-wormwood-feather-grass (*Stipa krylovii, Artemisia glauca* Pall. ex Willd., *Artemisia frigida, Artemisia scoparia* W. et K., *Artemisia dracunculus* L., *Festuca valesiaca, Koeleria cristata, Cleistogenes squarrosa*) steppes' on normal shallow chernozems. In winter snow here is blown off, and soils strongly freeze up to the depth of more than 2 m, while in the spring they thaw late. The soils of these steppes are characterized by a changeable hydroregime. In spring and summer they are dry; in late summer and in autumn they become humid. The developmental rhythm of this steppe vegetation is closely connected with the moisture conditions. These dry and cold steppes of the Mongolian type (Gerasimov et al. 1964) differ drastically from the true steppes of the Russian and Western Siberian Plains, but they are similar to the steppes of Khakasia.

Previously meadow steppes occupied large areas of watersheds in the southern forest-steppe. They were remarkable for their great species richness, thick grass cover and rich colors (Cherepnin 1961). To date these steppes are almost entirely tilled. Meadows are confined to flood plains and terraces above flood-plains of low river valleys. While in the northern subzone of forest-steppes there are considerable areas of natural hayfields (floodplain, forest, watershed, and steppificated meadows on slopes), in the southern subzone considerable areas of meadows have a hummocky topography, are salinized and are of little use.

In the southern subzone of forest-steppes forests are confined to northern slopes and occupy small areas. In saucer-shaped hollows amid arable lands there occur isolated birch (*Betula pendula* Roth. and *B. pubescens* Ehrh) forest stands. In addition to birch woods, pine (*Pinus sylvestris* L.) and pine-larch (*Pinus sylvestris* L.-*Larix sibirica* Ledeb.) forests occur here.

On the whole, in the vegetation cover of the forest-steppe the following series of complexes of different origin and age are distinguished: (1) true, dry and stony steppes of the Mongolian type; their occurrence has an island character (within the Kanskaya and Krasnoyarskaya forest-steppes they from isolated patches); (2) the complex of salt desert vegetation; (3) park pine (*Pinus sylvestris* L.) and birch (*Betula pendula* Roth. and *B. pubescens* Ehrh) tall-grass forests, in their species composition and developmental rhythm similar to the mountain forests of the Sayan and the Kuznetsk Alatau; (4) forb meadow steppes and birch (*Betula pendula* Roth. and *B. pubescens* Ehrh) forest stands; this is likely to be the youngest complex of vegetation of the forest-steppes. Its developmental rhythm completely corresponds to the current climatic conditions (Gerasimov et al. 1964).

16.3 Processes

The total of natural processes of movement, exchange and transformation of matter and energy gives an insight into the geosystem's functioning. To learn about the functioning of a geosystem as an integral system necessitates to study individual functional links, namely, the hydrological cycle, mineral metabolism in the plants – soils system (the biogeochemical cycle) and other elementary processes. They provide an idea of the dynamics of a geosystem (Isachenko 1991).

Here I present a spatial and temporal analysis of the present state of the soils of forest-steppe geosystems within the Nazarovskaya Depression in the context of climate changes.

16.3.1 The Setting

The forest-steppe of the Nazarovskaya Depression (Photos 16.1 and 16.2) is situated at the interface of two geomorphological provinces, namely, the Altai-Sayan mountain region and the Western Siberian Plain. In the south the low Solgonsky Ridge (the Eastern Sayan) separates it from the Chulymo-Yeniseiskaya Depression, whereas in the north the Arga Range (spurs of the Kuznetsk Alatau) separates it from the Western Siberian Plain. The Nazarovskaya Depression stretches east–west for 180 km, north–south up to 70 km. Morphologically, the Depression is divided into three parts: the southern cuesta-ridge (actual elevations of 400–600 m), the central hill and the northern plain (200–300 m).

Its climate is determined by its inland position at the interface of the mountain systems of Southern Siberia, the Central Siberian Plateau and the Western Siberian Plain and, of course, by the atmospheric circulation pattern. The thermic regime is continental and highly variable in time. Mean January air temperature is −16°C to −20°C, July 17–18°C, while the yearly mean temperature is below zero at all weather stations (except for Sharypovo). Temperature at the open soil surface in January is −18°C to −21°C, in July 20–22°C, and the yearly mean approaches 0°C. The soil freezes up to a depth of 1.5–2.0 m; continuous freezing of the upper horizons starts in late October, while its complete thawing occurs only mid July (Bufal et al. 1983).

The precipitation pattern is intricate. On the plain it averages 450–520 mm annually and it tends to increase east-, south- and south-eastwards to the foot of the spurs of the Kuznetsk Alatau and the Eastern Sayan (Drozdov and Grigorieva 1963). The south of the Depression has the lowest precipitation values. Accordingly, two types of climate are distinguished: type III 3 C for the central and southern parts, characterized by moderately insufficient moistening, a moderately warm summer and a moderately severe dry winter, and type II 3 D for the mountainous surroundings, characterized by optimum moistening, a moderately warm summer and a moderately severe winter with plenty of snow (Grigoriev and Budyko 1959).

The landscapes of the Nazarovskaya Depression and its mountainous surroundings are heavily disturbed. Within the low-hill territory the natural vegetation cover is preserved at only 5–10% of the area, while in the rest of the territory it is substantially transformed as a consequence of felling, fires, overgrazing, and recreation. The gently rugged plain surface of the depression belongs to the forest-steppe belt, while the mountainous surroundings belong to the forest belt of the lower hills. The forest-steppe belt is dominated by the communities of the 'birch (*Betula pendula*

Photo 16.1 Forest-steppe of the Nazarovskaya Depression (forb-feather-grass facies), Solgonsky Ridge (Eastern Sayan) (Photo Irina B. Vorobyeva)

Photo 16.2 Forest-steppe of the Nazarovskaya Depression (grass-*Artemisia* facies), Solgonsky Ridge (Eastern Sayan) (Photo Irina B. Vorobyeva)

Roth. and *B. pubescens* Ehrh) steppificated sedge-gramineous-leguminous-tall grass association group' in combination with steppe on the slopes of the ridges or on the 'plakor' surface. The low-hill belt is dominated by the forests of the 'birch (*Betula*) and larch (*Larix*) sedge-gramineous-tall grass association group' on the southern slopes and of the 'tall grass group' on the northern ones. Almost the entire territory is affected by agricultural activities; therefore, the remaining natural vegetation is characterized by various degradation stages.

The soils of the Nazarovskaya Depression and its mountainous surroundings are cryogenic soils, namely, chernozems, grey wood, sod forest, podzolic, sod-calcareous, meadow-chernozem, bog, meadow-bog, alluvial meadow, saline, and alkali soils. Ordinary and leached chernozems are frequent on steppificated sites. Grey wood soils are located on the northern and eastern slopes with birch (*Betula pendula* Roth. and *B. pubescens* Ehrh) and mixed tall grass forests. Meadow-chernozem soils of different thickness occupy bottoms of small ravines and areas adjacent to the slopes. As a whole the soil cover is modified by agricultural activity, and is subjected to wind and water erosion to a different extent.

Antropogenic transformation of the landscapes of southern Middle Siberia proceeded non-uniformly. It is strongest in the forest-steppe and steppe zones. The degree of agricultural development of lands (63%) exceeds several times the value for Russia, and the percentage of tillage (43%) is one of the highest in the country. Moreover, the fuel and energy resources of southern Siberia had their effects: it provides the major part of standard quality coal (85%), including 74.4% of coal suitable for industrial use (the Kansk-Achinsk basin). Cities, production localities and industrial facilities are distributed irregularly, and are connected by roads and railways, making a giant technogenic network that affects the environment.

16.3.2 Measurements

I monitored the role of the hydrothermal conditions of typical biogeocoenoses (facies 2, 3, 5 and 7 in Fig. 16.1) in the mountainous surroundings of the Depression, The hydrothermal conditions strongly determine the processes of organic matter transformation in the soils and are one of the main characteristics of geosystem functioning. Thus, they are a good indicator of climate change in this area. The facies of the northwest-facing slope include trans-eluvial 'forb-sedge-gramineous-meadows' with leached shallow chernozem (2) and trans-accumulative 'forb-feather-grass' at the site of felled open forest with a dark-grey wood soil (3). The slopes are rather steep (25–30°), stony and have a shortened soil profile (often not exceeding 50 cm), often with rock outcrops.

Southeast facing facies are trans-eluvial with 'meadow-steppe forb-gramineous' vegetation on weakly-leached chernozem of middle thickness (5) and eluvial with 'forb-feather-grass' vegetation on ordinary calcareous chernozem (7). Here slopes are more gentle (10–15°), less stony and with a thicker soil profile. There are pronounced fissures at its surface as a result of strong freezing of soils with a heavy texture. They are 3–5 cm wide at the surface and fade at a soil depth of 40–60 cm. In this zone

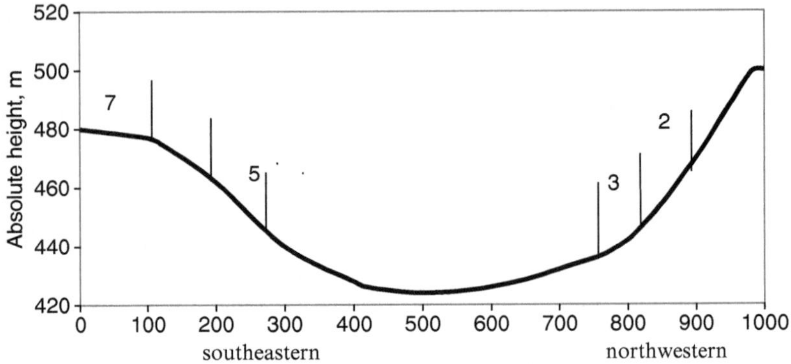

Fig. 16.1 Scheme of the Ashpansky experimental landscape-geochemical profile (in m)

of fissures a peculiar hydrothermal regime is being created. In spring an additional quantity of melting water enters these fissures and bring heat, which contributes to earlier and faster soil thawing. In summer better aeration conditions are created here. Monthly mean temperatures at the soil surface from September till April are below zero, and only from May till August they are above zero. Soil surface temperatures are highest in July and lowest in February.

Hydrothermal observations were made in July. By that time, the soil organic matter at different locations acquires relatively stable values. On a heterogeneous surface the soil temperature regime depends on slope exposure. Differences in moisture content of the soils of the facies are determined by the distribution of snow cover, peculiarities of spring snow melting on differently facing slopes and, correspondingly, of soil unfreezing, and their saturation with winter-spring moisture.

16.3.3 Results

The identical curves of the temperature response of the soils to solar radiation indicate that, during the past 20 years, the temperature regime in the 0–20 cm soil layer at different locations in the middle of the vegetative season is formed in the same way. In the period 1986–2005 the amplitude of temperature fluctuations in the cold period of the year (November–March) was higher than in summer. Mean air temperatures increased in all seasons, except in winter, when the monthly mean air temperature had an insignificant negative trend. The total precipitation showed the reverse tendency. Precipitation was generally close to normal (444 mm) with deviation towards both an increase (up to 668 mm) and a decrease (up to 276 mm); thus, the amount of precipitation may differ more than twofold (Fig. 16.2). The bulk of precipitation falls in the warm season (April–October).

The warming of the last decades manifests itself in the middle latitudes of the Northern hemisphere mainly in the cold season (summer temperatures do not show

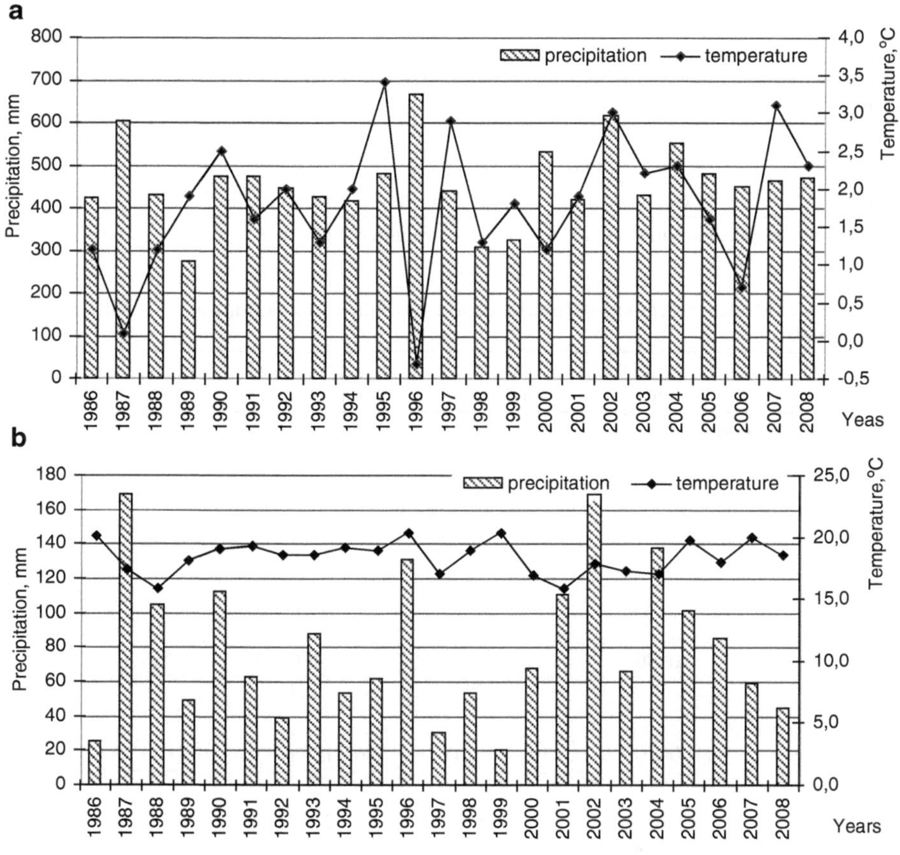

Fig. 16.2 Yearly mean air temperature and total precipitation (**a**), and monthly mean air temperature and precipitation in July (**b**) according to data from the station Sharypovo

tendencies for warming), whereas in the previous period of warming (1910–1940) it occurred simultaneously in winter and in summer. Climatic monitoring showed that the years 1998–2008 were the warmest ones in the Northern hemisphere for the entire period of instrumental meteorological observations (Vyrkin and Nechayeva 2007).

In the view of the heavy practical use of the forest-steppe territories, investigations into the deposition of productive moisture in soils both in space (on differently facing slopes) and in time are gaining special relevance. These parameters are of the utmost interest not only as an average for a vegetative period, but also for each season (Table 16.1). In spring moisture deposition in the 0–50 cm soil layer is very variable here but large enough for the development of plants, even on the dry upper parts of southern and western slopes. In summer moisture deposition is also sufficient in all locations, though less than in spring. In autumn deposits increase as compared to summer.

Table 16.1 Moisture deposition in the 0–50 cm soil layer of differently facing slopes

	Northwestern slope				Southeastern slope			
Seasons	Facies 2		Facies 3		Facies 5		Facies 7	
1987–1989	a	b	a	b	a	b	a	b
Spring	1,114	22	2,580	52	1,433	29	1,723	35
Summer	1,170	24	2,211	44	1,257	25	1,223	24
Autumn	697	14	1,988	40	1,094	23	1,276	25

Note: Moisture deposition (ton/ha): *a* – total, *b* – weighted mean at a depth of 0–50 cm

Table 16.2 Moisture deposition (mm) and water-soluble ions (kg/ha) in soils for the period 1987–1989

Facies	Soil layer (cm)	Moisture	Mineral carbon	Organic carbon
Northwestern slopes				
2	0–20	$\frac{28-88\ (60)}{53}$	$\frac{7-16\ (9)}{10}$	$\frac{79-331\ (252)}{204}$
	0–50	$\frac{105-177\ (72)}{132}$	$\frac{29-98\ (69)}{57}$	$\frac{823-1{,}690\ (867)}{1{,}181}$
3	0–20	$\frac{83-99\ (16)}{93}$	$\frac{54-229\ (175)}{131}$	$\frac{841-1{,}740\ (899)}{1{,}154}$
	0–50	$\frac{153-216\ (63)}{192}$	$\frac{123-460\ (337)}{291}$	$\frac{1{,}575-3{,}055\ (1{,}480)}{2{,}097}$
Southeastern slope				
5	0–20	$\frac{39-70\ (14)}{55}$	$\frac{50-124\ (74)}{86}$	$\frac{515-1{,}119\ (604)}{1{,}122}$
	0–50	$\frac{96-158\ (62)}{128}$	$\frac{83-210\ (127)}{151}$	$\frac{932-2{,}212\ (1{,}280)}{1{,}563}$
7	0–20	$\frac{34-59\ (25)}{45}$	$\frac{120-172\ (52)}{154}$	$\frac{420-1{,}325\ (905)}{780}$
	0–50	$\frac{101-144\ (43)}{367}$	$\frac{467-593\ (126)}{526}$	$\frac{933-3{,}062\ (2{,}129)}{1{,}752}$

Note: The numerator gives the range of values with in brackets the values' interval and the denominator gives mean values

Maximum moisture accumulation in the dark-grey wood soil of facies 3, formed at the site of felled forest on the northwestern slope, points to a different development here (forest type) than in the steppe facies 5 and 7 on the southeastern slope. This points to a potential return to the native forest, perhaps with the preservation of some elements of the steppe type of environment.

During the period 1986–2008 moisture deposition on differently facing slopes, in the 0–20 and 0–50 cm soil layers of facies 3 and 7, increased in total and in fluctuation range (Tables 16.2 and 16.3). Location is important too as a comparison of the data for facies 2 and 5 show. Soil moistening is of course related to precipitation and reflects long-term precipitation cycles.

Table 16.3 Moisture deposition (mm) and water-soluble form of elements (kg/ha) in soils for the period 2006–2008

Facies	Soil layer (cm)	Moisture	Mineral carbon	Organic carbon
Northwestern slopes				
2	0–20	$\frac{30-45\ (15)}{38}$	$\frac{20-23\ (3)}{22}$	$\frac{447-818\ (371)}{633}$
	0–50	$\frac{111-147\ (36)}{129}$	$\frac{70-122\ (52)}{96}$	$\frac{1{,}414-1{,}965\ (551)}{1{,}690}$
3	0–20	$\frac{82-130\ (48)}{112}$	$\frac{49-102\ (53)}{81}$	$\frac{2{,}262-3{,}096\ (834)}{2{,}771}$
	0–50	$\frac{149-235\ (86)}{204}$	$\frac{127-172\ (45)}{154}$	$\frac{3{,}938-4{,}636\ (598)}{4{,}250}$
Southeastern slope				
5	0–20	$\frac{28-72\ (44)}{55}$	$\frac{48-123\ (75)}{82}$	$\frac{731-797\ (66)}{766}$
	0–50	$\frac{70-131\ (61)}{110}$	$\frac{112-288\ (176)}{199}$	$\frac{1{,}195-1{,}547\ (352)}{1{,}414}$
7	0–20	$\frac{28-74\ (46)}{58}$	$\frac{84-203\ (119)}{151}$	$\frac{1{,}023-1{,}684\ (661)}{1{,}246}$
	0–50	$\frac{88-164\ (76)}{133}$	$\frac{431-559\ (128)}{480}$	$\frac{1{,}703-3{,}094\ (1{,}391)}{2{,}382}$

Note: The numerator gives the range of values with in brackets the values' interval and the denominator gives mean values

Soil temperature is influenced by the amount of heat near its surface and by diffusive factors of available heat. The main factor of heat diffusion by soil is moisture. If on the soil surface there is a large amount of moisture, the bulk of absorbed heat energy is spent on its evaporation. Since the surface temperature in this case does not increase, only an insignificant gradient is formed, which hardly contributes to the heat flow into the soil. If the soil surface is dry, absorbed energy heats it up and generates a significant temperature gradient that provokes considerable heat flow into the soil (Hanks and Ashcroft 1985).

Correlation of air and soil temperatures is one of the crucial conditions for the growth and development of plants. Their normal functioning requires roots to be situated in a cooler environment than aboveground organs. When the root layer is warmer than the air, respiration dominates over assimilation. Plants growing under the conditions of the soil temperature being warmer than the air temperature are able to survive; they then develop a deeper root system and reduce their height growth (Bessolitsina et al. 1991).

In the forest-steppe zone during most of the vegetative period plants function under a favorable temperature regime with the soil temperature at a depth of 15 cm in July at 5–7°C below the air temperature. Situations when, under the excessive heating of the soil surface and its 0–10 cm layer, the soil temperature exceeds the air

Table 16.4 Correlation coefficients between air temperature (*T*) at a height of 1.5 m and soil temperature at different depths

Facies	Index	T of air	Surface	At a depth of 5 cm	At a depth of 10 cm	At a depth of 15 cm
2	T of air	1.00				
	Surface	0.91	1.00			
	At a depth of 5 cm	0.83	0.89	1.00		
	At a depth of 10 cm	0.84	0.89	1.00	1.00	
3	T of air	1.00				
	Surface	0.82	1.00			
	At a depth of 5 cm	0.68	0.86	1.00		
	At a depth of 10 cm	0.56	0.80	0.88	1.00	
5	T of air	1.00				
	Surface	0.80	1.00			
	At a depth of 5 cm	0.77	0.80	1.00		
	At a depth of 10 cm	0.78	0.62	0.88	1.00	
	At a depth of 15 cm	0.80	0.62	0.83	0.92	1.00
7	T of air	1.00				
	Surface	0.76	1.00			
	At a depth of 5 cm	0.88	0.70	1.00		
	At a depth of 10 cm	0.77	0.65	0.91	1.00	
	At a depth of 15 cm	0.75	0.60	0.93	0.93	1.00

temperature, are likely to occur on the southern steppificated slopes. In facies 3 the air temperature in July is a factor 1.5 higher than in the upper soil layers (Table 16.4).

During the past 20 years mean annual air temperature fell below the norm in 1987, 1996 and 2006. In 2008 it was 2.3°C, which is the result of increasing air temperatures in winter, late autumn and early spring.

Changes in moisture deposition and heat supply occur in accordance with the dynamics of hydrothermal conditions and the cyclic development of the principal parameters of the soil climate. As regards yearly total precipitation the identification of long-term cycles is important. Most distinct are 7- and 9-year cycles in precipitation variability. Here a secondary cyclicity is observed between the nearest years of maximum and minimum values with an average duration of 3–4 years. In the mid-1980s precipitation values in summer time were below the norm. It resulted in the reduction of the deposits of moisture in the 0–20 and 0–50 cm layers in the soils of the south-eastern slope (Figs. 16.3 and 16.4). In those years, the annual mean air temperature amounted to 0.9°C.

Thermal and water regimes of soils are considered as ecological factors regulating the organic matter metabolism, which is a part of the biogeochemical carbon cycle and one of the leading landscape-forming processes. I observed indices of the final metabolic stage: the destruction block. It includes production of mobile organic compounds, which then migrate in the composition of soil solutions. The number of these compounds at the time of observation serves as a manifestation of an important

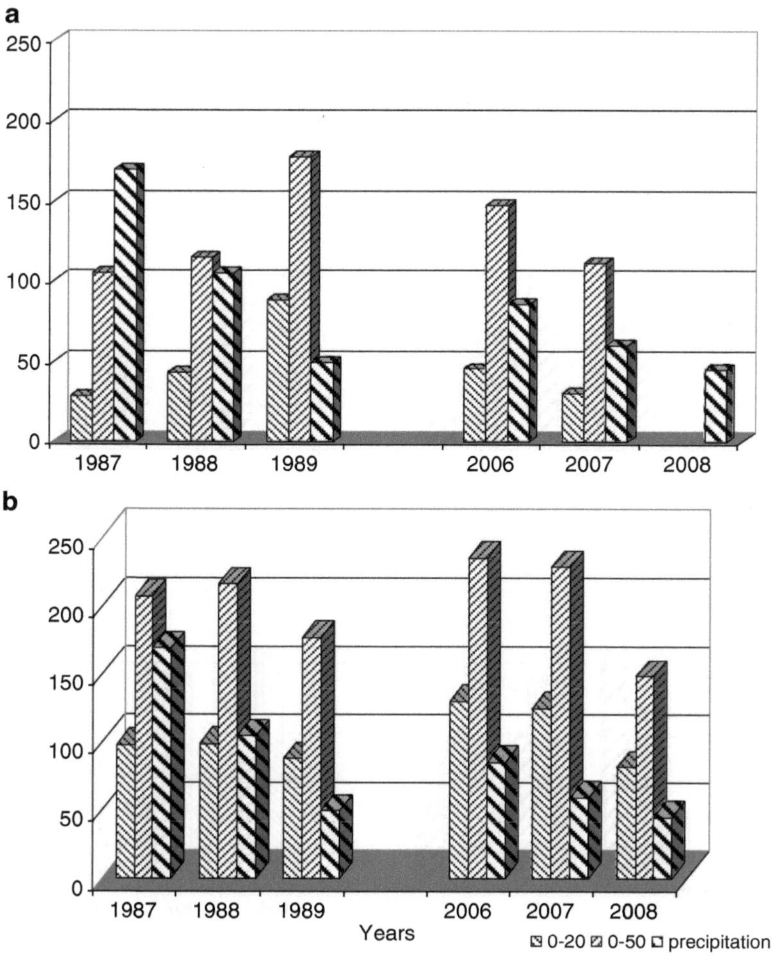

Fig. 16.3 Changes in precipitation (mm) and deposits of moisture (mm) in the 0–20 and 0–50 cm soil layers in facies *2* (**a**) and *3* (**b**) on the northwestern slope

soil property, namely, 'soil moment' (Sokolov and Targulyan 1976) and is an indicator of the dynamic state of a geosystem in that point in time (Martynov 1985; Nechayeva et al. 2004).

Observations since 1986 of the mobile form of carbon compounds showed the tendency in the development of the organic matter metabolism in the soils (Figs. 16.5 and 16.6). This metabolism is getting more active the last years, as suggested by an increase in the amount of water-soluble organic matter in the soils of facies 2 and 3 on the northwestern slope and a decrease in the soils of facies *5* and *7* on the southeastern slope.

Thus, temporal changes in the mobile organic matter reflect the dynamic state of forest-steppe geosystems in the context of global and regional changes of external

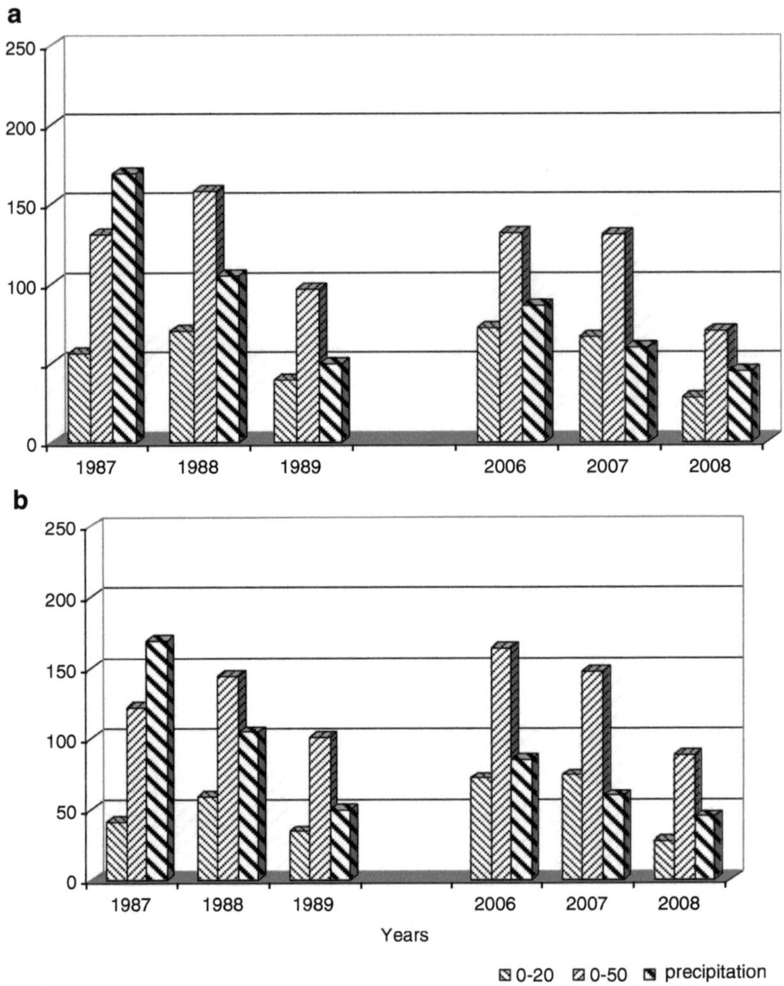

Fig. 16.4 Changes in precipitation (mm) and deposits of moisture (mm) in the 0–20 and 0–50 cm soil layers in facies *5* (**a**) and *7* (**b**) on the southeastern slope

factors as well as of soil climate (temperature and moisture deposits) on slopes of different exposure.

As the soil develops over time, along with the 'soil moment', the historically formed property of a 'soil memory' becomes clear, in which humus substances play a special role, due to the fact that humus acids (humic acids (HA) and fulvic acids (FA)) are the products of the humification process, which proceeds under any hydrothermal conditions in the presence of dead plant material.

The structure and composition of humus compounds are unsteady (variable) but stable enough within a certain framework set by the hydrothermal conditions under which they are formed and they correlate with climate characteristics (Grishina 1986).

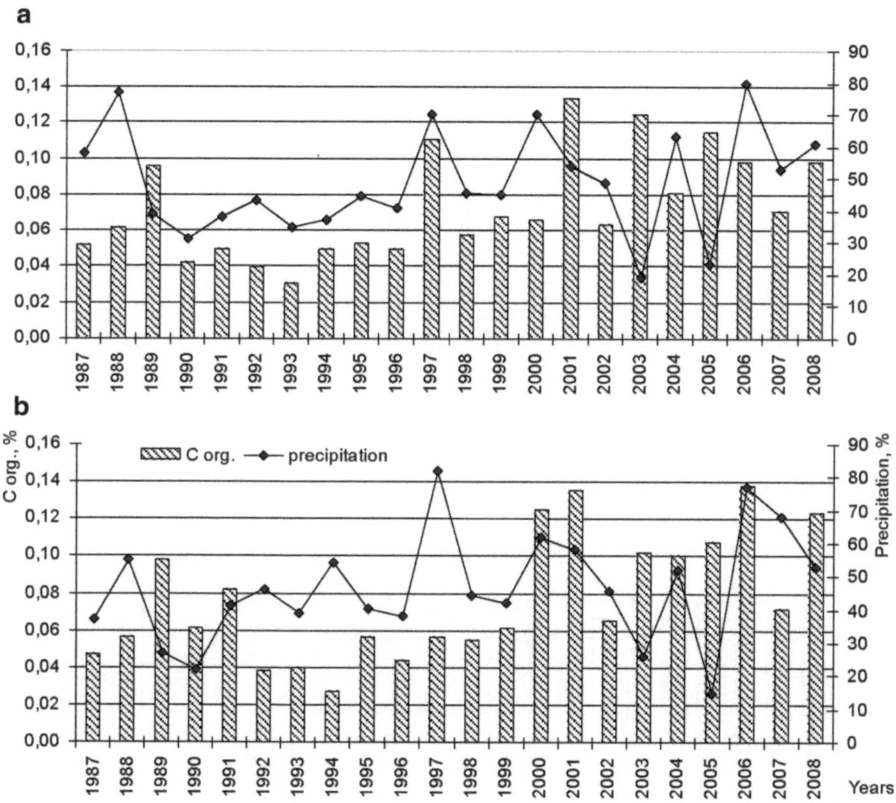

Fig. 16.5 Dynamics of soil humidity and water-soluble carbon (C org.) in soils of facies *2* (**a**) and *3* (**b**) on the northwest slope

Throughout, the correlations between climate indicators and values of the ratio HA: FA were significant. Furthermore, HA correlated significantly with thermal conditions (air temp., $t > 10°C$, soil temp., etc.), and FA with precipitation (Targulyan and Goryachkin 2008).

HA represent an accumulative component consolidated by the mineral part of the soil, while FA partially form complexes with HA, and partially migrate deeper into the soil.

In the Nazarovskaya Depression at facies 2 on the northwestern slope, carrying forest-steppe, the soil is characterized by a high content of humus, an abruptly decreasing distribution of carbon in the soil profile, a very high concentration of nitrogen in the organic matter, a medium degree of humification, and a humate-fulvate type of humus. The soils of facies 5 and 7 on the southeastern slope have a high content of humus, a gradually decreasing carbon content down the profile, a high concentration of nitrogen in the organic matter, a high degree of humification, and a humate or fulvate-humate type of humus (Vorobyeva 2004, 2005).

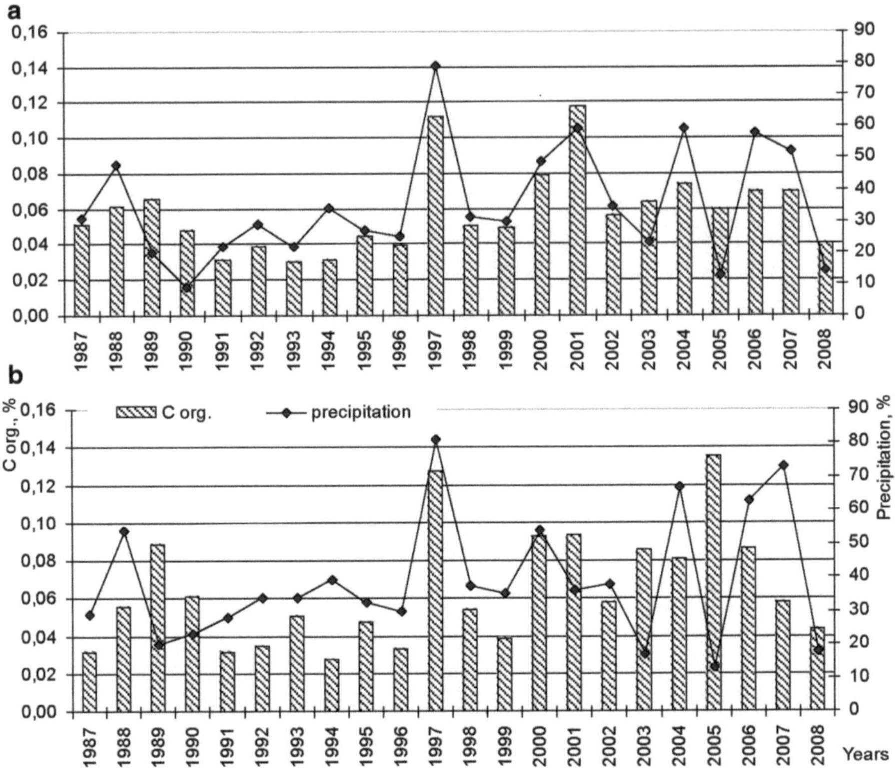

Fig. 16.6 Dynamics of soil humidity and water-soluble carbon (C org.) in soils of facies *5* (**a**) and *7* (**b**) on the southeast slope

Qualitative and quantitative characteristics of humus vary greatly in soils of different facies as a consequence of differences in the composition of the vegetation and the ecological conditions of decomposition of dead organic remains.

Humus accumulation in soils is affected by the annual input of organic material and its loss as a result of mineralization. In years that differ in hydrothermal conditions this effect changes. In steppe soils black humic acids prevail; they are water-soluble and in moist periods migrate into the deeper soil layers, resulting in the formation of thick humus horizons. In forest soils brown humic acids predominate; they are weakly-soluble in water, and thus their migration is restricted and humus horizons are thin (Kobak 1988).

Under the strong continental climate of the area the intensity and pace of the processes differ compared to those within the European territory: they are slower in spring and increase only by mid-summer. The organic remains decompose during a short summer period, concentrating in a thin layer in the top soil.

In 2007, after a period of more than 20 years, the content and reserves of humic acids, fulvic acids, and the unhydrated residue had increased by a factor of 1.5–5, 1.2–3.5, and 1.2–2, respectively (Fig. 16.7). Such changes can be seen in all soils,

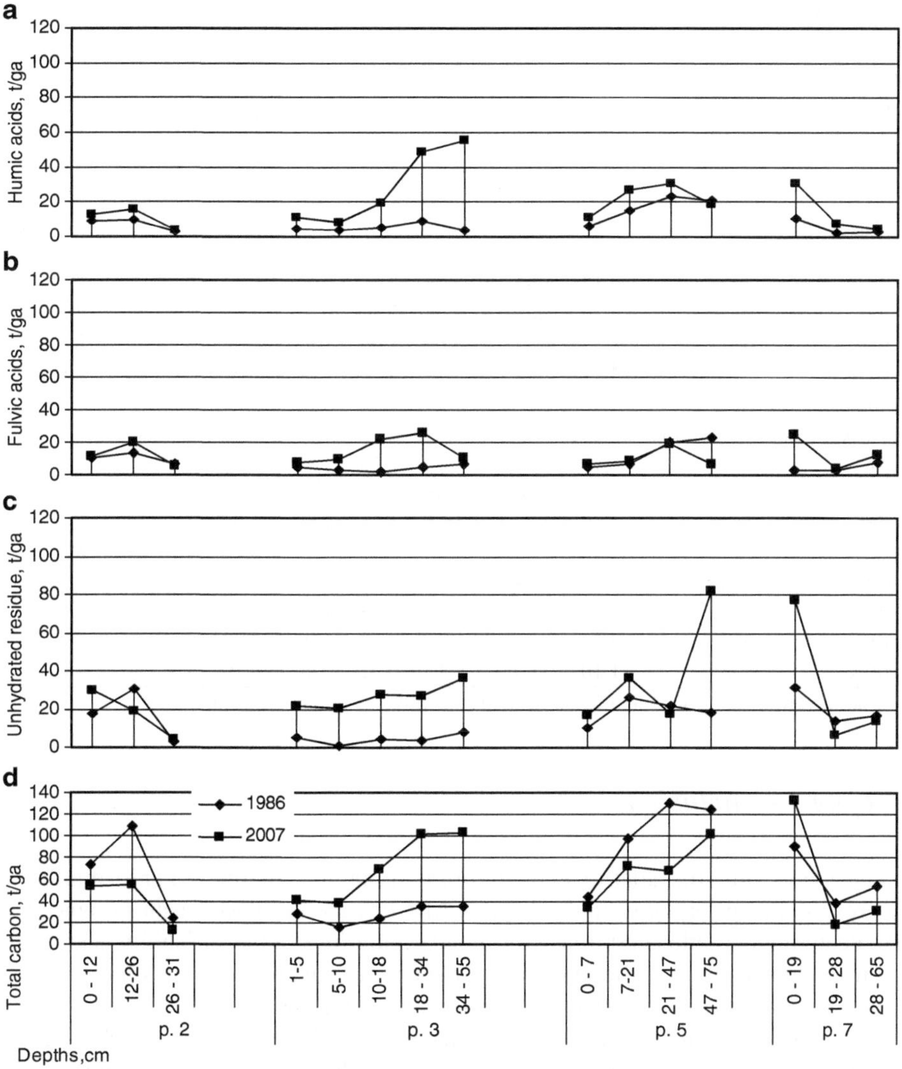

Fig. 16.7 Comparison of deposits of carbon of humic acids (**a**), fulvic acids (**b**), unhydrated residue (**c**) and total carbon (**d**) at different depths of soils of facies *2*, *3*, *5*, and *7* in 1986 and 2007

but with different intensity. Thus, in the soils of the northwestern slopes a low degree of humification of the organic matter (the proportion of humic substances in the composition of the organic matter) in leached chernozem changed to a medium degree over time, while the humate-fulvate and fulvate humus types were replaced by a fulvate-humate type. In the dark-grey wood soil the low degree of humification increased up to the medium degree, and the humate-fulvate humus of the upper part of the soil profile transformed into a fulvate-humate humus, and still deeper into humate humus (Table 16.5).

Table 16.5 Types of soil humus of the facies on the slopes, %

Horizon	Depth (cm)	C_{org}	$C_{ha}+C_{fa}$	C_{ha}	C_{fa}	$C_{ha/fa}$	C_{ins}	$C_{ha/org\cdot 100\%}$	$C_{ins/org\cdot 100\%}$
Northwest slopes									
Leached chernozem of trans-eluvial facies 2									
A	0–12	10.05	4.55	2.35	2.20	1.07	5.50	23	55
AB	12–26	6.00	3.90	1.75	2.15	0.81	2.10	29	35
BC	26–31	4.00	2.70	1.05	1.65	0.64	1.30	26	33
Dark-grey wood soil of trans-accumulative facies 3									
Ad	1–5	11.31	5.15	3.10	2.05	1.51	6.16	27	54
A1	5–10	10.14	4.85	2.25	2.60	0.87	5.29	22	52
AB	10–18	8.48	5.05	2.35	2.70	0.87	3.43	28	40
B	18–34	6.21	4.55	3.00	1.55	1.94	1.66	48	27
BCg	34–55	4.79	3.10	2.60	0.50	5.20	1.69	54	35
Southeast slopes									
Weakly-leached chernozem of trans-eluvial facies 5									
Ad	0–7	6.70	3.40	2.15	1.25	1.72	3.30	32	49
A1	7–21	6.24	3.10	2.35	0.75	3.13	3.14	38	50
AB	21–47	3.65	2.70	1.65	1.05	1.57	0.95	45	26
B	47–75	5.10	0.95	0.60	0.35	1.71	4.15	12	81
Ordinary calcareous chernozem of eluvial facies 7									
Ak	0–19	9.32	3.90	2.15	1.75	1.23	5.42	23	58
Bk	19–28	2.67	1.75	1.15	0.60	1.92	0.92	43	34
BCk	28–65	1.10	0.60	0.17	0.43	0.40	0.50	15	45

Note: *C* Carbon, C_{org} organic, C_{ha} humic acids, C_{fa} fulvic acids, C_{ins} insoluble residue

16.3.4 Conclusion

The study into the spatial and temporal dynamics of the organic matter in forest-steppe geosystems in the south of Middle Siberia showed a clear connection to changing climatic conditions. We found regularities in the variability, and determined changes in the differentiation of the organic matter. Dynamics in soil moisture deposition vary as a result of long-term precipitation cycles.

Temperatures in the root soil layer depend on a host of factors, including the structure of the vegetation cover and land forms (slope steepness, exposure).

Temporal changes in the mobile organic matter reflect the dynamic state of the forest-steppe geosystems in relation to changes in the global and regional soil climate (temperature and moisture deposition) on differently facing slopes. And the soil water regime determined differences in humus composition.

My experimental data proved that the humus state of the soils changes both in space and time. An increase of the mean annual air temperature over the past two decades under a sufficient amount of precipitation caused the strongest changes in the humus state of dark-grey wood soil and weakly-leached chernozem on the slope with forest-steppe. The increase in the amount of humic acids in the soils indicates that the steppe geosystem changes.

During the more than 20 years of observation the contents of total carbon, humic acids and insoluble residue increased up to a factor 3. These signs are evidence of a tendency in the present development of the southern Siberian forest-steppe towards the formation of islands of natural steppe. And this can be considered a landscape response to global and regional changes in the factors governing the natural environment.

References

Bessolitsina EP, Kakareka SV, Krauklis AA, Kremer LK (1991) Geosystems of the taiga–steppe contact: the south of Central Siberia. Nauka. Sib. otd-nie, Novosibirsk, 217 p

Bufal VV, Savelyeva IL, Turuschina LA (1983) Nature and economy of the region of top-priority formation of the KATEK. Nauka, Novosibirsk, 261 p

Cherepnin LM (1956) The vegetation cover of the southern part of the Krasnoyarsk territory and problems of its study, vol 5. Uch. Zap. Krasnoyar. Ped. In-ta, Krasnoyarsk

Cherepnin LM (1961) The vegetation of the Krasnoyarsk territory. In: The natural conditions of the Krasnoyarsk territory. Izd-vo AN SSSR, Moscow, pp 160–187

Drozdov OA, Grigoryeva AS (1963) Hydrological cycle in the atmosphere. Gidrometeoizdat, Leningrad, p 316

Gerasimov IP, Preobrashensky VS, Richter GD, Abramov LS (1964) Middle Siberia. Nauka, Moscow, 479 p

Grigoryev AA, Budyko MI (1959) Classification of climates of the USSR. Izv. AN SSSR. Ser. geogr., No. 3: 3–19

Grishina LA (1986) Humus formation and humus state of soils. Izd-vo MGU, Moscow, p 240p

Hanks RJ, Ashcroft GL (1985) Applied soil physics. Soil temperature and humidity. Gidrometeoizdat, Leningrad, p 150

Isachenko AG (1991) Landscape study and physical and geographical division into rayons. Vysshaya Shkola, Moscow, 366 p

Khotinsky NA (1961) Distribution of vegetation in the Kansk forest-steppe in the form of islands. Geogr Rep Moscow 2:93–94

Kobak KI (1988) Biotic components of the carbon cycle. Gidrometeoizdat, Leningrad, p 247p

Martynov AV (1985) Dynamics of substance in geosystems under intensifying technogenic impact: author's abstract of Cand. Sc. (Geogr) degree dissertation, Irkutsk, 20 p

Nechayeva EG, Davydova ND et al (eds) (2004) Trends of landscape-geochemical processes in geosystems of the south of Siberia. Nauka, Novosibirsk, 184 p

Richter GD (1960) Natural areas of Eastern Siberia. In: Development of production forces of Eastern Siberia. Agriculture. Izd-vo AN SSSR, Moscow, pp 7–16

Sokolov IA, Targulyan VO (1976) Soil and environment interaction: soil-memory, soil-moment. In: Study and development of the natural environment. Nauka, Moscow, pp 15–164

Targulyan VO, Goryachkin SV (eds) (2008) Soil memory: soil as a memory of biosphere-geosphere-antroposhere interactions. Izdatelstvo LKI, Moscow, 692 p

Vorobyeva IB (2004) Humic state of chernozems on the differently facing slopes of the Nazarovskaya forest-steppe. In: Humic substances in biosphere. Proceeding of II international conference. Izd-vo MGU, Moscow, pp 229–232

Vorobyeva IB (2005) Humic state of the steppificated slopes of forest-steppe geosystems of the southern Middle Siberia//Naukoviy Visnik Chernivetskogo Universitetu: Zbirnik Naukovikh Prats. Issue 252: Biology. Chernovets: Ruta, pp 42–47

Vyrkin VB, Nechayeva EG (eds) (2007) Geographical research of Siberia. Landscape forming processes, vol 2. Akademicheskoye izd-vo, Novosibirsk, «Geo», 317 p

Yerokhina AA, Lyubimova EL (1960) Influence of the long-term seasonal frost on soils and vegetation (exemplified by Middle Siberia). In: Proceedings of the international congress of soil scientists in New York, vol 39, pp 143–159

Chapter 17
Effects of Climate Warming and Vegetation Cover on Permafrost of Mongolia

Natsagdorj Sharkhuu and Anarmaa Sharkhuu

Abstract Long-term monitoring of permafrost in Mongolia showed that permafrost under the influence of climate warming is ubiquitously degrading at different rates. The average trends of the increases in active layer thickness and mean annual ground temperature are 5–20 cm and 0.1–0.3°C per decade, respectively. These trends of the last 15–20 years are higher than those of the previous 15–20 years. Furthermore, the degradation of permafrost in the Hövsgöl mountainous region is more intense than in the Hentei and Hangai mountainous regions. Experimental results have shown that vegetation cover, especially moss cover, forest, and dense grass, are natural insulators that maintain lower soil temperature and higher soil moisture by lowering the evaporation rate. Hence, intense grazing may accelerate the degradation of permafrost. Since permafrost plays an important role in the maintenance of ecosystems in this region, the recent degradation of permafrost may lead to significant changes in ecosystems, especially in the soil thermal state and moisture content.

17.1 Introduction

The Fourth Report by the International Panel on Climate Change documented recent changes in permafrost and seasonally frozen ground due to climate change in all northern latitudinal regions (Solomon et al. 2007). In addition to climate warming,

N. Sharkhuu
Institute of Geography, Mongolian Academy of Sciences, Ulaanbaatar, Mongolia
e-mail: sharkhuu_n@yahoo.com

A. Sharkhuu (✉)
Department of Earth and Environmental Sciences, University of Pennsylvania, Philadelphia, PA, USA
e-mail: anarmaa@sas.upenn.edu

M.J.A. Werger and M.A. van Staalduinen (eds.), *Eurasian Steppes. Ecological Problems and Livelihoods in a Changing World*, Plant and Vegetation 6,
DOI 10.1007/978-94-007-3886-7_17,

anthropogenic disturbances, including overgrazing, land cultivation, forest clear-cutting, mining, engineering construction, and forest fires, adversely influence permafrost, thus having impact on the ecosystem (Cannone et al. 2003; Cheng and Li 2003). Disturbance of the soil surface, especially of the vegetation and snow cover, is of main influence on the permafrost because of the thermal insulation effect of vegetation and snow cover on the permafrost systems. And this is considerably more important in the southern fringe of permafrost zones than in high-latitudinal zones (Kudryavtsev et al. 1974). Permafrost in semi-arid regions like Mongolia plays an important role in the hydrology. It acts as an impenetrable layer for ground water, thus making the ground water available to vegetation (Price 1971; Harris 1998; Walker et al. 2003, 2004).

The recent degradation of permafrost under the influence of climate warming has a direct impact on permafrost ecosystems. For example, permafrost degradation may cause shifts in the boundary of an ecosystems (Jorgenson et al. 2001), enhance the release of CO_2 (Lee et al. 2010; Schuur et al. 2009), emit nitrous oxide (Elberling et al. 2010) from previously frozen soil, enhance methane release from a thermokarst lake (Desyatkin et al. 2009), promote or inhibit plant growth (Jarvis and Linder 2000), and decrease surface soil moisture in arid and semi-arid regions (Cui and Graf 2009).

Permafrost monitoring is of considerable scientific and practical significance, because without knowledge about the permafrost degradation process, it is difficult to assess and predict the impacts of permafrost degradation on ecosystems. Accordingly, since the early 1990s, most countries with permafrost have conducted long-term permafrost monitoring within the framework of international and national programs (Brown et al. 2000; Burgess et al. 2000; Nelson et al. 2004).

In Mongolia, the effects of climate warming and human activities on permafrost have been monitored within the framework of several international projects in the recent years. The long-term (10–38 years) monitoring of permafrost in Mongolia has been conducted within the framework of the Circum-arctic Active Layer Monitoring (CALM) and the Global Terrestrial Network for Permafrost (GTN-P) programs since 1996 (Sharkhuu 1998, 2003; Sharkhuu et al. 2008). In addition, the monitoring of changes in permafrost under the influence of climate change and human activities, and studies on the thermal insulation effect of the vegetation cover in six valleys along the north-eastern shore of Hövsgöl Lake, Hövsgöl region, northern Mongolia, was carried out within the framework of the Hövsgöl Global Environment Facility/World Bank (GEF/WB) Project in the period of 2002–2007 (Goulden et al. 2005; Etzelmüller et al. 2006; Sharkhuu et al. 2006, 2007). Long-term monitoring of the ground temperature regime in the Nalayh depression and Terelj Valley near Ulaanbaatar in the Hentei region, central Mongolia, has been conducted by Japanese and Mongolian permafrost researchers within the framework of the Institute of Observational Research for Global Change (IORGC) project since 2002 (Jambaljav et al. 2008; Ishikawa et al. 2005). In order to extend the monitoring of permafrost in Mongolia, the Institute of Geography, Mongolian Academy of Sciences, drilled and instrumented several 10 m deep boreholes in 2009 and started monitoring these.

Here we summarize and review the results of the above studies on the long-term monitoring of permafrost under the influence of climate warming in Mongolia. In addition, we present preliminary results of observations and experiments on the insulation effects of different vegetation covers on soil temperature, moisture content, and active layer at the Dalbay and Hustai study sites in the Hövsgöl and Hentei regions (Sharkhuu et al. 2007, 2010; Sharkhuu and Anarmaa 2008).

17.2 Climate Warming and Grazing

According to climate change studies (Natsagdorj et al. 2000), the mean annual air temperature in Mongolia has increased by 1.56°C during the last 60 years. While the winter temperature increased by 3.61°C, the spring-autumn temperature increased by just 1.4–1.5°C. The mean annual air temperatures increased by 1.8°C in western and northern Mongolia and by 1.4°C in central Mongolia. In contrast, the mean annual air temperature of southern and eastern Mongolia increased by only 0.3°C in the period of 1940–1990.

A greater increase in air temperature in northern latitudes is predicted by IPCC models. While the increase of 1.6°C in annual air temperature in Mongolia is greater than the global average temperature increases (IPCC 2007), the increase of mean annual air temperature in central and eastern Yakutia (higher latitudinal regions) is three times as strong as in Mongolia (Gavrilova 2003).

Most of the observation sites for permafrost monitoring in Mongolia are located around the weather stations of Hatgal (on the southern shore of Hövsgöl Lake in the Hövsgöl region) and Buyant-Ukhaa (near Ulaanbaatar in the Hentei region). There, our studies on the insulation effects of vegetation cover on permafrost and seasonally frozen ground are also conducted. Figure 17.1 shows the linear trends of increase in the mean winter (MWT), summer (MST), and annual (MAT) air temperatures at these weather stations. The trends of increase in the mean annual air temperatures at the Hatgal and Buyant-Ukhaa weather stations are 0.35°C and 0.30°C per decade, respectively, but these trends change over time: the mean annual air temperatures at the Hatgal weather station increased by 0.61°C in the period of 1963–1989, and by 0.84°C in the period of 1990–2003 (Nandintsetseg and Goulden 2005).

In addition to a warming climate, Mongolia faces major environmental concerns related to human activities, such as pollution due to mining, overgrazing near settlement areas near water sources, abandonment or mismanagement of cultivated land, and clearance of forests (Corsi et al. 2002). These activities, along with engineering construction in certain areas, have caused the destruction of the vegetation cover, thus causing permafrost degradation. Among those activities overgrazing is the major issue because livestock husbandry is very common and important agricultural practice in Mongolia. The sub-sector of livestock husbandry employs 47.9% of the total population and accounts for 30% of the country's exports (Batima et al. 2005). Livestock

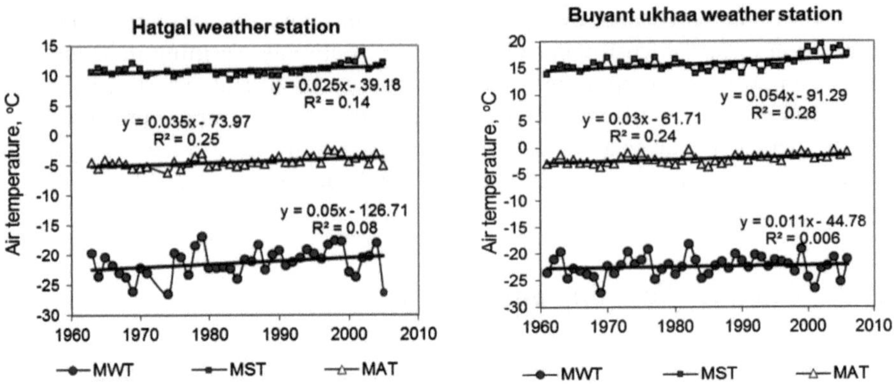

Fig. 17.1 Mean winter (*MWT*), summer (*MST*) and annual (*MAT*) air temperatures at the Hatgal, (Hövsgöl region) and Buyant-Ukhaa (Hentei region) weather stations

numbers never exceeded 23 million during the period 1921–1990 (FAO 2006), but it reached 33.6 million in 1999 and approximately 50 million in 2009, causing a reduction in pastureland and desertification (compare other chapters in this volume).

Destruction of the vegetation cover, a shift in the biome boundaries of forest and steppe regions, and desertification will affect the ground temperature regime, possibly causing permafrost degradation (Jorgenson et al. 2001; Yang et al. 2004). The thermal insulation effect of vegetation and snow covers on permafrost in the southern fringe of permafrost zones is more important than in high-latitudinal zones (Kudryavtsev et al. 1974). Hence, small changes in vegetation and/or snow cover may cause bigger changes in permafrost conditions.

To properly investigate this, the insulation effects of vegetation cover on permafrost and seasonal frozen ground were studied in Hövsgöl National Park in the Hövsgöl region and Hustai National Park in the Hentei region. Since both areas were designated as national parks in 1992, limited grazing was permitted there in a restricted zone only. This forced several families to either move out of the Hustai park area or decrease the number of their livestock, which resulted in a decrease of 46.1 km^2 of agricultural land and an increase of 166.5 km^2 of mountain steppe area (Bayarsaikhan et al. 2009). In contrast, grazing has become a more serious issue in the Hövsgöl park area due to the increased numbers of livestock and population. Nomad herders traditionally move approximately 20–25 km per season from place to place to ensure the recovery of pastureland between grazing periods. This movement is restricted to within 5 km in the valleys along the eastern shore of Lake Hövsgöl and results in overlapping pastureland between seasons, and between families. Moreover, a steppe plant community encroached into natural communities due to clear cutting, forest fires, and the draining of previously marshy areas; therefore, more families have moved into the area of the Hövsgöl park (Goulden et al. 1999). The Dalbay and Borsog Valleys in the Hövsgöl region are included in the strictly protected area of the park; thus, grazing is prohibited there. Nevertheless, three families settled in Dalbay and the impact of grazing has become apparent in the recent years. Other valleys, where permafrost monitoring has been conducted within the framework of the Hövsgöl GEF/WB project, are Turag, Shagnuul (both heavily grazed valleys), Noyon, and Sevsuul (both moderately grazed valleys); these valleys are not located in the restricted zone of the park (see Table 17.8).

17.3 Permafrost Conditions

Territory with underlying permafrost occupies almost two-thirds of Mongolia, predominantly in the areas surrounding the Hentei, Hövsgöl, Hangai and Altai Mountains, and it is located on the southern fringe of the Siberian continuous permafrost zone (Fig. 17.2). The territory is characterized by mountain and arid-land permafrost. It is sporadic to continuous in its extent. The temperature of the permafrost is close to 0°C in most areas. This makes it thermally unstable under the influence of climate change and human activities. In continuous and discontinuous permafrost zones, talics are found on steep south-facing slopes, under large river channels and deep lake bottoms, and along tectonic fractures with hydrothermal activity. In sporadic and isolated permafrost zones, frozen ground is found only on north-facing slopes and in fine-grained and moist deposits. The lowest limit of sporadic permafrost is found at an altitude between 600 and 700 m above sea level, while the lower limit of continuous permafrost on south-facing slopes ranges from 1,400 to 2,000 m in the Hövsgöl and Hentei mountains and from 2,200 to 3,200 m in the Altai and Hangai mountains. The average thickness and mean annual temperature of the permafrost in the continuous and discontinuous zones are 50–100 m and −1 to −2°C in valleys and depressions, and 100–250 m and −1 to −3°C in mountains. Permafrost in Mongolia has a low to moderate ice content in unconsolidated

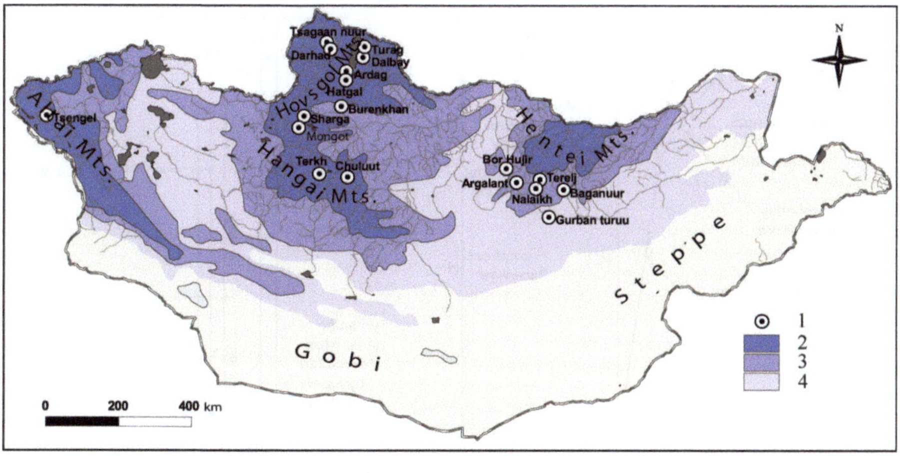

Fig. 17.2 Permafrost map and monitoring sites of Mongolia: *1* Permafrost monitoring sites, *2* Continuous and discontinuous permafrost, *3* Islands of permafrost, *4* Sporadic permafrost

sediments. Ice-rich permafrost (>20% by weight) is characteristic of lacustrine and alluvial sediments in valleys and depressions. Ice wedges are found in the Darhad depression, Hövsgöl region. The thickness of the active layer is 1–3 m in fine-grained soils and 4–6 m in course material. Cryogenic features such as frost heaves, cracks, icings, thermokarst terrain, solifluction lobes and sorted polygons are abundant in Mongolia (Gravis et al. 1974; Sharkhuu 2000, 2003).

17.4 Permafrost Monitoring and Studies of Insulation Effect of Vegetation Covers

17.4.1 Methodology of Permafrost Monitoring and Vegetation Cover

The primary parameters being monitored are the thickness of the active layer (ALT) and the mean annual permafrost temperature at the level of zero annual amplitude (10–15 m deep) (MAGT), as well as the temperature gradient (TG) of the permafrost (see Fig. 17.3a). Temperature measurements in the all boreholes were made using identical thermistors at corresponding depths, and were carried out on approximately the same dates of each year. We used moveable thermistor strings, made at the Geothermal Laboratory of the Melnikov Permafrost Institute, Siberian branch, Russian Academy of Sciences. The accuracy of the temperature measurements by Russian-calibrated thermistors (MMT-4 model) is 0.02°C. The thermistor resistance was measured by a multimeter (MB-400 model). In all boreholes the thickness of

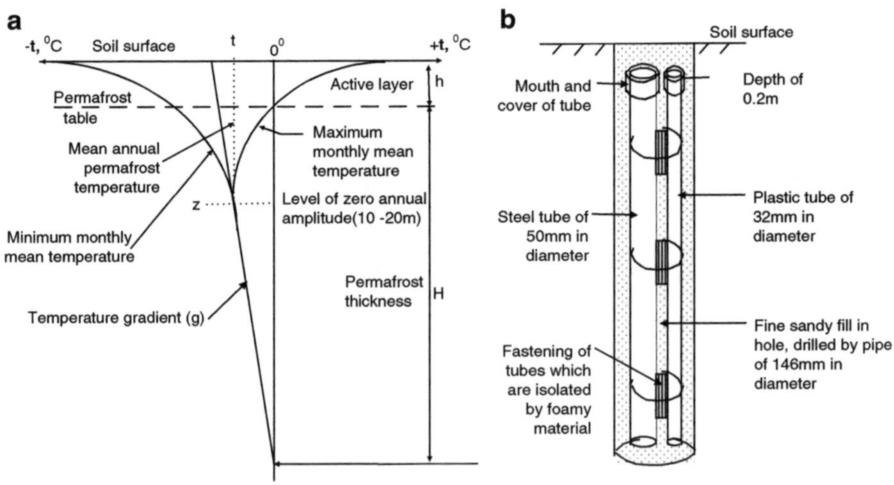

Fig. 17.3 Methodology of permafrost monitoring: (**a**) parameters of permafrost monitoring and (**b**) borehole instrumentation

the active layer was determined by interpolation of ground temperature profiles, obtained by borehole temperature measurements usually in late September. Four-channel temperature data loggers HOBO U12 were installed in 25 monitoring boreholes. In addition, Stow-Away or UTL-1 miniature data loggers were placed in seven boreholes. Accuracies of measurements of ground temperatures by using HOBO U12 and miniature data loggers are 0.03°C and 0.25°C, respectively. The interval time of temperature recordings by the data loggers was 90 min. Besides, six boreholes have permanent thermistor strings. Boreholes near Ulaanbaatar have permanent thermistor strings, and ground temperature measurements were made monthly. Monitoring of frost heaving, thermokarst and icing was based on leveling measurement with Russian leveling equipment of the model HB-1.

Each borehole was designed to prevent air convection and was protected against damage by passing people. All re-drilled boreholes with depths of 10–15 m were cased by parallel steel and plastic pipes of 3–5 cm diameter with the mouth of the pipe at depths of about 15–20 cm and covered by soil (see Fig. 17.3b). The empty space outside of the casing pipe in all the boreholes was filled with sand.

The primary parameters being measured for studying vegetation cover were plant biomass, soil temperature and soil water content. Live (green) and dead (litter) plant biomass was sampled from a $0.5 \times 0.5\ m^2$ and weighed after drying at a room temperature for more than 48 h. Soil temperatures down to 50 cm with intervals of 10 cm were measured by a digital thermometer with an accuracy of 0.1°C in specifically prepared holes, and in others soil temperatures at 10 and 50 cm depths were recorded by miniature data loggers (UTL-1 or StowAway). Soil surface temperatures under different vegetation covers were monitored by data loggers. Soil samples were collected for determining soil water content, using either a hand auger or by a trowel in specifically prepared holes. Active layer depths at observation sites were usually determined by using a 1.8 m long steel rod-probe. More detailed descriptions of methods used for insulation effects of vegetation cover are given by Sharkhuu et al. (2007, 2010) and Sharkhuu and Anarmaa (2008).

17.4.2 Monitoring Sites for Permafrost and Vegetation Cover

17.4.2.1 Monitoring Sites for Permafrost

At present, there are 16 permafrost monitoring sites, consisting of 46 boreholes of GTN-P and CALM in Mongolia (see Fig. 17.2). According to the TSP borehole depth classification, there are 19 surface (SU) boreholes with depths of 5–9 m, 17 shallow (SH) boreholes with depths of 10–25 m, 6 intermediate (ID) boreholes with depths of 25–125 m, and 4 deep (DB) boreholes with depths of 130–200 m. In the last 5–15 years, additional boreholes with depths of 10–15 m were drilled in locations near old boreholes in which temperature measurements were made 15–40 years ago. Most of the DB and ID geological boreholes are located at the Buren-khan and Ardag monitoring sites in the Hövsgöl region. Except for seven DB and ID boreholes, all of the boreholes were prepared by dry drilling. Currently, all boreholes are situated in natural conditions without thermal disturbance. In addition, we monitor changes in plant biomass, soil temperature and moisture content at the sites of the permafrost monitoring boreholes.

We selected six typical boreholes, two from each mountainous region, as examples for presenting the initial results of permafrost monitoring. The characteristics of the boreholes are presented in Table 17.1. Detailed information on all the boreholes for permafrost monitoring can be found at borehole metadata, hosted by the GTN-P and CALM websites (GTN-P 2010; CALM 2007). Boreholes are referred to by their borehole name, which includes CALM coding (e.g., Baganuur M1A), while sites are referred to by their names without additional specifications (e.g., Baganuur). One site may contain several boreholes or observational/experimental plots.

17.4.2.2 Study Sites for Vegetation Cover

The key experimental sites chosen for studying the thermal insulation effect of the vegetation cover were Dalbay Valley on the eastern shore of Hövsgöl Lake and Bayan-Spring Valley in Hustai National Park, 100 km west of Ulaanbaatar in the Hentei region.

Dalbay Valley is located in the mountain forest-steppe in the Taiga sub-province of the highlands of the eastern shore of Lake Hövsgöl and has three ecosystems in close proximity: (1) semi-arid mountain steppe on a permafrost-free south-facing slope, (2) riparian area with marshy tussocks and shrubs in the valley bottom, underlain by permafrost, and (3) boreal forest dominated by *Larix sibirica* on a north-facing slope, underlain by permafrost. The Hustai National Park also is located in mountain arid steppe and the sporadic permafrost zone. The landscape consists of mountains with altitudes of 1,400–1,842 m and river valleys. A small fraction is covered by a birch forest dominated by *Betula platyphylla* and *Populus tremula* (Bayarsaikhan et al. 2009).

The studies on the effects of vegetation cover on permafrost can be divided into two main studies: (a) comparative studies on the insulation effects of different vegetation covers (moss, grass, shrub, and forest cover in Dalbay Valley, and grass and forest cover in Hustai area) and (b) experiments with vegetation cover manipulation in both Dalbay and Hustai. In addition, we compared the insulation effects of grass and/or sedge-dominated vegetation in different aspects at both Hövsgöl and Hustai.

Comparative Plots for Studying Insulation Effects of Different Vegetation Covers

Four types of plots (a–d) were established for recording soil surface temperatures in the Dalbay valley area:

a1: a plot of which the vegetation cover was clipped,
a2: under a dense grass cover,

Table 17.1 Characteristics of the selected permafrost monitoring boreholes

Region	Hentei		Hangai		Hövsgöl	
Borehole name	*Baganuur M1A*	*Argalant M3*	*Terkh M6A*	*Chuluut M7A*	*Sharga M8*	*Burenkhan M4A*
Latitude, N	47-41-40	47-55-25	48-04-55	48-02-28	49-29-25	49-47-17
Longtitude, E	108-17-40	106-32-50	99-22-42	100-23-30	98-39-22	100-02-03
Elevation, m	1,350	1,385	2,050	1,870	1,855	1,705
Landform and terrain	Dry steppe plain in depression	Dry steppe valley bottom	Flood plain of wide valley	10 m high pingo top	Flood plain of wide valley	Northern slope of mountain grassland
Predominant soil profile	Sandstones covered by 3 m sand	0–6 m loam, 6–12 m clay	0–45 m sand, 45–90 m clay	0–5 m silt, 5–32 m ice, >32 m sand	0–12 m sand, 12–56 m loam	Limestones covered by 1 m thick debris
Ice content	Medium	High	Medium	High	Medium	Low
Thickness of permafrost, m	38	15	100	32	135	50

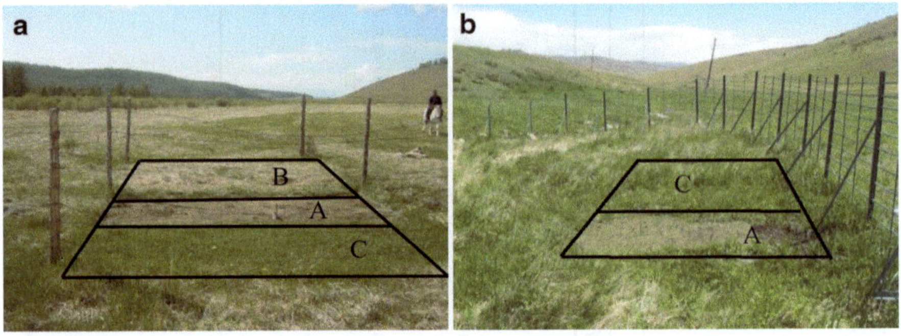

Photo 17.1 The Dalbay (**a**) and Hustai (**b**) experimental sites for studying the thermal insulation of grass cover (Photos Anarmaa in 2006 and 2007)

b1: under 0.5 m high dense shrub (*Salix glauca*),
b2: under 1.8 m high dense shrub (*Salix* sp.),
c1: under sparse *Larix* forest,
c2: under dense *Larix* forest,
d1: in forest without a moss cover,
d2: in forest under a 10 cm thick moss cover of *Rhytidium rugosum*,
d3: ditto but a 20 cm thick moss cover of *Rhytidium rugosum*,
d4: ditto but a 30 cm thick moss cover of *Rhytidium rugosum* (Sharkhuu et al. 2007).

In Hustai, three types of plots (a–c) were established on the north-facing slope of Hoshoot:

a1: a plot from which the vegetation cover was removed,
a2: under dense grass,
b: under dense shrub,
c: under *Betula* forest.

Manipulating the Vegetation Cover

Experimental plots (2×2 m) with different treatments were established on south-facing and north-facing slopes and riparian areas in both Dalbay and Hustai. Each experimental site in Dalbay Valley consisted of three treatments: control (C – vegetation cover left intact), clipped (B – green vegetation cover was clipped monthly but litter was left intact), and cleared (A – green vegetation and litter cover were removed monthly). Bayan-Spring Valley of Hustai consisted of only two treatments: control (C) and cleared (A) (see Photo 17.1).

In addition, we included some data from observations conducted in the Borsog, Shagnuul, and Turag Valleys near Dalbay Valley as additional data. These valleys vary in insulation effects of their vegetation cover due to different grazing pressures.

17.5 Findings

17.5.1 Recent Degradation of Permafrost Under the Influence of Climate Warming

17.5.1.1 Active Layer

The thickness of the active layer in Mongolia varies over the wide range of 1–1.5 to 5–8 m, depending on ground texture and its temperature-moisture regime. However, on average the ALT is 2.5–4.0 m. The thickest active layer has been observed in January-February in boreholes, such as M1a and M3, located at sites at which permafrost is characterized by a high MAGT and a low-medium ice content or at sites with bedrock. The full refreezing of such a thick active layer continues until February-April. A relatively shallow active layer is characteristic of boreholes (e.g. M6a, M7a, and M8) located at sites at which permafrost is characterized by a relatively low MAGT or an ice-rich fine sediment. The shallow active layer thoroughly refreezes in November-December.

ALT in all CALM boreholes of Mongolia increased at different rates, depending on changes in the mean annual air temperature, the environmental and ground conditions of regions, and human activities, especially the grazing regime (see Fig. 17.4 and Table 17.2). For instance, no substantial change of ALT was observed in the Chuluut, Terkh, and Sharga boreholes, which have ice-rich and fine-grained sediments and a shallow thawing depth. In contrast, the active layer depth in the Burenkhan and Argalant increased at rates of 25–40 cm per decade, and this rate appeared to be characteristic of areas with a deep active layer. Regional trends exist regardless of these differences in ground characteristics; the estimated average increase in ALT varied over the range of 0.2–1.5 cm $year^{-1}$ in the Hangai and Hentei regions and 0.3–2.4 cm $year^{-1}$ in the Hövsgöl region (Sharkhuu and Anarmaa 2005). The ALT measurements at the Hatgal borehole, which is composed of sandy gravel with relatively low ice content, suggest that the degradation of permafrost during the last 15–20 years was relatively more intense than in previous years. More specifically, the ALT at the Hatgal site was 3.6 m in 1969, 4.0 m in 1983, and 4.7 m in 2004 according to the interpolation of ground temperature values.

As stated above, we observed changes in the ALT. However, ALT in most boreholes in 2009 was less than in the previous 2 years. Moreover, ALT in the Hövsgöl region varies greatly from year to year due to a deep seasonal freezing and thawing of the ground. Hence, we cannot conclude that ALT in all of the monitored boreholes has been increasing significantly in recent years. In addition, we lack long-term monitoring data. Likewise, no significant trend of increase in active layer thickness was observed in Alaska and Siberia (Brown et al. 2000; Romanovsky et al. 2003).

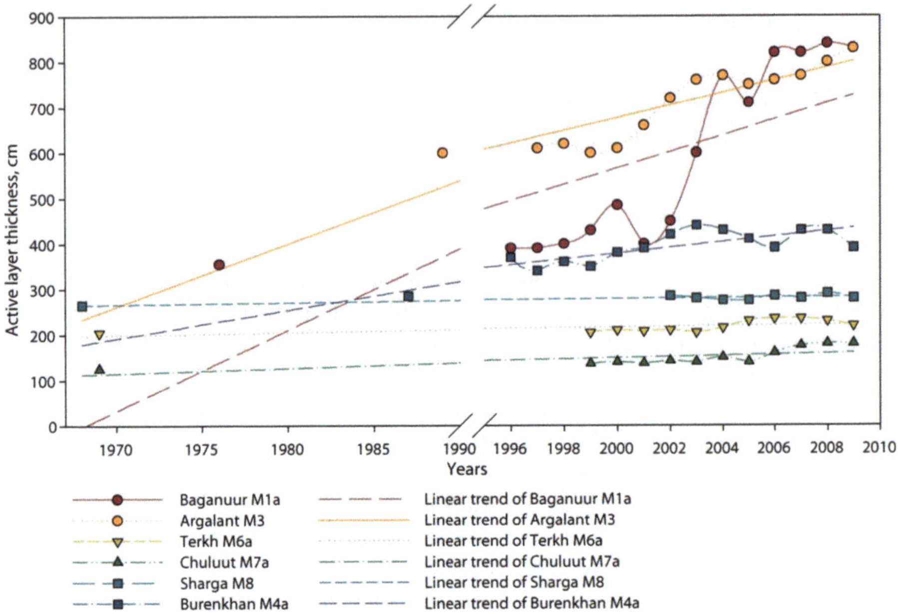

Fig. 17.4 Trends of increase in the active layer thickness in the selected boreholes

Table 17.2 Changes in ALT, MAGT, and TG in the selected boreholes

Region	Borehole name and number	Measured years	ALT, cm	MAGT at 10–15 m depth, °C	TG at 15–50 m depth, °C m^{-1}
Hentei	Baganuur M1a	1976	355	–0.45	0.020
		1996	390	–0.07	0.005
		2009	830	–0.06	0.005
	Argalant M3	1989	600	–0.48	
		1999	600	–0.33	
		2009	830	–0.19	
Hangai	Terkh M6a	1969	205	–2.04	0.022
		2002	210	–1.55	0.011
		2009	220	–1.35	0.007
	Chuluut M7a	1969	125	–0.72	0.038
		2002	142	–0.51	0.025
		2009	180	–0.43	0.019
Hövsgöl	Sharga M8	1968	265	–2.35	0.026
		2002	285	–1.67	0.010
		2009	280	–1.54	0.007
	Burenkhan M4a	1987	285	–1.00	0.026
		1996	370	–0.75	0.020
		2009	390	–0.40	0.014

17.5.1.2 Mean Annual Ground Temperatures

The mean annual ground or permafrost temperatures (MAGT) in Mongolia vary from 0 to −0.5°C in sporadic permafrost zones and from −1 to −3°C in continuous permafrost zones. As an example, the ground temperature regime down to the depth of yearly zero amplitude in the Terkh borehole M6a is presented in Table 17.3. The mean annual ground temperature gradually increases with depth. The mean winter (average of December, January and February) ground temperature increases until a depth of 4–5 m and then decreases between 6 and 14 m. The mean summer (average of June, July and August) ground temperature decreases until a depth of 6 m and then increases deeper than 6 m. The depth of the yearly zero amplitude temperature is approximately 14–15 m. The time lags for the mean summer soil surface temperature to penetrate to depths of 1, 4, and 10 m are 1, 5 and 7 months, respectively. Meanwhile, the time lag for the mean winter soil surface temperatures to penetrate to depths of 1, 4, and 10 m are 1, 3 and 5 months, respectively. The relatively short time lag for achieving a winter temperature at a deeper depth is caused by the high thermal conductivity of the frozen state of the active layer.

An increase in MAGT is a direct indicator of the warming and degradation of permafrost. During the last 10–40 years, the MAGT in all boreholes increased at different rates, varying from 0.01°C to 0.04°C year^{-1}. This wide range is caused by variations in local landscape, topography, and ground conditions, such as texture and thermal conductivity (see Fig. 17.5 and Table 17.2). For example, the highest rates are observed in the Ardag and Burenkhan boreholes on mountain watersheds and slopes, composed of bedrock with a high thermal conductivity. In addition, the rates of increase in MAGT were 0.27°C per decade on south-facing slopes, 0.19°C on north-facing slopes, 0.23°C in the upper watershed, and 0.11°C in the valley bottom of the Burenkhan Mountain area (Sharkhuu 1998). These changes show that the rate of increase in MAGT in bedrock or sandy sediments with high thermal conductivities is higher than that in unconsolidated and/or ice-rich fine sediments with low thermal conductivities, and the rate on south-facing slopes is higher than that on north-facing slopes. On a regional basis, the average trend in the increase in MAGT in different regions follows the pattern of the increase in mean annual air temperature in those regions. The increase in MAGT at a depth of 10–15 m in the Hövsgöl Mountain region reaches 0.02–0.03°C year^{-1}, but in the Hangai and Hentei Mountain regions it does not exceed 0.01–0.02°C year^{-1}. As compared to the Tsagaan nuur borehole (measured since 1989), the relatively low rate of increase of temperature in the Sharga borehole (measured since 1968) shows that the permafrost is degrading more intensively during last 15 years than during the previous 15–20 years (1970–1980s).

The average increases in the MAGT of permafrost in Mongolia are similar to those in central Asian and European mountain territories (Harris et al. 2003; Zhao et al. 2008a, b). However, increases in the MAGT of permafrost in Mongolia are lower than in eastern Siberia and Alaska (Osterkamp and Romanovsky 1999; Gavrilova 2003; Pavlov and Perlshtein 2006). For instance, the increases for the period of the late 1970s to the middle 1990s are 0.05–0.08°C year^{-1} in Siberia

Table 17.3 Mean winter, summer and annual ground temperatures, and its yearly amplitude (°C) at different depths in the Terkh borehole M6a (September 2009–2010)

Borehole depth, m	0	1	2	4	6	10	14
Mean winter T	−22.72	−7.53	−1.70	−0.69	−0.73	−1.04	−1.16
Mean summer T	15.00	3.11	−0.60	−1.53	−1.82	−1.41	−1.24
Mean annual T	−2.91	−2.15	−1.55	−1.41	−1.36	−1.21	−1.19
Yearly amplitude T	40.78	15.23	5.19	2.94	1.74	0.50	0.17
Max/Min T month	Jul/Jan	Aug/Feb	Sep/Mar	Dec/Apr	Jan/May	Feb/Jun	Apr/Aug

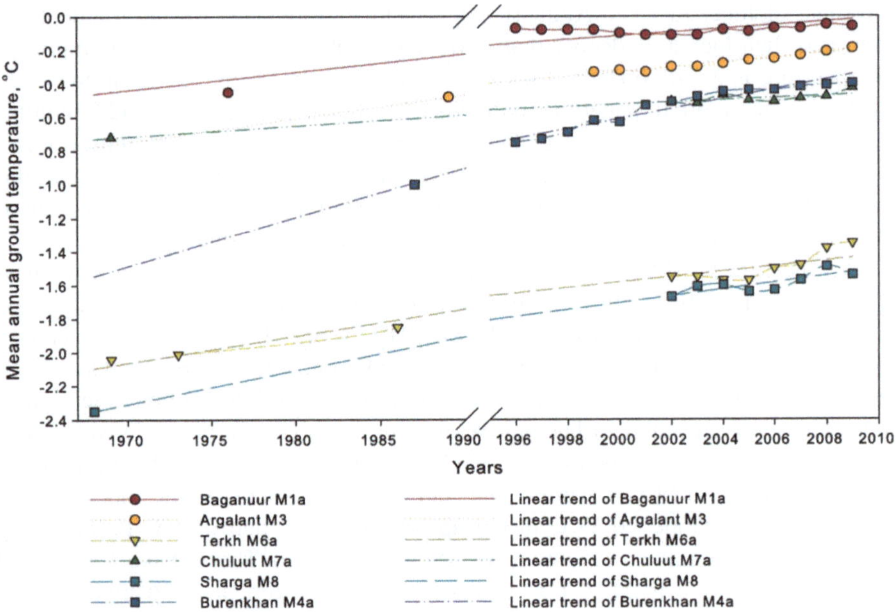

Fig. 17.5 Trends in increase in the mean annual permafrost temperatures in selected boreholes

(Pavlov and Perlshtein 2006), but the increases in permafrost table temperatures between 1987 and 2001 are 0.14°C year^{-1} at the West Dock and Franklin Bluffs sites and 0.21°C year^{-1} at the Deadhorse site in northern Alaska (Romanovsky et al. 2003). These trends are consistent with observed higher mean annual temperatures in high-latitude areas and suggest that continuous permafrost in those regions is degrading more rapidly than permafrost at lower latitude.

17.5.1.3 Ground Temperature Gradients

In a hypothetical situation with homogenous earth material, ground temperature is linearly related to depth. A distortion from this linear relationship can arise from changes in thermal conductivity of different earth materials and also from changes in soil surface temperature fluctuation. Any change in ground surface temperature penetrates slowly downwards. Based on this downward penetration, an occurrence time of a ground surface temperature anomaly can be inferred from the depth of a perturbation in a ground temperature gradient, which can be obtained from borehole temperature measurements. Theoretically, perturbations of homogenous earth material at the depth of 25, 80, and 250 m correspond to surface temperature pulses that occurred 10, 100, and 1,000 years ago (Harris and Chapman 1997). We compared the temperature gradients of different depth intervals and assumed that the temperature gradient in the upper part of a thermal profile reflects the recent history of the ground surface temperature. A decrease in the temperature gradient in the upper part indicates an increase in ground surface temperature, and a corresponding recent degradation of permafrost.

Changes in the temperature gradients of permafrost in two 200 m deep boreholes, M5b and M5c at the Ardag Mountain watershed site in the Hövsgöl region, which are composed of dolomite and limestone, are presented here as an example. Permafrost thickness in these boreholes is 210 m in M5b and 160 m in M5c. The temperature gradients, estimated from temperature measurements made in 1987 in borehole M5b, are 0.009°C m^{-1} for the range of 20–80 m depth, and 0.007°C m^{-1} for the range of 80–200 m depth. The temperature gradient in the upper depth of 20–80 m was 0.002°C m^{-1}, according to temperature measurements that were made in 2008 in the same borehole. The temperature gradients in the borehole M5c were 0.003°C m^{-1} in the upper 20–60 m and 0.009°C m^{-1} in the lower 60–200 m in 2008. The similar temperature gradients in the upper parts of these two boreholes, which were estimated from 2008 measurements, and the comparable temperature gradients in the lower parts in both boreholes, which were estimated from old and recent measurements, support our assumption that a distorted temperature gradient of the upper part would represent a recent surface temperature change. The assumption could be challenged by potential changes in thermal conductivity of the ground and complications related to it. However, the reduced temperature gradient in the upper part is a ubiquitous phenomenon in all the monitored boreholes. In addition, a simple calculation of how fast a temperature anomaly would propagate down to a given depth was done for the Burenkhan and Darhad deep boreholes, which are also located in the Hövsgöl region (Sharkhuu et al. 2007). According to this calculation, the temperature change at 30–40 m depth can reach its maximum within 15–30 years after a positive surface temperature pulse in the Darhad borehole. Similarly, the maximum temperature change at 80–100 m depth can be reached within 20–40 years in the Burenkhan boreholes. The temperature gradient in the upper part of the Darhad borehole profile would then correspond to a warming trend, which occurred in the early 1980s, and in the upper part of Burenkhan borehole profile, would represent a warming trend of the early 1960s (Sharkhuu et al. 2007). The difference in the periods of the warming trends was caused by climate change at those sites and thermal conductivity of the sediments in the boreholes. The deeper penetration of the ground surface temperature change in the Burenkhan borehole is caused by dolomite-dominated bedrock with high thermal conductivity. Nonetheless, the estimated years for the soil surface temperature anomaly to reach the depth of 40–80 m correspond to the observed decrease in the temperature gradient in the

Ardag boreholes. In general, the temperature gradients in the upper part of permafrost decreased in almost all monitored boreholes of Mongolia. The permafrost or ground temperature gradients at a depth of 15–50 m of the selected boreholes decreased by 0.01–0.02°C m^{-1} during the last 40 years (see Table 17.2).

17.5.1.4 Some Cryogenic Processes and Phenomena

Ongoing thermokarst and thermo-erosion processes in the Hövsgöl, Hangai, and Hentei mountain regions are direct indicators of the degradation of ice-rich permafrost under the influence of recent climate warming. We present some land forms of thermokarst and thermo-erosion processes in the Darhad depression (Hövsgöl region, Photo 17.2), Chuluut valley (Hangai region, Photo 17.3), and Nalayh depression (Hentei region, Photo 17.4) as examples.

The thermokarst depressions and the thermoerosional riverbanks with an average vertical extent of 3–7 m are characteristic of the Darhad depression. Some can reach 15–25 m (Photos 17.2a, b). Due to recent climate change, active thermokarst processes, such as thermokarst lakes and deep hollows with a vertical extent of 1–5 m, are observed everywhere in the Darhad depression. Newly formed cracks due to thaw settlement of ice-rich sediments are sometimes observed along a thermokarst lake shore, but dates for those formations have not yet been determined. In addition to ice-rich sediments, ice wedges are observed in steep exposures along the thermoerosional banks of the Sharga-ganga River and ice wedge polygons on the land surface are widespread here. The melting of ice wedges along polygonal hollow channels resulted in 2–3 m deep subsidence cavities on the permafrost table, and incidents of large herds (yaks and horses) falling into the cavities have occurred recently (Sharkhuu et al. 2007).

As shown on a river bank exposure and a 36 m deep borehole profile, the Chuluut River valley bottom is composed of ice-rich lacustrine sediments consisting of clay with mostly 5–20 cm, maximally 50–80 cm thick, ice layers, and nets (Photo 17.3a). The volumetric ice content of the lacustrine clay is about 20–50%. The permafrost thickness in the valley is 15–30 m. Data from measurements over the period 1969–1987 show that the rate of collapse of the 6–8 m high Chuluut River banks due to thermo-erosion was estimated to be in the range of 15–30 cm $year^{-1}$ (Sharkhuu 1998). There are numerous pingos and thermokarst lakes of different sizes and evolutionary stages in the valley bottom. The uneven distribution of ground ice leads to a highly uneven thaw settlement and surface subsidence of active thermokarst, both in spatial and temporal dimensions (see Photo 17.3b). Leveling measurement data taken at one of the active thermokarst sites in the Chuluut valley is presented in Table 17.4. According to a communication with local people, the thermokarst processes at this site began in the late 1980s and led to maximum subsidence in the mid and late 1990s. The data further showed that a maximum subsidence of up to 25–40 cm $year^{-1}$ was observed during the formation of an incipient thermokarst lake in the period of 1996 to 1998. The rate of thermokarst subsidence decreased from 15–25 cm $year^{-1}$ in 1999–2001 to 0–8 cm $year^{-1}$ in 2005–2009.

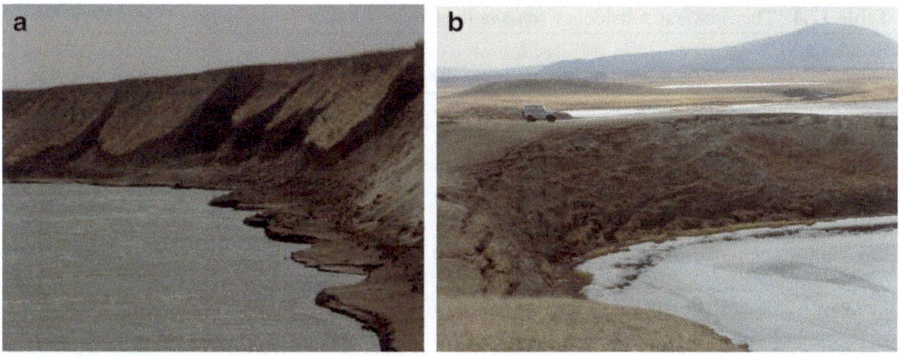

Photo 17.2 Thermoerosion bank (**a**) and thermokarst lake (**b**) in the Darhad depression, Hövsgöl region (Photos Anarmaa in 2004 and 2005)

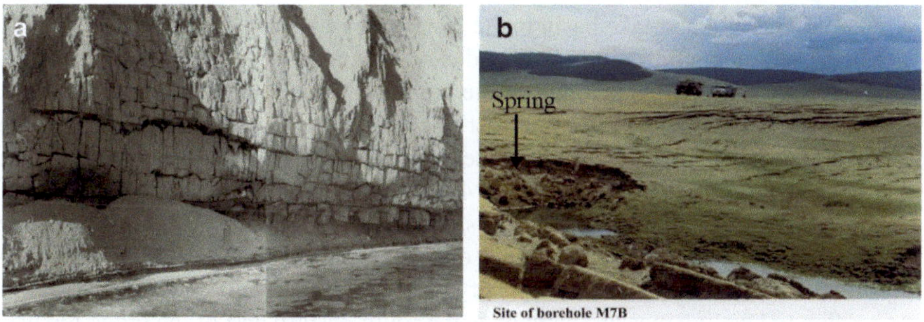

Photo 17.3 Thermoerosion bank (**a**) and thermokarst spring and pond (**b**) in the Chuluut valley, Hangai region (Photos Sharkhuu in 1969 and 2002)

Photo 17.4 Thermokarst lake (**a**) and pingo (**b**) in the Nalayh depression, Hentei region (Photos Sharkhuu in 1999 and 2005)

Table 17.4 Thermokarst subsidence (cm) in the Chuluut Valley in the period 1999–2009

Benchmark points	1999	2000	2001	2002	2003	2004	2005	2006	2009	Total	Depth
Bank of ravine slope	19	21	23	18	13	15	8	4	2	123	251
Spring channel	18	23	26	17	14	11	9	3	1	122	526
Top of rest-mound	21	19	13	9	6	3	3	1	0	75	338
Dry hollow bottom	25	23	19	22	16	9	8	5	2	129	600
Level of water pond	19	24	22	18	15	13	9	6	3	129	712

Its average rate in the period of 1999–2009 was 10–15 cm year^{-1}. Water discharge in a channel from the thermokarst pond was 1.2 L sec^{-1} in 1999 and 0.9 L sec^{-1} in 2002. In addition, water discharge in a new spring channel from the newly subsiding ravine was 0.3 L sec^{-1} in 2002.

A small thermokarst lake basin in the Nalayh depression is composed of ice-rich lacustrine sediment consisting of silt and clay with mostly 5–20 cm thick ice layers (Photo 17.4). The estimated value of the volumetric ice content in the sediment is approximately 20–40%. The permafrost thickness is approximately 40 m, and the active layer thickness is 1.3 m. The water level in the lake is decreasing: from 3 m in 1999 to 1.8 m in 2009. Until 2002, a 4–5 m high pingo, located in the middle of the lake, was surrounded by lake waters (Photo 17.4a). In late August 2004, thaw slumping of a 1.3 m thick active layer on the permafrost table (a pingo ice core) occurred on the south-eastern slope of this degrading pingo (see Photo 17.4b). According to leveling measurement data, the south-eastern part of the pingo top was subsided by 50 cm in 2005, 40 cm in 2006, and 45 cm between 2007 and 2009. This subsidence was caused by melting of the pingo ice core from its table.

17.5.2 Insulation Effect of Vegetation Cover on Permafrost and Seasonally Frozen Ground

17.5.2.1 Insulation Effect of Vegetation Cover in Discontinuous Permafrost Zone in Taiga – Forest-Steppe (Hövsgöl National Park)

Insulation Effects of Different Vegetation Covers

The results of the observation on the insulation effect of different vegetation covers in Dalbay Valley in the Hövsgöl region showed that the mean summer surface temperatures under different vegetation covers were reduced significantly, compared to the plots without vegetation cover. Moreover, taller dense shrub and forest cover greatly reduced summer surface temperature compared to shorter dense

Table 17.5 Averaged data from year-round temperature records under different vegetation covers in the Dalbay Valley: Mean winter (*MWT*), summer (*MST*), and annual (*MAT*) soil surface temperatures, the yearly amplitude (*A*) and differences in summer surface temperatures between a plot without vegetation cover and plots under different vegetation covers (*STD*), °C

Vegetation cover		MWT	MST	MAT	A	STD
Air temperature		*−23.9*	*10.8*	*−5.3*	*18.8*	–
No vegetation cover	a1	−13.7	12.4	1.1	14.7	0
Dense grass	a2	−11.1	10.2	0.9	11.9	2.2
50 cm high dense shrub	b1	−11.4	8.8	0.7	11.3	3.6
1.8 m high dense shrub	b2	−7.4	7.4	0.6	7.8	5.0
Sparse forest	c1	−14.8	8.6	0.7	13.0	3.8
Dense forest	c2	−11.7	7.5	0.6	10.5	4.9
No moss cover	d1	−11.5	11.9	−0.6	27.1	0.0
10 cm thick moss cover	d2	−8.7	8.9	−0.7	21.8	3.0
20 cm thick moss cover	d3	−7.2	6.9	−0.8	18.9	5.0
30 cm thick moss cover	d4	−5.3	2.7	−1.3	14.3	9.2

shrub, sparse forest, and grass cover (Table 17.5). Although the mean winter temperature under dense grass, shrub, and forest are warmer compared to sparse or no vegetation cover due to accumulation of a relatively loose and thick snow cover, the insulation effect of snow against cold temperature cannot compensate for the insulation effect of vegetation against warm temperature in summer (see MAT in Table 17.5). Different from regions with thicker snow covers, in this region the insulation effect of vegetation is more important than that of snow. The data suggest that permafrost may thaw if grassland encroaches into forest, and this is also predicted by the HadCM3 model simulation of climate change in Mongolia (Dagvadorj et al. 2009). Comparisons of summer surface temperatures between plots without a moss cover and plots with moss covers of varying thickness show that moss is also an important insulator. The differences in mean annual temperatures under moss cover with a thickness of 10, 20, and 30 cm are due to a thermal offset caused by a difference between frozen and thawed moss covers (Table 17.5). In general, moss is very susceptible to grazing and drought, and Van der Wal and Brooker (2004) reported that moss thickness is affected by grazing. These authors argued that grazing may reduce a moss cover and thus increase soil temperature, and this may positively influence grass growth. The authors also suggested that this shift in vegetation composition may occur as a result of grazing pressure and of increased global temperature. As a result, the arctic may shift from a carbon sink to a carbon source owing to permafrost degradation and enhanced decomposition.

Manipulation of Vegetation Cover

The experiment in the valley bottom of Dalbay Valley showed that the removal of either only litter or green plants and litter caused considerable increases in soil temperature and active layer thickness and decreases in soil moisture. Compared to the control plot in riparian area, the soil moisture content (down to 1 m in depth) decreased by 6.1%

Table 17.6 Soil temperature, moisture, and active layer thickness, measured 7 October 2006, in three plots differing in vegetation cover, in the riparian zone of Dalbay Valley

Treatments	Biomass ($g\ m^{-2}$)	Active layer thickness (cm)	Ground temperature, °C 10 cm	50 cm	100 cm	Soil water content, % 10 cm	50 cm	100 cm
Control (C)	625	128	0.8	1.7	1.0	61.3	50.2	30.2
With only litter (B)	375	138	0.9	2.0	1.3	55.2	43.0	38.9
Without green plant and litter (A)	135	173	1.1	2.5	1.5	51.1	40.3	33.6

in the B plot (only green plants were clipped), and by 10.2% in the A plot (green plant and litter cover were removed). This suggests that vegetation covers decrease evaporation. Another study conducted in the same study area suggested, however, that the effect of vegetation on soil moisture varies seasonally (Liancourt et al. 2012). In addition, the active layer thickness increased by 10 cm in B and 45 cm in A (Table 17.6). The differences in soil surface temperature among treatments were only 0.1–0.2°C, which is relatively small, but the measurements were made in October. The increases in soil temperatures at 100 cm depth are 0.3°C in B and 0.5°C in A, compared to control (Table 17.6).

Compared to the control plot, the active layer thickness increased by 11 cm in B and 51 cm in A on a north-facing slope (Table 17.7), while ALT increased 10 cm in B and 45 cm in A in the riparian area. These differences are due to sandier texture of soil on north facing than soil on riparian sites. The soil surface temperature differences between the control plot and A and B plots were less on the north-facing slope than those on the riparian site and south-facing slope. These differences can be explained by certain biotic and abiotic conditions such as solar radiation, plant biomass, and surface wetness, and show the importance of microclimatic and topographical differences.

Table 17.8 shows the differences of the active layer thickness, mean annual ground temperature, and estimated permafrost thickness among valleys that differ in grazing pressure. Although soil and topographic conditions are slightly different, relatively shallow active layers and low permafrost temperatures in the Borsog and Dalbay valleys, compared to the Shagnuul and Turag valleys, might be caused by the insulation effect of the vegetation cover. This suggests that overgrazing, as it reduces the vegetation cover, has an adverse impact on permafrost.

17.5.2.2 Insulation Effect of Vegetation Cover in Sporadic Permafrost Zone in the Arid Steppe (Hustai National Park)

Insulation Effects of Different Vegetation Types

Table 17.9 summarizes the initial results of observations on the thermal insulation effect of the forest on the Hoshoot north-facing slope, Hustai National Park.

Table 17.7 Active layer thickness and mean summer soil surface temperatures, measured 7 October 2006, in three plots in Dalbay

Location of the treatment plots	Active layer thickness, cm			Surface temperature, °C		
	Control (C)	With litter only (B)	Bare (A)	Control (C)	With litter only (B)	Bare (A)
South-facing slope	–	–	–	12.7	–	15.8
Riparian area	128	138	173	8.2	11.5	13.4
North-facing slope	102	113	153	8.8	10.8	10.7

Table 17.8 Active layer thickness, MAGT and estimated permafrost thickness in four valleys differing in the numbers of livestock and plant biomass

Valley	Borsog	Dalbay	Shagnuul	Turag
Livestock number, sheep units	51	530	3,001	3,330
Plant biomass, g m^{-2}	660	585	48	116
Active layer thickness, m	2.1	1.4	4.3	4.8
Mean annual ground temperature, °C	–0.91	–1.25	–0.56	–0.42
Estimated permafrost thickness, m	35	45	25	20

Table 17.9 Thermal insulation effect of forest on the Hoshoot north-facing slope: mean winter (*MWT*), summer (*MST*), and annual (*MAT*) soil surface temperatures, yearly amplitude (*MYA*), snow thickness and snow density

Vegetation cover	Plant biomass, g cm^{-3}	MWT, °C	MST, °C	MAT, °C	MYA, °C	Snow thickness, cm	Snow density, g cm^{-3}
Sparse *Betula* forest with dense grass cover	510	–10.7	16.2	0.7	14.9	6	0.18
Sparse grass without forest	230	–11.4	18.4	1.3	16.6	4	0.22
Difference	280	0.8	–2.3	–0.6	–1.7	2	–0.04

Compared to the plots with sparse grass, the mean summer (MST) and annual (MAT) soil surface temperatures, and the yearly amplitude (MYA) in the plots under sparse *Betula* forest with dense grass were less by 2.3°C, 0.6°C, and 1.7°C, respectively. However, the mean winter (MWT) soil surface temperature was warmer by 0.8°C due to the accumulation of a relatively thicker and looser snow cover in the forest than in the sparse grassy plots without forest. The temperature differences between the plots with and without forest were least during February, March, October, and November. In addition, the yearly and daily temperature amplitudes at the plot without forest were larger than those at the plot with forest.

The average soil surface temperatures measured in May and August under different vegetation cover on the Hoshoot north-facing foot slope are presented in

Table 17.10 Soil surface temperatures (°C) under different vegetation covers and their temperature insulation effects compared to a plot without vegetation cover at Hoshoot north-facing slope

Vegetation cover	May 12–31	Temperature difference	Aug 1–7	Temperature difference	Average difference
No vegetation cover	15.7	0	20.0	0	0
Under grass (biomass is 280 g/m^2)	9.2	–6.5	15.5	–4.4	–5.5
Under dense shrub	6.2	–9.4	14.1	–5.9	–7.7
Under dying birch forest	11.6	–4.0	14.3	–5.7	–4.9

Table 17.10. Compared to the plot without vegetation cover (the grass cover was clipped monthly), the soil surface temperatures in the plots under dying *Betula*, dense shrubs, and a grass cover with a biomass of 280 g m^{-2} were cooler by 4.9°C, 7.7°C, and 5.5°C, respectively. The thermal insulation effect of the vegetation cover might have been higher during June and July. Even though the *Betula* forest is dying due to soil moisture limitation (Bayarsaikhan et al. 2009), its thermal insulation effect was comparable to that of coniferous forest in Dalbay Valley. In contrast, the thermal insulation effects of dense shrub and grass cover were higher than of those in Dalbay.

Manipulation of the Vegetation Cover

The average January soil surface temperature of the control plot with grass cover of 402 g m^{-2} biomass in the Bayan meadow was 7.6°C warmer than the A plot without grass and litter cover (Table 17.11). However, it was colder by 5.4°C and 5.0°C in April and May. In contrast to the control plot, which had thick and loose snow cover, the plot without vegetation cover had little or no snow cover. Thus, the thermal insulation effect of both a thick snow and a vegetation cover in winter increased the soil surface temperature in the control plot and strongly protected the soil from deep freezing. Compared to the plot without vegetation cover, the soil temperature at 50 cm depth in the control was 2.4°C warmer in winter but by 1.5°C cooler in summer. During the spring and fall there was no great difference.

In Bayan Valley, the summer soil surface temperatures of the control plots with vegetation cover were 2.9°C cooler on the north-facing slope, and 1.5°C on the south-facing slope than those in plots of which the vegetation cover was removed monthly (Table 17.12). In general, vegetation biomass positively correlates to the summer thermal insulation effect of the vegetation cover (Sharkhuu 2007).

The average monthly subsurface summer soil temperatures and moisture contents (measured at 10 cm intervals from the surface down to 50 cm depth) of the control plots with vegetation cover (C) and the plots without vegetation cover (A), on south-facing and north-facing slopes and at the meadow of Bayan Valley, are

Table 17.11 Thermal insulation of 402 $g\,m^{-2}$ vegetation cover at Bayan meadow site, °C

Date of temperature recording	Soil surface			At 50 cm depth		
	Control (C)	Cleared (A)	Difference	Control (C)	Cleared (A)	Difference
Jan 5–20	−7.3	−14.9	7.6	−3.9	−6.3	2.4
Apr 7–9	0.5	6.0	−5.4	−1.1	−1.1	0
May 11–Jun 6	8.5	13.5	−5.0	1.0	2.4	−1.5

Table 17.12 Thermal insulation of grass cover at the Bayan observation sites on north and south slopes

	North-facing slope		South-facing slope	
Soil surface temperatures, °C	With grass (biomass 383 $g\,m^{-2}$)	Without grass	With grass (biomass 173 $g\,m^{-2}$)	Without grass
May	9.63	13.14	12.72	14.43
June	13.17	16.59	16.67	18.16
July	16.34	18.61	18.32	19.49
August	15.11	17.33	16.63	18.34
Average	**13.56**	**16.42**	**16.09**	**17.61**
Max	16.34	18.61	18.32	19.49
Min	9.63	13.14	12.72	14.43
Amplitude	6.71	5.47	5.60	5.06
Difference in average	2.86		1.52	

presented in Table 17.13. The average soil temperatures were at a minimum in June and a maximum in July. The average subsurface soil temperatures of the control plots in Bayan meadow were cooler by 2°C than those in the controls on a north slope and by 3.6°C compared to the controls on a south slope. Soil temperatures of plots without grass cover were always higher than the soil temperatures of plots with natural vegetation cover, independent of slope in this region. The differences in average summer soil temperatures between control plots and plots without vegetation cover were approximately 1.3°C on both south-facing and north-facing slopes. The differences were smallest in June and largest in August.

The average subsurface soil moisture content at the south-slope control plots at Bayan was 10.5% less than that of the control plots in the meadow and 3.2% less than that of the north-slope control plots. The subsurface soil moisture contents of the plots without grass cover on both slopes were always lower than that of the plots with natural grass cover. This difference was highest in June on the south slope, and in July on the north slope. As the north slope contained more soil moisture, the difference between the controls and the plots without vegetation cover was twice as large here than on the south slope (Table 17.13).

Table 17.13 Subsurface soil temperature and moisture content at Bayan Valley sites with and without grass cover, °C

Parameters	Site	June		July		August		Average	
		Control	Cleared	Control	Cleared	Control	Cleared	Control	Cleared
Soil temperature, °C	South slope	14.8	15.2	17.0	18.5	16.0	18.0	15.9	17.2
	Meadow	10.3	–	13.8	–	13.1	–	12.3	–
	North slope	12.8	13.5	16.0	16.7	14.2	16.9	14.3	15.7
Soil moisture content, %	South slope	7.9	10.0	8.1	9.0	9.8	10.4	8.6	9.8
	Meadow	12.0	–	20.5	–	24.8	–	19.1	–
	North slope	8.9	9.8	12.5	18.0	14.2	15.2	11.8	14.3

17.6 Conclusions

Increase in active layer thickness and mean annual ground temperature, decrease in the permafrost temperature gradient, and some active thermokarst and thermo-erosion processes indicate that permafrost in Mongolia is degrading under the influence of recent climate warming. The trends of the increase in the average ALT and MAGT are 0.5–2.0 cm $year^{-1}$ and 0.01–0.03°C $year^{-1}$, respectively. Recent climate warming leads not only to the increases in ALT and MAGT, but also to a significant decrease in soil moisture content, which is a main controlling factor in semi-arid ecosystems. Permafrost acts as an impermeable layer to ground water and makes it available to vegetation; thus, the degradation of permafrost leads to more arid conditions in certain areas.

The rate of permafrost degradation varies depending on local conditions. According to borehole measurements, the degradation rate in bedrock is more than that in unconsolidated sediments, is greater for permafrost in ice-poor substrates than that in ice-rich ones, and quicker on south-facing slopes than on north-facing slopes. Meanwhile, permafrost degradation under the influence of climate warming in Mongolia has been more intense during the last 15–20 years than during the previous 15–20 years (1970–1980). Permafrost in the Hövsgöl Mountain region (at a higher latitude) is degrading more intensely than in the Hangai and Hentei Mountain regions (at a lower latitude). The recent degradation of permafrost in Mongolia is similar to that in central Asia and European mountain territories. However, the recent degradation of permafrost in Alaska and eastern Siberia is more intense than in Mongolia.

Vegetation cover is a natural thermal insulator which reduces summer soil temperature and evaporation, thus maintaining shallow active layers in a continuous and discontinuous permafrost zone. The thermal insulation effects of moss, forest, shrub, and dense grass covers are more important to maintain a low ground temperature and shallow active layer in the continuous and discontinuous permafrost zone in the taiga Ð forest-steppe zone of Mongolia than in high-latitudinal zones of the North. Even a sparse steppe vegetation cover has a significant thermal insulation effect on subsurface soil temperature, moisture and seasonal freezing regimes of the arid steppe zone of Mongolia. In addition to climate warming, overgrazing by livestock in Mongolia is a major factor for soil surface disturbance, adversely affecting permafrost. Therefore, it is important to protect the vegetation cover, particularly forest, in order to prevent further degradation of permafrost and ecosystems under the influence of climate warming. As a result of permafrost degradation during the last 40 years, the southern boundary of the permafrost zone in Mongolia retreated by 80–120 km, and the altitudinal belt of permafrost distribution increased by 150–200 m. This estimate is based on data of altitudinal and latitudinal changes in MAGT in Mongolia (Sharkhuu 2000).

Acknowledgements Within the framework of the CALM project, the University of Cinne (Ken Kenith 2000), and the University of Delaware (F. Nelson and I. Shiklomanov 2003–2007) provided financial support and data loggers for the 2000 and 2003–2007 field trips for monitoring work.

Furthermore, within the framework of the GTN-P Project, the University of Alaska Fairbanks (K. Yoshikawa and V. Romanovsky) provided financial support and 20 HOBO U12 data loggers for the 2008 field trip for monitoring. This financial support and the necessary devices within the framework of the international CALM and GTN-P projects were helpful to the researchers who financed the preparation of boreholes and permafrost monitoring work in Mongolia. Monitoring of permafrost and experimental studies in the Hövsgöl region was supported by the GEF/WB Hövsgöl Project. In addition, the monitoring of permafrost in the Terelj and Nalayh sites were conducted within the framework of the IORGC project.

We would also like to thank Drs. Alain Plante and Marinus J.A. Werger for their help in improving the manuscript.

References

Batima P, Natsagdorj L, Gombluudev P, Erdenetsetseg B (2005) Observed climate change in Mongolia, AIACC, p 25

Bayarsaikhan U, Boldgiv B, Kim KR, Park KA, Lee D (2009) Change detection and classification of land cover at Hustai National Park in Mongolia. Int J Appl Earth Obs Geoinform 11:273–280. doi:10.1016/j.jag.2009.03.004

Brown J, Hinkel KM, Nelson FE (2000) The Circumpolar Active Layer Monitoring (CALM) program research designs and initial results. Polar Geogr 24:165–258

Burgess MM, Smith SL, Brown J, Romanovsky V, Hinkel K (2000) Global Terrestrial Network for Permafrost (GTNet-P): permafrost monitoring contributing to global climate observations. Geological Survey of Canada, Current Research, 2000-E14, pp 1–8

CALM (2007) CALM summary data table. Circumpolar Active Layer Monitoring. http://www.udel.edu/Geography/calm/data/north.html. Accessed Mar 2011

Cannone N, Guglielmin M, Hauck C, Muhll DV (2003) The impact of recent glacier fluctuation and human activities on permafrost distribution, Stelvio Pass (Italian Central-Eastern Alps). In: Philips M, Springman SM, Arenson LU (eds) Eight international conference on permafrost. A.A. Balkema Publishers, Zurich, pp 125–130

Cheng GD, Li X (2003) Constructing the Qinghai-Tibet railroad: new challenges to Chinese permafrost scientists. In: Philips M, Springman SM, Arenson LU (eds) Eight international conference on permafrost. A.A. Balkema Publishers, Zurich, pp 130–134

Corsi A, Dore G, Nagpal T, Whitten T, Iyer R, Mearns R, Lyngdoh YR, Rivera S, Ykhanbai H (2002) Mongolia environment monitor. The World Bank, Ulaanbaatar, p 40

Cui X, Graf H-F (2009) Recent land cover changes on the Tibetan Plateau: a review. Clim Chang 94:47–61. doi:10.1007/s10584-009-9556-8

Dagvadorj D, Natsagdorj L, Dorjpurev J, Namkhainyam B (2009) Mongolia assessment report on climate change 2009, Hiimori Printing, Ulaanbaatar, Mongolia, pp 228

Desyatkin AR, Takakai F, Fedorov PP, Nikolaeva MC, Desyatkin RV, Hatano R (2009) CH_4 emission from different stages of thermokarst formation in Central Yakutia, East Siberia. Soil Sci Plant Nutr 55:558–570

Elberling B, Christiansen HH, Hansen BU (2010) High nitrous oxide production from thawing permafrost. Nat Geosci 3:332–335

Etzelmüller B, Heggem ESF, Sharkhuu N, Frauenfelder R, Kääb A, Goulden C (2006) Mountain permafrost distribution modelling using a multi-criteria approach in the Hövsgöl area, northern Mongolia. Permafrost Periglac Process 17:91–104. doi:10.1002/ppp. 554

FAO (2006) Country pasture/Forage resource profiles – Mongolia. http://www.Fao.org/ag/AGP/AGPC/doc/Counprof/Mongol2.htm. Accessed Mar 2011

Gavrilova MK (2003) Recent climatic change in Asia along Yakutia-Mongolia transect. Pap Meteorol Hydrol 25:11–17

Goulden CE, Tsogtbaatar J, Chuluunkhuyag, Hession WC, Tumurbaatar D, Dugarjav C, Cianfrani C, Brusilovskiy P, Namkhaijantsen G, Baatar R (1999) The Mongolian LTER: Hovsgol

National Park. Paper presented at the 3rd international conference on Long-Term Ecological Research (LTER) in the East Asia-Pacific region, Seoul, Korea, 11–16 Oct 1999

Goulden CE, Ariuntsetseg L, Nandintsetseg B, Avirmed O, Batkhishig O, Sharkhuu A, Sharkhuu N (2005) Permafrost thermal properties and thaw and its relationship to soil and plant cover, Lake Hövsgöl, Mongolia. In: AGU fall meeting, San Francisco, pp C31A-1121

Gravis GF, Zabolotnik SI, Sukhodrovsky VL, Gavrilova MK, Lisun AM (1974) Geocryological conditions in the People's Republic of Mongolia. Nauka Publishing, Moscow

GTN-P (2010) Borehole metadata: Mongolia. Global Terrestrial Network for Permafrost. http://www.gtnp.org/inventory/mongolia_e.html Accessed Mar 2011

Harris SA (1998) Effects of vegetation cover on soil heat flux in the Southern Yukon territory. Erdkunde 52:265–284

Harris RN, Chapman DS (1997) Borehole temperatures and a baseline for 20th-century global warming estimates. Science 275:1618–1621. doi:10.1126/science.275.5306.1618

Harris C, Vonder Mühll D, Isaksen K, Haeberli W, Sollid JL, King L, Holmlund P, Dramis F, Guglielmin M, Palacios D (2003) Warming permafrost in European mountains. Glob Planet Change 39(3–4):215–225. doi:10.1016/j.gloplacha.2003.04.001

IPCC (2007) Climate change 2007: synthesis report. Contribution of working groups I, II and III to the fourth assessment report of the Intergovernmental Panel on Climate Change, IPCC, Geneva, Switzerland, p 104

Ishikawa M, Sharkhuu N, Zhang Y, Kadota T, Ohata T (2005) Ground thermal and moisture conditions at the southern boundary of discontinuous permafrost, Mongolia. Permafrost Periglac Process 16:209–216. doi:10.1002/ppp. 483

Jambaljav Y, Dashtseren A, Solongo D, Saruulzaya A, Battogtokh D, Iijima Y, Ishikawa M, Zhang Y, Yabuki H, Kadota T (2008) The temperature regime in boreholes at Nalaikh and Terelj sites, Mongolia. In: Ninth international conference on permafrost, Fairbanks, Alaska, 29 June–3 July 2008. pp 821–826

Jarvis P, Linder S (2000) Botany: constraints to growth of boreal forests. Nature 405:904–905

Jorgenson MT, Racine CH, Walters JC, Osterkamp TE (2001) Permafrost degradation and ecological changes associated with a warming climate in central Alaska. Clim Chang 48:551–579. doi:0.1023/A:1005667424292

Kudryavtsev VA, Garagulya LS, Kondratieva KA, Melamed VG (1974) Osnovi merzlotnogo prognoza pri ingenerno-geologicheskih issledovaniyah [Fundamentals of frost forecasting in geological engineering investigations]. Moskovskogo Universitet, Moscow

Lee H, Schuur EAG, Vogel JG (2010) Soil CO_2 production in upland tundra where permafrost is thawing. J Geophys Res Biogeosci 115. doi:G0100910.1029/2008jg000906

Liancourt P, Sharkhuu A, Ariuntsetseg L, Boldgiv B, Helliker B, Plante A, Petraitis P, Casper B (2012) Temporal and spatial variation in how vegetation alters the soil moisture response to climate manipulation. Plant Soil 351(1):249–261. doi:10.1007/s11104-011-0956-y

Nandintsetseg B, Goulden CE (2005) Recent climate change in Northern Mongolia and its impact on ecosystem. In: First international symposium on terrestrial and climate change in Mongolia, 2005, pp 19–22

Natsagdorj L, Dagvadorj D, Batima P, Tumurbaatar D (2000) Climate change and its impacts in Mongolia. JEMR Publishing, Ulaanbaatar

Nelson FE, Shiklomanov NI, Hinkel KM, Christiansen HH (2004) The Circumpolar Active Layer Monitoring (CALM) workshop and the CALM II program. Polar Geogr 28:253–266

Osterkamp TE, Romanovsky VE (1999) Evidence for warming and thawing of discontinuous permafrost in Alaska. Permafrost Periglac Process 10:17–37

Pavlov AV, Perlshtein GZ (2006) Monitoring and modeling the permafrost dynamics under climate changes. Paper presented at the Asian conference on permafrost, Lanzhou, 7–9 Aug 2006

Price LW (1971) Vegetation, microtopography, and depth of active layer on different exposures in subarctic alpine tundra. Ecology 52:638–647

Romanovsky VE, Sergueev DO, Osterkamp TE (2003) Temporal variations in the active layer and near-surface permafrost temperatures at the long-term observatories in Northern Alaska. In: Philips M, Springman SM, Arenson LU (eds) Eight international conference on permafrost. A.A. Balkema Publishers, Zurich, pp 989–994

Schuur EAG, Vogel JG, Crummer KG, Lee H, Sickman JO, Osterkamp TE (2009) The effect of permafrost thaw on old carbon release and net carbon exchange from tundra. Nature 459:556–559

Sharkhuu N (1998) Trends of permafrost development in the Selenge River basin, Mongolia. In: Lewkowicz AG, Allard M (eds) Seventh international conference on permafrost, Yellowknife, 1998. Collection Nordicana, pp 979–985

Sharkhuu N (2000) Regularities in distribution of permafrost in Mongolia, Scientific transitions of the Geo-ecology Institute, MAS, pp 217–232

Sharkhuu N (2003) Recent changes in the permafrost of Mongolia. In: Philips M, Springman SM, Arenson LU (eds) Eight international conference on permafrost. A.A. Balkema Publishers, Zurich, pp 1029–1034

Sharkhuu A (2007) Effects of snow and vegetation cover on ground surface temperature and active layer thickness in the Hovsgol Mountain taiga zone, Mongolia. Master thesis, National University of Mongolia, Ulaanbaatar

Sharkhuu N, Anarmaa S (2005) Recent degradation of permafrost in Mongolia. In: First international symposium on terristrial and climate change in Mongolia, Ulaanbaatar, 26–28 July 2005. Admon, pp 99–101

Sharkhuu N, Anarmaa S (2008) Ground temperature, soil moisture and seasonal freezing regime in the Hustai National Park, Mongolia. Takhi Trans Sci Articles 9:71–96

Sharkhuu A, Sharkhuu N, Goulden CE (2006) Permafrost condition and changes at eastern shore Hövsgöl Lake, Mongolia. Mongolian Geosci 29:86–93

Sharkhuu A, Sharkhuu N, Etzelmuller B, Heggem ESF, Nelson FE, Shiklomanov NI, Goulden CE, Brown J (2007) Permafrost monitoring in the Hovsgol mountain region, Mongolia. J Geophys Res 112:F02S06. doi:10.1029/2006JF000543

Sharkhuu N, Sharkhuu A, Romanovsky VE, Yoshikawa K, Nelson FE, Shiklomanov NI (2008) Thermal state of permafrost in Mongolia. In: Ninth international conference on permafrost, Fairbanks, Alaska, 29 June–3 July 2008, pp 1633–1638

Sharkhuu N, Anarmaa S, Batbyamba (2010) Thermal insulation of soil surface cover in the Hustai National Park, Mongolia. "Takhi" Transitions of Scientific Articles 9:30–45

Solomon S, Qin D, Manning M, Chen Z, Marquis M, Averyt KB, Tignor M, Miller HL (eds) (2007) Contribution of working group I to the fourth assessment report of the intergovernmental panel on climate change. Cambridge University Press, Cambridge UK/New York

Van der Wal R, Brooker RW (2004) Mosses mediate grazer impacts on grass abundance in arctic ecosystems. Funct Ecol 18:77–86

Walker DA, Jia GJ, Epstein HE, Raynolds MK, Chapin FS III, Copass C, Hinzman LD, Knudson JA, Maier HA, Michaelson GJ, Nelson F, Ping CL, Romanovsky VE, Shiklomanov N (2003) Vegetation-soil-thaw-depth relationships along a low-arctic bioclimate gradient, Alaska: synthesis of information from the ATLAS studies. Permafrost Periglac Process 14(2):103–123. doi:10.1002/ppp. 452

Walker DA, Epstein HE, Gould WA, Kelley AM, Kade AN, Knudson JA, Krantz WB, Michaelson G, AP R, Ping C-L, Raynolds MK, Romanovsky VE, Shur Y (2004) Frost-boil ecosystems: complex interactions between landforms, soils, vegetation and climate. Permafrost Periglac Process 15:171–188. doi:10.1002/ppp.487

Yang M, Wang S, Yao T, Gou X, Lu A, Guo X (2004) Desertification and its relationship with permafrost degradation in Qinghai-Xizang (Tibet) plateau. Cold Reg Sci Technol 39(1):47–53. doi:10.1016/j.coldregions.2004.01.002

Zhao L, Marchenko SS, Sharkhuu N, Wu T (2008a) Regional changes of permafrost in Central Asia. In: Ninth international conference on permafrost, University of Alaska Fairbanks, Fairbanks, Alaska, 29 June–3 July 2008, pp 1627–1632

Zhao L, Wu T, Ding Y, Xie C (2008b) Monitoring permafrost changes on Qinghai-Tibet Plateau. In: Ninth international conference on permafrost, University of Alaska Fairbanks, Fairbanks, Alaska, 29 June–3 July 2008, pp 2071–2076

Part IV
Livelihoods

Chapter 18
Degradation and Management of Steppes in China

Xiaoyong Cui, Ke Guo, Yanbin Hao, and Zuozhong Chen

Abstract China has 400 million ha of grassland, 72% of which is temperate grassland, including meadow steppe, typical steppe, and desert steppe. Some 90% of the grassland is degraded to different degrees. We present an overview of the indicators of various stages of degradation. The degradation is caused by long-term inappropriate use of the grassland in interaction with climate change. The most important causes are the explosive growth of the human population and livestock numbers in the Chinese grassland domain, together with the growing demand for economic improvement. We briefly review the effects of overgrazing, frequent mowing, transformation into cropland, and overharvesting of useful plants, and discuss the laws installed and measures taken to address these problems and manage grassland properly. Though the past decade has shown promising improvements, it is still a long way for China to achieve the goal of sustainable economic development and grassland conservation.

18.1 Overview of Chinese Grassland

Grassland in China covers an area of 400 million ha, taking up approximately 41% of the national territory. The dominant type is temperate grassland, which mainly occurs in northern China. It covers an area of 287 million ha, accounting for 72% of total grassland area. Alpine grassland and tropical - subtropical grassland account for the remaining 28% (Chen 2008; Squires et al. 2010; Suttie et al. 2005).

X. Cui • Y. Hao
Graduate University, Chinese Academy of Sciences, Beijing 100049, China
e-mail: cuixy@gucas.ac.cn; ybhao@gucas.ac.cn

Z. Chen • K. Guo (✉)
State Key Laboratory of Vegetation and Environmental Change, Institute of Botany, Chinese Academy of Sciences, Beijing 100093, China
e-mail: chenzuozhong@yahoo.com.cn; guoke@ibcas.ac.cn

M.J.A. Werger and M.A. van Staalduinen (eds.), *Eurasian Steppes. Ecological Problems and Livelihoods in a Changing World*, Plant and Vegetation 6,
DOI 10.1007/978-94-007-3886-7_18, © Springer Science+Business Media B.V. 2012

Temperate grassland in China is part of the great Eurasian grassland area. It extends from the Northeast Plain, across the Mongolian Plateau and the Loess Plateau, to mountainous regions in Xinjiang. Along this line, grassland changes from meadow steppe, typical steppe, to desert steppe, principally due to a decline in precipitation.

Meadow steppe is a zonal vegetation type in temperate semi-humid regions. The dominant species are semi-xeric, perennial tussock grasses and rhizomatous grasses, as well as semi-xeric and mesic herbage. Some semi-xeric dwarf shrubs often occur scattered in the community. The meadow steppe lies within the forest/grassland transition zone. It is the most humid grassland type, and is concentrated in the east of the grassland belt of China, such as the low mountains and highlands on the piedmonts of Daxinganling and some upper mountain grasslands. The Hulunbeier grassland and the Xilingol grassland consist for a large part of meadow steppe.

Typical steppe is formed under a temperate semi-arid continental climate. The main species are xeric and eurytopic xeric perennial bunch grasses, with, under certain conditions, shrubs and semi-shrubs. Temperate typical steppe is the most widely spread and also the most representative grassland type in China. It is mainly located in western parts of the Hulunbeier Plateau, most of the Xilingol plateau, the northern slopes of the Yinshan Mountain and the southern hills of Daxinganling (Photos 18.1, 18.2 and 18.3).

Photo 18.1 Typical steppe at Xilingol, Inner Mongolia (Photo Z. Chen)

Photo 18.2 Slightly degraded steppe at Xilingol, Inner Mongolia (Photo Z. Chen)

Photo 18.3 Severely degraded steppe at Xilingol, Inner Mongolia (Photo Z. Chen)

Desert steppe is a common vegetation type in temperate arid areas. Dominant species are often small perennial xeric tussock grasses, together with some xeric and super-xeric shrubs and semi-shrubs. It is constrained to the narrow area westward from typical steppe zone, between 75–114°E and 37–47°N in China. This includes mid-western Inner Mongolia, northern Ningxia, mid-Gansu, and Xinjiang. Desert steppe is under the strong influence of the Mongolian high pressure system and weakly touched by the south-east monsoon blowing from the Pacific Ocean to inland. It has a continental climate with an average annual precipitation of 150–250 mm. The soil is brown calcic with a low fertility.

Temperate steppe constitutes the main body of Chinese grassland. It covers more than half of northern China, showing clear zonal patterns along latitude, longitude, and altitude.

Latitudinal zonality. The temperate grassland extends across 17° of latitude in China. This leads to marked differences in thermal conditions and plant responses between the north and the south. Species in the north are generally cold-resistant, such as *Stipa grandis*, *Stipa krylovii, Leymus chinensis, Filifolium kitamura*, etc. In the south, however, plants are generally thermophilic, such as *Stipa bungeana, Bothriochloa ischaemum*, etc.

Longitudinal zonality. The temperate grassland stretches over 44° of longitude. Consequently, the distance from the oceans varies dramatically. Therefore, continentality increases and precipitation decreases from east to west, and thermal conditions changes accordingly. Vegetation zones, however, do not run strictly parallel to longitude but vary along a gradient in southeast-northwest direction, due to the influence of terrain relief. From east to west, the meadow steppe zone, typical steppe zone, and desert steppe zone appear sequentially.

Altitudinal zonality. There are several mountain ranges located in the temperate grassland area of China, including the Altai Mountains, the Helan range, the Daqing range, and so on. The altitudinal zonation of the vegetation is obvious in these ranges, especially on high mountains. From the bottom up there appears mountain grassland, mountain deciduous broad-leaved forest, mountain cold temperate coniferous forest, subalpine shrub, subalpine meadow, alpine meadow, alpine scree, and the glacier and nival zone, sequentially.

Temperate steppe, as the main part of Chinese grassland, is concentrated in North China, where environmental conditions are harsh. These regions are inhabited by ethnic minorities. The grassland provides an important environmental asset in their livelihood as it harbors essential resources for the social and economic development of the people, and they are increasingly concerned about the degradation of the temperate grassland.

Alpine grassland occurs in alpine regions and on high plains with elevations generally above 4,000 m and the climate varying from arid, sub-arid, to semi-humid. In China, it is mainly located on the Tibetan Plateau and some high mountains, such as the Tianshan and Kunlunshan ranges in Xinjiang, and the Qilianshan range in

Qinghai and Gansu provinces. Semi-xeric and xeric-mesophytes cold-resistant perennial herbage and small semi-shrubs are common. Physical conditions are harsh in alpine meadow areas, characterized by a cold climate. Annual average temperature is about 0–4°C, and annual rainfall around 300–400 mm. The soil is mainly an alpine steppe soil, usually with a thin loose sod layer and a light-colored humus layer. Soil organic matter content is high, usually more than 3%.

Tropical and subtropical grassland occurs in the regions south of the Yangtze River in China. It is formed after deforestation or abandonment of croplands, with types of secondary *herbosa* dominated by perennial herbs, secondary shrub *herbosa* scattered with shrubs or trees, and secondary savanna in some hot-dry valleys in Southwest China. Annual average temperature is generally 14–22°C. In most of the area there is abundant precipitation, from 800 mm on average to as high as 2,000 mm. The soil types include yellow earth, yellow brown earth, and mountain red earth. They are shallow and of low fertility.

18.2 The Degradation of Chinese Temperate Grassland

18.2.1 Degradation Indicators of Chinese Temperate Grassland

Some scientists in China argue that the grassland degradation is a systemic deterioration of both the biological communities (plants, animals, and microbial communities) and their environments under the interactive effects of human activities and natural forces, which is demonstrated by damage of the ecosystem structure and functions. This means, that grassland degradation refers not only to degeneration of the vegetation, but also to land deterioration (Chen 1990). Other scientists, emphasizing the process of succession, suggest that grassland degradation means a retrogressive process, which leads the ecosystem away from its climax under the interference of human activities such as grazing, land reclamation, and mowing (Li 1990). During the degrading succession, the ecosystem structure and functions gradually change, as indicated by a lower energy flow, an unbalanced material cycle, and an increase of entropy. The grassland then may reach a deflexed climax at a lower energy flow rate or lose its capability of maintaining ecosystem functions (Li 1997). The degree of grassland degradation, therefore, is denotable in terms of environment, biological composition and ecosystem function. Table 18.1 shows the indicators of different degradation stages for typical steppe ecosystems (Chen and Wang 2000).

About 90% of the Chinese grassland has been or is under degradation, according to a report of more than a decade ago by the National Environmental Protection Bureau of China (1999). The degradation of temperate grassland has attracted a lot of attention since it concerns most of the Chinese grassland area and plays a fundamental ecological role.

Table 18.1 Degradation levels and corresponding indicators for typical steppe ecosystems

Degradation level	Plant composition	Aboveground biomass and coverage	Ground cover	Indicative rodent	Indicative locust	Soil properties	Indicative soil animal	Indicative soil microbe	Ecosystem structure	Restorability
I Slight degradation	No significant change, with a slight decrease of *Leymus chinensis* and *Stipa grandis* and increase of *Artemisia frigida*, *Agropyron cristatum*, and other small grasses	Decreased by 20–35%	Significantly reduced	*Ochotona daurica*	*Chorthippus fallax*	No significant change, with slight increase of surface hardness and decrease of organic matter content	Enchytraoidae	*Bacillus subtilis, Phoma, Trichothecium*	No obvious change	Rapidly recoverable after enclosure
II Moderate degradation	Dominated by *A. frigida* with a large proportion of *L. chinensis* and *S. grandis*	Decreased by 35–60%	No ground cover	*Microtus brandti*	*Chorthippus dubius*	2-fold increase of surface hardness and obvious decrease of organic matter content	*Coleoptera* and mites	*Bacillus coagulans, Cephalotrichum*	Decrease of carnivores, and increase of rodents	Naturally recoverable after enclosure

III	Severe degradation	Dominated by *A. frigida*, *Potentilla acaulis* and small grasses, with disappearance of most endemic species and decrease of species richness	Decreased by 60–85%	Bare ground	*Microtus brandti*	*Angaracris barabensis*, *Myrmeleotettix palpalis*	2-fold increase of surface hardness, marked decrease of soil organic matter content, increase of coarse-grained particle fraction, salinization and appearance of alkali spots	*Hymenoptera*	*Absidia*	Dysfunctioning of the ecosystem as shown by shortened food chain and simplified structure	Difficult to recover naturally. Artificial improvement needed
IV	Extreme degradation	Disappearance of vegetation or with sparse *P. acaulis* and annuals	Decreased by more than 85%	Bare sandy soil, or with saline spots	*Meriones unguiculatus*		Loss of forage production value			Collapse of the system	Rehabilitation needed

18.2.2 Causes of Temperate Grassland Degradation in China

Grassland degradation is the result of interactive effects of climate change and long-term inappropriate human activities. The latter, including overgrazing, improper mowing, cultivation, overharvesting and so on, are the main cause.

The explosive increase of the population, coupled with a growing demand for economic improvement, are the underlying forces of irrational human activities. For example, the Baiyinxile Ranch occupies an area of 3,730 km^2 in Xilingol, in a typical steppe area. It had only 20 employees in 1950 upon establishment. This figure increased by more than 200 times in 12 years and reached 5,139 in 1962. The peak value was in 1982, with a population of 12,959. In 2000 it decreased to 10,210, about 510 times that in 1950. With population growth, livestock numbers rose as animal husbandry was the principal source of income. The total number was 1,023 in 1950, and reached a high of 252,248 in 1999. This is an increment of almost 240 times. At the same time, livestock composition shifted in response to market demand. This imposed further pressure on the grassland. Before 1975, large animals, such as horses and cattle, increased rapidly. The total number of horses got up to 17,261 in 1975, and declined thereafter. In contrast, the fraction of small animals, such as sheep and goats, especially goats, grew quickly after the 1980s. For instance, the number of goats was only 2,283 in 1962, but it rose by 13 times to 29,674 in 1997, owing to a large demand for cashmere, which due to sharply increasing prices, made goat raising profitable.

18.2.2.1 Overgrazing

It has been proved by plenty of experiments that overgrazing has serious negative effects on the vegetation (Photos 18.1, 18.2 and 18.3). Besides, overgrazing also has adverse effects on grassland soil (Jia et al. 1997). The top 0–5 cm soil layer was impacted most severely. Animal trampling led to compacted surface soil with a higher bulk density and hardness. For example, after 6 years of continuous heavy grazing, i.e. 6.67 sheep ha^{-1}, soil bulk density was 1.286 $g\,cm^{-3}$, 1.06 times of that in a non-grazed plot (1.209 $g\,cm^{-3}$). Surface hardness under heavy grazing was 34.71 kg cm^{-2}, compared to 30.93 kg cm^{-2} in a lightly grazed plot with a stocking rate of 1.33 sheep ha^{-1}.

The temperate grassland in China has long been overgrazed, in fact since the explosive growth of the human and livestock populations. More animals consequently led to less grassland area per animal. At Baiyinxile Ranch, one animal occupied an average area of 350 ha of grassland in 1950 but this had become only 1.5 ha in 2000. In the Inner Mongolia Autonomous Region, each sheep unit had 4.1 ha of pasture in 1947, a rather low grazing intensity. Till 1965, together with the large increase in livestock number, the figure declined to 0.97 ha per sheep unit. This stocking rate exceeded the carrying capacity of the grassland. That situation did not change in the following two decades, as the livestock number oscillated around 70 million sheep units. A survey showed that the grassland was overgrazed in 65 out of 108 banners, thus in approximately 60.2% of all pasture banners in the temperate grassland region at that time (Wu 1995). Since 1999 the overstocking problem accelerated in many areas (Squires et al. 2009). In 2006, the average rate of overstocking was about 34% across China, whereas it was about 40% in Gansu and even 70% in some areas in Xinjiang (Han et al. 2008; Jin and Zhou 2009). It was estimated that presently the overloading rate averages about 20–30% above the carrying capacity of natural grassland in China (Liu 2010).

18.2.2.2 Frequent Mowing

The principal land use regimes of the Chinese temperate grassland area are grazing and mowing. Both have similar influences on the grassland in terms of material cycling. Experiments demonstrated that the mowing frequency affected the forage production greatly (Zhong et al. 1991). After 5 years, the aboveground biomass was 113.8 $g.m^{-2}$ under an annual mowing treatment and this was only 76% of that of a biennially mown plot. Mowing at high frequency did not prove to be a sustainable way of grassland management. However, due to the increase in livestock numbers, particularly the stock in winter, rotational mowing was gradually replaced by annual mowing as the mowing field area became limited. Consequently, the nutrient output exceeded the natural input and the nutrient deficiency resulted in a decrease in grassland productivity.

18.2.2.3 Cultivation

Flat and fertile land with a deep soil and flourishing vegetation has long been considered as reclaimable wasteland. Since 1949, the Chinese grassland area has experienced several surges of cultivation to address food shortage. Up to the 1960s, a total of 6.67 million ha of grassland were reclaimed. An area of 6.67×10^4 ha of grassland was transferred into cropland during the three periods of large-scale reclamation in northwest China between the 1950s and 1970s. Till the early 1990s, there were several periods of large-scale grassland reclamation again throughout China. According to a survey of the Chinese Academy of Sciences, in one decade around 1988, as much as 970,851 ha of grassland became cultivated in 33 banners in eastern Inner Mongolia (Photo 18.4).

Food production in grassland areas consisted mostly of extensive cultivation. Little input and extensive management led to a fast loss of soil organic matter after land conversion. Furthermore, low rainfall and strong winds, especially in spring when cropland was hardly covered, resulted in severe soil erosion. Fine soil particles were blown away and land was severely degraded. Different from overgrazing-induced grassland degradation, cultivation generally caused fast and patchy sandification and degradation.

Photo 18.4 Large areas of steppe have been lost due to cultivation; here at Hulenbeier, Inner Mongolia (Photo Z. Chen)

Photo 18.5 Grazing in spring should be intermittently (Photo Z. Chen)

18.2.2.4 Overharvesting

There are a large number of medicinal plants and economic biological resources, including liquorice (*Glycyrrhiza* spp.), ephedra (*Ephedra sinica*), Anemarrhena (*Anemarrhena asphodeloides*), Astragalus (*Astragalus mongholicus*), mushrooms (*Agaricus campestris*), Nostoc (*Nostoc commune*), in the temperate grassland in China. Long-term digging, raking and cutting activities seriously damaged the grassland vegetation and soil, leading to grassland degradation. This was striking on the Ordos Plateau, where grassland vegetation was severely impaired due to long-term harvesting of liquorice and ephedra. Collecting 1 kg liquorice caused damage on 0.53–0.67 ha of grassland. It was estimated that harvesting damaged 26,700 ha of grassland per year in this region. Collection of Nostoc damaged 1.3×10^7 ha of grassland per year in Inner Mongolia, accounting for 19.5% of its total grassland area, of which 4×10^6 ha were totally destroyed.

18.2.2.5 Other Causes

The temperate grassland regions in China are rich in mineral resources, including coal, oil, natural gas, salt, saltpeter, soda and so on. Exploitation of mineral ores greatly promoted local economic development in pastoral areas in recent years. Nevertheless, it also negatively affected grassland through compacting by vehicles, tailings, mine dumps and so on. That induced grassland degradation and desertification around mineral sites. Once degraded, it was extremely difficult and slow to recover.

Grassland degradation brought about a series of ecological, environmental, and economic issues, such as reduction of biodiversity, productivity, and farmer income and living standards, as well as deterioration of ecosystems and the environment. Hence, there is an urgent need to restore degraded grasslands.

18.3 Management of Degraded Temperate Grassland in China

The Chinese government has taken a series of measures to protect rangeland and deal with grassland degradation. The most important were legal measures, imposed by the implementation of 'The Law on Grasslands of the PRC'. Financial and infrastructural measures were taken by carrying out large engineering projects to attract investments and improve roads, markets, and information necessary for efficient grassland management. Technical measures were adopted, making use of new technologies to prevent degradation and facilitate rehabilitation. Degenerated grasslands are being categorized and managed accordingly.

18.3.1 Managing Grassland in Accordance with the Law

The Chinese government attaches great importance to temperate grassland management. It promoted rangeland management according to the law so as to guarantee long-term sustainability of animal husbandry and the grassland ecosystem. As a milestone, on June 18, 1985, the NPC Standing Committee passed the "Law on Grassland of the People's Republic of China". It was put into force on October 1st of the same year. Around the beginning of the twenty-first century, the temperate grassland attracted public attention again with the increasing frequency and intensity of dust and sandstorms, as well as other ecological and environmental concerns in northern China. In order to strengthen the conservation of grassland and ensure its sustainable use and development, in September 2002, the State Council issued "Some recommendations on promoting grassland protection and construction". The "Law on Grassland" was amended by the Standing Committee of the National People's Congress in December 2002, further solidifying the legal foundation for grassland conservation and management.

18.3.2 Stimulating Investment by Key Projects

Since 1978, the Chinese government has paid more attention to grassland management. A series of key engineering projects were established and accomplished to promote investment and technological development in grassland management. Most of the projects were implemented since 2000, including the Grassland Revegetation Project, the Forage Grass Seed-base Construction Project, the Grassland Fencing Project, the Returning Cultivated Land to Grassland Engineering Project, the Beijing-Tianjin Sand Source Control Engineering Project, the Grassland Fire Control Engineering Project, and the Grassland Rodent and Pest Control Engineering Project, etc. A huge sum was invested through these projects. The central state alone already invested more than 16.4 billion RMB, of which over 10.0 billion RMB was given to the Returning Cultivated Land to Grassland Engineering Project. The engineering measures significantly improved grassland conservation and prevented further degradation in some areas. In 2007 grassland monitoring showed that the vegetation coverage increased by 16% on average in engineering areas. The eco-environment had improved to a certain degree (Liu 2008; Duan et al., 2011).

18.3.3 Promoting Scientific Grassland Management

18.3.3.1 Balancing the Grassland and Livestock

With the legal basis and financial support, advanced science and technology becomes the key for successful grassland management. The root cause of grassland degradation is the imbalance between forage demand by livestock and that available through grassland production and other ways at both farm and regional levels. Therefore, farmers and herdsmen are obliged to keep the fodder-livestock balance in order to maintain the grassland in good shape. To accomplish this goal, the Ministry of Agriculture of the People's Republic of China promulgated "Measures for the Balance of the Fodder and Livestock" in 2005. It ordained that the following principles should be adhered to in fodder-livestock balancing work: (1) strengthening grassland protection and promoting local development; (2) determining the number of household animals according to the availability of fodder and increasing the number of livestock only when fodder availability increases; (3) making a categorization of the grassland's status and manage the grasslands accordingly; and (4) proceeding in an orderly and gradual way. To avoid overloading with animals, specific technical measures should be taken by farmers and herdsmen or governments, including promoting the forage base construction, purchasing forage, applying pen-feeding to release grazing pressure, shortening the animal production cycle, optimizing the livestock structure, enlarging the grassland areas contracted by individual farmers through legally transferring the rights for grassland use (Decree of the Ministry of Agriculture of the People's Republic of China 2005).

18.3.3.2 Implementing Measures of Returning Grazing Land to Protected Grassland and Intermittent Grazing

Under the provision of "Recommendations on strengthening the management of Returning Grazing Land to Protected Grassland Engineering" promulgated by China's Ministry of Agriculture in April 2005, measures, including no grazing, intermittent grazing, rotational grazing, fencing and artificial grassland construction, should be taken to restore the vegetation, improve grassland productivity, and promote the equilibrium between ecological protection and socio-economic development.

Returning grazing land to protected grassland means that grazing should be temporarily prohibited in grasslands that are unable to endure further grazing. This refers to seriously degraded grasslands under long-term over-grazing, and/or under harsh physical conditions and with serious soil erosion. These grasslands are rehabilitated for one or more years and permitted to be used again when they fulfill certain criteria.

Intermittent grazing means no grazing during a short period per year, particularly in early spring when the grass just turns green. At this stage, overgrazing is extremely harmful to forage recovery and growth. Normal grazing is allowed after the vegetation is well developed. Temporary prohibition of grazing takes advantage of the self-renewal and self-repair capacities of grassland ecosystems to improve the environment and the growth of plants. These measures have achieved much success in the rehabilitation of degraded grassland in China. For example, intermittent grazing is practiced from early April to late June in the Xilingol steppe, Inner Mongolia (Photo 18.5). The vegetation had remarkably improved after a 2-months grazing stop, as compared to continuously grazed areas. No grazing at the early growing stage promoted biomass accumulation which provided a good foundation for grass growth later in the season. As a consequence, the final forage production was also higher (Li et al. 2005). Even better effects were obtained when intermittent grazing was practised during several years. Aboveground biomass, vegetation coverage and height had increased by 86%, 87.5% and 93% after respectively 1, 2 and 3 years of no grazing in spring in the Hulun Buir grassland (Zhu et al. 2008). In addition, carbon sequestration was promoted, due to a reduction in ecosystem respiration. In a *Kobresia humilis* alpine meadow, ecosystem respiration was 20.7% lower in areas that were not grazed in spring as compared to those under continuous grazing (Li et al. 2007).

18.3.3.3 Promoting Rotational Grazing

Rotational grazing is a planned and systematic grazing strategy. The grassland is divided into different units and animals graze through these demarcated plots in different seasons or years. Rotational grazing can reduce defoliation and animal trampling, breaks the life cycle of parasites and infectious diseases, and allows for regrowth of the pasture grasses, thus benefitting vegetation production. It can provide significant benefits for the animals by means of improved plant composition of the vegetation, thus earlier meeting the nutritional requirements of livestock. Due to greater utilization, in terms of more plant species being 'cropped' by the animals, stocking rates can usually be higher and animal production greater. This method can also provide for a longer grazing season, reducing the use of stored feeds (Demers and Clausen 2002). It has been shown that rotational grazing had very good results in grasslands in China. In a desert steppe, the proportion, importance value index, and production of both the abundant species *Stipa breviflora* and the dominant species *Cleistogenes songorica* were higher in rotationally grazed plots than in the continuously grazed area (Li et al. 2003). It was also found that warm-season rotational grazing facilitated the recovery of *Stipa bungeana* grassland (Wu et al. 2005).

18.3.3.4 Employing Categorized Management of Degraded Grasslands

In China, 90% of the grasslands are degraded to various degrees. They should be classified and different measures should be taken for their restoration (see also Table 18.1). For grasslands in the category of slight and moderate degradation, the principal technique is determining livestock number according to fodder availability, so as to keep the balance between feed and livestock. Rotational grazing should be adopted to prevent overloading. To prohibit grazing or utilization of this type of grassland completely is irrational, however, because grazing may also have beneficial effects to the grassland ecosystem. The effect of grazing on grassland is largely dependent on grazing intensity. Heavy grazing may cause damage to the grassland. Moderate grazing, on the contrary, stimulates grass growth, e.g. by the "over-compensation effect". Hence, the core is to control the stocking rate. Moreover, measures are recommended to improve forage production in grasslands of this category under suitable conditions. These include approaches to improve the soil physical properties and fertility, such as soil loosening, shallow ploughing, and fertilization, especially with nitrogen. Re-seeding native forage species is also acceptable to improve the plant community structure and facilitate the recovery of the vegetation.

For severely degraded grasslands, artificial assistant measures are hardly effective or economic. Aerial seeding and other measures are not encouraged, as the soil seed bank is abundant and consequently the effect is unsatisfactory. One suitable countermeasure is to fence and enclose such areas to allow natural recovery. Experiments indicate that these grasslands are completely repairable given sufficient time for natural rehabilitation under constant environmental conditions. Recovery was remarkable after 3-year fencing of typical steppe in Inner Mongolia, where "fencing and transferring" is employed as a strategy to conserve the severely degraded grassland. "Fencing and transferring" was accepted as a strategy in the Xilingol League of Inner Mongolia in 2000. It implies joint measures of not only fencing and prohibition of grazing, but also reduction of livestock numbers, transferring farmers out of their deteriorated habitats, as well as transforming grazing into intensive husbandry.

The extremely deteriorated regions that can not support human living and livestock raising are fenced and prohibited from grazing for the sake of the natural recovery of the vegetation. Meanwhile, the scattered housing is abandoned and people are moved to towns or newly established villages with sufficient supplies of water and electricity, convenient transportation, and high accessibility of broadcasting and telecommunication services. The immigrants are trained and engaged in labor-intensive and high-efficiency industries that provide employment opportunities and a steadily increasing income to the farmers (Gu and Chen, 2001). A huge amount of money is needed to fulfill the strategy in the whole grassland area. Thus, national investment is of fundamental importance. Great progress has been made, but there are still many problems to be solved. It is still a long way for China to achieve the goal of sustainable economic development and grassland conservation.

References

Chen Z (1990) Degradation of natural grassland ecosystem and its regulation in China. Control of land degradation in China. China Science and Technology Press, Beijing

Chen Z (2008) Into the grasslands. China Forestry Publishing House, Beijing, pp 5–13

Chen Z, Wang S (2000) Typical steppe ecosystem in China. Science Press, Beijing

Demers C, Clausen R (2002) Managing cattle or timberlands: forage management. University of Florida, Institute of Food and Agricultural Sciences (UF/IFAS)

Duan Q, Xin X, Yang G, Chen B, Zhang H, Yan Y, Wang X, Zhang B, Li G (2011) Current situation and prospect of grassland management decision support systems in China. In: Computer and computing technologies in agriculture, vol 347, IV IFIP advances in information and communication technology., pp 134–146

Gu Y, Chen Y (2001) Situation and development strategy of Xilingol League. Inner Mongolia Education Press, Huhhot

Han JG, Zhang YJ, Wang CJ, Bai WM, Wang JR, Han CD, Li LH (2008) Rangeland degradation and restoration management in China. The Rangeland J. 30:233–239

Jin G, Zhu J (2009) Chapter 14. Case Study 8: Northern Xinjiang. In: Squires VR, Lu X, Lu Q, Wang T, Yang Y (Eds.) Rangeland degradation and recovery in China's pastoral lands. CAB International, London. pp197–215

Jia S, Cui X, Li S, Chen Y, Wang F (1997) Soil physical properties along a grazing pressure gradient. In: Grassland ecosystem research. Science Press, Beijing, pp 12–16

Li B (1990) China's grasslands. Science Press, Beijing

Li B (1997) Degradation and countermeasures of grassland in the Northern China. Sci Agri Sin 1:1–5

Li Q, Han G, Ao T, Peng S (2003) Approach on restoration mechanism of rotational grazing system on desert steppe. Trans Chin Soc Agric Eng 19:224–227

Li Q, Zhao G, Zheng M, Amin B (2005) Delay grazing: a practical approach to grassland environment protection and sustainable animal production in grassland of Northern China. Acta Agrectir Sin 13 (suppl):53–56

Li Y, Qin X, Li W, Lin E, Gao Q, Wang Y, Shi F (2007) Impacts of no grazing in summer on greenhouse gas emissions from Kobresia humilis alpine meadow. Trans Chin Soc Agric Eng 23:206–210

Liu J (2008) Reform and development of Chinese grassland in 30 years. Acta Agrectir Sinica 16:547–551

Liu J (2010) Continuously developing grassland animal husbandry of China. Pratacult Sci 27:1–3

Liu X, Feng Q, Liang T, Long R (2010) Spatial-temporal dynamic balance between livestock carrying capacity and productivity of rangeland in Gannan of Gansu province, China. Chinese Journal of Grassland 32: 99–106

Ministry of Agriculture (2005) Decree of the Ministry of Agriculture of the People's Republic of China (No. 48)

National Environmental Protection Bureau of China (1999) State of the environment in China in 1998

Squires VR, Lu X, Lu Q, Wang T, Yang Y (2009) Rangeland degradation and recovery in China's pastoral lands. CAB International, London

Squires VR, Hua L, Zhang D, Li G (2010) Towards sustainable use of rangelands in North-West China. Springer, New York

Suttie JM, Reynolds SG, Batello C (2005) Grasslands of the world. FAO, Rome

Wu J (1995) Study on grassland degeneration in China. Ecol Econ 5:1–6

Wu X, Wu F, Chen W, Luo Y (2005) Performance of bungeana based on warm-season rotational grazing. Heilongjiang Anim Sci Vet Med 7:56–57

Zhong Y, Bao Q, Sun W (1991) A study on reasonable mowing system of natural mowing steppe in Baiyinxile farm, Inner Mongolia. Acta Ecol Sin 11:242–249

Zhu L, Zeng Z, Zhao B, Wang X, Hu Y, Hai T (2008) Effect of spring rest-grazing on steppe vegetation. Acta Agrectir Sinica 16:278–282

Chapter 19
Land Use Modernization and Agrarian and Conservation Prospects in the Russian Steppe

S.V. Levykin, A.A. Chibilyov, and G.V. Kazachkov

Abstract Agrarian and conservational problems facing the Russian steppes at the beginning of the twenty-first century are interrelated. We discuss the background and causes of the agrarian landscape degradation in the Russian steppe zone, including the virgin land campaign and other steppe exploitation policies, the agrarian developments in the steppe zone, the recent regeneration of secondary steppes on fallows, and a system of steppe land use modernization based on a restorative and adaptive land management concept, as an answer to the present challenges in the Russian steppe zone, and serving both ecological and rural economical goals. We emphasize that the Russian steppe zone has an excellent potential for pasturing animal husbandry and that it allows the development of an adaptive meat industry in the Russian steppe areas adjacent to Kazakhstan.

19.1 Introduction

Russia of recent years achieved some success in social and economical development. The crisis of the early post-Soviet years is met; good positions in the production of raw materials and in the transport industry, weaponry production and space exploration have been maintained. However, characteristic of Russia is that at the same time other spheres, the rural economy for instance, remains on average at the level of developing countries. The positive trend in the development of agriculture in regions with a favorable climate and other natural conditions is off-set by a series of crises in the agriculturally peripheral regions, especially in the south-east. At the same time, the conservation of land resources which are necessary to save the base

S.V. Levykin (✉) • A.A. Chibilyov • G.V. Kazachkov
Institute of Steppe, Urals Branch of the Russian Academy of Sciences, Orenburg, Russia
e-mail: stepevedy@yandex.ru; orensteppe@mail.ru; tsvikaz@rambler.ru

M.J.A. Werger and M.A. van Staalduinen (eds.), *Eurasian Steppes. Ecological Problems and Livelihoods in a Changing World*, Plant and Vegetation 6,
DOI 10.1007/978-94-007-3886-7_19,

of the rural economy, and the restoration of steppe ecosystems, that were almost totally lost in the twentieth century, is still beyond governmental priorities. Thus, the Russia of nowadays belongs to that majority of developing countries, that do not pay due attention to soil resources conservation and ecosystem restoration. Russian agriculture is still strongly depending on the quantity of ploughed land and is directly related to weather favorability. This dependence of the gross yield output of agrarian primary products on the quantity of agrarian land and on weather favorability seems inexorable in Russia, something like a "land needle", analogous to what Russians call the "oil needle", the nearly compulsory reliance of our national budget on oil.

With a system on this basis, the effects of risk factors are very significant, and that has been convincingly demonstrated in economically difficult periods and years of droughts, for example in the period 2008–2010. The tasks of modernization and diversification of the industry, and of the gradual liberation from the "oil needle", are set at Government level, but plans on agricultural development do still not provide room for measures to liberate the agricultural policy from the "land needle", which captured the agriculture in Soviet days and still enjoys a stable lobby in the Russia of today. The recent state of the Russian agricultural belt is the result of its historical specificity. Situated in the special geographical position between the agricultural west, which was short of land resources and created its own civilization of intensive tillage agriculture, and the steppic orient, that supported itself by extensive nomadic animal husbandry on boundless pasture ranges, Russia was not short of land resources and had natural opportunities to develop its agricultural economy. It resulted in the implementation of an extensive agriculture instead of an extensive animal husbandry.

19.2 The Historical Setting

Russian agriculture was born in the forest zone. In the early 2nd millennium A.D. agriculture gradually spread to the woodlands and steppes of European Russia. However, it was ejected back to the forests for several centuries by the military pressure from steppe peoples. Only in the eighteenth century it began to spread to the steppe zone again, and this zone was considered to be the obtained 'wild field', "dikoye pole" in Russian – the inexhaustible source of promising fertile arable lands. At the same time, a rather more negative than positive attitude to the virgin steppes developed, considering it the homeland of war-like nomads. On the one hand, there was the rapid development of science and culture and industrial growth in Russia, on the other serfdom and technological backwardness remained the basis of the rural economy. New lands were increasingly ploughed up for crops and land resources were so abundant that there was no economical need to apply advanced technologies.

Russian science of those days was conscious of this dangerous tendency to unlimited consumption of soil resources. A.T. Bolotov, the founder of Russian agricultural science, was very skeptical about the opinion that the profitability of an estate could be increased by increasing the area of ploughed fields, which was commonly advocated. He stated that if an estate gets too large an area of ploughed fields, and has not enough draft power and hands at its disposal, then the land will be worked badly and not in time, there will not be enough dung to fertilize the fields, and therefore its crop capacity will be low. He considered that increasing the soil fertility would be a much more reliable way to increase the profitability of an estate. For instance, he stated in 1784 that "keeping animal husbandry in due proportion to crop agriculture is most important for the rural economy. These two things are so interrelated that if one is falling out of attention the other is inescapably damaged." (Berdyshev 1988).

Ideas on a rational agriculture were proposed also by other progressive figures in eighteenth century Russia. For example, the great Russian military commander A.V. Suvorov wrote literally in his instructions to the peasants of his estate: "The abundance of fields evokes laziness, which leads to poverty. This means: raise livestock, fertilize fields with dung, plough only the fields well supplied with fertilizer, turn all the rest into meadows and pastures, and use these to produce fodder" (Sovremennik 1992).

Russian agriculture nevertheless stuck to the extensive way of development even in the nineteenth century. The construction of a continuous railway network and Black Sea harbors promoted the export of crops, the demand of which always grew. According to contemporaries, this evoked a genuine "wheat fever" over the steppe in the south of European Russia. In Ukraine, all plots that could be used for crops were ploughed up during this fever (Formozov 1962).

In the nineteenth century the steppes of European Russia were completely ploughed up, but in the Southern Urals, for instance, arable lands were sown for only 20–45%, and that gave some potential to save some steppe and allowed some steppes to regenerate (The State Archive of Orenburgskaya oblast 1895). Until the Soviet days the rotational fallow system dominated in Russian agriculture. As land resources decreased it evolved from a seven-field rotational system to a three-field system (The Handbook of the Russian Farmer 1993). But it should be noted that even under so modest a share of sown area the agroecological crisis developed and the catastrophic failure of crops in 1891 is commonly considered to be its reference point. Crop yields in those days strongly depended on how favourable the weather conditions during the growing and harvesting seasons were and on the quantity of sown lands. The low crop capacity of those days sharply fluctuated from 50 to 1,500 kg per ha (Collected Statistics about changes in population, cattle and harvests over Kazakh SSR from 1880 to 1922 1925). As dry crop farming in the steppe area expanded, land use optimization of the steppe became more topical than ever before. Famous scientists, such as V.V. Dokuchaev, D.I. Mendeleev, K.A. Timiryaziev, A.I. Voyeikov, P.A. Kostychev, A.N. Engelgardt, A.A. Izmailskiy, D.N. Pryanishnikov tried to find a solution for the problem of how to adapt the Russian rural economy to the landscape diversity of Russia.

At the dawn of the Collectivization D.N. Pryanishnikov criticized the increasing trend to shift the crop production to the South-East: "In the urge of gratuitous fertility we left regions that are not subject to droughts almost without agriculture, and we not only occupied all area suitable for dry farming, but also begin to plough up lands in regions where agriculture is a-priory a gamble and where there is evidently no place for farm expansion" (Pryanishnikov 1965, p. 343). He considered droughts the main threat to agricultural stability, and noted their disastrous effects in 1891, 1911, 1921. As an alternative to virgin land campaigns in the steppes, he proposed the development of intensive agriculture in the non-black soil areas to guarantee the food security for our country. He was convinced that agricultural lands in non-black soil areas would provide large enough yields to form crop reserve stocks of up to 17 million metric tons (Pryanishnikov 1965).

Land use optimization of the steppe was the research theme of V.V. Dokuchaev and his school and they wrote works that became considered to be the classics of Russian steppe science. A system approach to reforming steppe land management was proposed, and a main element proposed was to temporarily take the most degraded plots of ploughed lands out of production and conserve these. Later, other elements of Dokuchaev's system became widely applied: afforestation melioration, irrigation, etc. But even such a famous scientist did not succeed in shaking the extensivation paradigm of rural economical development, and Russia entered the twentieth century with a burden of unsolved agrarian problems. Also at the time of the achievements of Stolypin's agrarian reform and large governmental investments into rural economical development, the old extensivation paradigm remained unrevised. However, it should be noted that in the governmental agricultural recommendations of those days, in addition to stimulating the ploughing up of virgin lands for crops in the East, also fodder grass cultivation as the fodder base for the animal husbandry was promulgated for the intensely ploughed-up regions of European Russia (The Handbook of the Russian Farmer 1993).

19.3 The Soviet Years

The Soviet government, that declared fighting all the vestiges of monarchy, for unknown reasons not only left the extensivation principle of agricultural development without revision, but even cultivated it. During the Soviet days the new lands "pioneering syndrome" only deepened. The Soviet period became the triumph period of the "Transformism" approach in land use and it found its zenith in the 'virgin lands campaign' of 1954–1963. In 1953 a series of governmental programs to develop kolkhoz and sovkhoz meat livestock was brought to their close; from now on one counted on crops. These changes in agrarian policy won approval from specialists in earth science. The typical view of those days is reflected, for example, by A.I. Perelman's "The task is the maximal productive use of every plot on the earth's surface; the task is: no fallow land, no virgin land, no waste plot, no mire and other 'bad land' any more. The task is the extermination of all landscape features harmful for people, and the maximal development of the useful features of the landscape" (Perelman 1956, p. 61). The consequent ploughing up of huge areas of land of insufficient soil fertility or with too dry a climate led to the situation that low crop capacities were considered normal in those days and it caused crop yields to widely fluctuate from year to year in strong dependence on weather conditions.

At that time, in Russia, against the background of regular failures to factually meet planned crop yields and capacity plans, which had as a consequence that increasing the sown area was considered as the only way to increase gross grain yield, the idea grew that decreases in sown area should be utterly rejected, and this attitude still remains at present. This explains why the occurrence of fallow lands is commonly considered as evidence of trouble in the rural economy in most of the post-Soviet countries.

By 1990, the really sown areas in the agricultural regions of Zavolzhia and the Southern Urals exceeded the quantity of officially registered arable lands by 10–15%. This impeded objective assessments of the degree to which the steppe really was transformed, as well as of the real crop capacity. In the 1980s approximately 6.4 million ha were officially registered as arable lands in the Orenburgskaya oblast, but each year the sown area was registered at 7.2 million ha, the 0.8 million ha more being the so-called "lands for principal improvement" which were low-quality lands that normally were ploughed once in 5–7 years and in other years lay fallow and grew perennial wild grasses, but now were taken into long time use for crop growing. Furthermore, over 10,000 ha of unregistered ploughed lands were sown in territories of the Ministry of Defense and in conservation areas. Unfavorable ecological and economical phenomena, as observed in the steppe zone in the late 1980s, acquired the character of a challenge to the late Soviet system. The USSR did not have time to adequately answer this challenge and handed these problems to Russia and the post-Soviet states.

The system crisis that emerged rooted in the virgin lands campaign of 1954–1963. The characteristic feature of this campaign was that it automatically transformed the stage of ploughing up the land into a stage of governmental support of the applied land use system: Rather large quantities of grain widely dispersed over huge fields could be harvested only with the help of hands from outside, which did not require adequate reward (such as students, soldiers, freed prisoners, etc.) This problem was aggravated by the circumstance that most of the virgin areas were ploughed up in just 3 years (1954–1956) and this was followed by the additional ploughing up of millions of potentially low-productive hectares instead of providing the necessary facilities for the territories already taken into cultivation. Clearly this campaign was not well considered, and it was carried out in a frantic rush similar to wartime combat. Therefore, instead of using the steppe land in accordance with its natural conditions, which in the south-east could have been achieved by gradually evolving the rural economy, an a-priory unsustainable system was built by administrations and this was permanently supported by the national government. The system was evidently unviable without this support. The state itself became a hostage of this system, socially and economically, because the system involved millions of people, tens of millions of hectares of agrarian lands and some entire regions of the country.

The late Soviet approach to land management itself manifests the dependence of the government of the system. The national economic planning went beyond the objective potential of the lands. One of the most important tasks within the Soviet land management policy was turning non-used lands into an instrument of rural economy. If the properties of a piece of land did not allow it to give a certain economic return, then this land should be surely transformed and its properties should be improved (Kuznetsov and Proshlyakov 1977). Thus, in declaring the necessity to increase crop production capacity, the government virtually considered the increase of gross agricultural production output impossible without the involvement of new lands. No arguments to conserve pieces of poor land or take low-productive ploughed lands out of production were heard. Plans with permanently increasing gross grain yield levels required legislative confirmation that ploughing further steppe area was a priority, and it stimulated the permanent search for reserves to be turned into ploughed hectares. The inviolable ploughed field is the original product of the Soviet epoch, a monument of Soviet land management, which now successfully stifles the development of a sustainable rural economy.

19.4 The Post-Soviet Years

The weather conditions of the first post-Soviet years, 1992–1994, were favorable for crops, especially in the south-eastern periphery of the agricultural zone. During several years, high gross yields of grain were successfully taken in, also thanks to the stock of Soviet farming machines. At the same time, very inconsequent and contradictory agrarian reforms were implemented under the absence of a common concept of agrarian reform in Russia. In the non-black soil zone processes that were destructive for the rural economy and unordered fallowing of ploughed fields began in the late Soviet period and accelerated briefly during the first post-Soviet years (Luri et al. 2007). The degradation of the rural economy in the historical center of Russia started to acquire a stage that would make a reversion difficult. The intensive encroachment of bushy and woody vegetation on productive agricultural lands in the historical core of the Russian state, which grew for centuries around administrative centers, should be considered a serious agrarian and social challenge.

The decrease in sown area in the steppe zone began after a series of droughts in 1995–1996 and after the Soviet farming machines got worn and torn. The extent of the sown areas of farms became to a substantial degree determined by their opportunities to obtain fuel and lubricants, especially on favorable terms. After the drought of 1998, when farms lost their seed resources, governments all over all the post-Soviet countries raised the question of the transformation of ploughed fields to the interests of meat livestock raising. For example, in the Orenburgskaya oblast a governmental program of meat livestock raising was developed. But it was not followed by real activities because the government virtually abandoned the control of the rural sector.

However, since 2000 Russia supports a rural economy without a principal revision of land management and land use practices. The reincarnation of crop priority in the steppe zone encouraged the re-ploughing of fallow lands. While annual re-ploughing became an important criterion for rural economical reconstruction, earlier governmental programs of transformation of ploughed fields into pastures or reserves were virtually annulled. In 2009, after a series of favorable agricultural years, the Crop Forum in St Petersburg discussed how to achieve Russian leadership in crop export again by, amongst others, returning to plough fallow lands. Virtually, it amounted to a new virgin lands campaign being started with governmental financial support. On the other hand, with intensive governmental control, the reporting system on ploughed and sown hectares seems more acceptable and controllable than one based on indices of efficiency and profitability. While in the Soviet period the government directed the rural economy in such details as the "agronomist-in-chief", now regional executive authorities, applying the instruments of economical influence, rather than a land user himself, determine the specialization of a farm. There is still no free agrarian market and an agrarian reform is realized only partly.

Despite the rather liberal Russian Federation Land Code (The Russian Federation Land Code 2010), the real transformation of a ploughed field into a pasture virtually requires an official permission by an executive authority, which depends on agricultural lobbies and agricultural heavyweights, the main apologists of the radical crop approach. Consequently, the position of steppe scientists is weakened under this practice; it is virtually impossible to convince the agricultural lobby that ecological limitations should be respected and that the land use structure should be optimized. Degradation of the soil, which concerns the very base of agriculture, is the only argument taken into consideration when deciding on the transformation of ploughed fields into pastures. And based on this criterion, already in the late 1980s and early 1990s Russian scientists developed a theoretical base for a series of approaches to agrarian landscape optimization and to adapt agricultural systems to landscape features (Mirkin 1994; The scientific and technical bulletin... 2002; Volodin 2000).

In the Orenburgskaya oblast, one of the most ploughed up Russian regions, at least four concepts relating to the transformation of low-productive ploughed fields were developed. Klimentiev (1997; Klimentiev and Tikhonov 2001) based his arability criterion on the rate of soil formation to soil erosion; where erosion prevailed the land was considered non-arable. He estimated that more than 1.2 million ha of such lands, mainly in sloping terrain, were ploughed.

Rusanov (1998, 2003) assessed arability on the basis of its suitability to be ploughed. Lands were classified as suitable, damaged, or degraded. The latter category, considered non-arable, comprises 694,700 ha in the Orenburgskaya oblast. Rusanov's system is officially accepted.

Both authors consider loamy carbonate-alkaline dark chestnut soils to be sufficiently suitable for ploughing. However, their criteria did not include an economical assessment of the biological potential for agricultural production of the land. This limits the applicability of their systems largely to regions where the land value is high and the expenses for contouring the land (either straight or curvilinear) are easily repaid, i.e. conditions as found in the central black-soil regions of Russia and Ukraine.

The landscape-based criteria for the optimization of steppe land use, as developed in the late 1980s by Chibilyov (1992, 2003) showed that in the Orenburgskaya oblast low-productive ploughed fields increased from 10% in the typical black soil subzone to 30% in the dark chestnut soil subzone, and that at least 1.8 million ha of ploughed fields should be transformed into hayfields and pastures.

In 2000–2005, within the framework of the project "Conservation of Russian steppe biodiversity for sustainable agriculture", the Institute of the Steppe of the Urals Branch of the Russian Academy of Sciences elaborated new approaches to the ecological and economical assessment of the biological potential for agriculture of land resources in the steppe in view of the developing agrarian land market in Russia (Chibilyov et al. 2003; Levykin et al. 2005). 'Possible capitalized income' was an important criterion to determine optimal land use. Agriculture, animal husbandry, and ecosystem services were considered; soils were assessed on the basis of a soil ecological index, the biological potential for crop production, and the potential fodder productivity of the land. The data were evaluated in a cost-benefit balance assuming certain crop prices (Levykin 2006). It turned out that if the biological potential for crop production is less than 1,000–1,200 kg per ha dry farming is unprofitable even on soils in steppe watersheds.

Practical implementation of the approach, in the subzone of the southern black soils, proved that it is hardly effective, mainly because agricultural farming with curvilinear contours, that looks so easy on paper, proved cumbersome. Straight-contoured fields, the basis of the agrarian landscape structure, seem appropriate for the agrarian landscapes of the steppe (see also Nikolayev 1999).

19.5 Coping with the Future

The problem of low-productive ploughed fields aggravated due to recent climate change. The steppe regions of Zavolzhia and the Southern Urals have undergone disastrous June-droughts in 2006, 2009, 2010. In 2010 the Orenburg hydrometeocenter reported that in the south of the Orenburgskaya oblast the summer maximum was more than 5°C above average. Over 1.8 million ha of spring-sown fields were written off, and gross grain yield was 0.6 million metric tons, which was 5 times less than expected. The steppe land use system proved itself to be completely unready to climate change. This furthered the aggravation of the agrarian landscape crisis, which presently is a challenge in virtually the entire steppe region in the southeastern periphery of the Russian agricultural zone. This challenge can be described in its four constituents: economic structure, agroeconomics, soil resources, and ecological constraints.

1. **Economic structure**. The late Soviet centralized structure of the agrarian economy is still supported despite market requirements. There is no class of effective

land owners, no agrarian land market, no market price for agrarian land. The lobby of crop and livestock industries blocks the conditions necessary for the development of an adaptive fodder production and an adaptive animal husbandry. Radical agrarian initiatives and new land re-allotments adopt the character of a new virgin lands campaign.

2. **Agroeconomics**. Expenditures to support dry farming, independent of their sizes, are not repaid. Under a changing climate, dry farming becomes especially risky and partly resembles gambling. It is not clear what the minimum crop capacity for cost-effective crop farming is.
3. **Soil resources**. Despite the local application of "resource saving technologies", the crop sector in the South-East is extremely land-demanding and soil-wasting. The most important means of agrarian production are being destroyed.
4. **Ecological constraints**. Under the new virgin lands campaign, the chances that key biological objects of the steppe are conserved or restored decreased below the critical level. Sustainable development becomes problematic. The activities of the extensive crop sector accelerate soil erosion, and this in turn increases the greenhouse effect.

Presently, we need a governmental declaration affirming that the proper answer to this challenge requires first of all a scientifically based restoration of part of the semi-natural steppe ecosystems. It is necessary to officially declare that the steppe should not only be a crop field but also a sustainable fodder base for meat animal husbandry; it should have conservation areas within the system of Nature Areas of Preferential Protection (NAPP); it provides ecosystem services; it harbors aesthetic, spiritual, moral, and cultural values; it is an object of game economy, and a base for tourism, etc. The opportunity for a complex solution of the problem is not yet lost: through paying our debts to steppe nature by restoring groundlessly destroyed landscapes we can diversify steppe land use. Leading world experts in nature conservation suggest that 10% is the necessary minimum area to conserve the ecosystems of a natural zone. We consider that conservation embedded in a rational steppe land use should be based on the principle, which we named *caespesarium* (from Latin *caespes* – sod). This form of land use allows and promotes sod restoration and sustainable use of the steppe through regulated management of the vegetation and other bioresources.

In all post-Soviet countries, the system of NAPP in the steppe zone still does not concentrate on typical, zonal steppe ecosystems and landscapes, but mainly covers intrazonal elements, i.e. ecosystems and landscape elements that are restricted to special habitats, such as saline or rocky or deep-sandy soils, or lakes and ponds in the steppe zone. At the same time, old fallow lands, which are the main and almost only lands on which key steppe biological objects (taxa or ecosystem components) can survive climate change, are intensively re-ploughed. In this new virgin lands campaign, in the years 2008–2010, we observed this on a large scale in the Orenburgskaya oblast and adjacent oblasts in Kazakhstan. If this continues unlimited, and with the rapid northward expansion of a drier climate, several of the typical steppe biological objects can become extinct, because the steppe climate will persist

Photo 19.1 Secondary steppe regenerated in the period 1996–2011 on fallow land, Orenburgskaya oblast, Poduralskoye plateau, June 2010 (Photo by S.V. Levykin)

in the zone where the potential of steppe restoration is completely gone, while the remaining patches of zonal steppe will be under a desert climate. Therefore, now it is already necessary to form a system of representative steppe genetic reserves based on virgin and secondary steppes over the entire steppe area. Furthermore, Dzybov (2010) has shown that within European Russia the agrosteppe method suffices to preserve steppe species, but in Zavolzhia and Kazakhstan it is necessary to prevent the further ploughing up of old fallow lands as they are secondary steppes in their ecology (Photo 19.1). Prohibitions are not sufficient. The experience with the existing steppe NAPPs shows that successful steppe conservation and restoration is possible only by managing the annual phytomass production. This makes it possible to use the natural features of steppe production in a favorable combination of steppe conservation and steppe use (Levykin and Kazachkov 2008).

Based on present knowledge on the specific requirements of steppes conservation and the increasing risks of dry farming, we think that the challenge of degradation of the agrarian landscape can be adequately answered by a consistent implementation of the restorative and adaptive land management concept in steppe land management; management should be directed at the restoration of semi-natural, grass-based ecosystems on extensive stretches of land, and they should be used for scientifically based, sustainable, rotational grazing and haying systems, instead of (curvilinear) contour ploughing.

The payback time in pasture animal husbandry is longer than in agriculture or in livestock industry, and this is why pasture animal husbandry is less attractive to private investors. However, pasture animal husbandry on natural steppe fodder provides meat and dairy products of unique quality, which are far better than the products of the livestock industry. Under proper marketing, these products of high quality can be sold at higher prices. Pasture and industrial livestock products do not fiercely compete because, in a civilized market where the customer has a choice, they serve different customers. The government should support such developments in the steppe region. Kazakhstan produces the unique "marmoreal" meat that enjoyed a great demand before large stretches of virgin lands in the steppe regions of Zavolzhia and Northern Kazakhstan were ploughed up. After the virgin lands campaign, the rural economy of these regions was determined by crop production, the sustainability of which is questionable. We find an example in the USA, where a stable demand for meat grown on natural grasses was created. The main source of this product is American bison, which was on the verge of extinction. As result of the demand for ecologically pure products, the number of American bison grew during the last 15 years 10-fold and reached or even exceeded 500,000 (Gates et al. 2010; Redford and Fearn 2007; The National Bison Association 2010).

Russian steppe regions have a good potential for animal husbandry pasturing, but this potential cannot be realized until a sustainable fodder source based on grass ecosystems is in place (Dzybov 2010; Trofimov et al. 2010). The problem of a sustainable fodder base can be solved with restorative and adaptive land management. Thus, the two agrarian paradigms of steppe land use should be revised.

First of all, the forest shelter belt principle in agrarian landscape organization should be exchanged for the grass shelter belt principle. We observed forest shelter belts in the steppic South-east, grown with high expenses, but nowadays in poor conditions and virtually not performing its expected functions. An alternative network of steppe grass shelter belts, that virtually does not require any care, could for indefinitely long periods of time support ecological corridors and thus optimally promote the sustainability of steppe biodiversity, and it could serve as sanitary barriers against spreading of vermin, and as hayland.

In the second place, steppe land management should abandon the "inviolable ploughed field" paradigm and the permanent search for new lands for crops. The steppic South-east needs new land management planning on the district level, directed with priority at a rural economy of adaptive meat animal husbandry, aimed at forming a south-eastern "meat belt" (as has been proposed many times by economical and political figures of our country). In the Orenburgskaya oblast this belt should embrace its southern and south-eastern districts as well as the districts Svetlinskiy, Yasnenskiy and Dombarovskiy in the east. This would form a single economical territorial complex of over 900,000 ha (Fig. 19.1).

A network of collaborating plant melioration stations and seed farms is required for an effective restoration of the fodder base for an adaptive animal husbandry. But a stable sale of primary products is also necessary and requires the development of meat processing enterprises within the "meat belt" as well as the establishment of droves. The proximity to Kazakhstan of existing and future livestock processing

Fig. 19.1 The "meat belt" in the Orenburgskaya oblast

centers in and near the "meat belt" of the Orenburgskaya oblast could be used to process livestock grown in adjacent districts in Kazakhstan.

In addition to the problem of a sustainable rural economy, the steppe NAPP system needs modernization. The present system is not adequate given the specific features of existing steppe ecosystems and also does not cover the most typical varieties of steppes (Chernova 2010). Steppe conservation has always been running behind steppe ploughing; intrinsically so. In practice, only the remnants left after the agricultural colonization of flatland and other plain steppes, in other words the agrarian badlands, were declared conservation areas.

The global character and severity of the disastrous disappearance of the steppes of the Northern Hemisphere yet attracted, with a delay of 15 years, the attention of the important international conservation organizations. So, the UNDP and the Global Environment Facility (GEF) approved in 2009 and started in 2010 the project "Improving the Coverage and Management Efficiency of Protected Areas in the Steppe Biome of Russia". This program was started because, despite the evident economical growth and financial achievements by Russia, steppe land use problems still remain to be solved virtually at the level of developing countries. The severity of the steppe problem in the Northern Hemisphere was such that it was included in the priorities of the most authoritative international conservation organization, the IUCN. In 2010 this organization promoted a series of large international conferences on contemporary steppe problems, and a thematic group of steppe specialists, including Russian scientists and conservationists, was formed (Proceedings Mongolia 2010). Explicitly including the Russian steppe problems in an international declaration is in itself, of course, not a solution but it may serve as an additional argument for a scientifically based approach towards developing schemes for the optimal land use of the steppe and their practical implementation.

The efforts by the scientific and conservation communities presently have, to a certain extent, got attention at the regional level. For example, the Government of the Orenburgskaya oblast has seriously thought about the raised issues concerning the streamlining of steppe land use in the oblast. Firstly, the regional government admitted that, in the oblast, a series of unsolved steppe land use problems indeed remains, and that these problems stifle the conservation and the restoration of steppe ecosystems. Secondly, the Ministry of agriculture, food and processing industries of the Orenburgskaya oblast offered in turn concrete directions for the optimization of the agrarian land structure: the decrease of lands officially considered arable by the transformation of all low-productive plots into haylands and pastures; the sowing of plots that are being transformed with drought-resistant perennial herbs and herb mixtures; the maintenance of crop priority on only the most fertile lands; the creation of a land reserve for the restoration of land quality apart from the large araes of land that remain to be used and managed according to the directed purpose (crop production, pasture animal husbandry); the development of afforestation melioration. The official declaration of this intention to consider the above-mentioned problems of steppe land use and offer ways of solving these within the target programmes for the development of the Orenburgskaya oblast evokes a certain optimism.

Acknowledgment This work is supported by the Urals Branch of the Russian Academy of Sciences, project # 09-T-5-1027 "The geographical premises of sustainable land use in mountain and subarid (steppe) regions of the Urals and Zavolzhia".

References

Berdyshev AP, Bolotov AT (1988) Andrey Timopheevich Bolotov [Андрей Тимофеевич Болотов]. Agropromizdat, Moscow, 143 pp [Бердышев А.П. Андрей Тимофеевич Болотов. – М.: Агропромиздат, 1988. – 143 с.]

Chernova OV (2010) The increase of natural soil representativeness as the mechanism of Russian protected nature territories system optimization and development. 4(112):53–56 [Чернова О.В. Повышение репрезентативности естественных почв как механизм оптимизации и развития системы охраняемых территорий России//Использование и охрана природных ресурсов России. – 2010. – Вып. 4(112). – С. The use and conservation of natural resources in Russia [Использование и охрана природных ресурсов в России], 53–56.]

Chibilyov AA (1992) The ecological optimization of steppe landscapes. Yekaterinburg, Nauka, 172 pp [Чибилёв, А. А. Экологическая оптимизация степных ландшафтов/А. А. Чибилев. – Екатеринбург: Наука, 1992. – 172 с.]

Chibilyov AA, Levykin SV, Akhmetov RSh (2003) Geoecological aspects of sustainable rural economy modeling for Zavolzhia-Urals steppe subregion. Geography and environment. St. Petersburg, pp 86–94 [Чибилев А.А. и др. Геоэкологические аспекты создания модели устойчивого сельского хозяйства в Заволжско- Уральском степном субрегионе/ А.А.Чибилев, С.В.Левыкин, Р.Ш.Ахметов//География и окружающая среда.- СПб, 2003.- С.86– 94.]

Collected statistics about changes in population, cattle and harvests over Kazakh SSR from 1880 to 1922 (1925) Orenburg, 5 state print shop of Orenpoligraph, 97 pp [Сборник статистических сведений о движении населения, скота и урожаев по КССР с 1880 по 1922 г. – Оренбург, 5-я гос. тип-я Оренполигр., 1925. – 97 с.]

Dzybov DS (2010) Agrosteppes. Agrus, Stavropol. 256 pp [Дзыбов Д.С. Агростепи./Д. С. Дзыбов. – Ставрополь: АГРУС, 2010. – 256 с.]

Formozov AN (1962) Changes in natural conditions over the steppic south of USSR during the last 100 years and some features of modern steppe fauna. Researches in the geography of natural resources of plant and animal kingdoms (for the 60th anniversary of A.N. Formozov). Geographical institute of the RAS, Moscow, pp 114–160 [Формозов А.Н. Изменения природных условий степного юга Европейской части СССР за последние сто лет и некоторые черты современной фауны степей//Исследования географии природных ресурсов животного и растительного мира (к 60-л со дн. рожд. А.Н. Формозова). – М.: ИГ РАН, 1962. – с. 114–160.]

Gates CC, Freese CH, Gogan PJP, Kotzman M (eds) (2010) American bison: status survey and conservation guidelines. IUCN, Gland, 134 pp

Klimentiev AI (1997) Soil ecological criteria for steppe land use. Urals branch of the Russian Academy of Sciences, Yekaterinburg, 248 pp [Климентьев, А.И. 1997. Почвенно-экологические основы степного землепользования. Екатеринбург, УрО РАН. 248 с.]

Klimentiev AI, Tikhonov VYe (2001) The ecological and hydrological analysis of agrolandscape resistance to soil erosion. Soil Sci 6:756–766 [Климентьев А.И., Тихонов В.Е. Эколого-гидрологический анализ эрозионной устойчивости агроландшафтов//Почвоведение. – 2001. - №6. – с. 756–766.]

Kuznetsov GA, Proshlyakov VP (eds) (1977) Land management and rational land use. Kolos, Moscow, 247 pp [Землеустройство и рациональное использование земли. М.: Колос, 1977. – 247 с.]

Levykin SV (2006) The theory of agroecosystems land resources management based on the steppe landscape- and biodiversity conservation and rehabilitation. Author's abstract of the geography doctoral dissertation. Astrakhan, 2006, 34 pp [Левыкин С.В. Теория управления земельными ресурсами агроэкосистем на основе сохранения и реабилитации ландшафтно-биологического разнообразия степей. - Автореферат дисс. на соиск. уч. Степени доктора географических наук. – Астрахань, 2006. – 34 с.]

Levykin SV, Kazachkov GV (2008) The resource restorative approach in the theory of modern steppe science. Povolzhskiy ecologicheskiy zhurnal 4:379–385 [Левыкин С.В., Казачков Г.В. Ресурсовосстановительный подход в теории современного степеведения// Поволжский экологический журнал. – 2008. – Вып.4. – с. 379–385.]

Levykin SV et al (2005) The land: how to value an invalue. In: Levykin (ed) The land: how to value an invalue. Methodical approaches to economical evaluation of the biological potential of land resources in the steppe zone. [Земля: как оценить бесценное. Методические подходы к экономической оценке биопотенциала земельных ресурсов степной зоны]/С.В.Левыкин [и др.]; под общ. ред. С. В. Левыкина. – Новосибирск: Сиб. экол. центр, 2005. – 170 с.]

Luri DI, Nefedova TG, Konyushkov BD (2007) Russian agrarian lands: mythes and reality. Ecologicheskoye planirovaniye i upravleniye 2(3):54–63 [Люри Д.И., Нефёдова Т.Г., Конюшков Б.Д. Сельскохозяйственные земли России: мифы и реальность//Экологическое планирование и управление. – 2007. – № 2(3). – с. 54–63.]

Mirkin BM (1994) Sustainable agrosystems: dream or reality? Priroda 10:53–62 [Миркин Б.М. Устойчивые агросистемы: мечта или реальность?/Природа. – 1994. - № 10. – с. 53–62.]

Nikolayev VA (1999) The adaptive spatial and temporal organisation of agrarian landscapes. Vestnik MGU, Series 5. Geographies 1, pp 22–26 [Николаев В.А. Адаптивная пространственно-временная организация агроландшафта//Вестник Московского университета, сер. 5. География. – 1999. - №1. – с.22-26.]

Perelman AI (1956) The history of atoms and geography. State publishing house for geographical literary, Moscow, p 61 [Перельман А.И. История атомов и география. – М.: Гос. изд-во геогр. Лит-ры, 1956. – с. 61.]

Proceeding of the international conference on "Eurasian steppes: status threats and adaptation to climate change". 2010. Mongolia, 9–12 Sept 2010, Hustai National Park, Mongolia

Pryanishnikov DN (1965) The popular agrochemistry. Nauka, Moscow, 397 pp [Прянишников Д.Н. Популярная агрохимия. – М.: Наука, 1965. – 397 с.]

Redford KH, Fearn E (eds) (2007) Ecological future of bison in North America: a report from a multi-stakeholder, transboundary meeting. American Bison Society working paper no. 1, 64 pp

Rusanov AM (1998) The basis of the arability concept. In: Rusanov AM, Kononov VN (eds) The optimization of nature use and conservation in Southern Urals: materials of Russian theoretical and practice conference. Orenburg, pp 70–73 [Русанов А.М. Основные положения концепции пахотнопригодности земель/А.М.Русанов, В.М.Кононов//Оптимизация природопользования и охрана окружающей среды Южно-Уральского региона : материалы Рос. научно-практ. конф. – Оренбург, 1998. – С. 70–73.]

Rusanov AM (2003) The land arability concept: substance and role. Economical, legal and ecological problems of land use under market economy of Russia and CIS (methods, theory and practice of management): materials of international theoretical and practice conference, vol 2. Orenburg agrarian state university, Orenburg, pp 114–118 [Русанов А.М. Концепция пахотопригодности земель: содержание и значение//Экономико-правовые и экологические проблемы землепользования в условиях рыночной экономики России и стран СНГ (методология, теория и практика хозяйствования). - Материалы международной научно-практической конференции. – Ч.2.- Оренбург: ОГАУ, 2003. – с. 114–118.]

Sovremennik (1992) Sowers and protectors. In two volumes, vol 1. Sovremennik, Moscow, 415 pp [Сеятели и хранители. В двух книгах. Кн.1. – М.: Современник, 1992. – 415 с.]

The Handbook of the Russian Farmer (1993) Priboy, Moscow, 704 pp [Настольная книга русского земледельца. – М.: АО«Прибой», 1993. – 704 с.]

The National Bison Association (2010) URL: http://www.bisoncentral.com/index.php?s=&c=63&d=64&a=1017&w=2&r=Y. Accessed 19 Apr 2010

The Russian Federation Land Code//Consultant plus (2010) URL: http://base.consultant.ru/cons/cgi/online.cgi?req=doc;base=LAW;n=102874. Accessed 1 Nov 2010

The scientific and technical bulletin on the problem "The optimization of agrolandscapes and of landscape adaptive agriculture systems" (2002) Issue 1(70), Russian research institute for agriculture and soil protection from erosion. Kursk, 88 pp [Научно-технический бюллетень по проблеме«Оптимизация агроландшафтов и адаптивно-ландшафтных систем земледелия», Вып. 1(70)./ВНИИ Институт земледелия и защиты почв от эрозии. - Курск, 2002. – 88 с.]

The State Archive of Orenburgskaya oblast. Fund 164, list 1, unit of storage 125 . The statistics of land use in Orenburgskaya province (1895) [Государственный архив Оренбургской области. Фонд 164, опись 1, ед. хр. 125б. Статистические сведения о землепользовании в Оренбургской губернии. 1895 г.]

Trofimov IA, Trifimova LS, Yakovleva EP (2010) Grass ecosystems in the Russian rural economy. Ispolzovanie i okhrana prirodnykh resursov v Rossii 4 (112):37–40 [Трофимов И.А., Трофимова Л.С., Яковлева Е.П. Травяные экосистемы в сельском хозяйстве России//Использование и охрана природных ресурсов в России. – 2010 г. – Вып. 4(112). – с.37-40]

Volodin VM (2000) The ecological grounds of soil fertility assessment and the soil fertility use. Russian research institute for agriculture and soil protection from erosion. Moscow, 334 pp [Володин В.М. Экологические основы оценки и использования плодородия почв/ВНИИ Институт земледелия и защиты почв от эрозии. – М. – 334 с.]

Chapter 20
Rethinking Pastoral Risk Management in Mongolia

Stephan Baas, Savina Tessitore, and Tessa Jelley

Abstract Risk, in particular climate risk, is inherent in extensive pastoralism, mainly because it is carried out in environmental conditions where other forms of agriculture are not suitable. Different factors within environmental and socio-economic systems compound each other, affecting herders' exposure to risks and their capacity to cope with and recover from shocks and stresses.

In Mongolia, the 2009/2010 zud revealed the limited coping capacity of herders and institutions, and insufficient preparedness at all levels, questioning the effectiveness of the current disaster risk reduction and management systems. This article reviews Mongolia's current pastoral risk management (PRM) framework, assessing its strengths and weaknesses. It then discusses opportunities to strengthen the PRM framework by better linking disaster risk reduction, climate change adaptation and social protection in order to tackle vulnerability from different angles and with a long-term perspective. Opportunities to strengthen social protection and livelihood diversification correspond to a plurality of pathways for pastoralist livelihoods which sees them diversifying, becoming more commercialized, or for some herders even moving away from the pastoral system through well-prepared exit strategies. The proactive use of new opportunities which open up because of climate change are also discussed, such as funding for PRM through proactive soil carbon sequestration projects, as a necessary complement to existing PRM actions.

This paper concludes that the basic know how and technology options for PRM are already known. It is essential though, to consolidate and replicate what has already been tested successfully. Proactive measures are needed to ensure the

Background paper prepared for the IUCN Conference on "Eurasian Steppes: Status, Threats and Adaptation to Climate Change" Hustai National Park, Mongolia, September 9–12, 2010.

S. Baas (✉) • S. Tessitore • T. Jelley
Climate, Energy and Tenure Division FAO, Rome, Italy
e-mail: Stephan.Baas@fao.org; S.Tessitore@alumni.ids.ac.uk; tessatess@hotmail.com

M.J.A. Werger and M.A. van Staalduinen (eds.), *Eurasian Steppes. Ecological Problems and Livelihoods in a Changing World*, Plant and Vegetation 6,
DOI 10.1007/978-94-007-3886-7_20, © Springer Science+Business Media B.V. 2012

necessary outreach, and provide more sustainable capacity building and policy support to disaster risk reduction, hazard preparedness and livelihoods diversification measures.

Glossary

Aimag Primary administrative unit: "province".
Bag Lowest level of the administrative hierarchy, rural "sub-districts".
Bod Cattle-based animal unit. 1 bod = 1 horse/1 bovine/0.8 camels/8 sheep/8 goats. The definition is not standardized and may vary slightly by user.
Bog Sheep-based animal unit.
Carbon sequestration "the process of removing carbon from the atmosphere and depositing it in a reservoir"; (UNFCCC 2010)
GDP Gross Domestic Product
Khot Ail society formed of customary groups of two to ten cooperating households. "Fluid group of herding households that cooperate in livestock and pasture management notably to take advantage of economies of scale"; (Mearns 2004: 133)
Khural Representative assembly
NEMA National Emergency Management Agency
Otor Movement of livestock to make use of distant pasture to escape drought or zud or to prepare animals for winter.
PRM Pastoral Risk Management
Social protection includes "all initiatives that transfer income or assets to the poor, protect the vulnerable against livelihood risk and enhance the social status and rights of the marginalised"; (Devereux and Sabates-Wheeler 2004)
Sum Administrative unit below province, rural "district".
Zud A winter disaster that affects the welfare and food security of pastoral communities through large-scale debilitation and death of livestock, caused by a variable combination of summer drought, heavy winter snow, lower than normal temperatures, drifting windstorm and a dangerous spring thaw. Herders' main criteria to define it are not so much meteorological conditions as its consequences, particularly in terms of animal mortality.

20.1 Introduction

The dry summer of 2009 in Mongolia was a prelude to what is locally known as a *zud*, a natural disaster triggered by a complex series of extreme meteorological conditions and causing high rates of livestock mortality. The *zud* unfolded in the course of the following winter and spring, as heavy snowfalls, extreme cold, spring thaw and successive freezing of the snow cover preventing access to grazing (Photos 20.1 and 20.2) compounded each other in causing the death of animals already weakened

Photo 20.1 Animals exposed to snow and extreme cold; no fodder (Photo Ovorhangay Province Office 2010)

Photo 20.2 Icy snow cover prohibits access to standing hay (Photo Jim Suttie)

by the summer drought, and putting herders' food security and livelihoods at severe risk. By May 2010, 7.8 million head were recorded to be dead, with long-term consequences on the 769,106 people affected, or 28% of the population, for whom livestock is the major source of income and food. Hardest hit were 8,711 households who lost all their animals (Photos 20.3 and 20.4).

Photo 20.3 Herders camp during Tiger zud 2010 (Photo Ovorhangay Province Office)

Photo 20.4 High animal losses during Tiger zud 2010 (Photo David Hadrill)

Risk, and in particular climate risk, is inherent in extensive pastoralism, mainly because this type of livestock rearing is carried out in environmental conditions where other forms of agriculture are not suitable. Different factors within dynamic environmental and socio-economic systems compound each other, affecting herders' exposure to risks as well as their capacity to cope with and recover from shocks and long-term stresses. Pastoralists are regularly exposed to various inter-related types of environmental and socio- economic risks, including political and institutional ones (Baas et al. 2001). The high risk exposure and the central role of pastoralism in Mongolian culture has kept pastoral risk management high on Mongolia's policy agenda.

However, in the 1990s changed framework conditions caused by Mongolia's transition to a market economy called for the development of a new strategy for risk management. The preparation of this new strategy was promoted by two FAO Technical Cooperation Projects (TCPs) in 1996–2003. Linked to a string of hazard (*zud*) events between 1999 and 2001, the projects proposed the implementation of an articulated pastoral risk management (PRM) framework, which included a variety of capacity building measures and technical interventions as part of a broader strategy to reduce vulnerability to the risk of natural disasters in Mongolia. PRM was then taken forward and further fine-tuned by many actors, spearheaded by the government itself with assistance from the World Bank, UNDP and other donors. A number of the components of PRM are presently mainstreamed into the Government's livestock, land tenure, desertification, decentralization and other policies.

The evolution of the 2009/2010 crisis, with its heavy toll on households and high animal losses, however, revealed limited coping capacity of herders and institutions, and insufficient preparedness at all levels. It signalled shortcomings in hazard preparedness and contingency planning, as well as in prevention and emergency measures, which are known to be effective, and which should have been routinely checked and as a matter of course ready for activation when needed. The apparent limited capacity of these mechanisms to mitigate the full-blown consequences of the 2009/2010 drought and *zud* events questions the effectiveness of the original framework and the way it was implemented.

20.2 Scope for Enhancing the Pastoral Risk Management Framework

This article will review the PRM framework and try to assess the strengths and weaknesses of the original design. Harder to gauge precisely, on the other hand, is the progress and impact of the many related projects and programmes which, throughout the last 10 years, have built on the seminal experience of the two FAO TCPs. In this sense, three considerations are worth making: in the first place, an important constraint to full effectiveness of measures is the timeframe over which the different components were implemented. Probably more time is necessary for many of the new arrangements to take root, especially when the direct involvement and active participation of households and communities is a precondition to their effectiveness, as is the case of measures regulating access to natural resources or land tenure. In the second place, though found effective by initial evaluations, some components are only now being scaled up by the Government of Mongolia, and this presently limits their potential mitigating and recovery capacity.[1] Finally, external changes due to global geopolitical, environmental and economic trends, and notably the recent financial crisis, have influenced national and local scenarios in ways that could not be accounted for at the time, with consequences on resilience, coping modalities and capacity to respond to extreme weather conditions and other risks.

Predicted climate change scenarios throughout the country's regions require a broadening of the strategic PRM framework to include impacts of global warming on ecosystems more firmly in the picture. Not only do different factors within dynamic environmental and socio-economic systems make livestock-based livelihoods increasingly vulnerable, the very nature of vulnerability is changing, as new challenges emerge. The many uncertainties regarding the actual evolution and magnitude of change of weather-related phenomena, and how they will interact with each other and the wider physical, social and economic contexts, call for novel and flexible responses. Greater resilience and robustness of livelihood systems needs to be built up and solutions that are sustainable for both environmental and pastoral systems, also taking advantage of the positive externalities resulting from a changing climate, need to be envisaged.

In the meantime, reliance on national, regional and global income flows has rendered households' livelihoods more dependent on the expansions and contractions of markets. Environmental degradation, pollution, epizootics, urbanization, competing claims on land and water resources by agriculture and the mining industry represent further challenges to the viability of pastoral systems. To face these and other endogenous risks which have negatively affected the overall coping capacity of systems, an extension and articulation of the timeframe over which measures are implemented is necessary, combining short-term emergency interventions with medium and long term ones, as was in fact the original intent of the PRM strategy. Over the long-term, herder communities' resilience needs to be reinforced by diversifying the portfolio of income-generating activities, within and beyond livestock-based livelihoods. Diversification is critical to relieve the present unsustainable pressure of livestock on the rangeland ecosystem, spread risk, enhance commercial production towards processing and trade, and make the exit from pastoralism an acceptable option for the most vulnerable households.

Opportunities to further enhance the PRM framework in this sense include: (i) better grazing management practices and land use planning to enhance production, (ii) provision of environmental services, stewardship and protection of biodiversity and ecosystems (iii) the expansion of ecotourism services, (iv) the promotion of

[1] For example the innovative Index Based Livestock Insurance Project, supported by the World Bank with multi-donor funding, see the IBLIP web site http://www.iblip.mn/inpro/news.php?vnewsid=120

farming activities, fodder production, value-added livestock processing and quality products and (v) improved access to niche markets for livestock products. These options correspond to a plurality of pathways for pastoralist livelihoods, in the sense of "stepping up" towards a more commercialized livestock-based system, "stepping out" by diversifying activities, or even "moving away" from a pastoral system through well- prepared exit strategies.[2] Many of these opportunities have already been piloted by a number of projects and programmes that agencies have been developing in the past 10 years.

Besides its expected negative impacts, climate change may also create new opportunities: rangelands, for example, at present contain approximately 8% of the world's soil carbon stocks, and with appropriate management this figure has an estimated potential to increase by 6% in temperate regions. Carbon sequestration could benefit households, not only by increasing production potential, but also through the payments envisaged for this kind of programmes (Neely et al. 2009). This paper will therefore also look into carbon sequestration activities as an aspect of an enhanced PRM cycle. By working towards removal of greenhouse gases from the atmosphere on the one hand, and increasing land productivity, producing knock-on environmental benefits and opening up opportunities of tapping into the carbon market on the other, carbon sequestration has a potential to contribute to both adaptation and mitigation. In terms of a framework for reducing pastoral risk, these actions correspond to long-term strategies to enhance livelihood resilience, and to recover from the impact of a disaster by offering alternative livelihood options.

Disaster risk reduction, climate change adaptation and social protection[3] all have the aim of reducing vulnerability, and a more explicit link between these three approaches would allow for a greater integration in design, implementation and targeting of vulnerability-reducing interventions. Linking the disaster risk reduction and climate change adaptation frameworks to social protection is proposed in this paper in order to tackle vulnerability from different ends and with a long-term perspective. Accounting for climate change introduces a longer timeframe in the design of social protection and risk reduction measures, and it highlights the role of social protection interventions to possibly reduce dependence from climate sensitive livelihoods activities in order to maximize their long-term resilience to climate shocks (Davies et al. 2009). Indeed, some of the components of PRM developed over the years correspond to preventive measures to forestall deprivation and promote enhanced income opportunities and capabilities, as envisaged by the social protection framework. Building on these components, the paper will take a further step in the same direction, to indicate ways in which pastoral risk can be prevented and mitigated and alternative livelihoods promoted through social protection instruments.

[2] These scenarios have been proposed as a way forward for pastoralists in the different regional setting of the Greater Horn of Africa by Devereux and Scoones (2007).

[3] Social protection includes "all initiatives that transfer income or assets to the poor, protect the vulnerable against livelihood risk and enhance the social status and rights of the marginalised" (Devereux and Sabates-Wheeler 2004).

Photo 20.5 Unpleasant wake-up after snow storm (Photo Ovorhangay Province Office 2010)

20.3 Pastoralism in Mongolia

Mongolia is a landlocked country with a territory of 1.56 million km^2 making it one of the least densely populated countries in the world, with 2.76 million inhabitants, 57% of which are settled in urban areas. It is located in the temperate steppe zone of Central Asia, with the Gobi desert traversing it along the Southern border with China, and many mountain ranges throughout the country. It is generally characterised by an extremely cold and dry winter, a dry cold and windy spring (Photo 20.5) and short summers. Overall precipitation is low. Most of it falls in summer, between mid-June to the end of August (Johnson et al. 2006; Dagvadorj et al. 2009). Extreme weather is characteristic of Mongolian seasons. Although lakes, streams and rivers are plentiful in the Northern part of the country, they are concentrated in this region, and almost 65% of the territory as a whole has no open water sources.

Mongolia's six distinct rangeland ecosystems – alpine tundra, forest-steppes, swamp steppes, grass steppes, desert steppes and desert – present different rainfall distribution, soils, elevation, temperatures and vegetation patterns, though extended drought and winter snow or ice storms can occur in all (Fig. 20.1).

About 82% of its total land area is made up of rangeland (FAO 2007), and 76% is considered to have potential for agricultural – primarily extensive pastoralist – production (Johnson et al. 2006).

The rangelands stretch through a continuum from moister equilibrium environments to the non-equilibrium ones of the desert zones (Swift and Baas 1999). The ecological variation induces a vegetation gradient with big differences between

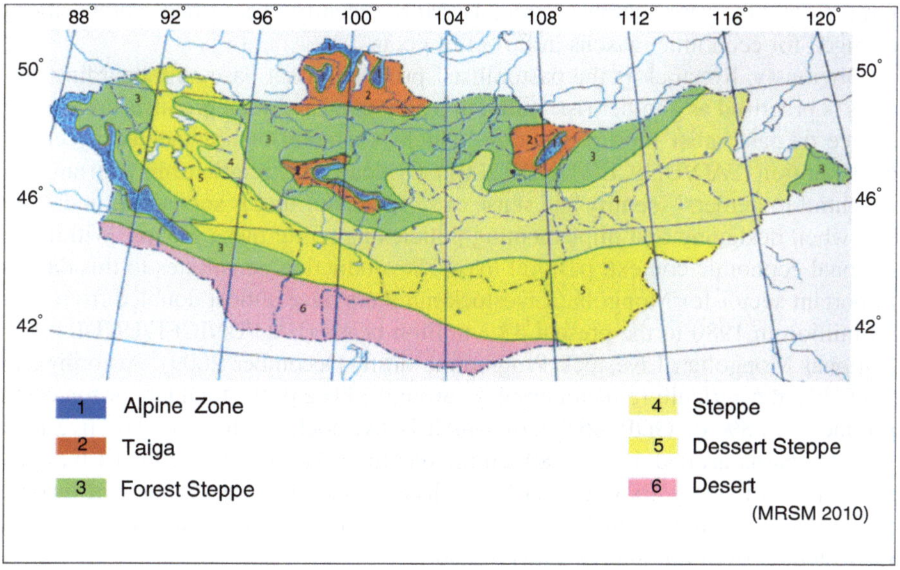

Fig. 20.1 Mongolia's ecological zones

North and South and widely differing rangeland productivity. In line with the ecological regions, Johnson et al. (2006) classified six main vegetation zones: (i) Alpine Tundra (3% of total area) (ii) Mountain Taiga, (4.1%) (iii) Mountain Steppe and Forest-Steppe (25.1%) (iv) Grass Steppe 26.1%, (v) Desert Steppe (27.2%) and (vi) Desert (14.5%). Large inter-annual and seasonal changes in temperature and rainfall, and location-specific diverse grazing pressures make it difficult to provide operationally useful data on large scale rangeland productivity. Johnson et al. (2006) presented the following ranges of estimated standing crop yields for the following vegetation zones: Alpine Tundra 1,050–1,500 kg/ha; Forest-Steppe: from 1,150 to 1,940 kg/ha; Grass Steppe: 650–1,300 kg/ha; and 290–380 kg/ha in Desert steppes. A more recent study on ecological trends in Desert and Forest-Steppe pastureland (Sheehy et al. 2010), which compares range conditions and productivities at the same sites in the years 1997 and 2008, revealed a significant decline in ecological rangeland condition over the last decade..

The long-standing environmental differences resulted in the development of different models of pastoral mobility in the rangeland landscapes, and distinct grazing management strategies suitable to different combinations of species according to seasonal patterns of vegetation growth. The mobility of herds and flexible exploitation of different patches are key to risk spreading in the face of extreme weather variability. Goats and camels are best adapted to the deserts and desert steppe, and sheep and horses to the steppes. Cattle make up a greater proportion of herds in the steppe and forest-steppe region, and in high mountains, cattle are replaced by yak (Johnson et al. 2006). However, most herders keep multi-species herds in all ecosystems as a rangeland and risk management strategy, while pastureland management requires grazing segregation of animals. More recently overall herd compositions changed for economic reasons in favour of goats.

Obviously, livestock is the pastoralists' principal asset base, and hoarding animals is practised as another risk-mitigation strategy. Livestock products are the key source of pastoralist food production, which is usually complemented by cereal consumption (FAO/UNICEF/UNDP 2007). Critically, the relationship with animals is central to herders' identity and sense of self, a factor that is sometimes undervalued when designing and implementing policies targeting these groups. Within the national economic context, pastoral livestock production constitutes to this day an important sector for Mongolia: Livestock numbers have almost doubled from over 23 million in 1980 to the present 43.3 million heads (FAO/UNICEF/UNDP 2007; National Mongolian Livestock Programme draft, December 2009). According to 2009 data the agriculture sector employs around 34.6% of the total labour force and produces 18.8% of GDP, 86.9% of which is livestock production. The livestock industry earns around 10% of all export income (National Mongolian Livestock Program draft 2009). Export of cashmere has become the largest source of income for herders with more than 300 goats, while the poorest households gained 20% of their income from this source (Kerven 2006).

Grazing is accomplished by cooperation between households specialising in different herding tasks (Mearns 1991). Cooperation is also required to build hay reserves and accomplish other risk reduction and mitigation tasks. The seasonal rangeland productivity determines mobility patterns, and annual and seasonal grazing rotation and emergency reserve pastures are managed accordingly. Migratory movements may be classified by characteristics of the animals moved, duration and seasonality. The forest-steppe environments induce predictable seasonal oscillatory or transhumant *otor* across shorter distances, and inter-annual movements only occur exceptionally. In the wetter steppe and in the mountain steppe regions, movements are more frequent within the same season, while inter-annual movements are needed only in bad years. Finally, in the Southern desert steppe and Gobi desert, non-equilibrium conditions push herders to cross greater distances in search of patch, following more chaotic and opportunistic strategies (Ojima and Chuluun 2008).

Extreme weather is characteristic of Mongolian seasons, and pastoral systems have evolved accordingly, preparing for the harsh and long winter by fattening animals during the short summer season to build resilience and making hay for winter fodder. Consecutive drought, sandstorms, and *zud* events weaken animals progressively, and have ratcheting effects on livestock morbidity and mortality. Particularly devastating for herders' food security and livelihoods are *zuds*, complex events resulting from a series of successive weather conditions such as long-lasting or sudden heavy snowfall, extremely low temperatures and windstorms, leading to high livestock mortality due to hunger, freezing or exhaustion in the following spring, when the consequences of the *zud* are most apparent (Batima et al. 2006). In all areas winter *otor* may be undertaken as an emergency measure to escape *zud* and poor winter pasture (FAO 2003). Successful transition through grazing terrains, while maintaining spatial separation of pastures used in different seasons, calls for negotiations between groups regarding ordinary, seasonal and emergency access to pastureland.

During the socialist period, until the early 1990s, land was collectively owned, and pasture use seems to have been informally regulated by customary norms as well as by formal regulatory institutions (Mearns 2004). Under the centrally planned system, pastoral collectives (*negdels)*, provided services, social security and most importantly, were crucial in assisting herders to cope with and minimize risk by transporting animals between pastures, storing emergency feed reserves, clearing roads from snow, providing veterinary and breeding services, and mechanisms for the insurance and restocking of herds. With the end of collectivisation traditional kin and friendship networks (*khot ails*) resurfaced, partially substituting for the shrinking state presence. They acted as safety-nets to poorer households, pooling risk and coordinating pasture use, though some of the essential services critical to the security and sustainability of their livelihoods previously provided by the state, and notably transport of animals, maintenance of shelters, water sources and strategic fodder reserves, remained overall lacking. Critically, regulation of pasture use and allocation fell through the institutional vacuum created by the retreat of the state, with the deep-set traditional ethic of open access prevailing, to the detriment of sustainability and equality of access to natural rangeland resources. It has been suggested that the present land law, confirming public ownership and free access to most pastureland, while devolving control of time-partitioning and use of seasonal pastures to sub-district (*bag*) level, needs good implementation guidelines and rules, rather than further development (Mearns 2004). Attempts to devise secure collective land tenure arrangements and formalise group membership to coincide with customary use, while ensuring mobility and flexibility of access, have been part of the efforts of projects developing PRM, as described further below.

According to the commonly accepted view, the minimum number of heads to cover food and other basic requirements sustainably is 100. A household with less than 20/30 animals is considered poor, while engagement with the commercial sector begins with 150/200 heads (FAO/UNICEF/UNDP 2007).

20.4 Risk Exposure and Vulnerability of Mongolian Herders

Control over different forms of capital, and capacity to access financial, natural and social resources in the context of the broader social and economic environment, are general determinants of individual household vulnerability and coping capacity. Risks and shocks affect livelihoods differently, according to whether they are covariate, when they hit everybody in a community equally, as with climate-related events or a financial downturn; or idiosyncratic, when events such as illness, death or animal theft affect an individual or a single household. The exposure to recurrent shocks and stresses coupled with the capacity to successfully cope with uncertainties and recover from them, enhance or diminish underlying systemic, household and individual vulnerabilities.

In Mongolia, the most important current risks are snow disasters, or *zuds*. Herders rank them as the most serious risk, since they can kill several million animals in a few weeks. A *zud* is especially dangerous because it creates a covariate risk, affecting all herders in a given area, and making reciprocal help difficult. *Zud* events evolve throughout the year and have a progressive impact on herder livelihoods, as summer droughts prevent proper preparation of animals and feed reserves, and severe winter temperature, windstorms and snow further weaken livestock, who often end up dying in great numbers at the beginning of spring. The slow onset and cumulative nature of *zud* disasters calls for a sequence of preparedness and response measures in answer to its inter-seasonality.

The 2010 *Tiger zud* (Photos 20.1, 20.2, 20.3 and 20.4) was the latest drastic example. NEMA data indicate that almost 9,000 herder families lost all their livestock and around. 33,000 families lost more than half their animals. In total more than 9.2 million livestock perished nation-wide, effectively decimating rural communities with particularly devastating impacts on smaller-scale herders (with fewer than 250 animals) who lost more than 80% of their livestock and face the risk of losing even more animals within early summer. Annex 20.1 provides detailed data on animal losses by province (aimag) and species. In total 22.1% of the national herd was lost. In a normal winter, the average loss of livestock is generally no more than 2%. The tremendous loss of livestock in the 2010 *zud* is evident also in a comparison between the worst four *zuds* of the past 50 years (Table 20.1).

Key lessons learnt from the 1999–2002 *zuds* that struck Mongolia for three consecutive winters were that their impacts had long-lasting implications for herders' livelihoods and security, and particularly for women and children. During those three *zuds* 15,000 herders lost all their animals. Long-term impacts were deepened poverty, a lowering of the GDP and an increase in levels of chronic malnutrition and maternal mortality. Between 2001 and 2002 the number of herders fell by 4.2%, which precisely corresponded to the growth of Ulaanbaatar's population – the final destination for about 50,000 people. Those families that migrated to cities or provincial centres faced, and in many cases continue to face, a dearth of basic social services. They lack access to electricity, water for drinking and household use, sanitation, health care, and pre-schools and primary schools. Similar or even worse trends must be expected as long-term impacts of the 2010 *zud*.

Less important, but also dangerous in the overall pastoral risk management context, are droughts, predation, animal disease, animal theft, conflict over natural resources, market failure, human illness, torrential rain and floods, and bush fires. Climate change is likely to enhance the risk of *zud* in the future, probably creating a more variable climate, with more severe extremes.

Vulnerabilities need to be assessed broadly and comprehensively, their interactions understood and factored into any risk management framework and support plan, for greater effectiveness and in order to disaggregate interventions according to needs. The following section discusses socio-economic and environmental sources of vulnerability that affect Mongolian pastoral systems and environments, and herder households.

Annex 20.1 Animal losses during *Tiger zud* 2010 by June (Data received from NEMA)

Aimag	Numbers lost	Camels	Horses	Cattle/Yak	Sheep	Goats	% of total herd/ aimag
01 Arhangai	1,085,777	27	70,086	126,449	495,263	393,952	30
02 Bayan-Ulgii	144,508	154	6,365	11,378	46,975	79,636	11.1
03 Bayanhongor	543,763	693	13,936	31,230	134,849	363,055	20.3
04 Bulgan	197,987	7	8,764	11,908	101,299	76,009	7.6
05 Gobi-Altai	715,077	1,875	22,375	16,054	239,500	435,273	32.9
06 Dornogobi	38,357	119	507	1,508	7,639	28,584	3.6
07 Dornod	275,557	120	7,902	24,048	127,498	115,989	19
08 Dundgobi	803,546	682	15,245	10,590	339,969	437,060	37.4
09 Zabhan	986,855	293	28,399	66,887	487,459	403,817	34.9
10 Uburhangai	1,643,643	766	78,616	79,225	761,053	723,983	45.4
11 Umnugobi	597,819	5,773	11,810	4,874	130,559	444,803	34.1
12 Suhbaatar	28,423	58	997	1,805	11,388	14,175	1.6
13 Selenge	99,751	7	1,389	10,425	41,732	46,198	6.4
14 Tub	507,219	67	24,426	24,624	221,981	236,121	14.6
15 Ubs	616,229	1,257	14,873	26,044	315,458	258,597	26.6
16 Hobd	466,665	786	14,451	20,411	149,603	281,414	20.6
17 Hubsgul	653,860	46	9,470	51,932	305,104	287,308	16.7
18 Hentii	172,256	31	6,354	19,819	72,448	73,604	7.7
19 Darhan-Uul	45,094	32	403	4,100	18,363	22,196	12.5
20 Ulaanbaatar	63,966	10	1,209	3,716	29,837	29,194	16
21 Orhon	27,734	0	445	2,998	11,006	13,285	10.8
22 Oobi Sumber	12,509	2	209	106	4,402	7,790	7.4
Total losses	**9,726,595**	**12,805**	**338,231**	**550,131**	**4,053,385**	**4,772,043**	**22.1**

Table 20.1 Livestock losses during the past four worst *zuds*[a]

Year	No of livestock lost
1999–2000	3.34 million (9.9% of total national herd)
2000–2001	4.15 million (13.7% of total national herd)
2001–2002	2.18 million (8.3% of total national herd)
2010 (to June)	9.73 million (22.1% of total national herd)

[a]Data provided by National Emergency Management Agency (NEMA) 2010

20.4.1 Vulnerability to Socio-economic Risks

As discussed, pastoralism is by its very nature exposed to risky natural environments and adaptation strategies to environmental stresses have been developed over long time frameworks and are built into pastoral management practices.

The transition to a market economy in the early 1990s induced major changes and long-lasting impacts on the vulnerability patterns of pastoralists, as had already happened during the earlier transition to socialism. In terms of risk management, there was a major shift after the transition to a market economy, with the risk burden transferred from the state and the community to individual households. Previously state-owned herds were privatized, the state stopped providing essential services, and infrastructure fell into disrepair, while before the transition the state had provided supplementary feed during harsh winters, transport between pastures, infrastructures, and mobile education, health and veterinary services (Fratkin and Mearns 2003). Lack of employment opportunities saw new and often inexperienced families entering the livestock sector, which became a social safety net of last resort. A combination of lack of social capital, labour and skills of new herding households often resulted in diminished mobility and concentration of grazing near urban centres and roads, with consequent untenable pressure on the natural resource base and parallel under-use of less accessible but critical seasonal pastures. Herding households increased from 17% in 1990 to 35% in the mid 1990s (Mearns 2004). Major losses during the subsequent *zud* years, between 1999 and 2002, demonstrated the sector's high vulnerability.

Over the years, the viability of pastoral systems and their capacity to sustainably and equitably provide a living for growing numbers of herder families have been further exposed to the gradual erosion of customary pasture and rangeland management mechanisms and institutions, but also by the laissez faire state policies adopted after de-collectivisation. These were later relinquished, and state presence has been building up again in the last decade.

Developments were influenced by different levels of skills and interests between older and newer herding households. These contributed to mounting inequality and reduced incentives towards cooperation and traditional risk-spreading practices of asset redistribution and resource pooling. Starker social differences and the commercialization of the sector manifested themselves in a variety of ways. The traditional livestock and pasture management strategies of herd diversification were affected by the changing composition of households and groups, notably when poverty entailed

Table 20.2 Increasing inequality of animal ownership between herder households

	Poor <31 animals	Vulnerable 31–100 animals	Rich >200 animals
1991	48%	40%	2%
2002	30%	38%	12%
2007	19%	27%	29%

Source: FAO/UNICEF/UNDP (2007), Mongolia National Statistical Office (2007)

lack of labour, and by export prices of cashmere, which have encouraged a strong increase in goat numbers. Reliance on export of raw cashmere translates in an overall greater exposure to market fluctuations in terms of production and investment, spilling over into shrinking informal social transfer mechanisms.

Currently, poorer *khot ails* are unable to provide traditional informal safety nets. This has left their members more vulnerable to idiosyncratic shocks which affect single households or individuals. At the same time, the customary practices of assistance between wealthier and asset-poor households have often been transformed into wage-labour relationships, ultimately depriving the poorest from the benefits of the traditional safety nets provided though local risk-pooling institutions (Mearns 2004). Ownership of trucks also enhances social differences by giving richer herders access to better pastures regardless of previous informal arrangements (Blench 2007). Greater political clout is also often associated with bigger herds, manifested in the possibility of exercising informal pressure on local administrators regarding distribution of benefits and management of natural resources (Mearns 2004).

Lack of alternative income-generating activities enhances vulnerability especially of the poorest and less experienced households, and, together with the breakdown of social networks and mounting inequality, can lead them to drop out of livestock-rearing altogether, and join the growing number of the unemployed living at the margin of urban centres.

A Participatory Living Standards Assessment conducted in 2000 found increased economic, environmental, social and physical insecurity among Mongolian households, and evidence of growing poverty and inequality measured in increasing differences in herd size (Nixson and Walters 2004). The results were confirmed by a joint FAO/UNICEF/UNDP food security assessment mission to Mongolia, conducted in 2007 (Table 20.2).

Access to and information on markets is crucial in order to destock and restock profitably, and can make all the difference in terms of gains and losses when disasters strike. However, constrained marketing of livestock and livestock production remains a main obstacle to sustainable pastoral production, enhancing herders' vulnerabilities. The high dependence on cashmere markets was already mentioned. Exports of meat are also limited, and well below levels of the 1990s (FAO/UNICEF/UNDP 2007). Only 3% of the meat produced in the country is processed in the formal sector and obsolete technologies result in high losses along the meat food chain. At the end of the 1990s three quarters of Mongolia's processed milk and dairy products were imported, after the decline of the dairy industry following decollectivisation. In 2005 only around 7% of its milk was processed by formal domestic industry, and most milk consumed in urban areas was imported, though the industry is reviving also thanks to a Government/JICA/FAO project which recommends replicating and tailoring piloted commercial dairy units to other aimags and sums (FAO/UNICEF/UNDP 2007).

The recent financial crisis exposed the vulnerability of the Mongolian economy, whose export sector is strongly reliant on primary products. The global recession depressed prices for cashmere and other livestock off-take products, hitting the pastoralist economy hard, with rural families increasingly unable to afford school fees and essential medical care (Walker and Hall 2010). Lack of cash, defaults in loan payment and soaring food prices caused a reduction in the use of supplementary feed, and failure to prepare animals for winter during the summer preceding the 2009/2010 *zud*. Key social support mechanisms, which had generously been expanded up until 2008, depended on fiscal revenue from exports, and the Government responded to lack of revenue due to the contraction of commodity prices by cutting welfare. This weighed disproportionately on the poorest households. The collapse of financing for social assistance later triggered a reform, assisted by donors, to achieve fiscal sustainability and ensure targeting of social benefits to the poor (Walker and Hall 2010).

The pastoral way of life means herders are self-sufficient in terms of meat and milk products, but dependant on cash for wheat, totalling half of their energy intake, and other staples. Poorer families are often vulnerable to seasonal food insecurity in spring, after winter weakening or loss of animals, or following a *zud* event. Constrained access to cash also creates vulnerability to the inflationary effects on food prices provoked by seasonal lack of meat on the market, while flooding of meat markets before the winter results in unfavourable terms of trade. Exposure to hazards and lack of diversification therefore make poor herders in Mongolia the group most vulnerable to food insecurity in rural areas, though severe food insecurity is more widespread in urban and *aimag* settings where the poorest live (FAO/UNICEF/UNDP 2007).

20.4.2 Environmental Vulnerabilities

According to the National Action Plan for Combating Desertification in Mongolia, the great majority of Mongolia's territory is vulnerable to degradation and desertification. This is due to a combination of climatic variations, including changing summer rainfall patterns, increasing evapotranspiration, dust and sandstorm events and inappropriate natural resource management practices such as overgrazing, deforestation and mining activities, though the relative contribution of each is not measurable (National Action Program to Combat Desertification in Mongolia draft 2009).

A recent study comparing ecological conditions in desert and forest-steppe zones across a period of 11 years found changes in ground surface attributes, plant canopy cover and standing crop, and plant species composition of fodder, with deterioration or lack of improvement of pastures in the desert zone and shift from fair to poor of the ecological condition of the forest-steppe zone. The primary causes of degradation were identified in soil aridity due to changing and decreasing precipitation patterns and dominance in herd structure of goats, whose dietary plasticity and capacity to access preferred and desirable plant species causes unsustainable pastureland stress (Sheehy and Damiran 2009). Lack of water also hindered use of transitional spring and fall pastures, increasing the grazing pressure on summer and winter ones. Though limited to two zones only, this survey revealed a pattern of degradation that is highly symptomatic of the interaction between herd management, climate change and rangelands, and of the overall ecological conditions of Mongolian pastureland. The strain on pastureland ecosystems caused by overgrazing must be seen in the context of relinquished traditional risk management practices, increasing animal numbers and weak institutional regulation of grazing management practices.

Under the socialist system between 1960 and 1990 livestock numbers were kept at 23–25 million constantly (Mearns 1991). Thereafter the overall animal numbers soared to 33.5 million by the end of the 1990s (FAO/UNICEF/UNDP 2007) and the balance of the five major species (sheep, goat, horse, cattle/yak, camel) shifted substantially. The number of livestock has now reached 43.3 million, and herd composition has changed too. Goats are currently 46.1% of the total number of livestock, while cattle have decreased by 23.7% compared to 2000 (National Action Program to Combat Desertification in Mongolia draft 2009). Seeing the devastating impact of the increasing animal numbers on rangeland productivity and in particular the enhanced pressure caused by goats on ecosystem health, it is mandatory to find ways to reduce their numbers and rebalance herd structures.

Human-induced climate change is another factor observed that has been recently affecting the country. The ecosystems pastoralists inhabit are particularly sensitive to climatic variability and change. Impacts of climate change are predicted to increase in the future, with warmer winters, melting of mountain ice and snow caps and reduction of groundwater. Annual mean temperatures have already risen in Mongolia by 1.8° between 1940 and 2003, with a more pronounced warming of winters and an increase of heat wave durations and decrease of cold waves. Annual precipitation changes are site-specific. Melting of high mountain glaciers is increasing and permafrost is degrading intensively. Ground water table is decreasing in arid regions (Batima et al. 2006). The incidence of drought is also expected to increase through climate change in the future, directly affecting the vulnerability of households and animals to winters. Increased snowfalls are predicted, coupled with untimely melting of snow due to shorter cold waves. These will be followed by freezing of melt, preventing animals from grazing due to the ice sheet formed on the land cover, and likely to have a negative impact on plant growth and animals. These predicted trends are likely to enhance the severity of *zud* phenomena in terms of livestock mortality (Batima et al. 2006). Climate change is likely to further enhance existing vulnerabilities of Mongolian pastoralists due to an increase in climate variability and the occurrence of more extreme events, shifts in the growing seasons and changes in temperature and precipitation. Socio-economic knock-on effects, such as migration away from negatively affected areas are also likely, thus increasing the pressure on other ecological zones.

On the other hand, the inbuilt capacity of pastoralists to cope with the difficult ecological contexts where their way of life evolved is precisely what could make them more able than other human groups to cope with the uncertainty of human induced climate change. Not only have they been recognised as good managers of fragile ecosystems, but also the global importance of rangelands as carbon sinks opens new opportunities for them in the context of climate change, through carbon-sequestration rangeland management practices.

20.5 Pastoral Risk Management in Mongolia

The previous paragraphs have clearly shown how important systematic risk management is to maintain sustainable pastoral production and livelihood systems.

The 'pastoral risk management agenda' developed by FAO/TCP through two projects in Mongolia in the late 1990s and early 2000s intended to ensure that mechanisms were put in place to address short and longer term causes of risk: in the short term the aim was to mitigate potential impacts and enhance better preparedness to overcome extreme, sudden and unpredictable shocks. For the medium and long term, its components were designed to provide a viable and sustainable long-term strategy for pastoralist livelihoods. The most important components of the proposed framework, which are still very valid are briefly discussed below as well as some lessons learnt[4]:

Strengthening and Empowering Herders and Pasture User Groups

- The creation and strengthening of herder groups was identified as a key reform if risk management is to be effective. Herder groups can organise collective action and capture economies of scale not available to individual herding households. Current efforts at building field level herder institutions are advancing, but need to be continuously extended and reinforced. The current variety of organisational forms, although sometimes confusing on paper, can serve as a real life laboratory from which lessons can be drawn. This is especially important given that the widely varying ecological conditions in Mongolia make it likely that there should not be one standard type of herder group or association, but rather that organisational forms should be adapted to local ecological and economic circumstances and interests.
- Herders forming a group can choose between legal forms. Until 2007, herder groups preferred to register as NGOs. NGOs are legal bodies and can enter into legal relationships with other legal bodies including the *sum* government. The weakness of the NGO format becomes evident when an organization wants to undertake a larger volume of business but cannot, since their legal statute does not allow them to. The focus on creating and strengthening pasture user groups (PUG)

[4] For more details of the framework and specific areas of intervention proposed for herders, as well as sum, aimag and national levels please refer to FAO (2007).

has been pursued successfully by the Green Gold Pasture Management Project, as the institutional basis for pasture management; the World Bank's Sustainable Livelihoods Project (SLP phase ii), linked the PUG concept with pastoral risk management throughout the country. Discussions about the comparative advantages of different forms of herder organizations are going on, whereas the overall importance of herders' organizations to take a lead role at local level in pasture and risk management has become common understanding. However, neither herder groups nor PUGs have managed yet to systematically include poor households as a matter of course, neither were they successful in limiting animal numbers. The readiness of poor households to join such institutions needs to be further explored as does the willingness of richer households to include them, (in order to reduce their vulnerability), and of herder groups or associations to give them additional support. The associations of PUGs (APUG) federated at sum levels could be encouraged to address the issue of poverty and make these associations the main defence against impoverishment and the escape from poverty.

- Herders groups and PUGs should be further empowered to take an increasing role in grazing management, as well as in solving grazing conflicts, which have increased over time, and in reducing overall animal numbers where needed, to correspond with location specific and seasonally available grazing resources. The authorities together with PUGs should coordinate seasonal migrations so that (i) the interests of both parties can be taken into account; and (ii) the number of animals entering into certain territories is based on estimates of carrying capacities of the pastures. This negotiation task should be progressively taken over by the relevant PUGs, herders groups and herder associations.

Land Reform and Tenure Security

- More secure land rights, especially to winter-spring pasture, hayfields and camping sites, are critical to better risk management. Good progress has been made on this under the provisions of the Land Law. The FAO assessment in 2007 revealed that herders recognise that possession certificates could deliver a mechanism to control overstocking and degradation of pastures, and theoretically made it possible to adjust grazing pressure to pasture carrying capacity. As a result there is increasing willingness on the part of herders to take out possession certificates (i.e. sign formal lease certificates with the sum government) for pastures and hayfields.
- However, the legal environment is still not fully satisfactory in this area, and uncertainties persist about the status of pasture lease certificates and the relationship between them and mining permits. The MFA is drafting amendments to the existing Land Law in order to create a better legal basis for the possession of pastures by herders.
- Recent studies analysing the current land legislation have indicated that the best way to regulate the access to pasture land would be through community-based management based on herder groups (Schulze 2009). Further legislation is therefore required to connect the principles of land allocation with group formation so that pastoralists have more incentives to form herder groups and be responsible for the management of the land (Schulze 2009). The recent draft Law on Pasture

Land is a move towards better land tenure laws; however it does not include all of the principles required for clear legal land tenure rights (Schulze 2009). Better implementation, negotiations and enforcement mechanisms, and procedures based on community-based management through herder groups or PUGs could be beneficial for the development of soil carbon sequestration projects. Collective action would lower transaction costs and only that could make such projects viable (Cacho and Lipper 2007).

Pasture Management and Pasture Mapping

- Pasture management needs to be directed towards recovering and sustaining traditional best practices and by re-strengthening seasonal grazing rotation and emergency reserve pastures, supported by the new tenure arrangements, especially the possession certificates for winter campsites and winter spring pastures and hay fields, discussed above. The methodology initially developed by the Centre for Policy Research (CPR), which involves herders and PUGs drawing pasture maps and preparing seasonal grazing plans is instrumental to this. National policy should also, where and as appropriate, encourage herders to adopt alternative strategies (fencing, improved pastures), especially in areas where traditional seasonal rotation of grazing is constrained for any reason. New awareness-raising activities may be needed to revitalize commitment for sustainable pasture management as a priority for herders and local government.
- The systematic pasture mapping and management methodology promoted by the WB SLP project has demonstrated its utility, especially when undertaken in collaboration with local governments and in areas where land possession certificates have been issued. This methodology may be used nationally to improve the quality of management and reduce conflicts. Differences in ecology and land use mean there should not be a uniform approach to pasture management. Although the mapping approach has been widely accepted by herders and local governments, so far it has not catalyzed better adjustment of livestock numbers to overwinter carrying capacity.

Fodder Production

- Fodder preparation by households is an important component of winter preparation and risk management. Adequate animal feed stored within the household is an essential part of winter preparation. Households make hay, but on a small scale. Key issues are, first, the allocation and security of tenure of hayfields to households, and second, access to the appropriate technology – especially animal powered hay mowers and rakes. Tenure has improved and herder who wish to make hay have greater security of tenure. Access to appropriate tools is still difficult. Natural hay productivity is low, and there are few proven technologies for substantially raising yields.
- Fodder concentrates are not produced according to demands. Experiments with winter cereal varieties such as of winter wheat or winter barley (for fodder production) should be conducted to assess their potentials for enhancing vegetation periods and thus yields as basis for enhanced fodder production. Approaches to promote fodder concentrate production, through specialized enterprises or

possibly as a way of livelihood diversification among pastoralists were already proposed in the original PRM framework but not fully taken up. Constraints and new opportunities at this stage – also in light of increasing temperatures and numbers of frost-free days due to climate change – should be further investigated.

- Although they exist on paper, it seems that in practice there are at present no significant, operational emergency fodder funds. The World Bank project explored economic models of successful feed and fodder storage. Emergency grazing reserves are being rehabilitated.
- World Bank efforts to develop a viable commercial model of fodder production are worthwhile and should be extended to herders if there were successful outcomes. Efforts to re-establish inter-*sum* and inter-*aimag* emergency grazing reserves are important, as are efforts to enhance a genuine national emergency fodder fund.

Enhanced Access to Financial Services

- Micro-finance has a key role to play in pastoral risk management, though so far it has been under-exploited. Credit has become much more readily available in the countryside, as a result of the spread of normal banking and savings-and-loans operations, which are preferable to dedicated pastoral credit operations. Index-based livestock insurance (discussed further on), as introduced by the World Bank, offers a new approach to risk management. Tailor-made micro-finance products are needed for herders. Further awareness-raising campaigns should be conducted among herders to attract more active use of saving opportunities within the banking system, as well as of micro-credit options. Forms of savings other than "savings on the hoof" need to be promoted, since this would also have the important side effect of reducing further grazing pressure. An effort may be necessary to persuade herders to continue to buy index-linked insurance after two or three subsequent years of mild weather.

Consequent Winter Preparation and Contingency Planning at Different Levels

- Winter preparation by herders is an essential element of a risk management strategy at household level and above and local government can support herding households in this. So far, experience with the preparation of *sum* and *aimag* level winter preparedness and reports and risk forecasts has been positive. A start has been made in preparing contingency plans in conjunction with pasture mapping and management planning. *Sum* risk management committees have taken on these tasks, and have developed detailed guidelines for the preparation of annual risk management plans. Regular evaluation of contingency plans is crucial to share experiences between different ecological and economic zones.

Functioning Early Warning and Response System Reaching to the Local Level

- Current weather forecasting is of a high standard, but distribution and use by herders is still not fully satisfactory. Attempts have been made to improve this, and the resulting weather forecast bulletins are in use by officials. Early warning is an important part of risk management, and effective early warning bulletins, covering

Table 20.3 Examples of relevant actors and projects promoting PRM

Government of Mongolia, Ministry of Environment – National Action Plan for combating desertification
Government of Mongolia, MFALI – National Mongolian Livestock Programme
WB – Sustainable Livelihoods Project (phases 1 and 2)
SADC – Livelihoods project "Green Gold Ecosystem Pasture Management Programme"
UNDP – National Climate Risk Management Strategy and Action Plan for Mongolia
USAID – Development of Herder Alliances in the Gobi Region of Mongolia

weather forecasts and other indicators are key to it. It is crucial for warning stages to be linked to pre-defined response actions, triggered at defined thresholds and well known and agreed upon (as part of the system) by all stakeholders. In this regard little progress has been made yet in combining seasonal forecasts with adjusting animal numbers to estimated over-winter carrying capacities.

Well Targeted Restocking in Post Disaster Contexts

- A wide variety of restocking exercises were carried out in the wake of the bad *zuds* at the start of the 2000s, and some of these have been evaluated and their lessons tabulated. Potentially restocking has an important role to play as part of an anti-poverty strategy. However, it is only a part of such a strategy, useful in certain circumstances, and not a panacea. Some form of individual or index-linked insurance of restocked animals is essential, as restocked households risk being made destitute with large debts in case of another *zud* during the repayment period. Several restocking exercises carried out in the past ignored the basic lessons available from earlier experience. As a result they were unsuccessful, and there has been a backlash against restocking. If targeted and undertaken properly though, restocking remains a powerful tool in certain well-defined circumstances, and should not be ignored (on social protection see also 20.7.2 below).

Well Coordinated Institutionalisation of Risk Management in all Relevant Ministries and Agencies

- Progress has been made in institutionalising risk management within government agencies and programmes in a sustainable manner. Notable steps include the creation of the National Emergency Management Agency (NEMA), as well as several important ongoing projects and programmes on PRM, but also the creation of *sum* and *aimag* working groups on risk management bringing together all stakeholders, including herders, and preparing and approving local risk management and pasture plans. Good progress has also been made in including detailed responsibilities for risk management, in the form of clearly specified outputs, into the job descriptions and civil service contracts of officials at *aimag*, *sum* and *bag* level (Table 20.3).
- Linkages between the activities of different Ministries have improved, but are not yet consistently implemented: For example, links between programmes to build the capacity of herder associations and technical activities, such as well

drilling or repair, where the herder associations might turn out to be the best managers of the wells.

Technical Capacity Building for Risk Management

- The risk management agenda introduced a simple plan to manage risk. Nevertheless, this has no effect unless herders and government staff at all levels understand and trust the plan, and play their part in its implementation. The continued upgrading of capacity building at *aimag* and *sum* levels to disseminate the knowledge and skills is constantly necessary in risk management. Trainers teams will need to be supported until civil society organisations and the private sector can provide capacity-building.

The above components of the proposed PRM framework constituted an effort to assist government, pastoralist households and communities to restore and strengthen their resilience to extreme weather events. In general, the framework addressed the abilities to better manage socio-economic and environmental vulnerabilities and strengthen capacities to cope with uncertainty.

While recognizing the positive achievements and impact of several programmes and projects with a PRM agenda over the last decade, the lack of sufficient funding remained the key constraint to up-scaling these reforms for PRM into wide enough operation. Some form of fiscal decentralisation, including use of revenues generated from local taxes on pasture or livestock (possibly to be re-established for those animals in big herds above a maximum ceiling of tax exemption) may be one route to follow, though there are many calls and different views on this issue.

20.6 Lessons for Pastoral Risk Management Revealed by the 2009/2010 *Tiger Zud*

Two assessments conducted between March and May 2010 of the conditions leading to loss of animals and of the humanitarian response capacity following the *zud* give some indication of current shortcomings of the pastoral disaster management system in Mongolia (UN Mongolia Country Team 2010; Hadrill 2010). Some of the key observations are summarized below.

Even though the National Emergency Management Agency (NEMA) reacted quickly to assess needs and was effective in its effort to secure and distribute goods at provincial level, several bottlenecks were reported, limiting emergency response operations. A major constraint was faced in March, when all available high fibre animal feed had already been purchased by relief agencies and herders themselves, and no more was to be found on the market. Persistent gaps reported included: the lack of a common comprehensive needs assessment tool, no geographic mapping of interventions to ensure coordination, insufficient coordination between NEMA and line ministries, lack of an emergency database and other information management tools, the absence of an emergency response supply or contingency plan, absence of snow equipment and snow-removal mobiles – initial access to more remote areas was difficult due to snow – and overall communication and equipment shortages creating difficulties in communication with villages. Insufficient or conflicting information impeded prioritizing of the response actions was also reported. Particularly striking was the failure of comprehensive situation analysis, information management, and mapping that impeded the effectiveness of the humanitarian response, even though there had been long-term investments in terms of resources and time by a number of agencies towards the goal of mapping and information gathering for better preparedness.

The analysis of the 2009/2010 *zud* response presented above mainly highlights gaps in preparedness for response measures, but the severity of the hazard's impact was also exacerbated by a set of factors linked to the limited implementation of other long-term risk reduction, prevention and recovery interventions, included to different degrees in the original PRM framework. Among the PRM activities supposedly in place, there was a failure to fatten animals during to the previous summer drought; insufficient fodder preparation by government and *sums*; too little emphasis on the role of herders organizations and of locally responsible grazing management, including lack of incentives for de-stocking; limited use of traditional herding skills; changing herd structure and lower goat survival capacity in cold weather; overstocking and overgrazing; increased ecological vulnerability. These in turn were directly affected by policies on fiscal measures, regulations, incentives and subsidies for the livestock and other productive sectors and by market failures at different levels. The past years' food and financial crises may also have had a multiplier effect on the impact of the disaster, eroding asset bases and consumption patterns, and weakening resilience and coping capacity. A key challenge was certainly to maintain PRM high on the agenda over 10 years without the occurrence of extreme events.

Through its limited capacity to prevent the full-blown consequences of the 2009/2010 *zud* the efficiency of the original PRM strategy has to be put into question, while the possibility that provisions have not been implemented adequately, or for a sufficient period of time, should also be considered.

On the other hand, the lessons from the 2009/2010 *zud* reconfirmed the need for and the relevance of the PRM framework described above, and of most of its components. As in 2001, similar gaps were identified in 2010 as the main causes of the *Tiger zud* disaster, and similar strategies are now recommended again to strengthen PRM. The overall shortcomings in 2010 of preparedness and response measures indicate the need for even more capacity building as could be delivered over the past years by projects and programmes concerned with building up PRM. The conclusions of the assessment about the progress in implementation and streamlining of PRM in government planning undertaken in 2007 are still valid today: capacity needs to be firmly built both downstream and upstream, coordination needs to be ensured between institutions at different levels and longer-term vision and planning need to be in place (FAO 2007). There is a need for more consistent implementation and wider coverage. But the question of how to enrich and strengthen a PRM strategy further is also open to draw in more firmly government and herders alike.

20.7 Enhancing the Pastoral Risk Management Framework

Some of these broader issues indicate that a further effort to rethink the pastoral risk framework is called for, taking an evolving national and international economic and political scenario into account, and factoring in lessons learned from years of implementation.

This last section will confront the open question of how to enhance a PRM framework considering the unique case and conditions in Mongolia. The possibilities of strengthening the PRM framework in Mongolia are favourable. Unlike in many other national contexts, where pastoral systems' capacity to respond to stressful environments and confront the impacts of climate change is hindered by the political, social and economic marginalisation of pastoralists (Nori and Davies 2007), in Mongolia pastoralism is recognized as having a central role in forging the national identity. Pastoralism is a major economic factor and revenues from extensive herding constitute an important part of the national economy. The recognition of the importance of pastoralism and of the viability of its coping and adaptive strategies puts the country at an advantage to exploit them in the global market for environmental services and carbon-offset. It will therefore be crucial to value and reward the protagonists accordingly (Grahn 2008).

The recommended effort to strengthen the current PRM framework in the first place goes in the direction of further consolidating and multiplying what has already been done, but broadening its scope and providing more sustainable capacity building and policy support to disaster risk reduction, hazard preparedness and livelihoods diversification measures. On top of this, this section will consider additional factors, based on the broad need to further increase livelihood resilience of herders by strategically linking disaster risk reduction, climate change adaptation and social protection strategies.

This chapter will thus look at climate change as an opportunity despite persistent overstocking of animals, rangeland degradation, lack of markets and mounting inequalities, without significant progress in the reduction of income poverty. From a PRM perspective, despite expected negative impacts, climate change is also likely to offer a range of new opportunities, both through changing seasonal patterns towards enhanced productivity in some ecosystems, and possibly through payments coming from global carbon trade mechanisms.

The enhancement of a PRM strategy will also turn to social protection mechanisms, since a process of stratification has been ongoing and leading to starker social differentiation between pastoral households, reflected in livestock numbers and composition of herds, use of pastureland and natural resources, mobility patterns, spatial distribution, cooperation mechanisms, different needs, coping strategies and capacity, growing impoverishment and food insecurity, and migration to urban centres. A sound analysis of this process could offer the basis for a more diversified and evidence-based social protection response to reduce vulnerability of labour and asset-poor households and wealthier ones, promoting diversification within livestock-based livelihoods or away from them, according to needs.

20.7.1 Diversification of Livelihood Strategies as a Contribution to Enhanced PRM

Diversification of livelihood opportunities as a strategy to adapt to changing conditions is in line with the inbuilt capacity of pastoral systems to be flexible and opportunistic, temporarily or permanently shifting between options according to circumstances. If PRM strategies are to be effective, and to make sure that risk is not enhanced by compromising on the mobility and flexibility of livestock-herding, wider livelihood diversification should be accepted by herders themselves as a wholesale alternative (Kirkbride and Grahn 2008). This said, alternative and complementary income generating activities are crucial to curtail the common practice of risk spreading by hoarding animals, which severely degrades pastures and is increasingly unsustainable on the long run.

Poorer, less skilled and therefore more vulnerable households in particular stand to gain from "moving away" from the sector, while middle income families with viable herds can take advantage of restocking and insurance provisions, building on this kind of initiatives envisaged in the PRM framework in 2001 and supported by FAO, the World Bank and others over the years.[5]

20.7.1.1 Ecotourism Services, Stewardship, Protection of the Ecosystem and of Biodiversity

A long history of co-evolution with the natural environment means that traditional mobile and transhumant herding practices and knowledge can be vital for ecosystem health and sustainability. And while in the past wildlife conservation has been pitted against the presence of pastoralists in protected habitats, it has now been recognised that in fact in certain areas wildlife populations and grasslands rely on well-managed mobile pastoral systems as an integral part of their existence. Managed livestock grazing maintains and enhances biodiversity and the ecosystem services supported by biodiversity, including water and nutrient cycles, and maintenance of species habitats by preventing scrub encroachment and removing plant material without cutting

[5] For example, one of the key areas of the Country Assistance Strategy (CAS) 2009/2010 endorsed by the World Bank for Mongolia was "protecting the poor and the vulnerable", while the consultation with communities for the 2009–2011 CAS highlighted as one of its pillars "improving rural livelihoods and the environment" http://go.worldbank.org/91J9Y3DGF0

Table 20.4 Pastoral economics and ecosystem services

Examples of projects encouraging expansion of pastoral economics towards provision of ecosystem services, including:
UNDP Sustainable Land Management for Combating Desertification
Implemented over 2008–2012 in the central and desert steppe, one component supports community-based approaches to natural resource management, with a focus on grassland and water management and sylvo-pastoralism[a].
UNDP/GEF Biodiversity conservation in the Altai Sayan Ecoregion
The project supports a range of initiatives to promote the development of alternative and environment-friendly activities, including sustainable use of non-timber forest products, support for traditional economic pursuits and participation in tourism[b].
WWF Khar Us Nuur National Park
An assessment was carried out encouraging WWF to adopt a pro-active role in creating an enabling business environment for ecotourism development in KUNNP and its buffer zones, ensuring high local community benefits to achieve the twin objectives of rural development and conservation.

[a]See http://www.undp.mn/snrm-slmcd.html
[b]See http://www.altai-sayan.org/eng/projectoutputs/

and burning. Livestock mobility can also be regulated to increase soil cover. The goods and services provided by pastoralists (biodiversity, combating erosion and desertification as well as carbon sequestration as mentioned above) should be acknowledged and remunerated as contribution to an enhanced PRM framework. Incidentally, a focus on ecosystem services would also constitute an incentive towards recuperating good rangeland management practices and sharing them between more and less experienced herders (Table 20.4) (Hatfield and Davies 2006).

Lessons and operative proposals from these projects to generate benefits for herders and the environment should be proactively integrated into an enhanced PRM framework.

20.7.1.2 New Opportunities Arising from Carbon Markets

International climate change negotiations have opened up new possibilities for funding, especially for mitigation options. Tapping into mitigation funding is not yet part of Mongolia's PRM framework. Once it will be possible to choose win-win options, addressing mitigation, adaptation and production in a synergetic way, households could greatly benefit from these through carbon finance payment and increased production potential.

20.7.1.2.1 Scope

A suitable win-win option for the pastoral landscape from the carbon market is soil carbon sequestration which "is the process of removing carbon from the atmosphere and depositing it in a reservoir (the soil)" (UNFCCC 2010). Degraded and badly managed soils have depleted carbon stocks, and techniques such as improved rangeland management, improved crop management and agro-forestry have shown potential to reverse this depletion, though these practices need to be adapted to each specific local environment (Lal 2004a).

In the pastoral landscape improved rangeland management is the most effective way of sequestering carbon in the soil. The carbon sequestration potential of rangeland is high, due to the large global areas of rangeland and their current state of degradation, which means that they are far from saturated with carbon.[6] Different ecosystems have different carbon sequestration potentials, as this potential is dependent on variables such as climate, soil type, vegetation, and land use history (Tennigkeit and Wilkes 2008). Parton et al. (1995) provide a general indication that the sequestration rate in temperate steppe, the ecosystem of Mongolia, is about 180 tCO_2e per year per km^2. The pasture area of Mongolia in 2007 was 1,120,000km^2 (MRSM 2010) and thus the soil sequestration potential of Mongolia is estimated at about 200 $MtCO_2e$ per year. This is a huge potential, as it is five times more than Mongolia's overall greenhouse gas emissions in 2005 (36.5 Mt CO_2e) (World Bank 2007).

By undertaking specific rangeland management techniques, which either increase the carbon inputs in the soil or reduce the carbon losses from the soil, herders could tap this huge potential (see Table 20.5 for an overview of management practices). In Mongolia many of these techniques have already been adopted through the current PRM framework, since many of the practices which enhance the productivity and resilience of the land, reducing the vulnerability to natural disasters, also sequester carbon. Soil fertility is improved by increasing the organic matter accumulation in the soil which: (a) improves soil physical properties, for example by increasing its water holding capacity and aeration; (b) improves the soil's chemical properties, for examples by improving nutrient cycling; and (c) improves the soil's biological properties, for example by enhancing faunal activity. Increased decomposition of organic matter augments the carbon stored in soil. Increasing the levels of organic matter, and the carbon stocks that it entails, can also help enhance and maintain species' growth, biomass and diversity (Lal 2004a, b). These management practices thus bolster the resilience and reduce the risk and vulnerability of pastoralists to climate variability and disasters firstly by improving land productivity, which increases the economic returns from the land; and secondly by improving the resilience of the land itself, reducing the risks associated with unreliable rains and prolonged drought (Neely et al. 2009).

Many management practices that sequester carbon are also beneficial to other environmental processes (Table 20.5). Some studies have shown that they maximise vegetative cover, therefore reducing erosion by wind and water, while others have demonstrated that many of these practices can enhance the ecosystem water balance in arid/semiarid environments through enhanced water infiltration and retention;

[6]By improving rangeland management Smith et al. (2007) estimate that globally rangeland has the biophysical potential to sequester 1.3–2 $GtCO_2e$ up until 2030. CO_2e is "the unit of measurement used to compare the relative climate impact of the different greenhouse gases. The CO_2e quantity of any greenhouse gas is the amount of carbon dioxide that would produce the equivalent global warming potential" (Carbon Neutral 2010: Global Land Reserve website http://globallandreserve.com).

Table 20.5 Management practices, which have the potential to either increase carbon sequestration or decrease carbon losses from rangelands (Adapted from Tennigkeit and Wilkes 2008, p. 11)

Type of carbon sequestration	Management practice
Increases C inputs	Improved grazing management
	Optimising the stocking rates to not exceed the carrying capacity of the land
	Setting up a rotational, planned, or adaptive grazing system
	Fencing off pastureland from livestock grazing
Increases C inputs	Increasing biomass by adding nutrients, water to the soil or improving the soil composition
	Fertilisation
	Irrigation
	Introduction of grass species with higher productivity or deeper roots
Decreases C losses	Managing land use conversion to maintain land use with high carbon sequestration potential
	Converting agricultural land to pastureland
	Avoiding the conversion of pastureland to agricultural land
	Avoiding the conversion of forests the pastureland
Decreases C losses	Fire management
	To reduce the GHGs, smoke, and ozone forming substances released into the atmosphere
	To maintain shrub cover with higher carbon density
Decreases C losses	Introduction of alternative energy to replace dung/shrubs as fuel

there is also some evidence that they could improve biodiversity levels (Conant 2010). Soil sequestration in the form of rangeland management is thus a strategy that can be used both to adapt to climate change as well as to mitigate it, and because of this, it may be a way to link the three related UN conventions on climate change, desertification, and loss of biodiversity (Lal 2004a, b).

As mentioned, such rangeland management practices are not new to Mongolian herders and are already part of the PRM framework. The carbon sequestration aspect of these practices, however, is a new opportunity and could open up access to finance for herders or herder groups. Carbon sequestration projects owned by herders or herder groups could obtain support nationally, as Mongolia's national policy priorities promote reductions of greenhouse gases, improved rangeland management and sustainable livelihoods for herders.

20.7.1.2.2 Criteria and Barriers to Overcome

National support is one of the main criteria required for the development of carbon finance projects. Other important criteria include: institutional capacity to develop and support the implementation of these practices; clear legal rights for land tenure; project size capable of meeting transaction costs; and good scientific documentation of the impacts of specific management practices on carbon sequestration in the area in question (Tennigkeit and Wilkes 2008). Since strengthening rangeland management is already part of the PRM framework, the institutional capacity needed to include carbon sequestration within PRM should not be very different. More likely to be lacking is national expertise, though this should probably increase in the next few years through a capacity building programme on Clean Development Mechanism (CDM) which is currently being undertaken in Mongolia, funded by the Asian Development Bank. While recognizing that currently soil carbon sequestration is not recognised under the CDM mechanism – which implies that CDM is not an option to obtain funding – the institutional capacity needed for the development of soil carbon sequestration projects and CDM projects would be similar, so the CDM capacity building process could be beneficial for carbon finance projects too. There might also be a chance that soil carbon sequestration will be recognised under the CDM mechanism in the future.

Clear legal rights for land tenure were also identified as a key issue in the PRM framework from the very beginning, and the issue has not yet been fully resolved. However, recent studies indicated that the best way to regulate access to pasture land would be through community-based management centred on herder groups (Schulze 2009). This type of regulation would not only strengthen the PRM framework by achieving one of its initial objectives, it would also be beneficial for the development of soil carbon sequestration finance projects, as collective action would lower transaction costs, making the projects viable (Cacho and Lipper 2007), further strengthening the framework. Though this type of project might only be possible in certain areas of Mongolia where herder groups generally use one or two valleys for grazing, while it might not be feasible in areas such as the Gobi, where herders move over great distances.

In order to integrate carbon sequestration in the PRM framework the latter needs to be integrated with more national and regional information on the impacts of specific management practices on carbon sequestration. Pilot projects would be a good way of obtaining this information, and since Mongolia's national priorities support such projects, they could be promoted and funded by the government.

20.7.1.2.3 Soil Carbon Project Development and Possible Funding Sources

Projects for soil carbon sequestration could access mitigation finance through the voluntary carbon market. This market accepts credits generated from soil carbon projects, and its methodologies and procedures for project registration are not too complex, unlike the Clean Development Mechanism (CDM) of the regulatory market (Seeberg-Elverfeldt 2010). The development of soil carbon sequestration projects has many initial costs, and it can take more than 2–5 years for money to be received from the sale of carbon credits. It is therefore necessary to identify a project developer, who can assist in obtaining funding and in the development of the

Table 20.6 Carbon project document

Currently no standard methodologies exist for measuring soil carbon sequestration potentials from improved grassland management; however a project is being developed in Qinghai Province, in China for such a methodology for the Voluntary Carbon Standard (Wilkes and Tennigkeit 2009). Using such an approved methodology increases the likelihood that a project will be registered and can be prepared faster (Seeberg-Elverfeldt 2010). The project plan will be assembled in a Carbon Project Document, and the project developer needs to identify an accredited certifier to review and validate this document to ensure transparency. Credits from validated projects are then kept in a registry until they are bought (Seeberg-Elverfeldt 2010).

project. Funding for carbon sequestration projects is available from grant programs or NGOs. (Seeberg-Elverfeldt 2010)

In the voluntary market, carbon credits are generally purchased by the private sector. Most commonly the private sector is driven to buy credits by corporate social responsibility or public relations, and it can purchase the credits directly from the projects, from companies, or from funds (Seeberg-Elverfeldt 2010). In order to sell carbon credits, projects need to show additionality, which means that emission reductions or sequestration that has occurred is in addition to any that would have occurred without the project; and they need to show permanence, indicating that the carbon has been stored for a certain length of time and is not re-released into the atmosphere. In order to access the market a carbon sequestration concept needs to be developed by herder groups. This concept could be linked to the overall PRM framework, but it needs to be a separate project, able to demonstrate additionality.

A Project Idea Note (PIN) must be put together to provide the details of the project including: an overall project description, the estimated greenhouse gas emission removals, the environmental and social impacts, project ownership, host government laws and policies, and a plan for the project's financing. Once a PIN has been created a project developer needs to be identified who has sufficient experience to prepare the project for the market (companies such as Eco-Security, Carbon Neutral Company, World Bank Carbon Finance Unit can do this). In order to access the carbon market the project will have to comply with one of the approved standards (e.g. the Voluntary Carbon Standard) to show that the project is real, additional, measurable, permanent, independently verified, and unique (Seeberg-Elverfeldt 2010) (Table 20.6).

Linking carbon sequestration projects to the PRM framework should significantly strengthen the framework as it could open up funding opportunities for pastoralists, constituting an additional incentive for improved rangeland management. These projects are not only positive for pastoralists, they are also positive for the government of Mongolia, since besides assisting the country to reduce emissions, they also meet other national priorities such as decreasing herder vulnerability, reducing risk, enhancing pastureland and increasing productivity.

20.7.1.3 Livestock and Farm-Based Alternative Livelihoods

An Action Plan (2010–2015) for the implementation of the National Mongolian Livestock Program was recently launched to intensify future efforts in the development of the sector by a number of actions. Among these are improving animal breeding, raising the veterinary service standard, developing livestock production that is adaptable to climatic and ecological changes with strengthened risk management capacity, developing targeted markets for livestock and livestock products and establishing proper processing and marketing structures and increasing economic turnover (National Mongolian Livestock Programme draft, December 2009). Recognizing that weak implementation of past programmes has greatly hindered their effectiveness, and that current state funding is inadequate, the programme promises to invest greater and more targeted resources for programmes and policies supporting livestock. Though the trade-offs between greater productivity, intensification and commercialization of livestock production and ecological sustainability, are not explicitly addressed, this Action Plan could become another pillar to reduce herders' exposure to risk and minimize the impact of livestock on overgrazed and stressed pastureland. It ties in with the aim of diversifying pastoral livelihoods by adding value to livestock production, focusing on quality products and developing processing and trade.

A diversification option for herders with important knock-on effects is to engage in farming activities, including enhanced small-scale vegetable production, enhanced fodder and emergency fodder production. Cropping can benefit from distribution of improved winter cereal varieties exploiting the changing seasonal cycle induced by climate change. By growing leguminous fodder in rotation with cereals, the practice of cultivating much needed fodder would also go towards the enhanced PRM objective of sequestering carbon and mitigating the effects of climate change.

An important caveat in designing interventions encouraging partial or permanent transition into farming is that under-used and unused land should be identified to this end, with care not to exploit precious patches or use resources presently critical for herd transhumance or *otor* movements, in order to avoid further over-exploitation and conflict around access.

20.7.2 Social Protection Options to Complement Pastoral Risk Management

A framework to reduce herders' risk and build their resilience, tying into poverty reduction and food security aims and in the context of a changing climate impacting on rangeland and ecosystems that have already been greatly eroded by social, economic and environmental shocks and stresses, could benefit from a link to what has been called *adaptive social protection* (Davies et al. 2009) (Table 20.7). The concept is useful to bring under one label the actions and policies already

Table 20.7 Social protection for pastoral risk management

Social protection measure	Aim	Target group
Asset transfers	Provide post-disaster relief	Low income HHs with less than 100 heads
Seed distribution	Protect assets (livestock)	All in emergencies
Improved and emergency fodder	Diversify income-generating activities	Skilled HHs with at least a 100–200 heads base (restocking)
Restocking	Enhance productivity through appropriate seed and animal varieties adapted to CC	
Unconditional cash/ food transfers	Provide relief	Low income HHs with less than 100 heads
	Smooth consumption allowing flexible risk-taking and investment	All in emergencies
	Protect assets and prevent distress sales	
	Prevent food insecurity in times of crisis	
	Boost local markets (cash)	
Cash/food for work	Provide relief	Low income HHs with less than 100 heads
	Smooth consumption allowing flexible risk-taking and investment	All in emergencies
	Protect assets and prevent distress sales	
	Prevent food insecurity in times of crisis	
	Boost local markets (cash)	
	Build community assets (shelters, wells, roads, soil conservation)	
Index-linked insurance	Buffering livelihoods in moments of crisis	Middle and high income HHs with more than 100 heads
	Restocking	
	Replace animal hoarding as a risk-spreading strategy	
Employment schemes and unemployment benefits	Provide a safety net and alternative livelihoods pathways for poor herders who have lost all their animals	Poorest HH with unviable/no herds
		Ex pastoralist urban dwellers

implemented, as well as the ones proposed in this paper involving livelihood diversification and returns from carbon finance. But more than a matter of simply renaming what have been ongoing efforts by international aid agencies and the Government of Mongolia, the *adaptive social protection* approach has specific contributions to make: it stresses the importance of differentiating pastoral risk management interventions by mapping vulnerability and targeting groups accordingly; it points to the urgency of transforming productive livelihoods as well as protecting them from and helping them adapt to climate change; and it indicates the need for a long-term perspective for social protection policies, that takes the changing nature of shocks and stresses into account and ensures that predictable and sustainable social transfer mechanisms are in place (Davies et al. 2009).

One of the original background documents developing a PRM approach identified the role of social capital in pastoral economies as "a key determinant of the ability of pastoral groupings to survive extreme risk" (Baas et al. 2001). Weak social capital is both a cause and a consequence of unviable pastoral systems and of the erosion of customary practices like herd specialisation and sharing and informal agreements on rotational access to pastureland. Success in reconstructing it is therefore key to the success of PRM itself, not least for the role of social capital in activating informal safety nets capable of assisting households affected by idiosyncratic shocks.

A number of projects with a PRM component have been working towards the reconstruction of social capital. For example, the Swiss Agency for Development and Cooperation's (SDC) Green Gold project has been developing community-based pasture management activities by encouraging the formation of pasture-user groups (PUGs) between families traditionally using the same pasture. Establishment of PUGs is achieved through a deliberative process among herders themselves and ratified by the *sum* government.

Nevertheless, past projects' efforts have not proved sufficient to reactivate forms of cooperation that effectively protect herders, and especially the poorest households, against risk. Linking into the Government's present drive, assisted by donors and IFIs, to reform its welfare system to make it financially more sustainable and better targeted (Walker and Hall 2010), a range of social protection instruments could be developed with the goal of reducing pastoralist groups' vulnerability, and complementing the PRM's mitigation, relief and recovery objectives. Besides their role in smoothing consumption, facilitating long-term investments and building assets, predictable social protection interventions allow households to plan over a longer timeframe, particularly critical in a context of climate change where economic, social and ecological factors are in a high state of flux (Davies et al. 2009). The measures proposed here are examples of possible action in this sense, and only meant as indications. A comprehensive social protection strategy for herders could only be designed and adopted following a proper assessment of the problems, opportunities and constraints to their enactment, including financial ones. It also needs to be linked with broader policies adopted by the Government of Mongolia for livestock and for poverty reduction.

A set of predictable and sustainable social transfer mechanisms targeted according to needs, and linked to diversification strategies, would encourage a plurality of pathways for the future of pastoralists. By providing the poorest owners of unviable herds with alternative livelihood options and safety nets, it would assist them in "stepping out" of pastoralism altogether. Middle-income herders would see the opening up of possibilities related to a plurality of productive activities compatible and even associated with livestock rearing, including farming, possibly with a partial differentiation of income-generating activities within households. This would go in the direction of "moving away", and probably at least for some on the long term it would mean abandoning herding altogether. But it would give them the time and means to transition gradually and adapt to the new state. Benefiting from insurance and financial services, a smaller proportion of owners of large herds would feel secure enough to engage in risk-taking activities, “stepping up” into a more commercial sector for livestock and products. Enhancing commercial productivity would presumably also facilitate off-take of bigger herds, relieving their pressure on pasturelands (Table 20.7).

Social assistance is needed for vulnerable herders with less than 30 heads. It is also needed for primary consumption and to buy feed and veterinary medicine for those who have lost all their animals following a *zud* or other hazard. Therefore, it can both function as an ordinary measure to reduce long-term vulnerability, as well as providing short-term relief following a disaster. Safety nets can take the form of unconditional transfers, or food/cash for work programmes. A recent example of the latter in an emergency was a post-*zud* carcass removal programme launched by UNDP as part of the UN coordinated *zud* response and targeting 60,000 of the poorest herder households (UN Mongolia Country Team 2010).

Besides being a relief measure in emergencies, continuous **unconditional cash transfers** are a means to reduce the vulnerability of the poorest herders by raising their income, smoothing consumption, taking moderate economic risks and avoiding negative coping strategies, including protecting their assets from distress sales, in particular by preventing non-seasonal sale of animals with low returns. Cash transfers can also have multiplier effects on local economies. Seeing the strong seasonal variability of income of pastoral economies, it is especially critical that they be predictable and regular, to allow recipients to plan, invest and save over the long term.

If food security and nutrition are a concern and access to markets for a range of foodstuffs is not granted, food packages or a mixed cash and food system should be contemplated as a possibility, especially in the first phases of relief. In general, the decision of whether to provide cash or food depends on a number of factors. These should be assessed in the stages of designing social transfer interventions, and can make the balance shift in favour of food in determined circumstances: constrained access to markets, as is the case in many regions of Mongolia during the winter months; micro-nutrient deficiency among the targeted population, as has been found among Mongolian herder groups (FAO/UNICEF/UNDP 2007), requiring an input of fortified foodstuff; finally, price fluctuation and changing value of cash can diminish their purchasing power and lead to a shift in preference towards food transfers.

The wide stretches of land over which pastoral communities are dispersed pose a problem in terms of distribution. Ad hoc solutions would need to be envisaged, possibly linking delivery to mobile and delocalised education, health and veterinary service provision.

Cash for work programs have the additional benefit of contributing to adaptation and risk management by building community assets such as shelters, water sources and establishing soil conservation projects that enhance resilience, though in other settings the assets built have been of low quality, and not maintained following the completion of the scheme. A further limit is that they often end up being controlled by the better off, benefiting them exclusively. Lessons for the future about this kind of programs should be drawn from the World Bank's Mongolia Poverty Alleviation for Vulnerable Groups Project, and its mixed success in achieving the objective of creating gainful employment and income for poor people rural communities.[7]

Different **asset transfers** in the form of the provision of animals, seeds, goods and services more or less directly promote livelihoods by providing inputs that encourage diversification and risk-taking, including off-farm rural enterprise and industry. Assets for income-generating activities need to be suited to local conditions, since they have the potential to both reduce and exacerbate climate change. Appropriate inputs for the Mongolian context relevant to PRM include but are not limited to supplementary fodder and emergency fodder stocks, winter cereal seed varieties adapted to changing climate conditions and the longer growing season, restocking of locally adapted animal species, alongside training on their value, and on the impact on ecosystems of goat prevalence and of climate change. Restocking of herders following disasters needs to be undertaken following stringent selection criteria of households, as initially envisaged by the FAO projects piloting PRM. That includes proof of herding capacity, labour adequacy, residency away from urban centres, obligatory insurance of restocked animals, and a minimum number of animals transferred to make up a viable herd size (FAO 2007). Unfortunately, these criteria slipped over the years in the implementation of other restocking initiatives. Currently restocking programmes should be approached with extreme caution, seeing the unsustainable expansion of herd numbers: large-scale distribution of animals as a poverty-reduction strategy is to be avoided at any cost, and restocking should principally target herders with a viable herd for sustainable livelihoods.

In line with the Asian Development Bank's Social Security Sector Development Program implemented since 2001 in Mongolia, **Unemployment benefits and employment schemes** are necessary to assist poor herders who have lost all their assets after a crisis to recover, and enable them to take up another occupation, and to avoid inexperienced unemployed from urban areas to enter the sector. The ADB has provided a policy loan, and under its "Employment promotion and improvement of working conditions", it is pilot-testing six livelihood assistance projects, which involve conditional cash transfers, compensation packages and livelihood support. The ADB has also conducted surveys and made projections on the necessary funds, contribution rate, pension level and other factors required for the inclusion of herders into a basic social pension insurance scheme (Asian Development Bank n.d.). An option could be to fund an unemployment scheme through fiscal revenues from herds over a certain size, starting taxation above a certain number of animals.

---

[7] The project was implemented from 1995 to 2000, with the overall development objective to mitigate the adverse effects of Mongolia's economic transition on vulnerable groups. The public works component was successful in creating temporary employment and in halting the deterioration of economic and social infrastructure. Although the duration of employment was too short to have an impact on the livelihoods of the poor, and productivity was not enhanced on the long run, the infrastructure works themselves contributed to poverty reduction. http://go.worldbank.org/KYPG2LENE0

As mentioned in the original pastoral risk management framework, savings and credit are necessary seeing the seasonality of the livestock-based production cycle. Access to **financial services** might be a way to avoid the expansion of herds as the primary route to investment and insurance. Credit would also allow to buy supplementary feed when income is lacking, to promote and enhance livelihoods by investing in implements, veterinary services, education of children. Credit would produce multiplier effects on the pastoral economy, if some herding households, which engage in small-scale cultivation would produce supplementary feed as a complementary income-generating strategy, as mentioned in the previous section on diversification.

Herd insurance is a further critical measure in the direction of avoiding herd expansion as a risk spreading strategy, also serving the purpose of buffering livelihoods in moments of crisis and to restock. Spearheaded in 2005 by the World Bank's SLP and initially piloted for 3 years in 3 *aimags*, selected according to agro-geo-climatic variation and different exposure to risk of livestock loss, the Index Based Livestock Insurance Project (IBLIP) protects herders against disaster-induced livestock losses by index-linking animal deaths to historical mortality rates. Private insurance pays out to individual herders over a certain rate of deaths, while in case of a catastrophe affecting great numbers of animals and herder families, the state sets in. The IBLIP currently covers 14% of herders in 4 pilot *aimags*. An evaluation found that all main performance indicators had been achieved or exceeded and nation wide scaling up is expected by 2013 (www.iblip.mn). Though, in theory the possibility of choosing between insurance and disaster response products sold at different rates, makes the IBLIP accessible to herders with different herd size, in practice it might be advisable for the scheme to adopt criteria close to the ones mentioned for restocking, and mainly target middle and upper income households with viable herds.

20.8 Conclusion

The original PRM framework proposed in the early 2000s after the transition to a market economy, with critical FAO inputs, has been influential in creating new policy and project approaches to risk management. The framework included a variety of capacity building measures and technical interventions as part of a broader strategy to reduce vulnerability to the risk of natural disasters in Mongolia. The main risk management interventions proposed as part of the PRM framework have been adopted, and were further fine-tuned by the government as well as by many subsequent large donor-funded projects in the poverty and natural resource management area. Important lessons have been learned – not so much leading to changes in the PRM agenda itself, as to increasing sophistication in its implementation.

In view of the major impacts of the 2009/2010 *zud*, the question was raised of whether the existing PRM strategy has been sufficient to comprehensively address the environmental, social and economic challenges faced by Mongolian pastoral systems since the transition to a market economy. Shortcomings in the implementation and in particular the outreach of the PRM framework and risk reducing actions emerged. Gaps in operational linkages between long-term PRM activities and the mobilization of quick emergency response operations were particularly striking. In view of the importance of a functioning PRM system in Mongolia this paper concludes that the basic know how and technology options for PRM are actually in place. It is essential though, to further consolidate and replicate what has been tested already with success. Proactive measures are needed to ensure the necessary outreach, and provide more sustainable capacity building and policy support to disaster risk reduction, hazard preparedness and livelihoods diversification measures.

The proposals for enhanced social protection and livelihoods diversification, including a proactive use of new opportunities, such as funding for PRM through proactive soil carbon sequestration projects, are suggested as additional components of an enhanced background framework constituting a necessary complement to existing PRM actions.

Seeing the cross-cutting approach proposed here, enhanced cross-sectoral engagement will be crucial. This implies working across different line ministries and agencies in order to link the disaster risk reduction, climate change adaptation and social protection approaches. In Mongolia it requires that actions be consolidated and coordinated, with a close collaboration in this direction between NEMA and the Ministry of the Environment, the Ministry of Food, Agriculture and Light Industry and the Ministry of Social Welfare and Labour.

Fundamental to enhance the resilience of pastoralism, including the benefits deriving from an enriched PRM strategy, however, is the need to maintain the natural resource base at a sustainable productive level, including through regulated destocking and restocking of animals. Unless this will be accomplished foremost, the scope for positive impacts from PRM will remain limited. Improved framework conditions for sustainable production have to be addressed in parallel. Improving livestock marketing, which was not discussed in this paper, is a key challenge for Mongolia to succeed in balancing its stocking rates as well as to reduce the vulnerability of herders by making them less dependent on limited and malfunctioning markets. Enhancing the engagement of pastoralists in local, regional and international markets is, as well known, difficult but highly desirable.

References

Asian Development Bank (n.d.) Improving Mongolia's social security sector. Social protection project briefs. ADB, Manila

Baas S, Erdenbaatar B, Swift J (2001) Pastoral risk management for disaster prevention and preparedness in Central Asia – with special reference to Mongolia. In: Report of the Asia-Pacific conference on early warning, prevention, preparedness and management of disasters in food and agriculture, Chiang Mai, Thailand, 12–15 June 2001, FAO RAP Publication no. 2001: 4

Batima P et al (2006) Climate change, vulnerability and adaptation in the livestock sector of Mongolia, a final report submitted to Assessments of Impacts and Adaptation to Climate Change (AIACC), Project No AS 06. International START Secretariat, Washington, DC

Blench R (2007) Grasshoppers, mice and pashminas: ecology and economics and the future of pastoralism in the Gobi aimags of Mongolia. Unpublished report for the Centre for Policy

Research, Ulaan Bataar, mimeo. http://www.rogerblench.info/Pastoralism/Pastoralism%20 Asia/mice%20and%20grasshoppers%202004.pdf. Accessed July 2010

Cacho O, Lipper L (2007) Abatement and transaction costs carbon-sink projects involving smallholders. Fondazione Eni Enrico Mattei. http://www.bepress.com/cgi/viewcontent.cgi?article=1083&context=feem. Accessed May 2010

Conant RT (2010) Challenges and opportunities for carbon sequestration in grassland systems: a technical report on grassland management and climate change mitigation. Food and Agriculture Organisation of the United Nations, Rome. http://www.fao.org/fileadmin/templates/agphome/documents/climate/AGPC_grassland_webversion_19.pdf. Accessed July 2010

Davies M, Günther B, Leavy J, Mitchell T, Tanner T (2009) Climate change adaptation, disaster risk reduction and social protection: complementary roles in agriculture and rural growth? IDS WP 320. Institute of Development Studies, Brighton

Dagvadorj D, Natsagdorj L, Dorjpurev J, Namkhainayam B, Ulanbataar (2009) Mongolia assessment report on climate change. Ministry of Environment, Nature and Tourism, Government of Mongolia, UNEP, UNDP, Mongolia, 228 pp. ISBN/ISSN: 978999299343X

Devereux S, Sabates-Wheeler R (2004) Transformative social protection. IDS working paper 232. IDS, Brighton

Devereux S, Scoones I (2007) The crisis of pastoralism: a response. Future agricultures consortium discussion paper. Institute of Development Studies, Brighton

FAO (2003) Managing pastoral risk in Mongolia: a plan of action. Project TCP/FAO/MON/0066, prepared by the Rural Institutions and Participations Service, FAO in collaboration with IDS, Sussex. Food and Agriculture Organisation of the United Nations, Rome

FAO (2007) Case study. Institutionalizing pastoral risk management in Mongolia: lessons learned. Prepared by JJ Swift, under the overall guidance from the Rural Institutions and Participation Service, FAO. Food and Agriculture Organisation of the United Nations, Rome

FAO/UNICEF/UNDP (2007) Report of the joint food security assessment mission to Mongolia. ftp://ftp.fao.org/docrep/fao/010/j9883e/j9883e00.pdf. Accessed August 2010

Fratkin E, Mearns R (2003) Sustainability and pastoral livelihoods: lessons from East African Maasai and Mongolia. Hum Organ 62(2):112–122

Grahn R (2008) The paradox of pastoral vulnerability, from poverty to power. Background paper. Oxfam International, Oxford

Hatfield R, Davies J (2006) Global review of the economics of pastoralism. IUCN, Nairobi

Hadrill D (2010) Mongolia winter disaster (dzud). Assessment report. TCE/FAO, Rome

Johnson DA, Sheehy DP, Miller D, Damiran D (2006) Mongolian rangelands in transition. Sécheresse 17(1–2):133–141

Kirkbride M, Grahn R (2008) Survival of the fittest: pastoralism and climate change in East Africa. Oxfam briefing paper number 116. Oxfam International, Oxford

Kerven C (2006) Review of the literature on pastoral economics and marketing: Central Asia, Mongolia and Siberia. IUCN, Nairobi

Lal R (2004a) Soil carbon sequestration impacts on global climate change and food security. Science 304:1623–1627

Lal R (2004b) Carbon sequestration in dryland ecosystems. Environ Manage 33:528–544

Mearns R (1991) Pastoralists, patch ecology and perestroika: understanding potentials for change in Mongolia. IDS Bull 22(4):25–33. IDS, Brighton

Mearns R (2004) Decentralization, rural livelihoods, and pasture-land management in post-socialist Mongolia. Eur J Dev Res 16:38–63

Mongolia National Statistical Office (2007) Report on livestock. National Statistical Office, Ulaanbatar

MRSM (2010) Mongolian society for range management: livelihood study of Herders in Mongolia. Selengepress Co., Ltd, Mongolia

National Action Program to Combat Desertification in Mongolia (2010–2020) Draft December 4, 2009. Ministry of Nature, Environment and Tourism, Government of Mongolia, Ulanbataar

Neely C, Bunning S, Wilkes A (eds) (2009) Review of evidence on drylands pastoral systems and climate change – implications and opportunities for mitigation and adaptation. FAO, Rome. http://www.fao.org/uploads/media/LWdisc_paper8_temp.pdf. Accessed May 2010

Nixson F, Walters B (2004) Privatisation, income distribution and poverty: the Mongolian experience. Report submitted to UNDP Mongolia. http://soros.org.mn/res_mat/Privatization_eng.pdf. Accessed August 2010
Nori M, Davies J (2007) Change of wind or wind of change? Climate change, adaptation and pastoralism. IUCN, Nairobi
Ojima D, Chuluun T (2008) Policy changes in Mongolia: implications for land use and landscapes. In: Galvin KA et al (eds) Fragmentation in arid and semi-arid landscapes: consequences for human and natural systems. Springer, Dordrecht, pp 179–193
Parton WJ et al (1995) Impact of climate change on grassland production and soil carbon worldwide. Glob Chang Biol 1:13–22
Schulze A (2009) Land legislation and the possibilities for pastoral risk management and adaptation to climate change – the example of Mongolia. Food and Agriculture Organisation of the United Nations, Rome. http://foris.fao.org/static/data/nrc/Schulze2009_LandLegislationCCA_Mongolia.pdf. Accessed July 2010
Seeberg-Elverfeldt C (2010) Carbon finance possibilities for agriculture, forestry and other land use projects in a smallholder context. Food and Agriculture Organisation of the United Nations, Rome. http://www.fao.org/docrep/012/al060e/al060e00.htm. Accessed July 2010
Sheehy DP, Damiran D (2009) Time comparative study of Mongolian pastureland condition (1997–2008). World Bank, Washington, DC
Sheehy DP, Sheehy CM, Johnson DE, Damiran D, Fiemengo M (2010) Livestock and wildlife in the context of development in the Southern Gobi Region, with special attention to wild ass. Mongolia discussion papers, East Asia and Pacific Sustainable Development Department. World Bank, Washington, DC
Smith P et al (2007) Agriculture. In: Metz B et al (eds) Climate change 2007: mitigation. Contribution of working group III to the fourth assessment report of the Intergovernmental Panel on Climate Change. Cambridge University Press, Cambridge/New York
Swift J, Baas S (1999) Pastoral institutions and approaches to risk management and poverty alleviation in Central Asian countries in transition. SD Dimensions. Food and Agriculture Organisation of the United Nations, Rome
Tennigkeit T, Wilkes A (2008) An assessment of the potential for carbon finance in rangelands. ICRAF working paper 68. World Agroforestry Centre China, Beijing
UNFCCC (2010) Glossary of climate change acronyms. http://unfccc.int/essential_background/glossary/items/3666.php#C. Accessed May 2010
UN Mongolia Country Team (2010) Consolidated Mongolia Dzud appeal. Office for the Coordination of Humanitarian Affairs, New York
Walker W, Hall D (2010) Reforming social protection systems when commodity prices collapse: the experience of Mongolia. In: Bauer A et al (eds) Poverty and sustainable development in Asia: impacts and responses to the global economic crisis. Asian Development Bank, Manila
Wilkes A, Tennigkeit T (2009) Grassland restoration and conservation in the three rivers source area, Quinghai Province, China. ICRAF – China and Center for Mountain Ecosystem Studies, Beijing
World Bank (2007) Agriculture for development. World Bank, Washington, DC

Chapter 21
Mongolian Nomads and Climate Change – A Herder's View

Ts. Sanjmyatav

Abstract Mongolia is one of the few countries that practices nomadic livestock breeding on a large scale. Climate change has a strong impact on Mongolian environmental conditions and does affect nomadic livelihoods considerably, often catastrophically. Political changes also brought strong changes to nomadic livelihoods.

The herders near Hustai National Park are organized in herder groups to better conserve their environment, graze sustainably, generate alternative means of income, and improve their economical situation. The author heads such a group. The author reviews the changes affecting their livelihoods, and tells about the measures they take and alternatives they develop to cope with the changes and develop the economic basis of their semi-nomadic way of life.

21.1 Introduction

Mongolia is one of the few countries that practice nomadic livestock breeding on a large scale. Even today many rural citizens depend on nomadic livestock breeding. Long distance traveling while looking for good pastures has changed into a semi-nomadic practice which includes the prolonged stay at a winter camp during that season. While urban land has been privatized since the mid-1990s, pasture land is still commons, and long-term agreements on a winter camp and permissions to travel to good pastures in summer time are obtained from a local administration.

Ts. Sanjmyatav (✉)
Argalant Soum Herder and Head of Tost Herder Group, Tov aimag, Mongolia
e-mail: takhi@hustai.mn

M.J.A. Werger and M.A. van Staalduinen (eds.), *Eurasian Steppes. Ecological Problems and Livelihoods in a Changing World*, Plant and Vegetation 6,
DOI 10.1007/978-94-007-3886-7_21,

The impact of climate change is well noticeable in Mongolia, lying in a cool zone in central Asia, far away from oceans. The average annual temperature has increased by more than 2°C over the past 70 years. Due to climate change precipitation in winter increased, while precipitation in summer decreased. As a consequence vegetation yield decreased, the frequency of drought and extreme winter conditions increased, and many rivers, springs and lakes fell dry and the forest has contracted.

Herders are very vulnerable to climate change. Under adverse conditions they are unable to stick to their usual life as livestock breeders. In the extreme winter of 2009/2010 increased snow cover caused the loss of 6.5 million animals in the central and western aimags of Mongolia, and 8,711 herder households were left with no animals at all. Therefore, adaptation to climate change is an urgent matter for herders.

The Hustai Center helps herders in the bufferzone of Hustai National Park to adapt to climate change. Thanks to the support of the Hustai Center herders establish herder groups and jointly conserve nature and wildlife, use the natural resources sustainably, and create alternative means to generate income. The Tost herder group of Argalant soum, Tov aimag, is a very good example of this practice (Photos 21.1 and 21.2).

21.2 First Briefly About Mongolia

Mongolia is located in the cool zone of Central Asia, away from oceans, and has a harsh continental climate. Locally temperatures can be down to −50°C in winter and up to more than 35°C in summer. Annual precipitation is 250 mm. The mountains in the north west of the country are on average 2,500–3,500 m above sea level, while the steppes, hills and the Gobi in the south east are 1,000–1,500 m above sea level.

The country comprises 1,564,116 km^2 of land and, given its total population of just over three million (2009 data), it has the lowest population density of any country: 1.9 per km^2. This sparsely populated country comprises many different landscape zones, including the alpine zone, taiga, forest-steppe, steppe, desert steppe and desert (Fig. 21.1).

Nomadic livestock breeding has always been of great importance in Mongolia. Up to mid-twentieth century this was the main sector of the economy and it produced about 90% of the gross domestic product. But more recently mining is also becoming very important.

Nomadic livestock breeding is year round dependent on pastures. From October the winter season starts and the land is covered with snow up till next March. However, Mongolian animals are resistant to the harsh climate. In winter animals do not get supplementary fodder. They dig with their feet in the snow and graze the grass they find under it. Mongolian animals put on fat during summer and autumn and that fat provides them with energy in winter time, when there is not enough hay in the pasture. The animals are small in size and have a low productivity compared

Photo 21.1 Ger (nomadic felt tent) of a herdsman family in the bufferzone (Photo by M.A. van Staalduinen)

Photo 21.2 Felt making, rolling the moistened wool (Photo by P. Wit)

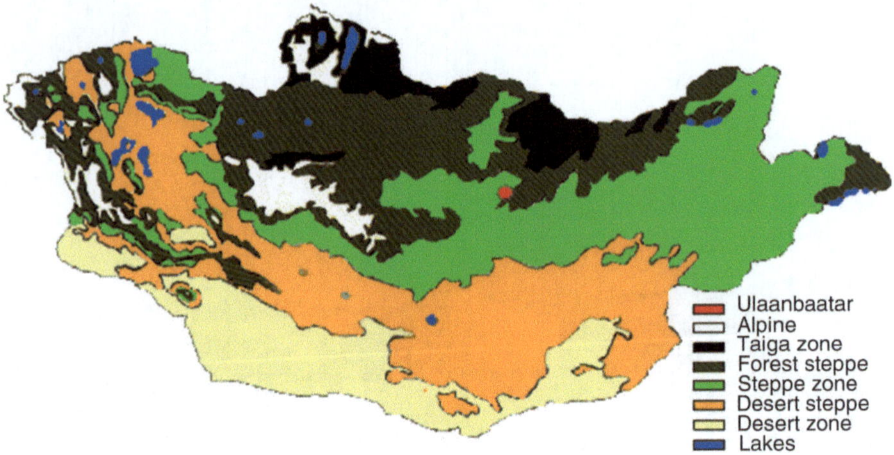

Fig. 21.1 Landscape zones of Mongolia

to other countries, but they are hardy. Main source of living for herders are the animals. In olden times herders made their living solely on animals: meat and milk was used for food, skin was processed and used for clothing and flooring of the ger (the dwelling of nomads) and dung was used as fuel.

Pasture is part of the commons and is property of the state. That means that a herder can use any pasture. In olden times herder families moved around in areas with good pastures, but as from the 1960s they started a semi-nomadic lifestyle, staying every year at the same winter camp and moving around in summer and autumn to put fat on the animals.

In recent decades (semi-)nomadic livestock breeding has decreased its share in the national economy. This partly is due to a decrease in pasture availability, but certainly also to an increase in the human population and the number of animals. According to the 2009 animal census there were 44.0 million animals, including 19.3 million sheep, 19.6 million goats, 2.6 million cattle, 2.2 million horses, and 0.3 million camels. At the same time there were a total of 171,000 herder families in Mongolia.

21.3 The Place Where I Live

We live in the bufferzone of Hustai National Park (HNP), some 100 km away from Ulaanbaatar, capital city of Mongolia. The Przewalskii horse (which is called 'takhi' by Mongolians), was near extinction, but was brought back to its natural habitat from The Netherlands, as from 1992, in an area of 50,600 ha in the Hustai range, which was designated a national park (see Chap. 13, this volume). HNP is rich in biodiversity: more than 450 species of vascular plants, 85 species of lichens, 90 species of mosses, 44 species of mammals, such as red deer, roe deer, Mongolian antelope, wild boar, wild sheep, wolf, fox, corsac fox, badger, lynx, wildcat, and 221 species of birds, such as vulture, eagle, kite, swan, crane, and bustard, have been recorded.

The success of the reintroduction of the takhi made HNP famous around the world, and it has boosted nature conservation and the study of steppe ecosystems. At HNP an average of about 10,000 tourists and researchers from about 50 countries visit every year, mainly to see the takhi and other wildlife.

Hustai National Center (in short Hustai Center) is a non-governmental organization that manages HNP on the basis of an agreement with the government of Mongolia. Hustai Center takes measures to improve the livelihood of herders and involve herders in conservation of the ecosystem. They try to explain to herders how they can adapt to the effects of climate change and desertification and how they can create a strategy for improved livelihood. As part of those efforts 39 herders groups were established in the buffer zone of HNP. Herders that live in each others neighbourhood join into such a herder group. Our group, for example, counts ten herder households, living in the bufferzone north of HNP. The aim of a herder group is to cooperate in nature conservation, improve their livelihood, facilitate the selling of their products to the market, and overcome natural hardship.

The soums of Altanbulag, Argalant and Bayanhangai in Tov aimag lie in the bufferzone of HNP. There are 6,600 people in 1,800 households in these three soums. From these, 1,085 are herder households and they had 411,300 animals on 1 January 2010.

21.4 Climate Change and Its Impact in Mongolia

The average annual temperature increased by 2.14°C in the past 70 years. In the last 10 years the increase has intensified. The number of hot days (more than 33°C) increased by 5–8 days over 10 years (Natsagdorj and Dagvadorj 2010). Researchers claim that events of extreme coldness in winter decreased but the amount of snow increased and summer heath increased. The amount of summer precipitation is stable but because of the increase in temperature there is a tendency towards an increase of drought. What other changes do these changes in climate bring about in Mongolia?

- *Rain characteristics*. Increase in temperature caused changes in rainfall: soft rain events decreased while heavy shower events increased; the latter in fact doubled in number over the past 20 years. Heavy shower rain provides less moisture to the soil and hardly promotes plant growth but harbors an increased risk of floods and associated damage.
- *Drought*. Higher temperatures in the warm season caused less precipitation and more droughts. Over the past 67 years there was a decrease in summer precipitation in the steppe and Gobi zones of 9.7–17.6%. That's why 25% of the total land surface of Mongolia experiences a drought once every 2–3 years and more than 50% of the country experiences it once in 4–5 years.

- *Water resources*. A census of surface water showed that 852 rivers out of a total of 5,128, 2,274 springs out of 9,306, 1,181 lakes out of 3,747, and 60 mineral springs out of 429 dried up.
- *Ice*. In the Mongolian Altai range the amount of ice decreased by 30% in the period from 1940 to 2002.
- *Dust storms*. Because the snow cover in spring thaws earlier and the start of the growth season is delayed by 1 month, the number of days with bare soil increased. As a result the loose topsoil is blown by the wind and built into more powerful 'yellow' dust storms that reach deep into China and all the way to Korea and Japan (Photo 21.3).
- *Desertification*. 78.2% of pasture land is affected by medium to intense desertification according to ground and satellite measurements (Mandakh et al. 2007). Also 2.8% of the land is covered by sand.
- *Vegetation*. In 2006 the area of bare land had increased by a factor 3, compared to 1992, and the area covered by forest decreased by one third according to satellite data. Pasture yield decreased by 20–30% (Bolortsetseg et al. 2010). As the spring precipitation decreased by 20%, the growing season of the vegetation was delayed by 1 month in the Gobi and steppe zones.
- *Forest and steppe fires*. The frequency of forest and steppe fires increased because of the drought and the length of the season of fire risk also increased. While it was common to have fires in spring, these days fires also occur in autumn, or even in summer (in 2002). During the last 10 years 590,000 ha of forest burnt.
- *Pasture pests*. The distribution area of Brandt's vole (*Laciopodomus brandtii*) increased and caused serious pasture degradation. Furthermore, the grasshopper *Eclipophleps tarbinskii* grew in numbers and increased its movement. It seems that the lifetime of those animals also increases.
- *Extreme winters (Zud)*. When there is a lot of snow on the pasture and animals cannot reach the plants below the snow cover and thus cannot graze we speak of extreme winter. In extreme winters several thousands to several millions of animals may die and cause devastation to the herder's livelihood and the economy of the state. Studies showed that the frequency of extreme winters increased since 2000.

21.5 Changes in My Living Environment

I live in the northern part of the steppe zone, bordering on the taiga zone. Mongolian researchers say that climate change is most noticeable in the steppe and forest-steppe zones. Let me mention some changes in my environment (Oyunbileg and Tserendulam 2010).

1. Less pasture yield every year. Even though pasture yield depends on the amount of precipitation pasture yield tends to drop. A research team has monitored the forest-steppe ecosystem at Hustai Center over a long period. Their studies show that the pasture yield in fenced areas is dropping (Fig. 21.2)

Photo 21.3 Yellow dust storm on the Mongolian steppe (Photo by Ts. Sanjmyatav)

Photo 21.4 Withered birch forest (mainly *Betula platyphylla* with some *Populus tremula*) at Hustai National Park (Photo by Ts. Sanjmyatav)

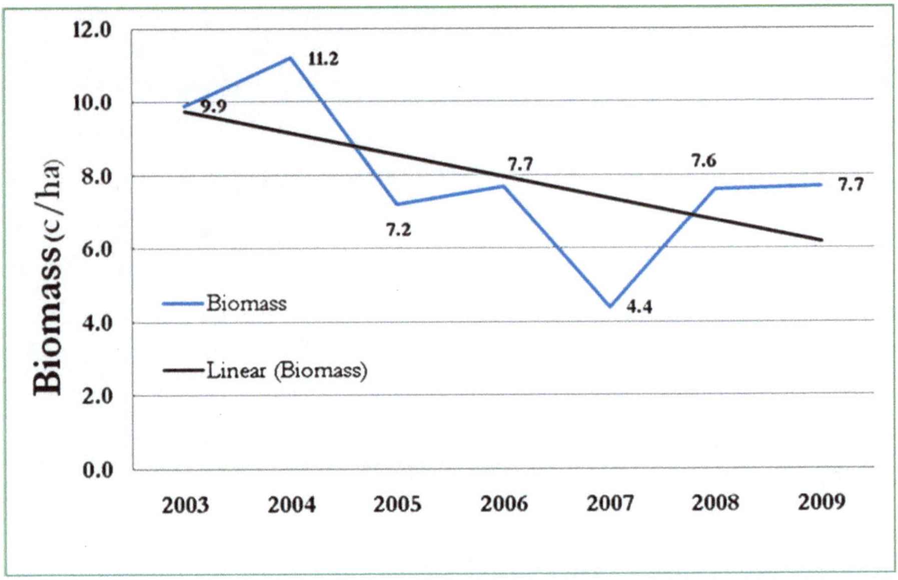

Fig. 21.2 Changes in aboveground biomass (in 100 kg per ha) over the period 2003–2009 in a fenced area in the forest-steppe zone at Hustai

2. HNP contains 1,850 ha of birch forest (*Betula platyphylla* with some *Populus tremula*). Seventy percent of that forest dried out during the last 10 years. Researchers say that melting of the permafrost caused the moisture deficiency (Photo 21.4).
3. Where there were 11 streams at HNP 10 years ago, seven of them or now dry. And the remaining streams carry much less water.
4. The Tuul River passes through HNP to Lake Baigali. The water flow of this river diminished and sometimes it has no water at all.
5. Our main source of living, the pasture, is degrading. In winter our herder group moves to the area north of HNP, crossing the Hustai range, and in spring time we pass the Hustai Range again to get to the Tuul River basin. The Tuul River basin carries large human and animal populations in summer and autumn. This causes serious degradation of the pasture land. Nutritious plant species are getting less, while weeds, such as species of *Chenopodium* and *Artemisia* expand.

The steppe zone (in Dundgobi aimag) experienced drought for many years and herders had to move far away. Some of them showed up in our territory and herders from western aimags that wanted to be close to the market also came to our territory. Therefore we have more animals than the carrying capacity allows for and this causes pasture degradation.

21.6 Our Tost Herder Group

Changes in climate and society force us to leave nomadic livestock breeding which is based on self-sufficiency. The Hustai Center provides advise in this process of change and encourages establishment of herder groups in the bufferzone of HNP.

One of these herder groups is 'Tost'. It was named after a nearby small mountain. At first, in 2004, the Tost herder group united five herder households. Presently it has expanded to ten households. We have a total of 4,650 animals. This includes 350 horses, 265 cattle, 2,200 sheep, and 1,835 goats. Traditionally, our living depended on animals. We used and still use animals for food and milk in summer and make various dairy products. We sell the wool and cashmere of the animals and with that money buy consumer products (Photos 21.5 and 21.6). But the productivity of the local animals is not high. For example, yearly 300 l of milk is produced and a goat produces 200 g of cashmere. Cashmere sells for 15,000–30,000 tugrug per kg (which amounts to about 12–24 US $ at the average 2010–2011 price level). But prices are not stable. Possessing a 100 goats a herder can generate 300,000–600,000 tugrug a year from cashmere. Consumer demands for cashmere increase every year. To meet that demand the number of animals should be increased. But an increase in

Photo 21.5 Women milking goats in the bufferzone (Photo by Bufferzone)

Photo 21.6 Meeting of pastoral women's collective (Photo by Bufferzone)

the number of animals will result in the degeneration of the pasture. And in extreme winters the animals will die and we will be poor. Therefore we do not want to increase the number of animals, but keep it below the carrying capacity and seek to generate alternative sources of income. Such alternatives are:

1. *Crop farming*. Herders never used to grow vegetables or cereals. Now we are growing vegetables. In our neighbourhood there are abandoned crop fields of 20 years ago. We started cultivating these fields 2 years ago. The herder group applied to the soum administration for permission to cultivate some parts of the fields for several years. In 2009 our group took 0.5 ha into cultivation and

harvested 10.5 tons of potatoes which generated 3.2 million tugrug. Some other groups grow cucumbers and tomatoes in a greenhouse to generate income. Hustai Center provides financial support to herders who want to buy seed potatoes and procure a greenhouse.

2. *Community based tourism* (CBT). Over the last years Hustai Center stimulates that herder groups in the bufferzone of HNP host tourists. Now, more than ten herder groups are involved in tourism. The lifestyle, traditions and culture of nomads is interesting for foreign guests. We show them:

   - The traditional way of making dairy products from cow, sheep, or goat milk (curds, butter, yogurt, cheese, dried curd, etc.). Mongolians make drinking bags from cow hide and fill these with airag (fermented mare's milk). It is a tasty drink and foreigners like it a lot.
   - Lamb and kid skin can be processed to make warm winter clothes and sheep skin can be processed to make outer covering for winter boots. With this covering you can herd your animals all day even at minus 20°C.
   - Cow, horse and camel hides can be processed into leather products, such as bridles, hobbles, halters, long ropes for tying livestock, or lassos.
   - Wives are good in sewing traditional clothing and make these for all family members. They show tourists how to sow such traditional cloths and some tourists venture to buy Mongolian traditional costumes.
   - We demonstrate how to play Mongolian games, including various games which can be played with ankle bones of sheep or goats.
   - Every year on 11 and 12 July we celebrate our national festival with wrestling, horse racing, ankle bone shooting, etc. Herder groups organize such a festival for tourists at other times.

3. When nomads move they use their carts, drawn by their camels, horses, or castrated bulls, to transport their ger (the dwelling of nomads) and other household furniture. Tourists are interested to see how this is done, and they like to help with the removal and the erection of the ger. They also like to stay in the ger.
4. Herders sell cloths and other souvenirs they produce in the souvenir shop of Hustai Center.

21.7 Our Goals for the Future

In order to adapt to the effects of a changing climate, we herders proposed the following goals:

1. Improve the organization of and cooperation with herder groups. At the moment there is no legal framework covering herder groups. Being legalized is important in order to conduct independent activities, to build contacts with other organizations, and getting loans from the bank. I hope that, in the near future, the Mongolian parliament will approve a law on herder groups.

2. Prioritize nature conservation. Mongolia's pristine nature is being devastated, wildlife is diminishing, and their distribution area decreases because of human impact and climate change. For example, there were 13 million marmots in 1990 against four million now, and at the same time the number of 4.7 million Mongolian antelopes was reduced to 1.6 million, 130,000 deer to 10,000, and 60,000 wild sheep to 15,000 (Samya and Mulenberg 2006). In HNP the numbers of deer, marmots, and Mongolian antelopes are increasing, but poaching, especially of marmots, is common.

   Under such circumstances herder groups should pay attention to conserve wildlife in their neighborhoods.
3. Keep the pasture in good condition. Herder groups should keep the pasture in a good condition; they should use it on a rotational basis, and should improve the pasture (e.g. by sowing seeds of perennial plants). And the state should take measures. Presently, pasture land is commons and property of the state. It is open for use to anyone and this causes over-exploitation and degradation of the pasture land. It is clear that nomadic livestock breeding will go on for some time to come and thus privatization of pastures is not possible. Therefore herder groups should possess rights for pasture use under long-term (at least 15 years) contracts. A draft law on pastures is now being broadly discussed and the state should formulate a decision soon.
4. Generate alternative sources of income. It is important to establish alternative sources of income, which are not fully dependent on variation in nature and altering climatic conditions. For example, small farms that are suited to Mongolian conditions could be established.
5. Produce not only traditional products but develop new products. At the moment especially new milk products seem promising. Small workshops of herder groups can be established that pasteurize milk and package it so that it has a long shelf life.
6. Establish a distribution network to the market.

21.8 Conclusion

The impact of climate change is profound in Mongolia. Even though the number of cold days decreased, the number of hot summer days increased, there is more snow in winter, and less rain in summer. As a consequence, it is getting drier, pasture yields decrease and late rains in spring delay the growing season by 1 month. Nomadic livestock breeding is strongly affected by these changes in climate and the survival of herders becomes critical.

In order to adapt to these changes herder households should join and cooperate. Then they can better cope with the natural hardships, improve their livelihoods, conserve nature, and find alternative sources of income.

Hustai Center – a non-governmental organization – implemented the program 'adapting and developing HNP and its bufferzone under a changing climate', an international project funded by The Netherlands' government. It also implemented the project 'developing sustainable livelihoods in the bufferzone of HNP', again funded by The Netherlands' government in the period 2004–2008. These projects helped the herders a lot in adapting to the effects of changes in climate.

References

Bolortsetseg B, Erdenetsetseg B, Bat-oyun Ts (2010) Changes in the development of pasture plants and pasture yield during the last 40 years. Research paper 24. Institute of hydrology and meteorology, Ulaanbataar, pp 108–114

Mandakh N, Dash D, Khaulenbek A (2007) Present status of desertification in Mongolia. In: Geoecological issues in Mongolia. Institute of Geoecology of Mongolian Academy of Sciences, Ulaanbataar, pp 63–73

Natsagdorj L, Dagvadorj D (2010) Adapting to climatic changes. Report, Ulaanbaatar

Oyunbileg Kh, Tserendulam Ts (2010) The vegetation condition of HNP in 2009. Research report, Ulaanbataar

Samya R, Mulenberg I (2006) Conservation in Mongolia, Mongolian Academy of Sciences, Ulaanbaatar

Index

M.J.A. Werger and M.A. van Staalduinen (eds.), *Eurasian Steppes. Ecological Problems and Livelihoods in a Changing World*, Plant and Vegetation 6,
DOI 10.1007/978-94-007-3886-7, © Springer Science+Business Media B.V. 2012

MIX
Papier aus verantwortungsvollen Quellen
Paper from responsible sources
FSC® C105338

If you have any concerns about our products,
you can contact us on
ProductSafety@springernature.com

In case Publisher is established outside the EU,
the EU authorized representative is:
Springer Nature Customer Service Center GmbH
Europaplatz 3, 69115 Heidelberg, Germany

Printed by Libri Plureos GmbH
in Hamburg, Germany